introduction to
SWITCHING THEORY
AND LOGICAL DESIGN

introduction to
SWITCHING THEORY
AND LOGICAL DESIGN

SECOND EDITION

Fredrick J. Hill

Professor of Electrical Engineering
University of Arizona

Gerald R. Peterson

Professor of Electrical Engineering
University of Arizona

JOHN WILEY & SONS

New York   London   Sydney   Toronto

Library of Congress Cataloging in Publication Data:

Hill, Frederick J.
   Introduction to switching theory and logical design.

   Includes bibliographies.
   1. Switching theory. 2. Digital electronics.
I. Peterson, Gerald R., joint author. II. Title. III.
Title: Switching theory and logical design.

TK7868.S9H5   1974   621.3819′58′2   72-21783
ISBN 0-471-39882-9

Printed in the United States of America

10 9 8 7 6 5 4 3 2

Preface

In the field of digital computers, there is an area somewhere between circuit design and system formulation which is usually identified as switching theory. Although this topic must be mastered by the digital computer engineer, it does not answer all the questions concerning layout of an efficient digital computing system. The remaining questions, which are less susceptible to formalization, may be grouped under the heading "Digital System Theory."

Our approach is to follow the basic framework of switching theory. We hope, however, to motivate the student by presenting examples of the many problems which appear repeatedly in the design of digital systems. It is not our intent to provide the student with a detailed knowledge of such specialized system topics as computer arithmetic. Instead, we attempt to provide the framework through which the student might develop a sound design philosophy applicable to any digital design problem. Thus, we do not feel that the use of the term "logical design" in the title is unjustified.

The usefulness of switching theory is not restricted to engineers actively engaged in computer design. In fact, it is our contention that almost every design engineer in the field of electronics will have some opportunity to draw on this subject. Applications occur whenever information in communication, control, or instrumentation systems, for example, is handled in other than analog form.

The above three paragraphs, which began the preface of the First Edition of this book, seem to us to be even more accurate today than they were six years ago. In these years, digital techniques have become virtually standard in many areas once considered the exclusive preserve of analog techniques. This "digitalization" of electrical engineering has become so pervasive that it is a rare electrical engineer who will not have some contact with digital design problems.

We undertook preparation of this second edition in order to strengthen some of the pedagogical weak spots of the first edition as well as to include new material made necessary by recent developments, particularly in the area of integrated circuits. As an example of the latter, we have rewritten

v

Chapter 5 and added Chapter 16 to reflect the general acceptance of integrated circuits and LSI throughout industry. We have completely reworked Chapters 9–13 and have added Chapters 14 and 15 on sequential circuits. The emphasis of these chapters has shifted toward creative design and away from formal minimization. This will be most apparent in Chapter 15, which describes a digital hardware design language. We found the original treatment in Chapters 2, 3, 4, 6, and 7 to be quite satisfactory and have made few changes except in the problem sets. Considerable effort has been devoted to updating and improving the problems throughout the book.

Chapters 1 and 2 were written with the goal of providing student motivation. We have endeavored to sustain this motivation throughout the book by treating with special care points which might otherwise become sources of confusion. It is our belief that, if questions concerning justifications behind manipulative procedures go unanswered, the student will be discouraged. Our intention to avoid such pitfalls is exemplified in Chapter 4 by a reliance on Huntington's original postulates and careful distinction between the algebra of the Karnaugh map and the two-valued algebra of truth values. (This is essentially the same as the distinction between Boolean forms and switching functions, which is made with great care in some more advanced texts.) Our treatment of Boolean algebra is formal to be sure, but we have found that it represents no conceptual difficulties to the student.

The introductory chapter should occupy no more than one day of class time, or it could be left as a reading assignment if class time is critical. The binary number system is taken up in Chapter 2 to provide a background for the binary notation of Boolean functions and maps in Chapters 4 and 6. The subject is allotted a separate chapter which is placed prior to the introduction of Boolean algebra to minimize the chance of confusing the two concepts. The additional material on binary arithmetic is essential only for examples and might be omitted, or referred back to, if desired.

Chapters 3 and 4 provide the basic logical and algebraic basis for switching theory and certainly have not diminished in importance in any way. We believe the benefits to the students will be great if the topics of these chapters are covered in order with no omissions. Chapter 5, on components, has been totally rewritten to reflect the almost total replacement of discrete logic circuits by integrated circuits. The coverage of this chapter may be adjusted to suit the purposes of the instructor.

Chapters 6 and 7 remain unchanged in the second edition except for new problem sets. Rapid developments in components, particularly integrated circuits, have greatly changed the relative importance of various cost factors in digital design, but the basic correlation between simple algebraic forms and economical designs remains undisturbed. And we believe the Karnaugh map and Quine-McCluskey minimization remain unchallenged as the most natural, powerful, and widely applicable tools for simplifying switching functions. These two chapters should be covered in detail.

From Chapter 8 on, the book has been almost totally rewritten. Chapter 8 has been expanded to include additional topics related to integrated circuit realization of certain specialized combinational circuits. Included is a section on the use of programmable read-only memories as generalized logic elements. Chapter 9 has been expanded to include a fairly extensive coverage of flip-flops and comparison between the various types of sequential circuits and their timing characteristics.

Chapter 10, on clocked sequential circuits includes an improved treatment of the problem of setting up the initial state table and a better discussion of state assignment. State assignment is certainly the most difficult aspect of sequential circuit design. The literature includes many papers dealing with various sophisticated and complex state assignment techniques which rarely produce results which can justify the time and effort required to learn and apply them. We present a more intuitive approach which relies heavily on the designer's natural *a priori* insight into the structure ultimately required and which will provide satisfactory assignments for a reasonable expenditure of effort.

The pulse mode circuits material has been moved from Chapter 12 in the First Edition to Chapter 11 and changed only slightly. This provides a smoother transition into incompletely specified circuit examples. The treatment of incompletely specified circuits, now in Chapter 12, has been simplified to avoid undue emphasis on state table minimization. The Grasselli-Luccio minimization procedure is now covered in Appendix A.

Chapter 13 is totally rewritten, to provide adequate coverage of the general class of "nonpulsed" circuits, which we have christened "level mode circuits" and of which fundamental mode is only a special case. In particular, we have expanded the treatment of race-free assignments and have covered circuits in which two or more inputs may change simultaneously.

Chapter 14 is entirely new and presents a method of hazard-free level mode design which has apparently been used by integrated circuit designers but which, to the best of our knowledge, has never before been treated in a textbook. Chapter 16, also new, introduces some of the special problems of logic design involving medium-scale and large-scale integrated circuits.

As noted earlier, it is generally recognized that the formal techniques of switching theory, or logic design, do not extend readily to the design of larger systems. In the area of system design, register transfer languages are gaining wide acceptance. Our companion volume, *Digital Systems: Hardware Organization and Design* (Wiley, 1973) provides a thorough treatment of the application of a register transfer language to the design of large-scale digital systems. Chapter 15 of this book attempts to "bridge the gap" through application of a simplified register transfer language to the design of circuits which are too large to be efficiently attacked by the logic design methods presented earlier but are not large enough to be considered full-scale digital systems in the usual sense. The reader who has used both texts will notice a difference in the primary hardware implementation of the control unit in the two books. In this book, we rely on the reader's familiarity with standard clock mode sequential circuits, while in the other book we introduced a special element called the control delay. In that book, we were primarily concerned with describing large digital hardware systems to students who may or may not have been expert in sequential circuits. The language used in Chapter 15 to describe hardware will be a subset of and completely consistent with the AHPL of the other book. Once a reader has become familiar with both developments, he will have no difficulty switching between the two hardware interpretations.

Chapters 17 and 18 (Chapters 14 and 15 of the First Edition) are little changed except for the addition of material on iterative circuits to Chapter 17. This is a topic of considerable importance in the design of integrated circuits, where the use of repetitive structures may lead to significant economies. Chapter 16 of the First Edition, on linear circuits and codes, was written because, at that time, there was no satisfactory undergraduate text on this subject. In the interim, excellent texts have appeared, so this chapter has become superfluous.

The diagram illustrates the prerequisite relationships between the various chapters, as an aid to instructors in planning courses based on this book. It can readily be seen from the chart that Chapters 2, 3, 4, 6, 9, and 10 would necessarily be included in most any course structure. Beyond this, the

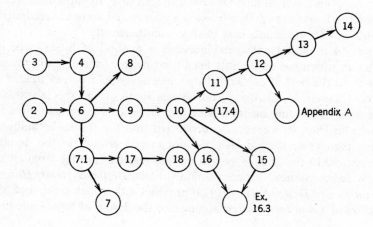

instructor has considerable freedom in structuring the course. Of the remaining restrictions, the most critical is that the level mode material of Chapters 13 and 14 has as a prerequisite the incompletely specified state table minimization of Chapter 12, which in turn requires Chapter 11. One perfectly satisfactory approach is to take up the chapters in numerical order. Another alternative which may be attractive to instructors wishing to expose their students to some system design problems would be to take up Chapters 15 and 16 after Chapter 10. These two chapters contain a number of system design examples of varying degrees of complexity. Where students already have a background in combinational logic, a course can be begun with any of Chapters 7, 8 or 9. There is more than enough material of practical value to constitute a one semester course beginning with Chapter 9.

Finally, we might note that for logic symbols we have retained the uniform shape symbols of standard IEEE No. 91-A.S.A. Y 32.14-1962. After this book went into production, a new standard IEEE No. 91-ANSI Y 32.14-1973 appeared. It is the first standard to be adopted by all concerned agencies, both civilian and military, in the United States, and is in substantial agreement with international standards. As such it represents a long overdue step toward standardization, which we heartily applaud. Unfortunately, it appeared too late to be included in this book, except for some special cases. The differences between this new standard and the symbols used in this book are explained fully in Chapter 3.

Obviously, many persons contribute to the development of any book and there is no way we could acknowledge them all. Our primary thanks must go to our families, to whom it must seem as though we will never stop writing and revising books.

TUCSON, ARIZONA

F. J. Hill
G. R. Peterson

Contents

1 | Introduction   1

1.1 Characteristics of Digital Systems   1
1.2 A Brief Historical Note   4
1.3 Organization of Digital Computers   5
1.4 Types of Digital Circuits   7

2 | Number Systems   9

2.1 Introduction   9
2.2 Conversion between Bases   10
2.3 Arithmetic with Bases Other than Ten   14
2.4 Negative Numbers   16
2.5 Binary-Coded Decimal Numbers   18
    Problems   19
    Bibliography   21

3 | Truth Functions   23

3.1 Introduction   23
3.2 Binary Connectives   26
3.3 Evaluation of Truth Functions   28
3.4 Many-Statement Compounds   29
3.5 Physical Realizations   30
3.6 Sufficient Sets of the Connectives   35
3.7 A Digital Computer Example   38
    Problems   41
    Bibliography   43

4 | *Boolean Algebra* — 45

4.1 Introduction — 45
4.2 Truth-Functional Calculus as a Boolean Algebra — 47
4.3 Duality — 48
4.4 Fundamental Theorems of Boolean Algebra — 49
4.5 Set Theory as an Example of Boolean Algebra — 53
4.6 Examples of Boolean Simplification — 56
4.7 Remarks on Switching Functions — 61
4.8 Summary — 64
     Problems — 65
     Bibliography — 67

5 | *Switching Devices* — 69

5.1 Introduction — 69
5.2 Switches and Relays — 69
5.3 Logic Circuits — 71
5.4 Speed and Delay in Logic Circuits — 77
5.5 Integrated Circuit Logic — 79
5.6 Comparison of IC Logic Families — 86
5.7 Logical Interconnection of NAND and NOR Gates — 88
5.8 Conclusion — 95
     Bibliography — 96

6 | *Minimization of Boolean Functions* — 97

6.1 Introduction — 97
6.2 Standard Forms of Boolean Functions — 98
6.3 Minterm and Maxterm Designation of Functions — 101
6.4 Karnaugh Map Representation of Boolean Functions — 103
6.5 Simplification of Functions on Karnaugh Maps — 114
6.6 Map Minimizations of Product-of-Sums Expressions — 126
6.7 Incompletely Specified Functions — 129
     Problems — 133
     Bibliography — 137

7 | *Tabular Minimization and Multiple-Output Circuits*   139

7.1   Cubical Representation of Boolean Functions   139
7.2   Determination of Prime Implicants   142
7.3   Selection of an Optimum Set of Prime Implicants   148
7.4   Multiple-Output Circuits   159
7.5   Map Minimization of Multiple-Output Circuits   161
7.6   Tabular Determination of Multiple-Output Prime Implicants   166
        Problems   168
        Bibliography   174

8 | *Special Realizations and Codes*   175

8.1   Introduction   175
8.2   The Binary Adder   176
8.3   Coding of Numbers   181
8.4   The Decoder   185
8.5   Code Conversion and Read-Only Memories   188
8.6   NAND and NOR Implementation   194
8.7   Parity   199
8.8   Error-Detecting-and-Correcting Codes   202
8.9   Hamming Codes   204
        Problems   207
        Bibliography   210

9 | *Introduction to Sequential Circuits*   211

9.1   General Characteristics   211
9.2   Flip-Flops   213
9.3   Why Sequential Circuits?   217
9.4   Shift Registers and Counters   221
9.5   Speed-Versus-Cost Trade-Off   223
9.6   A General Model for Sequential Circuits   226
9.7   Clock Mode and Pulse Mode Sequential Circuits   227
9.8   Timing Problems and Master-Slave Flip-Flops   230
9.9   Level Mode Sequential Circuits   235
9.10 Conclusion   236
        Problems   237
        Bibliography   239

10| *Synthesis of Clock Mode Sequential Circuits*   241

10.1 Analysis of a Sequential Circuit   241
10.2 Design Procedure   244
10.3 Synthesis of State Diagrams   245
10.4 Finite Memory Circuits   249
10.5 Equivalence Relations   256
10.6 Equivalent States and Circuits   258
10.7 Determination of Classes of Indistinguishable States   261
10.8 Simplification by Implication Tables   267
10.9 State Assignment and Memory Element Input Equations   275
10.10 Partitioning and State Assignment   289
10.11 Conclusion   299
      Problems   300
      Bibliography   305

11| *Pulse-Mode Circuits*   307

11.1 Introduction   307
11.2 Mealy Circuits and Moore Circuits   309
11.3 Design Procedures   311
11.4 Mealy Moore Translations   317
11.5 Counters Revisited   323
      Problems   326
      Bibliography   328

12| *Incompletely Specified Sequential Circuits*   329

12.1 Introduction   329
12.2 Compatibility   332
12.3 Completion of Design   350
      Problems   355
      Bibliography   361

13| *Level Mode Sequential Circuits*   363

13.1   Introduction   363
13.2   Analysis of a Fundamental Mode Circuit   366
13.3   Synthesis of Flow Tables   367
13.4   Minimization   370
13.5   Transition Tables, Excitation Maps, and Output Maps   376
13.6   Cycles and Races   379
13.7   Race-Free Assignments   382
13.8   Hazards in Sequential Circuits   394
13.9   General Level Mode Circuits   401
       Problems   409
       Bibliography   417

14| *A Second Look at Flip-flops and Timing*   419

14.1   Introduction   419
14.2   Clock Skew   420
14.3   A Flow Table for a J-K Master-Slave Flip-Flop   421
14.4   Another Approach to Level Mode Realization   424
14.5   Realizing the Standard J-K Flip-Flop   426
14.6   Analysis of Races and Hazards   429
14.7   Summary   435
       Problems   435
       Bibliography   436

15| *Description of Large Sequential Circuits*   437

15.1   Introduction   437
15.2   Clock Input Control   438
15.3   Extended State Table   440
15.4   A Program Description   442
15.5   Synthesis   443
15.6   Vector Operations   454
15.7   Logical Functions of Vectors   456
15.8   Applications   460
15.9   Summary   468
       Problems   468
       Bibliography   472

16| LSI-MSI   473

16.1   Introduction   473
16.2   Definitions   474
16.3   Design with Standard MSI Parts   476
16.4   Basic Economics of Integrated Circuits   484
16.5   MOS/LSI   489
16.6   Simulation   495
16.7   Test Sequence Generation   500
       Problems   502
       Bibliography   504

17| Combinational Functions with Special Properties   507

17.1   Introduction   507
17.2   Symmetric Functions   508
17.3   Boolean Combinations of Symmetric Functions   510
17.4   Higher-Order Forms   512
17.5   Simple Disjoint Decomposition   515
17.6   Complex Disjoint Decomposition   523
17.7   Iterative Networks   529
17.8   Ordering Relations   534
17.9   Unate Functions   536
       Problems   540
       Bibliography   542

18| Threshold Logic   545

18.1   Generalized Resistor-Transitor Logic Circuit   545
18.2   Linear Separability   548
18.3   Conditions for Linear Separability   553
18.4   Magnetic Threshold Logic Devices   560
18.5   The Realization of Symmetric Functions Using More than One
       Threshold Device   562
       Problems   568
       Bibliography   570

Appendix A/ Selection of Minimal Closed Covers   571

    Bibliography   579

Appendix B/ Relay Circuits   **580**

B.1   Basic Characteristics of Relay Circuits   580
B.2   Relay Realizations of Symmetric Functions   585
B.3   Relays in Sequential Circuits   586
     Bibliography   592

Index   593

1 Introduction

1.1 Characteristics of Digital Systems

The purpose of this book is to present the basic mathematical tools used in the design of digital systems. This statement raises the question: What is a *digital* system? Broadly speaking, a *digital* system is any system for the transmission or processing of information in which the information is represented by physical quantities (signals) which are so constrained as to take on only discrete values. If the signals are constrained to only two discrete values, the system is *binary*. These definitions can be best clarified by considering a specific example, the problem of transmitting the water level in some lake, river, or irrigation canal to some remote central control station.

This is a very practical problem in the control of large hydroelectric systems or irrigation and flood control systems. The sensing device will probably be some sort of float mechanism, as shown in Fig. 1.1. Here a float is suspended at the end of a cable which passes over a pulley to a counterweight. As the float moves up and down with the water level, the pulley turns, and this motion is conveyed through suitable gearing to an output shaft, which makes one revolution over the maximum possible range of water levels. The problem now is to devise a means of transmitting the position of this shaft to the remote control station.

One very simple system is shown in Fig. 1.2, in which the output shaft controls the position of a potentiometer. The resultant output voltage is transmitted by wire to a meter at the remote station. It would appear to be a

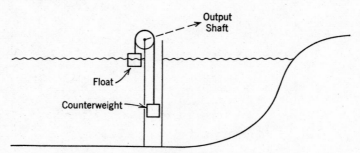

FIGURE 1.1. *Water-level sensing mechanism.*

simple matter to calibrate the meter in terms of water level. This type of system is an analog, or continuous, system, since the voltage can vary continuously between 0 and V. With suitable refinements, this system can work satisfactorily, but there are some problems. The main one is that the voltage at the receiving end is dependent on the condition of the transmission lines as well as on the position of the potentiometer. The transmission lines may be hundreds of miles long and their resistance thus quite sizeable. Furthermore, the resistance will be a function of temperature. For such reasons, the calibration of the meter in terms of water level may be quite unreliable.

A possible method of correcting the deficiencies of the analog system is shown in Fig. 1.3. We assume that the possible range of water levels is 10 feet and that it is sufficient to know the water level within 1 foot. We now let the output shaft control the position of a ten-position rotary switch which applies a voltage to one of ten lines running to the remote station, resulting in one of ten lamps lighting. It is apparent that this system is much more reliable than the analog system. The characteristics of the transmission lines will still affect the exact voltage received at the remote station, but now we need only discriminate between no voltage at all and enough voltage to turn on a lamp.

The system of Fig. 1.3 has the drawback of being quite expensive since it requires ten separate wires. An alternate scheme is shown in Fig. 1.4. Here the output shaft controls a serial pulse generator, which periodically puts

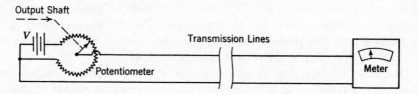

FIGURE 1.2. *Analog system for transmitting water level.*

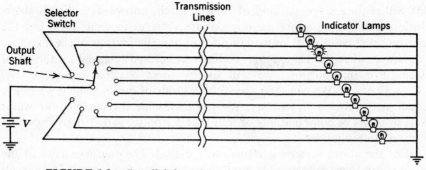

FIGURE 1.3.   *Parallel digital system for transmitting water level.*

out a train of one to ten pulses corresponding to the water level. The operator at the control station can then determine the water level by counting the number of flashes of the indicator lamp. Clearly, the operator would rather not be required to count pulses. Probably an electronic device would be included which would count each sequence of pulses or display this count by lighting one of ten lamps. Such a device would be a simple example of a sequential circuit. (Sequential circuits are the topic of Chapters 9, 10, 11, 12, and 13.)

Since the pulses arrive one at a time, in *series*, the system of Fig. 1.4 is known as a *serial* system, whereas the system of Fig. 1.3 is known as a *parallel* system. The pulse generator in the serial system will be more costly than the simple switch in the parallel system, but this will be more than compensated for by the saving of nine wires. The serial system will thus be less costly than the parallel system but has the disadvantage of being considerably slower. This "trade-off" between speed and cost in parallel and serial systems exists in almost any situation where digital techniques can be applied.

Further improvements in both serial and parallel systems can be made through the use of *coding*. In the parallel system, rather than restricting ourselves to a voltage on only one line, we can use various combinations of voltages on several lines. Since there are $2^4 = 16$ possible combinations of voltage and no-voltage on four lines, the necessary information about ten possible water levels can be carried on just four lines by suitable coding.

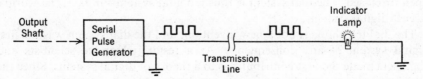

FIGURE 1.4.   *Serial digital system for transmitting water level.*

This will require a special kind of rotary switch, known as a digital shaft-position encoder, to apply the appropriate voltages to four lines, but the extra cost will be more than made up for by the elimination of six lines.

In a somewhat similar manner, coding can be used to speed up the serial system. For this purpose, we will specify three possible voltage levels, $+V$, $-V$, and 0. We can then specify two kinds of pulses, which we might denote as 0 pulses and 1 pulses. We can then code the ten possible water levels in terms of combinations of four 0 and 1 pulses rather than anywhere from one to ten pulses.

All of the above systems are obviously digital. The characteristics of the transmission lines will certainly have some effect on the exact voltage levels at the receiving end. Nevertheless, the voltages will be restricted to two or three widely separated ranges of voltage, and a voltage in one range will be clearly and unmistakably distinguishable from a voltage at one of the other levels. Thus, we see that digital systems are generally far more reliable than analog systems. It would appear from these simple examples that this reliability is expensive, but this is not necessarily the case. For just the reason that they need function only at a few discrete levels, digital devices are often simpler, and consequently less costly, than continuous devices.

1.2 A Brief Historical Note

The concepts of digital systems have found their widest and most dramatic application in recent years, but the ideas are far from new. The reader will recognize in the system of Fig. 1.4 the basic concept of the telegraph, probably the very first practical application of electricity. The telegraph uses a coding technique not mentioned here, the use of pulses of variable length, but it is unquestionably a digital system.

The principle employed in the system of Fig. 1.4 is directly applied in the dial telephone. When you turn the dial and then release it, the spinning dial activates a switch which puts out a number of pulses equal to the number dialed. These pulses are then counted and recorded by special switches at the central office. After all the numbers have been dialed and recorded, they control the operation of a number of switches which connect you to the desired party. A switch is a digital device since it has only two conditions, open or closed. The dial system is thus a digital system—in fact, the world's largest digital system.

The dial telephone is not a new invention, and the dial system was the first digital system of any consequence. As a result, telephone scientists and engineers made the first contributions to a theory of digital systems. Since the telephone system was largely constructed of switches, this theory became known as *switching theory*, the name by which it is still known today.

Shortly before World War II, it was recognized that the theory and technology developed by telephone engineers could be applied to the design of digital computing machines. In 1939, Harvard University was given a contract to develop a machine to compute ballistic tables for the U.S. Navy. The result of this work was Harvard Mark I, which went into operation in 1943. In the next few years, more machines were built in research laboratories around the world. The first commercially produced computer, Univac I, went into operation in 1951. Today, some two decades later, digital computers have revolutionized modern society. During this period of tremendous development, switching theory, or the theory of logical design, developed at a similar pace. Today the theory of logical design is one of the most comprehensive and well-developed in the entire field of engineering.

1.3 Organization of Digital Computers

Since digital computers form the most important single class of digital systems, it is appropriate to consider their basic organization. The basic block diagram of a typical general-purpose digital computer is shown in Fig. 1.5.

The input/output (I/O) section consists of such devices as card readers and punches, printers, paper-tape readers and punches, magnetic tape units, and typewriters. All data and instructions are fed to the computer through the input section, and all results are returned through the output section.

All data and instructions passing through the input section go directly to *Memory*, where they are stored. The memory consists of various devices,

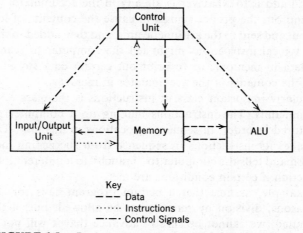

FIGURE 1.5. *Basic organization, typical digital computer.*

usually magnetic or electronic, by means of which information can be stored in such a manner as to be available to other sections of the computer at electronic speeds. All information fed to the computer is divided into basic units, called *words*, *bytes*, or *characters*, depending on the exact organization of the machine. The memory is in turn divided into individual locations, each capable of storing a single word, byte, or character. Each location has associated with it an identification number or *address* by means of which information can be directed to it or retrieved from it.

The information stored in memory consists of data and instructions. Within the machine there is no difference in form between the two; both are simply collections of numbers or characters. The difference lies in what is done with them. Data elements are sent to the *Arithmetic-Logic Unit* (ALU), where they are processed. The ALU basically consists of a number of registers, electronic units for temporarily storing information, and certain logical circuits for performing operations on the numbers stored in the registers. There is often a special register, known as the *accumulator*, which is permanently connected to the *adder* circuits. The basic arithmetic operation of the computer consists of adding a number taken from memory to whatever number is already in the accumulator. The accumulator thus performs a function similar to that of the dials of a desk calculator, accumulating the running sum of numbers sent to it.

Instructions are sent to the control unit, where they are decoded or interpreted. The control unit decides what the instructions call for and sends electronic signals to the various other parts of the computer to cause the appropriate operations to take place. A typical command might be ADD4000, which tells the computer to take the number stored in location 4000 in memory and add it to whatever is already in the accumulator. The control unit will send out the proper signals to cause the contents of location 4000 to be read out and sent to the arithmetic unit, and then added to the accumulator. Other typical instructions might tell the computer to read a card and store the data in memory, or to print out certain data stored in memory, or to store the contents of the accumulator in memory.

A particularly important class of instructions is the class of decision, or *branch*, commands. The instructions making up a computer program are normally stored in order, in sequential locations in memory. The computer normally takes its instructions in sequence from succeeding locations. The branch command tells the computer to "branch" to a different location for its next instruction if certain conditions are met.

As an example, assume that a certain program calls for division. For obvious reasons, division by zero cannot be allowed, but, if the divisor is itself computed, we cannot be sure in advance that it will not be zero. To guard against this situation, we will first load the divisor into the accumulator.

We will then use the *Branch on Zero* command, which tells the computer to inspect the number in the accumulator. If the number is not zero, the computer should take its next instruction from the next sequential location, in the normal manner. If the number is zero, the computer should branch to some other location, as specified, for its next command. At this branch location, the programmer will have stored a set of instructions for dealing with the contingency of a zero divisor.

1.4   Types of Digital Circuits

In this book, we will not be concerned with designing complete computers or other digital systems of comparable complexity. Such *systems design* is the subject of our other book, *Digital Systems: Hardware Organization and Design*. Here we will be concerned with the design of many types of circuits and subsystems which appear within computers and other types of digital systems. These circuits fall into two general categories, *combinational* and *sequential*. An example of a combinational circuit is one for making the branch decision described above. This circuit would have as its input the contents of the accumulator. Its output would take on one value if the accumulator contents were zero, another value if they were not. The name *combinational* conveys the idea that the output at any time is a function solely of the combination of all the inputs at that time. Another example of a combinational circuit in a computer is the adder; the inputs are the two numbers to be added, the output is the sum. Techniques for the design of combinational circuits are developed in Chapters 6, 7, and 8.

An example of a sequential circuit is the accumulator. Its output, i.e., its contents, at any given time will be a function not only of current inputs but also of some *sequence* of prior events, e.g., what numbers have been added into it since it was last cleared. Another sequential circuit in a computer is the program counter. Recall that the instructions making up a program are normally stored in consecutive (sequentially numbered) locations in memory and are executed in sequence unless a branch instruction is executed. The program counter is set to the address of the first instruction in the program and is incremented to succeeding values as instructions are executed sequentially, thus keeping track of progress through the program. It is a sequential circuit because its output (contents) is clearly a function of a series of prior events, i.e., its initial setting and how many times it has been incremented.

While digital computers certainly form the most prominent class of digital systems, digital techniques may be applied to many other areas, some quite unrelated to computers. Until fairly recently, digital techniques were primarily restricted to computing applications, where their high cost could be justified. This situation was changed by the development of the integrated circuit, which has reduced the cost of individual logic elements from dollars to fractions of a cent. As a result, digital techniques are now being applied to virtually every area of technology, and virtually every engineer will have occasion to use digital techniques no matter what his area of specialization.

One very important area for digital applications is instrumentation. For example, in a manufacturing plant, analog readings of various instruments could be connected to a multiplexer at a central point. The multiplexer, a sort of digital switch, would connect each line in turn to an analog-to-digital converter, which would convert each signal to digital form. Each digital signal would then be sent to an assembly register to be joined to an identifying signal generated by the multiplexer. The complete word, containing the identification of the signal and its value, could then be sent to a computer for processing or recorded for later processing. Such a system in which the data is recorded is often referred to as a *data logging* system.

Another important area of digital application is numerical machine tool control. In such systems, a sequence of desired positions of a cutting tool might be recorded in digital form on punched paper tape or magnetic tape. The machine tool would be equipped with sensors to indicate the position of the cutting tool, also in digital form. In operation, the desired next position of the tool will be read off the tape into a register, a sequential device. This position will be compared with the actual position in a comparator, a combinational device. If the desired and actual positions do not agree, signals will be issued to move the tool in the appropriate direction until they do agree.

Digital traffic controllers measure traffic flow on cross streets and adjust the period of traffic signals to maximize flow. Digital clocks measure time simply by counting the cycles of line current. Digital anti-skid controllers measure the speed of the wheels on trucks and adjust the braking force to prevent skids. Ignition controllers measure the operating parameters of automobile engines and adjust carburetion and timing to reduce emissions. Radios and other communications gear are tuned by digital frequency division and phase-locked loops. Digital meters are replacing conventional meters, digital calculators are replacing mechanical office machines, and digital wristwatches will quite likely replace mechanical watches in another ten years. Every day, ingenious designers figure out new ways to apply digital techniques to their problems.

All these various types of systems vary widely in function and complexity, but all are made up of interconnections of individual combinational and sequential circuits. The design of these individual circuits is the basic topic of this book.

Chapter 2 Number Systems

2.1 Introduction

A familiarity with the binary number system is necessary in this book for two primary reasons. First, binary numbers will appear frequently as a notational device. Second, digital computers generally work with binary numbers in some form, and many of the best examples of both combinational and sequential circuits are drawn from subsystems within computers. Some readers will already possess a working knowledge of the binary number system and, indeed, may have first encountered the topic in elementary school. Such readers may wish to skip this chapter and proceed directly to Chapter 3. However, this chapter should be covered by those readers who are not reasonably proficient in the binary number system. First, let us review what is involved in the definition of a number system. We write, for example:

$$7419$$

as a shortened notation for

$$7 \cdot 10^3 + 4 \cdot 10^2 + 1 \cdot 10^1 + 9 \cdot 10^0$$

As the decimal number system is almost universal, the powers of 10 are implied without further notation. We say that ten is the *radix* or *base* of the decimal number system. The fact that 10 was chosen by man as the base of his number system is commonly attributed to his ten fingers. Most any other relatively small number (a large number of symbols would be awkward) would have done as well.

Consider the number

$$N = 110101$$

and assume that the base of N is two. The number system with base-2 is called the *binary* number system. We may rewrite N as follows,

$$N = 1 \cdot 2^5 + 1 \cdot 2^4 + 0 \cdot 2^3 + 1 \cdot 2^2 + 0 \cdot 2^1 + 1 \cdot 2^0$$
$$= 32 + 16 + 0 + 4 + 0 + 1 = 53 \text{ (decimal)}$$

thus obtaining the decimal equivalent. Notice that the binary number system requires only two symbols, 0 and 1, while the familiar ten symbols are employed in the decimal system.

In contrast to men, digital computers do not have ten fingers and are not biased toward the decimal system. They are constructed of physical devices which operate much more reliably when required to switch between only two stable operating points. Some examples are switches, relays, data transmission lines, transistors, etc. One of the stable operating points may be assigned to represent 1 and the other 0. Thus, base-2 must play some role within any digital system.

Many examples throughout the book will be problems in digital computer design. This, however, is not the only reason for introducing binary numbers at this point. We shall find the binary number system indispensable as a notational device in the next few chapters.

The reader who is already familiar with arithmetic in other bases may skip rapidly over the next two sections.

2.2 Conversion between Bases

Suppose that some number N is expressed in base-s. It may be converted to base-r by the following sequence of divisions carried out in base-s. The digits A_i are the remainders from each division, so that $A_i < r$.

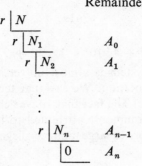

Alternatively, this division may be written as follows:

$$N = r \cdot N_1 + A_0$$
$$N_1 = r \cdot N_2 + A_1$$
$$\cdot$$
$$\cdot$$
$$\cdot$$
$$N_n = r \cdot 0 + A_n \tag{2.1}$$

or as

$$N = r(rN_2 + A_1) + A_0$$
$$= r^2N_2 + rA_1 + A_0$$
$$= r^2(rN_3 + A_2) + rA_1 + A_0$$
$$\cdot \qquad\qquad\qquad \cdot$$
$$\cdot \qquad\qquad\qquad \cdot$$
$$\cdot \qquad\qquad\qquad \cdot$$
$$= A_nr^n + A_{n-1}r^{n-1} + \cdots + A_1r + A_0 \tag{2.2}$$

Example 2.1

Convert 653_{10} (the subscript indicates base-10) to (a) base-2 and (b) base-5.

(a)

```
2 | 653        Check
2 | 326   1    1010001101₂ = 512 + 128 + 8 + 4 + 1
2 | 163   0                = 653₁₀
2 |  81   1
2 |  40   1
2 |  20   0
2 |  10   0
2 |   5   0
2 |   2   1
2 |   1   0
    0     1
```

(a) $1010001101_2 = 512 + 128 + 8 + 4 + 1$
$$= 653_{10}$$

(b)

```
5 | 653
5 | 130   3
5 |  26   0
5 |   5   1
5 |   1   0
    0     1
```

$10103_5 = 1 \times 5^4 + 1 \times 5^2 + 3 \times 5^0$
$$= 1 \times 625 + 1 \times 25 + 3$$
$$= 653_{10} \qquad \blacksquare$$

Example 2.2

Convert the number 1606_{10} to base 12: The decimal numbers 10_{10} and 11_{10} become one-digit numbers in base 12, which may be represented by α and β respectively.

$$
\begin{array}{r|l}
12 & 1606 \\
\cline{1-2}
12 & 133 \quad 10 = \alpha \\
\cline{1-2}
12 & 11 \quad 1 \\
\cline{1-2}
 & 0 \quad 11 = \beta
\end{array}
$$

Check

$\beta 1 \alpha_{12} = 11(144) + 1(12) + 10$

$\qquad = 1606_{10}$ ∎

In each of the above examples, arithmetic was carried out in base-10. This is convenient since we are familiar with base-10 arithmetic. If a number is to be converted from base-r ($r \neq 10$) to some base-b, we have the choice of doing the arithmetic in base-r or else first converting to base-10 (as shown in Sec. 2.1) and then to base-b. Arithmetic in bases other than 10 is the subject of the next section.

The numbers we have discussed so far have been integers. Suppose it is desired to convert numbers with fractional parts from one number system to another. Consider a number

$$N = N_I + N_F = A_n r^n + \cdots + A_1 r + A_0 + A_{-1} r^{-1} + A_{-2} r^{-2} \cdots \qquad (2.3)$$

where N_I and N_F are the integral and fractional parts, respectively. The fractional part in one base will always lead to the fractional part in any other base. Therefore, N_I may be converted as before and N_F converted separately as follows:

$$N_F = A_{-1} r^{-1} + A_{-2} r^{-2} + A_{-3} r^{-3} \cdots \qquad (2.4)$$

To determine the coefficients A_{-1}, A_{-2}, A_{-3}, etc., for the base-r, we note that each of these coefficients is itself an integer. We first multiply by r:

$$r N_F = A_{-1} + A_{-2} r^{-1} + A_{-3} r^{-2} \qquad (2.5)$$

Thus, the integral part of $r N_F$ is A_{-1}. We subtract A_{-1} and again multiply by r:

$$r(r N_F - A_{-1}) = A_{-2} + A_{-3} r^{-1} + A_{-4} r^{-2} \qquad (2.6)$$

thus determining A_{-2}. This process is continued until as many coefficients as desired are obtained. The process may not terminate.

Notice we have avoided the term decimal point or decimal places. For base-2, for example, the appropriate terms are binary point and binary places.

Example 2.3

Convert the number 653.61_{10} to base-2.

$2 \cdot (0.61) = 1.22$ $A_{-1} = 1$ Check

$2 \cdot (0.22) = 0.44$ $A_{-2} = 0$ $653.61_{10} = 1010001101.1001110 \cdots$

$2 \cdot (0.44) = 0.88$ $A_{-3} = 0$ $= 653 + \frac{1}{2} + \frac{1}{16} + \frac{1}{32} + \frac{1}{64} \cdots$

$2 \cdot (0.88) = 1.76$ $A_{-4} = 1$ $= 653 + 0.5 + 0.0625 + 0.03125$

$2 \cdot (0.76) = 1.52$ $A_{-5} = 1$ $+ 0.015625$

$2 \cdot (0.52) = 1.04$ $A_{-6} = 1$ $= 653.609375$ ■

$2 \cdot (0.04) = 0.08$ $A_{-7} = 0$

Base-8 (octal) and base-16 (hexadecimal) are of particular importance in the field of digital computers because of their close relationship to base-2. Since $8 = 2^3$, each octal digit corresponds to three binary digits (bits). The procedure for converting a binary number into an octal number is to partition the binary number into groups of three bits, starting from the binary point, and assign the corresponding octal digit to each group.

$$111001010011.010110011_2 = 7123.263_8$$
$$7 \quad 1 \quad 2 \quad 3 \quad 2 \quad 6 \quad 3$$

The hexadecimal equivalent of a binary number is obtained in a similar fashion, dividing the binary number into groups of four bits. One complication is that we need sixteen distinct symbols for a hexadecimal system. The most common convention is to use the first six letters of the alphabet plus the ten decimal digits, as shown in Fig. 2.1. With this specification of the hexadecimal system, the conversion shown below is straightforward.

$$1101101110000110.10100011_2 = DB86.A3_{16} \qquad (2.8)$$
$$D \quad B \quad 8 \quad 6 \quad A \quad 3$$

Decimal Number	0	1	2	3	4	5	6	7	8	9	10	11	12	13	14	15
Hexadecimal Digit	0	1	2	3	4	5	6	7	8	9	*A*	*B*	*C*	*D*	*E*	*F*

FIGURE 2.1

2.3 Arithmetic with Bases Other Than Ten

Our ability to perform arithmetic in base-10 depends on a set of basic addition and multiplication relations which are committed to memory. These operations become so familiar that we tend to forget the significance of these tables, particularly the addition table. Doing arithmetic in some other base similarly requires a certain familiarization with the counterparts of these tables for that base. We see in Fig. 2.2 the addition and multiplication tables for base-2 and base-5.

All further rules of arithmetic, such as the carry and borrow operations, apply in precise analogy for bases other than 10. We are now ready to consider nondecimal arithmetic.

Example 2.4

Add 321 to 314 and multiply 142 × 32 in base-5.

```
        1        1    ← Carry
               3 1 4                    1 4 2
             +3 2 1                   × 3 2
             -------                   -----
             1 1 4 0                      1    ← Carry
                                        2 3 4
                                      1 2 1    ← Carry
                                      3 2 1
                                    ---------
                                    1 1 1 4 4  ■
```

+	0	1
0	0	1
1	1	10

×	0	1
0	0	0
1	0	1

+	0	1	2	3	4
0	0	1	2	3	4
1	1	2	3	4	10
2	2	3	4	10	11
3	3	4	10	11	12
4	4	10	11	12	13

×	0	1	2	3	4
0	0	0	0	0	0
1	0	1	2	3	4
2	0	2	4	11	13
3	0	3	11	14	22
4	0	4	13	22	31

(a) Base-2   (b) Base-5

(*Note:* Where two digits are included in a square, the most significant may be regarded as a carry.)

FIGURE 2.2. *Addition and multiplication tables.*

Example 2.5

Convert 431_5 to base-2, performing any necessary arithmetic in base-5.

$$
\begin{array}{r|rl}
2 & 431 & \\
2 & 213 & 0 \\
2 & 104 & 0 \\
2 & 24 & 1 \\
2 & 12 & 0 \\
2 & 3 & 1 \\
2 & 1 & 1 \\
& 0 & 1
\end{array}
$$

Check:

$$N = 1 \cdot 2^{11} + 1 \cdot 2^{10} + 1 \cdot 2^4 + 0 \cdot 2^3$$
$$+ 1 \cdot 2^2 + 0 \cdot 2 + 0 \cdot 1$$
$$= (2^3)^2 + 2^2 \cdot 2^3 + 2^2 \cdot 2^2 + 2^2$$
$$= (13)^2 + 4(13) + 4 \cdot 4 + 4$$
$$= 224 + 112 + 31 + 4 = 431 \quad \blacksquare$$

$$N_2 = 1110100$$

It is not difficult to acquire a facility for base-2 arithmetic because of the simplicity of the base-2 addition and multiplication tables. We shall see in succeeding chapters that, for this same reason, physical implementation of base-2 arithmetic is economical.

In the decimal number system, the multiplication of a number by a power of 10 requires only a shifting operation. The same thing is true for powers of 2 in base-2. Consider, for example,

$$25_{10} \times 2^4{}_{10} = (11001 \times 10000)_2 = 110010000$$

Example 2.6

Carry out the following arithmetic operations in base-2. Convert the answers to base-10 as a check.

$$24_{10} + 11_{10} \qquad 24_{10} + 31.5_{10} \qquad 24_{10} - 11_{10} \qquad 24_{10} \times 11_{10}$$

$$
\begin{array}{r|rl}
2 & 24 & \\
2 & 12 & 0 \\
2 & 6 & 0 \\
2 & 3 & 0 \\
2 & 1 & 1 \\
& 0 & 1
\end{array}
\qquad
\begin{array}{r|rl}
2 & 31 & \\
2 & 15 & 1 \\
2 & 7 & 1 \\
2 & 3 & 1 \\
2 & 1 & 1 \\
& 0 & 1
\end{array}
\qquad
\begin{array}{r|rl}
2 & 11 & \\
2 & 5 & 1 \\
2 & 2 & 1 \\
2 & 1 & 0 \\
& 0 & 1
\end{array}
$$

$$24_{10} = 11000_2 \qquad 31.5_{10} = 11111.1_2 \qquad 11_{10} = 1011_2$$

$(24 + 11)$ $(24 + 31.5)$

Check

$$
\begin{array}{r}
11000 \\
+ \ 1011 \\
\hline
100011
\end{array}
\qquad 32 + 2 + 1 = 35
$$

$$
\begin{array}{r}
11000 \\
+ \ 11111.1 \\
\hline
110111.1
\end{array}
$$

Check

$$32 + 16 + 4 + 2 + 1 + \tfrac{1}{2}$$
$$= 55.5$$

Example 2.6   (*continued*).

(24 − 11)

```
    1111  ← borrows
   11000
 −  1011
   ─────
    1101 = 8 + 4 + 1 = 13₁₀
```

(24 × 11)

```
   11000
 ×  1011
   ─────
   11000
   11000
  11000
 ─────────
 100001000 = 256 + 8 = 264₁₀  ■
```

2.4 Negative Numbers

Negative numbers in base-2 may be treated in much the same way as in base-10. For example, in adding a positive number to a negative number, the number smaller in magnitude is subtracted from the larger, and the sign of the larger is assigned to the result. Thus, the mind must carry on a sequence of logical manipulations independent of the actual arithmetic. We shall see that these logical operations may be implemented in the circuitry of an electronic digital computer.

If a slightly different approach is used, the amount of such circuitry required may be greatly reduced, with the additional benefit that the computer need not distinguish between addition and subtraction. That is, the circuitry employed in the operation of addition may also be used for subtraction.

Most computers deal with numbers in the form

$$A \cdot 2^\alpha$$

where A is a number less than 1. Machines equipped with a *floating point* feature provide separate circuitry to accomplish shifting and to keep track of the exponent α. In some machines, the task of assuring that $A < 1$ is left to the programmer.

Suppose x is a negative number expressed in base-2 such that $x < 1$. We define the 2's *complement* of x to be the quantity

$$x_C = 2 - x \tag{2.9}$$

Example 2.7

Determine the 2's complement of $x = .110101100_2$

```
   11.111111     ← borrows
   10.000000000
 −    .110101100
   ─────────────
   01.001010100
```

Notice that the three least significant digits are the same for the number x as for its 2's complement. The remaining digits, however, are complemented. Any zero's at the right end of a fraction are nonsignificant and will have no effect on the complement, since they are subtracted from zero. The least significant 1 when subtracted from zero yields the underlined 1 in the difference. From there on, due to the repeated borrows, each digit is subtracted from a 1. Therefore, each of these digits is complemented in the answer. ■

From the above example, we may state the following rule for obtaining the 2's complement of a number less than 1.

Rule 2.1. The 2's complement of a number $x < 1$ may be obtained by (1) writing a 1 for the digit corresponding to the least significant 1 of x together with 0's for all less significant digits, and (2) by complementing (changing 0's to 1's and 1's to 0's) the remaining digits.

Implementation of Rule 2.1 may be accomplished within a computer without resorting to a separate addition (subtraction) cycle.

Notice, from Example 2.7, that negative numbers in 2's complement form will always be distinguished by a 1 immediately to the left of the binary point. This digit will be zero for properly shifted positive numbers. Thus, the information regarding the sign of a number is not lost in the complementing operation.

The subtraction of a positive number x from a positive number y may be expressed as the addition of the 2's complement of x as follows:

$$y + x_c = y + (2 - x) = (y - x) + 2 \qquad (2.10)$$

Suppose that $x < y$, in which case $y + x_c$ will exceed by two the proper difference $y - x$. This excess 2 will take the form of a carry to the second digit left of the binary point. If, as is actually done in a computer, we merely ignore all but the first digit left of the binary point, the result will be correct. This point of view will yield consistent results for any combination of signs and magnitudes of x and y.

If $x > y$, we may write

$$y + x_c = 2 - (x - y) \qquad (2.11)$$

Now the result is a negative number expressed in 2's complement notation. This case will always be distinguished from the former by the 1 which will appear left of the binary point. Such a 1 would never appear if $x < y$ because the result must necessarily be a positive number that is less than 1.

If two numbers of the same sign, each of magnitude less than one, are added, the magnitude of the result might be greater than one. Unless special steps are taken, a computer could not distinguish between the resulting one to the left of the binary and the 2's complement indicator just mentioned. When the magnitude of a result exceeds one, we have what is termed an *overflow* condition. Most computers provide for the detection of overflow and the generation of an overflow signal. This signal may, at the option of the programmer, initiate some corrective action, such as a right shift and adjustment of the exponent.

Example 2.8

Accomplish the following operations in 2's complement arithmetic.

$$\text{(a) } .1011 - .1001 \qquad \text{(b) } .1011 - .1101$$

(a) $(-.1001)_c$ is found by inspection to be 1.0111

$$
\begin{array}{ll}
1.111 \leftarrow \text{Carry's} & \\
.1011 & \text{Check:} \\
\underline{+1.0111} & .1011 - .1001 = (\tfrac{11}{16} - \tfrac{9}{16})_{10} = \tfrac{1}{8}_{10} \\
0.0010 = \tfrac{1}{8}_{10} &
\end{array}
$$

(b) $(-.1101)_c = (1.0011)$

$$
\begin{array}{ll}
11 & (.1110)_c = (.0010) \quad \text{The 1 to the left of the binary point indicates} \\
.1011 & \text{that this result is negative.} \\
\underline{1.0011} & \\
1.1110 & \text{Check:} \\
& .1011 - .1101 = (\tfrac{11}{16} - \tfrac{13}{16})_{10} = -\tfrac{1}{8}_{10} = -.0010 \quad \blacksquare
\end{array}
$$

2.5   Binary-Coded Decimal Numbers

Although virtually all digital systems are binary in the sense that all signals within the systems may take on only two values, some nevertheless perform arithmetic in the decimal system. Indeed, some digital computers are referred to as "decimal machines." In such systems, the identity of decimal numbers is retained to the extent that each decimal digit is individually represented by a binary code. As there are ten decimal digits, four binary bits are required in each code element. The most obvious choice is to let each decimal digit be represented by the corresponding 4-bit binary number, as shown in Fig. 2.3. This form of representation is known as the BCD (Binary-Coded Decimal) representation or code.

In computer systems, most input or output is in the decimal system since this is the system most convenient for human users. On input, the decimal numbers are converted into some binary form for processing, and this process is reversed on output. In a "straight binary" computer, a decimal number such as 75 would be converted into the binary equivalent, 101011. In a computer using the BCD system, 75 would be converted into 0111 0101.

Decimal Digit	Binary Representation $X_3\ X_2\ X_1\ X_0$
0	0000
1	0001
2	0010
3	0011
4	0100
5	0101
6	0110
7	0111
8	1000
9	1001

FIGURE 2.3.   *Binary-coded decimal digits.*

If the BCD form is used, it must be preserved in arithmetic processing. For example, addition in a BCD machine would be carried out as shown below.

$$
\begin{array}{r r}
37 & 0011\ 0111 \\
+24 & +0010\ 0100 \\
\hline
61 & 0110\ 0001 \\
\text{(Decimal)} & \text{(BCD)}
\end{array}
$$

The principal advantage of the BCD system is the simplicity of input/output conversion; the principal disadvantage is the complexity of arithmetic processing. The choice (between binary and BCD) depends on the type of problems the system will be handling. Computers are built both ways, and the proper choice is a subject of some debate.

The BCD code is not the only possible coding for decimal digits. Any set of at least ten distinct combinations of at least four binary bits could theoretically be used. Indeed, some writers use the term BCD to refer to any set of ten binary codes used to represent the decimal digits, but we shall use BCD only to refer to the code of Fig. 2.3. Some other codes which are commonly used will be discussed in Chapter 8.

PROBLEMS

2.1. Convert the following base-10 numbers to base-2 and check.
   (a) 13   (b) 94   (c) 356
2.2. Repeat Problem 2.1, converting to base-5.

2.3.   Repeat Problem 2.1 for base-8 and base-16.

2.4.   Convert the following base-10 numbers to base-2.

(a) .00625   (b) 43.32   (c) .51

2.5.   Find the base-8 equivalent of the following base-2 numbers.
(a) 10111100101   (b) 1101.101   (c) 1.0111

2.6.   Express (a) 734_8 and (b) 41.5_8 in base-2.

2.7.   Express the base-2 numbers of Problem 2.5 in base-4 without employing division.

2.8.   Construct addition and multiplication tables for base-4.

2.9.   Perform the indicated arithmetic operations in base-5.

(a) 143_5 (b) 124.22_5 (c) 413_5 (d) 404_5
      $+221_5$ $+ 21.34_5$ $\times 32_5$ -213_5

2.10.   Convert the base-5 number $4{,}433{,}214_5$ directly to base-12.

2.11.   Perform the indicated arithmetic operations after converting the operands to base-2.

(a) 659_{10} (b) 695_{10} (c) 272_{10} (d) $\dfrac{272_{10}}{23_{10}}$
      $+272_{10}$ -272_{10} $\times 23_{10}$

2.12.   Repeat Problem 2.11 for the following.

(a) 97_{10} (b) 131_{10} (c) 97_{10} (d) $\dfrac{586_{10}}{39_{10}}$
      $+83_{10}$ -219_{10} $\times 43_{10}$

2.13.   Carry out the following computations after first converting each negative operand to 2's complement notation. Check your results.

(a)   .11101   (b)   .10001   (c) $-$.10101
      $-$.10111 $-$.11011 $-$.11010

2.14.   Carry out the operations of Problem 2.11 in BCD, i.e., convert the operands to BCD and express the results in BCD.

2.15.   Repeat Problem 2.14 for the operations of Problem 2.12.

2.16.   The method of subtracting by adding complements may also be employed in the decimal number system.
(a) Determine the 10's complement of 373.
(b) Accomplish the subtraction $614 - 373$ by adding the 10's complement.

2.17.   One of the most common uses of the octal system is as a "shorthand" notation for binary numbers. Shown in Fig. P2.17 is a table listing a standard binary code for the letters of the alphabet. Produce a new table with the binary codes replaced by their octal equivalents. Note the comparative simplicity of the new table.

A	011000	N	000110
B	010011	O	000011
C	001110	P	001101
D	010010	Q	011101
E	010000	R	001010
F	010110	S	010100
G	001011	T	000001
H	000101	U	011100
I	001100	V	001111
J	011010	W	011001
K	011110	X	010111
L	001001	Y	010101
M	000111	Z	010001

FIGURE P2.17

BIBLIOGRAPHY

1. Ware, W. H. *Digital Computer Technology and Design*, Vol. I, Wiley, New York, 1963.

2. Chu, Y. *Digital Computer Design Fundamentals*, McGraw-Hill, New York, 1962.

3. Flores, I. *The Logic of Computer Arithmetic*, Prentice-Hall, Englewood Cliffs, N.J., 1962.

Chapter 3 Truth Functions

3.1 Introduction

The power of many digital systems, particularly computers, is largely dependent on the capability of their components to (1) make decisions and (2) store information, including the results of decisions. We shall defer consideration of the storage capability until Chapter 9 and first explore the problem of physical implementation of logical decision processes.

Digital computers are not the only physical systems which utilize electronic components to make logical decisions, although in digital computers this practice has reached its highest degree of sophistication. In order to explore the relation of electrical systems to logical decisions, let us consider a relatively simple design problem.

A logic network is to be designed to implement the seat belt alarm which is required on all new cars. A set of sensor switches is available to supply the inputs to the network. One switch will be turned on if the gear shift is engaged (not in neutral). A switch is placed under each front seat and each will turn on if someone sits in the corresponding seat. Finally, a switch is attached to each front seat which will turn on if and only if that seat belt is fastened. *An alarm buzzer is to sound when the ignition is turned on and the gear shift is engaged provided that either of the front seats is occupied and the corresponding seat belt is not fastened.*

These conditions are sufficiently simple that the reader could most likely work out a circuit design by trial and error. We shall take another approach requiring a formal analysis of the statements in the previous paragraph. In this way we shall lead up to formal procedures which can be applied to much more complicated systems.

The statements in the design specification may be listed as follows:

The alarm will sound:	A
The Ignition is on:	I
The Gearshift is engaged:	G
The Left front seat is occupied:	L
The Right front seat is occupied:	R
The Left seat Belt is fastened:	B_L
The Right seat Belt is fastened:	B_R

To each statement, we have assigned a *statement variable*, which will represent the statement and which may take on a *truth value*, **T** or **F**, according to whether the corresponding statement is true or false. In general, we define a statement as any declarative sentence which may be classified as true or false.

The mathematics of manipulating statement variables and assigning truth values is known as *truth-functional calculus*. The application of truth-functional calculus is not restricted to simple positive statements such as those above. First, for any positive statement, there is a corresponding negative statement. For example,

"The left seat belt is *not* fastened": $\bar{B}_L$

Since the statement $\bar{B}_L$ is true whenever B_L is false and is false whenever B_L is true, $\bar{B}_L$ is called the *negation* of B_L. (Other commonly used symbols for the negation of a statement A are $\sim A$ and A'.)

Consider next this statement:

"The left seat belt is not fastened

and the left front seat is occupied": $\bar{B}_L \wedge L$

This is a *truth-functional compound*, a compound statement, the truth value of which may be determined from the truth values of the component statements. The exact relationship between the truth of the component statements and the truth of the compound statements is determined by the *connective* relating the components. In this case, the connective is AND, indicating that the statement $\bar{B}_L \wedge L$ is true if and only if both the component statements, $\bar{B}_L$ AND L, are true. Since AND is a common relationship between statements, we shall assign it a standard symbol, $\wedge$, so that a compound of two statements, A and B, related by AND may be represented by $A \wedge B$.

A	B	$A \wedge B$	$A \vee B$	$A \oplus B$
F	F	F	F	F
F	T	F	T	T
T	F	F	T	T
T	T	T	T	F

FIGURE 3.1

There are only four possible combinations of truth values of the two component statements A and B. We may, therefore, completely define $A \wedge B$ by listing its truth values in a four-row table such as Fig. 3.1. A table of this form is called a *truth table*.

This table also defines $A \vee B$, which symbolizes a statement which is true if and only if either statement A is true, statement B is true, or both of these statements are true. Common usage of the word "or" is not always in agreement with the definition of $A \vee B$. Consider the sentence

"Joe will be driving either a blue sedan or a green convertible."

Clearly, it does not mean that Joe might be driving both cars. The latter usage of "or" is called the *exclusive-or* and is symbolized by $A \oplus B$. The truth values of $A \oplus B$ are also tabulated in Fig. 3.1. The connective $\vee$ is known by contrast as *inclusive-or*. Since this connective turns out to be more common in logical design problems than the exclusive-or, it is standard to refer to it simply as OR, leaving the "inclusive" understood.

One more connective, $\equiv$, must be defined before truth-functional calculus can be applied to the seat belt problem. By $A \equiv B$ it is meant that A and B always have the same truth value, i.e., A is true if and only if B is true. This connective is tabulated in Fig. 3.2.

Let us now attempt to represent the information contained in the verbal specification of the alarm system by a truth-functional equation. Rephrasing the specifications in terms of the simple statements, we see that *the alarm will sound* (A) if and only if *the ignition is on* (I) and *the gearshift is engaged* (G) and *the left front seat is occupied* (L) and *the left seat belt is not fastened*

A	B	$A \equiv B$
F	F	T
F	T	F
T	F	F
T	T	T

FIGURE 3.2

($\bar{B}_L$), or *the right front seat is occupied* (R) and *the right seat belt is not fastened* ($\bar{B}_R$). As a truth functional equation, all this is simply written as

$$A \equiv I \wedge [G \wedge ((L \wedge \bar{B}_L) \vee (R \wedge \bar{B}_R))]. \tag{3.1}$$

The design process is certainly not complete, but we have seen that the truth-functional calculus enables us to put the specifications in a form that is compact, unambiguous, and suitable for mathematical manipulation. It should be noted that the designer must be able to express the specifications in the form of statements related by clearly defined connectives. Not all declarative sentences are subject to such expression. Consider the sentence

"The alarm sounded because the right seat belt was not fastened."

This is not a truth-functional compound, since the truth of the statement cannot be determined solely from the truth of the component statements. For example, even if both components are true, the compound statement may not be true, since the right seat may have been unoccupied, in which case the latter half of the statement is true, but the alarm was set off by the driver not fastening his seat belt. Such sentences cannot be handled by truth-functional calculus. In the next four sections, it will become clear that the basic usefulness of electrical circuits in making logical decisions is limited to situations in which truth-functional calculus is applicable. Thus, the designer must avoid formulating a design statement in such a way that it does not constitute a truth-functional compound.

3.2   Binary Connectives

The connectives defined by the truth tables of Figs. 3.1 and 3.2 are *binary*, since they relate only two component statements. Each binary connective corresponds to a unique assignment of truth values to the four rows of the truth table, corresponding to the four possible combinations of truth values of two component statements. There are $2^4 = 16$ ways of arranging T's and F's on four rows, so there are sixteen possible binary connectives, as tabulated in Fig. 3.3. Each connective is numbered, and symbols are indicated for those already defined and for others of particular interest.

Note particularly connective 13, which may be written

$$A \supset B$$

and which represents the compound sentence:

"If A is true, then B is true."

A	B	0	$\wedge$ 1	2	A 3	4	B 5	$\oplus$ 6	$\vee$ 7	$\downarrow$ 8	$\equiv$ 9	$\bar{B}$ 10	11	$\bar{A}$ 12	$\supset$ 13	$\uparrow$ 14	15
F	F	F	F	F	F	F	F	F	F	F	T	T	T	T	T	T	T
F	T	F	F	F	F	T	T	T	T	F	F	F	F	T	T	T	T
T	F	F	F	T	T	F	F	T	T	F	F	T	T	F	F	T	T
T	T	F	T	F	T	F	T	F	T	F	T	F	T	F	T	F	T

FIGURE 3.3

This is similar to the *if and only if* statement discussed previously, except that B is not necessarily false if A is false. For example, consider the statement

$$(2 > X > 1) \supset (X > 0) \tag{3.2}$$

If $X = .5$, for example, the statement $X > 0$ is true; but the statement $2 > X > 1$ is false. There is no contradiction to statement 3.2, and it is true as defined in Fig. 3.3.

The truth values for the *if-then* statement may seem unnatural at first, but this concept is necessary in many mathematical arguments. The distinction between $\supset$ and $\equiv$ is very important. One is required to prove both

$$A \supset B \quad \text{and} \quad B \supset A$$

in order to assert that

$$B \equiv A \tag{3.3}$$

We shall leave it as a problem for the reader to prove that $(A \supset B) \wedge (B \supset A)$ means the same as $B \equiv A$. That is,

$$(B \equiv A) \equiv [(A \supset B) \wedge (B \supset A)] \tag{3.4}$$

Since the logic designer is often concerned with converting natural language specifications into precise mathematical form, he should keep in mind that, in conventional usage, *if-then* may be thoughtlessly used when *if and only if* is intended. For example, a carelessly written specification might say:

"If the switch is closed, then the circuit output will be 5 volts,"

although what is almost certainly meant is:

"The circuit output will be 5 volts if and only if the switch is closed."

The designer thus has a dual responsibility—to be precise in his own writing of specifications and to be careful in interpreting specifications written by others.

Also of special interest are connectives 8 and 14,* written

$$A \uparrow B \equiv \overline{A \wedge B} \qquad (3.5)$$

and

$$A \downarrow B \equiv \overline{A \vee B} \qquad (3.6)$$

$A \uparrow B$ is the negation of the AND relation, i.e., its truth values are precisely opposite those of $A \wedge B$. It is commonly known as the NAND (not AND) relation. $A \downarrow B$ is similarly the negation of $A \vee B$ and is known as the NOR relation. The usefulness of these two connectives will become clear in later sections.

3.3   Evaluation of Truth Functions

We are now ready to consider compounds related by more than one binary connective. Suppose, for example, we are interested in the truth value of

$$\bar{A} \vee (A \supset B) \qquad (3.7)$$

for the case where both A and B are false. We replace the statements A and B in the expression by the truth value **F**, yielding

$$\bar{\mathbf{F}} \vee (\mathbf{F} \supset \mathbf{F}) \equiv \mathbf{T} \vee \mathbf{T} \equiv \mathbf{T} \qquad (3.8)$$

As $A \supset B$ is defined in Fig. 3.3 to be true where A and B are both false, $\mathbf{F} \supset \mathbf{F}$ may be replaced by the value **T**. Similarly, $\bar{\mathbf{F}}$ may be replaced by **T**. Furthermore, two true statements related by OR also produce a true statement, completing the evaluation in expression 3.8. In each case, successive steps of the evaluation are connected by $\equiv$, indicating that the truth value is the same for the expressions on either side of the symbol. Expression 3.7 is evaluated for the three remaining combinations of values of A and B in Fig. 3.4.

A	B	$\bar{A}$	$A \supset B$	$\bar{A} \vee (A \supset B)$	$\bar{A} \vee B$
F	F	T	T	T	T
F	T	T	T	T	T
T	F	F	F	F	F
T	T	F	T	T	T

FIGURE 3.4

* Connective 14 in Fig. 3.3 is often designated by a simple stroke, $A \mid B$, and called *Sheffer's stroke* in honor of H. M. Sheffer. In 1913, Sheffer presented, using "$\mid$", a set of postulates defining both $\uparrow$ and $\downarrow$.

A	B	$\bar{A}$	$\bar{A} \vee A$	$(\bar{A} \vee A) \supset B$	B
F	F	T	T	F	F
F	T	T	T	T	T
T	F	F	T	F	F
T	T	F	T	T	T

FIGURE 3.5

Notice that the resulting truth value in each row is the same as that defined in Fig. 3.3 for the statement $A \supset B$. We may, therefore, write

$$\bar{A} \vee (A \supset B) \equiv (A \supset B) \qquad (3.9)$$

Note also the parentheses in expression 3.7; without them, this expression might be interpreted as

$$(\bar{A} \vee A) \supset B \qquad (3.10)$$

The truth values of this expression as compiled in Fig. 3.5 are not the same as for expression 3.7. We note instead that

$$(\bar{A} \vee A) \supset B \equiv B \qquad (3.11)$$

We may conclude, then, that the placement of parentheses is essential to the meaning of these expressions.

Any expression in terms of only two statements, A and B, related by any number of binary connectives with parentheses designating a unique order of operation can be reduced to a list of four truth values corresponding to each combination of values of A and B. As all sixteen such functions are listed in Fig. 3.3, the original expression may be replaced by A and B related by a single binary connective.

3.4   Many-Statement Compounds

Consider a statement Z, which can be symbolized by

$$Z \equiv (H \wedge \bar{R}) \supset D \qquad (3.12)$$

A truth table analysis of this sort of function is distinguished from those is the previous section only in that the truth tables have more than four rows. In this case, an eight-row truth table is required to provide for all possible combinations of truth or falsity of H, R, and D. This table with a derivation of truth values for expression 3.12, appears in Fig. 3.6. By reference

H	R	D	$\bar{R}$	$H \wedge \bar{R}$	$(H \wedge \bar{R}) \supset D$	$\overline{H \wedge \bar{R}}$	$\overline{(H \wedge \bar{R})} \vee D$
F	F	F	T	F	T	T	T
F	F	T	T	F	T	T	T
F	T	F	F	F	T	T	T
F	T	T	F	F	T	T	T
T	F	F	T	T	F	F	F
T	F	T	T	T	T	F	T
T	T	F	F	F	T	T	T
T	T	T	F	F	T	T	T

FIGURE 3.6

to Fig. 3.3, the reader can verify that

$$A \supset B \equiv \bar{A} \vee B. \tag{3.13}$$

It then follows immediately that

$$Z \equiv (H \wedge \bar{R}) \supset D \equiv (\overline{H \wedge \bar{R}}) \vee D \tag{3.14}$$

The validity of expression 3.14 is also demonstrated in Fig. 3.6. Notice that the right side of 3.14 employs only $\wedge$, $\vee$, and NOT. We shall see that any truth-functional compound, of any number of statements, may be expressed in terms of these three binary connectives. A proof of this assertion must await methods to be developed in Chapter 6.

Following the same procedure, an abbreviated truth table for the seat belt problem is determined in Fig. 3.7. Only the final 16 of the 64 rows of a six-variable truth table are shown. On the other 48 rows, I or G are false so that $A \equiv \mathbf{F}$.

In some problems, it may be less confusing to identify the statement variables and compile the truth table directly rather than trying to first formulate a truth-functional compound. This can be accomplished by first listing the 2^n combinations of truth values of the n variables. Then the truth value of the compound can be deduced from the problem statement for each row of the truth table.

3.5   *Physical Realizations*

At this point, a brief consideration of the physical realization of binary connectives may help clarify the relationship between the truth functional calculus and logic circuits. Although any of the connectives of Fig. 3.3

I	G	L	B_L	R	B_R	$\bar{B}_R$	$(R \wedge \bar{B}_R)$	$(L \wedge \bar{B}_L) \vee (R \wedge \bar{B}_R)$	A
T	T	F	F	F	F	T	F	F	F
T	T	F	F	F	T	F	F	F	F
T	T	F	F	T	F	T	T	T	T
T	T	F	F	T	T	F	F	F	F
T	T	F	T	F	F	T	F	F	F
T	T	F	T	F	T	F	F	F	F
T	T	F	T	T	F	T	T	T	T
T	T	F	T	T	T	F	F	F	F
T	T	T	F	F	F	T	F	T	T
T	T	T	F	F	T	F	F	T	T
T	T	T	F	T	F	T	T	T	T
T	T	T	F	T	T	F	F	T	T
T	T	T	T	F	F	T	F	F	F
T	T	T	T	F	T	F	F	F	F
T	T	T	T	T	F	T	T	T	T
T	T	T	T	T	T	F	F	F	T

FIGURE 3.7

might theoretically be realized physically, in practice only AND, OR, NAND, NOR, exclusive-OR, and negation are implemented in logic circuits. Circuits realizing the first five of these connectives are known as *gates*, and circuits realizing negation are known as *inverters*. A logic gate is an electrical or electronic device* with one output lead and an arbitrary number of input leads. The voltage potential, with respect to ground, of any input or output lead may take on one of only two distinct values. One of the voltages will represent true throughout the system, and the other will represent false.

In an AND gate, the output will be at the voltage representing true if all inputs are at that voltage and will be at the voltage representing false if any of the inputs are at that voltage. For an OR gate, the output will be true if any inputs are true and will be false only if all inputs are false. For other types of gates, the input and output voltages must correspond in the same manner to the truth values given in Fig. 3.3 for the corresponding connective. An inverter has one input and one output, with the output being false when the input is true, and vice-versa. A discussion of the circuit aspects of these logic elements will be deferred until Chapter 5.

---

* Although electronic gates are by far the most common, there are other possible forms of physical realization, e.g., hydraulic and pneumatic.

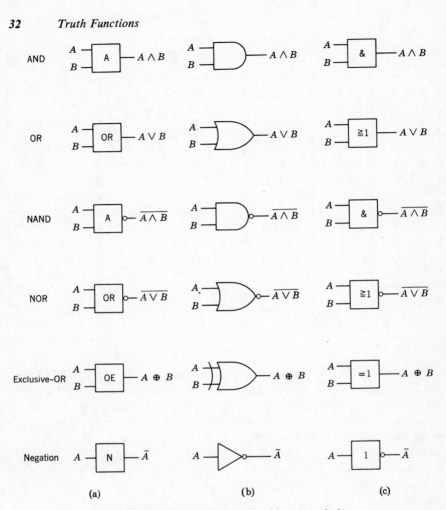

FIGURE 3.8.   IEEE Standard logic symbols.

The most common pictorial representation of logic circuits is the *block diagram*, in which the logic elements are represented by standard symbols. The standard logic symbols specified by IEEE Standard No. 91 (ANSI Y32.14, 1962) are shown in Figs. 3.8a and 3.8b. Although the symbols are shown with two inputs for the gates, AND, OR NAND, and NOR gates may have any number of inputs. Exclusive-OR, however, is defined only for two inputs.

The reader will note that there are two types of symbols, the *uniform shape* symbols and the *distinctive shape* symbols.

Note added in Proof: A new standard, IEEE Std 91-1973 (ANSI-Y32.14-1973) uses the same distinctive shape symbols but different uniform shape

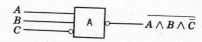

FIGURE 3.9. *Example of dot negation.*

symbols, as shown in Fig. 3.8c. We chose to use the uniform shape symbols of Fig. 3.8a to avoid possible confusion with MIL-STD-806B, which uses the same distinctive shape symbols, but with slightly different meaning in a few special cases. The new standard supersedes MIL-STD-806B and is consistent with international standards (IEC Publication 117-15), while the distinctive shape symbols are used only in the United States. Although the symbols of Fig. 3.8a are used in this edition we strongly recommend the adoption of the newer standard of Fig. 3.8c.

There are two ways of indicating inversion or negation. One is the inverter symbol as shown in Fig. 3.8. The second is a small circle on the signal line where it enters or leaves a logic symbol, sometimes referred to as *dot negation*. This system is used for NAND and NOR symbols, as shown in Fig. 3.8, and Fig. 3.9 gives an example of a more complex use of this notation. Generally, the dot negation should be used only if the inversion is actually accomplished within the circuitry of the associated gate. If the inverter is a physically distinct element, it is preferable to use the separate symbol.

Let us now return to the seat belt alarm problem as symbolized by expression 3.1, which is repeated as

$$A \equiv I \land [G \land ((L \land \bar{B}_L) \lor (R \land \bar{B}_R))] \tag{3.15}$$

As in the determination of truth tables, a physical realization of expression 3.15 may be obtained by first realizing the innermost statements and then working outward. Letting a positive voltage represent a true statement and 0 volts a false statement, we obtain Fig. 3.10. Consistent with the original problem statement, a statement variable is true when the corresponding switch is closed.

The reader should be able to verify that the output of the circuit Fig. 3.10 will correspond to true for those combinations of inputs indicated in the truth table of Fig. 3.7. As will be illustrated presently, it is also possible to determine a logical expression for a truth-functional compound directly from its truth table. With this in mind, we may phrase the design procedure for decision circuits as follows.

Step 1.   Obtain a statement of the design problem which can be symbolized as a truth-functional compound or translated directly to a truth table.

Step 2.   Obtain an expression for the output of the problem in terms of the connectives AND, OR, and NOT.

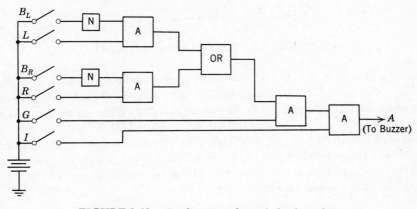

FIGURE 3.10.   *Realization of seat belt alarm logic.*

Step 3.* Obtain the logical expression, equivalent to that obtained in Step 2, which will result in the most economical physical realization.

Step 4. Construct the physical realization corresponding to Step 3.
This procedure will be illustrated in the following example.

Example 3.1

The flight of a certain satellite may be controlled in any one of three ways: by ground control, manually by the astronaut, or automatically by a computer aboard the satellite. The ground station, which has ultimate authority, will cause a voltage source, S, aboard the satellite to assume a true level when

"Control is aboard the satellite": S

is true. The astronaut then decides between manual and automatic control. He switches a voltage source, M, to the false level when

"Control is manual": M

if false. A check signal C generated within the computer will remain at the true level as long as the computer is functioning properly and is able to assume or remain in control.

As long as C is true, if S is true and M is false, then a panel lamp is lit both at ground control and aboard the satellite. Should this lamp go out, an alarm will result, and manual or ground control will be assumed immediately. This may be symbolized by

$$(S \wedge \overline{M}) \supset C \equiv L \tag{3.16}$$

We have not discussed a physical implementation of $\supset$. In fact, none is commonly available. We have already established, however, from the truth tabulations in

* This step is the topic of Chapters 6 and 7.

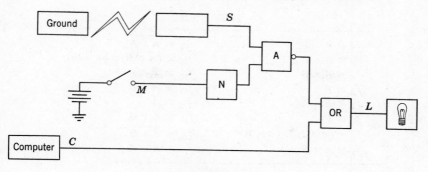

FIGURE 3.11

Fig. 3.6 that

$$(S \wedge \overline{M}) \supset C \equiv \overline{(S \wedge \overline{M})} \vee C \tag{3.17}$$

A physical implementation of the right side of Eq. 3.17 is shown in Fig. 3.11. The reader should verify that the lamp will light for the proper combinations of input truth values as given in Fig. 3.6.   ■

3.6   Sufficient Sets of the Connectives

We saw in the previous section that certain of the binary connectives in Fig. 3.3 could be expressed in terms of various other connectives. The question arises: Can a smaller set of connectives be found in terms of which all sixteen functions can be expressed? Let us search for such a smaller set.

For example, consider the set of connectives containing only $\wedge$ and $\vee$. The case where A and B are both false is critical. Clearly, $A \wedge B$ and $A \vee B$ must both be false. By way of induction, visualize a set of statements, arrived at by relating false statements A and B by $\wedge$ and $\vee$ in a fashion of arbitrary complexity, but such that they are all false. Now relate any two of these statements, or relate one of them to A or B, utilizing $\wedge$ or $\vee$. By definition, the result must be another false statement. As the set initially contained only false statements A and B, induction tells us that the set will always include only false statements. In Fig. 3.3, however, note that functions 8 through 15 must be true when A and B are both false. Therefore, these eight functions cannot be realized by any expression written in terms of only the connectives $\wedge$ and $\vee$.

Suppose the NOT operation is added to our set of connectives. The reader who intuitively expects that all functions can be written in terms of these three connectives is quite correct. That each of the functions in Fig. 3.3 can be expressed in terms of $\wedge$, $\vee$, and "–" may be verified by determining

TABLE 3.1

$f_0 \equiv A \wedge \bar{A}$	$f_8 \equiv \overline{A \wedge B}$
$f_1 \equiv A \wedge B$	$f_9 \equiv (\bar{A} \wedge \bar{B}) \vee (A \wedge B)$
$f_2 \equiv A \wedge B$	$f_{10} \equiv \bar{B}$
$f_3 \equiv A$	$f_{11} \equiv A \vee \bar{B}$
$f_4 \equiv B \wedge \bar{A}$	$f_{12} \equiv \bar{A}$
$f_5 \equiv B$	$f_{13} \equiv \bar{A} \vee B$
$f_6 \equiv (\bar{A} \wedge B) \vee (A \wedge \bar{B})$	$f_{14} \equiv \overline{A \wedge B}$
$f_7 \equiv A \vee B$	$f_{15} \equiv A \vee \bar{A}$

truth values for the corresponding expression listed in Table 3.1. This task is left to the reader.

The reader may wonder whether any list of fewer than three connectives might be sufficient. The answer is yes. Either AND and NOT or OR and NOT are sufficient sets. As will be apparent in Chapter 5, statements can usually be expressed most conveniently by using all three.

At this point, we pause briefly in our discussion of sufficient sets of connectives to verify, for truth functional calculus, two very important theorems called *DeMorgan's Theorems.* We shall prove these theorems again, algebraically, in Chapter 4. As they facilitate our discussion, we introduce them here.

THEOREM 3.1.

(1) $A \downarrow B \equiv \overline{A \vee B} \equiv \bar{A} \wedge \bar{B}$

(2) $A \uparrow B \equiv \overline{A \wedge B} \equiv \bar{A} \vee \bar{B}$

Proof: As elsewhere in the chapter, the proof consists of a truth table analysis as given in Fig. 3.12.

Using DeMorgan's Theorems, we see that $\wedge$, $\vee$, and NOT can be expressed in terms of $\uparrow$ and in terms of $\downarrow$. Clearly $X \wedge X \equiv X$, so we have an expression

A	B	$A \vee B$	$\overline{A \vee B}$	$\bar{A}$	$\bar{B}$	$\bar{A} \wedge \bar{B}$	$A \wedge B$	$\overline{A \wedge B}$	$\bar{A} \vee \bar{B}$
F	F	F	T	T	T	T	F	T	T
F	T	T	F	T	F	F	F	T	T
T	F	T	F	F	T	F	F	T	T
T	T	T	F	F	F	F	T	F	F

FIGURE 3.12. *Truth functional proof of DeMorgan's theorems.*

for $\bar{X}$ in Equation 3.18. Using the obvious

$$\bar{X} \equiv \overline{X \wedge X} \equiv X \uparrow X \tag{3.18}$$

theorem, $\bar{\bar{A}} \equiv A$, and DeMorgan's Theorems we have expressions for $X \wedge Y$ and $X \vee Y$ in Equations 3.19 and 3.20.

$$X \wedge Y \equiv \overline{X \uparrow Y} \equiv (X \uparrow Y) \uparrow (X \uparrow Y) \tag{3.19}$$

$$X \vee Y \equiv \overline{\bar{X} \vee \bar{Y}} \equiv \overline{\bar{X} \wedge \bar{Y}} \equiv \bar{X} \uparrow \bar{X} \equiv (X \uparrow X) \uparrow (Y \uparrow Y) \tag{3.20}$$

We similarly obtain expressions 3.21, 3.22, and 3.23 in terms of $\downarrow$.

$$X \equiv \overline{X \vee X} \equiv X \downarrow X \tag{3.21}$$

$$X \wedge Y \equiv \overline{\overline{X \wedge Y}} \equiv \overline{\bar{X} \vee \bar{Y}} \equiv \bar{X} \downarrow \bar{Y} \equiv (X \downarrow X) \downarrow (Y \downarrow Y) \tag{3.22}$$

$$X \vee Y \equiv \overline{X \downarrow Y} \equiv (X \downarrow Y) \downarrow (X \downarrow Y) \tag{3.23}$$

We now conclude that any truth-functional compound may be expressed in terms of a single connective, which may be either "$\downarrow$" or "$\uparrow$". An $\wedge$, $\vee$, "$-$" expression may be converted to a $\uparrow$ expression by replacing the former connectives and the statements to which they apply by the appropriate expression out of 3.18, 3.19, and 3.20. For instance, f_{13} in Fig. 3.3 becomes

$$\bar{A} \vee B \equiv (A \uparrow A) \vee B \equiv [(A \uparrow A) \uparrow (A \uparrow A)] \uparrow [B \uparrow B] \tag{3.24}$$

If the "obvious" theorem $\bar{\bar{A}} \equiv A$ (easily verified by truth table analysis) is applied, a much simpler expression will result. That is,

$$[(A \uparrow A) \uparrow (A \uparrow A)] \uparrow [B \uparrow B] \equiv (\bar{A} \uparrow \bar{A}) \uparrow (B \uparrow B)$$

$$\equiv \bar{\bar{A}} \uparrow (B \uparrow B)$$

$$\equiv A \uparrow (B \uparrow B) \tag{3.25}$$

Certainly, there are other theorems which might be used to simplify complex logical expressions. A systematic exposition of such theorems in the form of Boolean algebra is the topic of Chapter 4.

The observation that all logical expressions can be expressed in terms of either $\downarrow$ or $\uparrow$ previews an important notion which will be developed in Chapter 5. That is, all logic circuits can be constructed using only NOR gates or only NAND gates.

3.7   A Digital Computer Example

Before leaving this chapter, let us formulate one final example. For this, we have chosen a complex statement which must be realized physically in the design of most digital computers.

On most computers, there is a provision for the detection of *overflow*. Overflow occurs when an arithmetic operation produces a result which is too large to be correctly handled by the machine. When this occurs, a signal must be generated which may, typically, cause the computer to switch from the regular program to a special routine for dealing with this contingency. For this example, we shall assume that the operands and results are all placed in registers, or temporary storage locations, of the same fixed capacity of N digits. Then the addition of two N-digit numbers may produce an $N + 1$-digit number too large to be stored, in which case we have *additive overflow*. Multiplication and division can also cause overflow, but we shall restrict our attention in this example to additive overflow.

If only addition of the positive numbers is considered, overflow is always indicated by a carry from the most significant digit; but if subtraction or addition of negative numbers is also considered, the situation becomes more complicated. As discussed in Chapter 2, subtraction is often accomplished in computers by addition of complements, and negative numbers are often represented by their complements. When complement arithmetic is used, the relationship between overflow and carry from the most significant digit depends on the operation and the signs of operands. For details, we refer the interested reader to books on computer arithmetic [1, 2]. For this example, we ask the reader to accept the following statements as being true for a machine using complement arithmetic.

1. When the operation is addition and both addend and augend are positive, overflow is indicated by a carry from the most significant digit.
2. When the operation is addition and both addend and augend are negative, overflow is indicated by the absence of carry from the most significant digit.
3. When the operation is subtraction and the minuend is positive and the subtrahend negative, overflow is indicated by a carry from the most significant digit.
4. When the operation is subtraction and the minuend is negative and the subtrahend positive, overflow is indicated by absence of a carry from the most significant digit.

In order to symbolize the above statements as truth-functional compounds, let us assign variables to the various component statements as follows:

Statement	Truth variable
The operation is addition	A
The operation is subtraction	S
The sign of the augend is negative	W
The sign of the addend is negative	X
The sign of the minuend is negative	Y
The sign of the subtrahend is negative	Z
There is a carry from the most significant digit	C
An additive overflow error has occurred	E

Note that we have not included statements about the signs of the operands being positive. The reason for this is that all numbers in a computer are either positive or negative. Therefore, e.g., the statement "The augend is positive" is properly represented by $\overline{W}$. It would be equally valid to assign its truth variables to statements that the signs of the operands are positive and let the conditions of negative signs be represented by the negative truth variables.

To simplify discussion, let us designate the overflow in each of the above cases as E_1, E_2, E_3, and E_4 respectively. Now let us consider the first statement about overflow. This statement says that overflow exists, i.e., E_1 is true when statements A and C are true and $\overline{W}$ and $\overline{X}$ are true. Thus, the statement is represented by the truth-functional equation

$$E_1 = (A \wedge C) \wedge (\overline{W} \wedge \overline{X}) \tag{3.26}$$

If the two AND statements enclosed by parentheses in 3.26 are generated by physical AND circuits whose outputs are the inputs to a third AND circuit, E_1 will be the output of this third circuit. This is illustrated in Fig. 3.13.

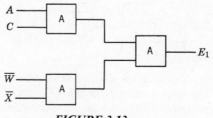

FIGURE 3.13

The following expressions for E_2, E_3, and E_4 are similarly developed:

$$E_2 = (A \wedge \bar{C}) \wedge (W \wedge X) \tag{3.27}$$

$$E_3 = (S \wedge C) \wedge (\bar{C} \wedge Z) \tag{3.28}$$

$$E_4 = (S \wedge \bar{C}) \wedge (Y \wedge \bar{Z}) \tag{3.29}$$

The physical implementations of these expressions are similar to the configuration in Fig. 3.13.

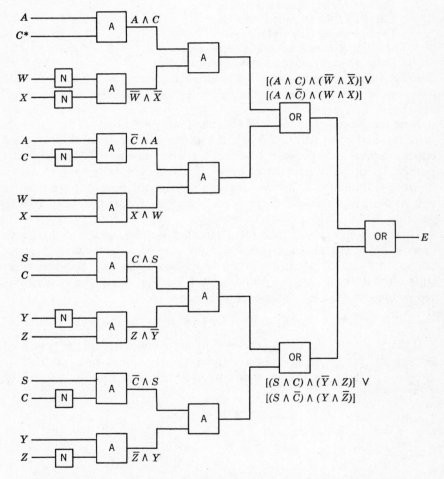

FIGURE 3.14. *Initial block diagram of overflow detector. *All lines labeled with the same switching variable (e.g., the four lines labeled C are assumed to be electrically common); to draw in all the interconnections would complicate the diagram unnecessarily.*

The statement E is true, i.e., overflow occurs in the machine whenever E_1 or E_2 or E_3 or E_4 is true. Relating expressions 3.26 to 3.29 in this manner results in

$$E = [[(A \wedge C) \wedge (\overline{W} \wedge \overline{X})] \vee (A \wedge \overline{C}) \wedge (W \wedge X)]] \vee$$
$$[[(S \wedge C) \wedge (\overline{Y} \wedge Z)] \vee [(S \wedge \overline{C}) \wedge (Y \wedge Z)]] \qquad (3.30)$$

Implementing the three OR connectives together with the circuits for E_1, E_2, E_3, and E_4 results in the circuit in Fig. 3.14. The output of this circuit will be the voltage representing true whenever overflow occurs in the machine.

The reader may have noticed that the procedure followed here is an example of the design philosophy described in the introduction. A mathematical model, consisting of the truth-functional expression 3.30 and the corresponding physical realization in Fig. 3.14, was developed for a problem, which was first specified as a set of natural language statements. It remains to simplify this expression and thereby the physical realization, utilizing an algebra of logic to be defined in the next chapter.

PROBLEMS

3.1. Symbolize the truth-functional compound describing the following design problem. Determine the truth values of the compound.

The flow of water into the brine solution to be used in a chemical process will be turned off if and only if
1. the tank is full; or
2. the output is shut off, the salt concentration does not exceed 2.5%, and the water level is not below a designated minimum level.

3.2. Symbolize the following statement and determine its truth values.

If it is hot in Arizona and it is raining outside or demonstrators are in the streets, then it is hot in Arizona, demonstrators are in the streets, and it is snowing in Argentina.

3.3. A burglar alarm system for a bank is to be operative only if a master switch at the police station has been turned on. Subject to this condition, the alarm will ring if the vault door is disturbed in any way, or if the door to the bank is opened unless a special switch is first operated by the watchman's key. The vault door will be equipped with a vibration sensor which will cause a switch to close if the vault door is disturbed, and a switch will be mounted on the bank door in such a way that it will close whenever the bank door is opened.

Symbolize the above system as a truth functional compound and construct the corresponding logic block diagram.

3.4. In many cars, the seat belt alarm buzzer is also used to warn against leaving the key in the ignition or leaving the lights on when the car is unoccupied. The following statement describes how such a system might operate.

The alarm is to sound if the key is in the ignition when the door is open and the motor is not running, or if the lights are on when the key is not in the ignition, or if the driver seat belt is not fastened when the motor is running, or if the passenger seat is occupied and the passenger seat belt is not fastened when the motor is running.

Symbolize the above system as a truth functional compound and construct the corresponding logic block diagram. You may assume that switch contacts are available which indicate the occurrence of the various individual conditions.

3.5. Using a truth table, verify that

$$(A \supset B) \supset [A \lor \overline{(\overline{A} \land B)}] \equiv A \land \bar{B}$$

3.6. Express $X \oplus Y$ in terms of only the connective $\downarrow$.

3.7. Prove that no connective in Fig. 3.3 other than $\downarrow$ or $\uparrow$ will serve as the only binary connective in expressions for every function in that figure.

3.8. Tabulate the truth values of:
(a) Example 3.1.
(b) The example of Sec. 3.7.

3.9. Verify Expression 3.3 by a tabulation of truth values.

3.10. Suppose three one-digit binary numbers, a, b, and c, are to be added together to form a two-digit number whose digits are denoted by $s_1 s_2$. For each of the above five binary digits, define the corresponding capital letter to be a statement variable which is True whenever the small letter digit is 1 and False otherwise. Determine a truth functional expression for S_2 such that S_2 will be true whenever $s_2 = 1$. Determine a similar expression for S_1. Construct the diagram of logic circuits realizing S_2 and S_1.

3.11. An argument consists of several compound statements regarded as *premises* with one additional compound statement called the conclusion, which is asserted to follow logically from the premises. We say that the argument is valid if the conclusion is true whenever the conjunction of the premises is true. Determine by using truth tables the validity of each of the following arguments:

(a) If I work then I earn money, and if I don't work then I enjoy myself. Therefore, if I don't earn money I enjoy myself.

(b) The governor of California hails from either Los Angeles or San Francisco. Mr. Jones doesn't come from San Francisco. Therefore, if Mr. Jones is not from Los Angeles, then Mr. Jones is not governor of California.

(c) If Harvard wins the Ivy League football championship, then Princeton will be second; and if Princeton is second, then Dartmouth will place in second division. Harvard will win or Dartmouth will not finish among the top four teams. Therefore, Princeton will not take second place.

BIBLIOGRAPHY

1. Flores, I. *Logic of Computer Arithmetic*, Prentice-Hall, Englewood Cliffs, N.J., 1963.

2. Hill, F. J. and G. R. Peterson. *Digital Systems: Hardware Organization and Design*, Wiley, New York, 1973.

3. Copi, I. M. *Symbolic Logic*, Macmillan, New York, 1954.

4. Quine, W. V. *Mathematical Logic*, Harvard University Press, Cambridge, Mass. 1955.

Chapter 4 *Boolean Algebra*

4.1 Introduction

Associated with the physical realization of any logic element is a cost. Similarly, a cost may be assigned to any logic circuit equal to the total cost of all the logic elements in the circuit. For example, the cost of the circuit in Fig. 3.14 is the cost of 3 OR gates plus the cost of 12 AND gates plus the cost of 6 inverters.

Where it is necessary to construct a physical realization of any truth function, it is highly desirable that this be done at minimum cost. It is possible to verify that a given circuit realizes the proper truth function by checking its output against the truth table for each combination of inputs. Similarly, two functions may be shown equivalent by this method. Truth-functional calculus, however, offers no convenient way for finding the most economical realization of a given function.

For this purpose, we introduce Boolean algebra. Perhaps the most convenient set of postulates for a Boolean algebra is the one set forth by Huntington in 1904.[1] The first postulate can be thought of as establishing the system under study.

1. *There exists a set of K objects or elements, subject to an equivalence* relation, denoted "=," which satisfies the principle of substitution.*

---

* A formal definition of an equivalence relation may be found in Chapter 10. In the present chapter, the reader will encounter no difficulty in the use of the standard notion of equality from ordinary algebra.

By substitution, it is meant: if $a = b$, a may be substituted for b in any expression involving b without affecting the validity of the expression.

The remainder of Huntington's postulates are:

IIa. *A rule of combination "$+$" is defined such that $a + b$ is in K whenever both a and b are in K.*

IIb. *A rule of combination "$\cdot$" is defined such that $a \cdot b$ (abbreviated ab) is in K whenever both a and b are in K.*

IIIa. *There exists an element 0 in K such that, for every a in K, $a + 0 = a$.*

IIIb. *There exists an element 1 in K such that, for every a in K, $a \cdot 1 = a$.*

IVa. $a + b = b + a$
IVb. $a \cdot b = b \cdot a$ (commutative laws).

Va. $a + (b \cdot c) = (a + b) \cdot (a + c)$
Vb. $a \cdot (b + c) = (a \cdot b) + (a \cdot c)$ (distributive laws)

VI. *For every element a in K there exists an element $\bar{a}$ such that*

$$a \cdot \bar{a} = 0$$

and

$$a + \bar{a} = 1.$$

VII. *There are at least two elements x and y in K such that $x \neq y$.*

Notice that nothing has been said which would fix the number or type of elements which make up K. As a matter of fact, there are many systems which satisfy these postulates. A few will be illustrated a little later.

The reader may have observed a similarity between these postulates and those of ordinary algebra. Note, however, that the first distributive law—that is, distribution over addition—does not hold for ordinary algebra. Also, there is no $\bar{a}$ defined in ordinary algebra.

If a set of postulates is to be of any use, it must be consistent. That is, none of the postulates in the set may contradict any other postulate of the set. To verify consistency, we might attempt a postulate-by-postulate examination to ascertain that no postulate contradicted any possible group of postulates. This is rather awkward; and, in fact, a much easier way is available. It is merely necessary to find an example of a Boolean algebra which is known independently to be consistent. If such a consistent system satisfies all of Huntington's postulates, then the postulates themselves must be consistent.*

The simplest example of a Boolean algebra consists of only two elements, 1 and 0, defined to satisfy

$$\bar{1} = 0 \qquad \bar{0} = 1$$
$$1 \cdot 1 = 1 + 1 = 1 + 0 = 0 + 1 = 1$$

---

* This matter of the consistency of a set of postulates is a rather subtle mathematical concept. The interested reader should consult Huntington's paper for a full discussion.[1]

and

$$0 + 0 = 0 \cdot 0 = 1 \cdot 0 = 0 \cdot 1 = 0$$

We see that postulates I, II, III, and VII are satisfied by definition. Consider, for example, IIIb. If $a = 1$, we have

$$a \cdot 1 = 1 \cdot 1 = 1$$

If $a = 0$, we have

$$a \cdot 1 = 0 \cdot 1 = 0$$

The satisfaction of both of the above equations results from our definition of the "$\cdot$" function for all possible combinations of arguments. Satisfaction of the commutative laws is evident, and verification of the distributive laws requires only a truth table listing of the values of each side of both equations for all combination of values of a, b, and c. This task is left to the reader.

We observe also the satisfaction of Postulate VI upon letting a take on the values 1 and 0. That is, letting $a = 1$,

$$a \cdot \bar{a} = 1 \cdot \bar{1} = 1 \cdot 0 = 0$$
$$a + \bar{a} = 1 + \bar{1} = 1 + 0 = 1 \tag{4.1}$$

and letting $a = 0$,

$$a \cdot \bar{a} = 0 \cdot \bar{0} = 0 \cdot 1 = 0$$
$$a + \bar{a} = 0 + \bar{0} = 0 + 1 = 1 \tag{4.2}$$

In addition to consistency, the question of *independence*[1] of the postulates has been of some interest. By independence, it is meant that none of the postulates can be proved from the others. The postulates presented here are, in fact, independent. However, a demonstration of this fact would be lengthy and is not essential to our discussion.

It is not necessary that we begin with an independent set of postulates. In fact, some authors save effort by including some of the theorems which we shall develop later as postulates. It is our opinion, however, that the proof of these theorems constitutes the best possible introduction to the manipulation of Boolean algebra.

4.2   *Truth-Functional Calculus as a Boolean Algebra*

There is a one-to-one correspondence between the truth-functional calculus of Chapter 3 and the two-element algebra introduced in the previous section. Table 4.1 interprets each truth value or logical connective as an element or a rule of combination, respectively, in a Boolean algebra. These interpretations imply the truth table correspondence shown in Fig. 4.1.

TABLE 4.1

Truth Calculus		Boolean Algebra
$\wedge$	$\longleftrightarrow$	$\cdot$
$\vee$	$\longleftrightarrow$	$+$
F	$\longleftrightarrow$	0
T	$\longleftrightarrow$	1
$\bar{A}$	$\longleftrightarrow$	$\bar{A}$

A precise distinction between the two columns of Table 4.1 is not always made. Generally, "$\cdot$" is identified in the literature as AND while "$+$" is called OR. We shall follow this convention throughout the book.

AB	ab	$A \wedge B$ $a \cdot b$	$A \vee B$ $a + b$
FF $\longrightarrow$ 00		F $\longrightarrow$ 0	F $\longrightarrow$ 0
FT $\longrightarrow$ 01		F $\longrightarrow$ 0	T $\longrightarrow$ 1
TF $\longrightarrow$ 10		F $\longrightarrow$ 0	T $\longrightarrow$ 1
TT $\longrightarrow$ 11		T $\longrightarrow$ 1	T $\longrightarrow$ 1

FIGURE 4.1

In the remainder of this chapter, Boolean algebra will be developed into a convenient manipulative tool. As this is accomplished, the results will be immediately applicable to truth functions and, therefore, to logic circuits.

4.3   Duality

Notice that Huntington's postulates are presented in pairs. A closer look reveals that in each case one postulate in a pair can be obtained from the other by interchanging 0 and 1 along with $+$ and "$\cdot$". For example:

$$a + 0 = a$$
$$\downarrow \ \downarrow$$
$$a \cdot 1 = a \qquad (4.3)$$

and

$$a + (b \cdot c) = (a + b) \cdot (a + c)$$
$$\downarrow \quad \downarrow \qquad \downarrow \quad \downarrow \quad \downarrow$$
$$a \cdot (b + c) = (a \cdot b) + (a \cdot c) \qquad (4.4)$$

Every theorem which can be proved for Boolean algebra has a dual which is also true. That is, every step of a proof of a theorem may be replaced by its dual, yielding a proof of the dual of the theorem. In a sense, this doubles our capacity for proving theorems.

4.4 Fundamental Theorems of Boolean Algebra

In this section, we prove the theorems necessary for the convenient manipulation of Boolean algebra. Some of these are labeled Theorems while others are labeled Lemmas. Our criterion for making this distinction is that only those equalities which are valuable tools for working problems are designated as Theorems. Intermediate results necessary in the proof of theorems and results, which are included only for the sake of logical completeness, are called Lemmas.

LEMMA 4.1.   *The 0 and 1 elements are unique.*

Proof: By way of contradiction, assume that there are two zero elements, 0_1 and 0_2. For any elements a_1 and a_2 in K we have

$$a_1 + 0_1 = a_1 \quad \text{and} \quad a_2 + 0_2 = a_2$$

(Post. IIIa)

Now let $a_1 = 0_2$ and $a_2 = 0_1$. Therefore,

$$0_2 + 0_1 = 0_2 \quad \text{and} \quad 0_1 + 0_2 = 0_1$$

Thus, using the first commutative law and the transitive property of equality, we have

$$0_1 = 0_2$$

As an example of duality, let us make the following change of symbols in our proof:

$$a_1 + 0_1 = a_1 \qquad a_2 + 0_2 = a_2$$
$$\downarrow \downarrow \qquad\qquad \downarrow \downarrow$$
$$a_1 \cdot 1_1 = a_1 \qquad a_2 \cdot 1_2 = a_2$$

and

$$0_2 + 0_1 = 0_2 \qquad 0_1 + 0_2 = 0_1$$
$$\downarrow \downarrow \downarrow \quad \downarrow \qquad \downarrow \downarrow \downarrow \quad \downarrow$$
$$1_2 \cdot 1_1 = 1_2 \qquad 1_1 \cdot 1_2 = 1$$

and

$$0_1 = 0_2$$
$$\downarrow \qquad \downarrow$$
$$1_1 = 1_2$$

Because we were aware that the dual of each postulate exists, it was unnecessary to cite postulates as the dual of each step was compiled. In fact, we could have merely stated that the dual of the theorem was true by the principle of duality. This approach will be followed in the remaining theorems of this section.

LEMMA 4.2.   *For every element a in K, $a + a = a$ and $a \cdot a = a$.*

Proof:	$a + a = (a + a) \cdot 1$	(Post. IIIb)
	$a + a = (a + a)(a + \bar{a})$	(Post. VI)
	$a + a = a + a\bar{a}$	(Post. Va)
	$a + a = a + 0$	(Post. VI)
	$a + a = a$	(Post. IIIa)
	$a \cdot a = a$	(Duality)

LEMMA 4.3.   *For every a in K, $a + 1 = 1$ and $a \cdot 0 = 0$.*

Proof:	$a + 1 = 1 \cdot (a + 1)$	(Post. IIIb)
	$a + 1 = (a + \bar{a})(a + 1)$	(Post. VI)
	$a + 1 = a + \bar{a} \cdot 1$	(Post. Va)
	$a + 1 = a + \bar{a}$	(Post. IIIb)
	$a + 1 = 1$	(Post. VI)
	$a \cdot 0 = 0$	(Duality)

LEMMA 4.4.   *The elements 1 and 0 are distinct and $\bar{1} = 0$.*

Proof: Let a be any element in K.

$$a \cdot 1 = a \qquad \text{(Post. IIIb)}$$
$$a \cdot 0 = 0 \qquad \text{(Lemma 4.3)}$$

Now assume that $1 = 0$. In this case, the above expressions are satisfied only if $a = 0$. Postulate VII, however, tells us that there are at least two elements in K. The contradiction can be resolved only by concluding that $1 \neq 0$. To prove the second assertion, we need only write

$$\bar{1} = \bar{1} \cdot 1 \qquad \text{(Post. IIIb)}$$
$$= 0 \qquad \text{(Post. VI)}$$

LEMMA 4.5.   *For every pair of elements a and b in K, $a + ab = a$ and $a(a + b) = a$.*

Proof:	$a + ab = a \cdot 1 + ab$	(Post. IIIb)
	$= a(1 + b)$	(Post. Vb)
	$= a \cdot 1$	(Lemma 4.3)
	$= a$	(Post. IIIb)

and

$$a(a + b) = a \qquad \text{(Duality)}$$

LEMMA 4.6.   *The $\bar{a}$ defined by Postulate VI for every a in K is unique.*

Proof: By contradiction. Assume that there are two distinct elements, $\bar{a}_1$ and $\bar{a}_2$, which satisfy Postulate VI, i.e., assume that

$$a + \bar{a}_1 = 1, \qquad a + \bar{a}_2 = 1, \qquad a\bar{a}_1 = 0, \qquad a\bar{a}_2 = 0$$

$$
\begin{aligned}
\bar{a}_2 &= 1 \cdot \bar{a}_2 && \text{(Post. IIIb)} \\
&= (a + \bar{a}_1)\bar{a}_2 && \text{Assumption} \\
&= a\bar{a}_2 + \bar{a}_1\bar{a}_2 && \text{(Post. Vb)} \\
&= 0 + \bar{a}_1\bar{a}_2 && \text{Assumption} \\
&= a\bar{a}_1 + \bar{a}_1\bar{a}_2 && \text{Assumption} \\
&= (a + \bar{a}_2)\bar{a}_1 && \text{(Post. Vb)} \\
&= 1 \cdot \bar{a}_1 && \text{Assumption} \\
&= \bar{a}_1 && \text{(Post. IIIb)}
\end{aligned}
$$

LEMMA 4.7.   *For every element a in K, $a = \bar{\bar{a}}$.*

Proof: Let $\bar{\bar{a}} = x$. Therefore,

$$\bar{a}x = 0 \qquad \text{and} \qquad \bar{a} + x = 1 \qquad \text{(Post. VI)}$$

but

$$a\bar{a} = 0 \qquad \text{and} \qquad \bar{a} + a = 1 \qquad \text{(Post. VI)}$$

Thus, both x and a satisfy Postulate VI as the complement of $\bar{a}$. Therefore, by Lemma 4.6,

$$x = a$$

LEMMA 4.8. $a[(a + b) + c] = [(a + b) + c]a = a.$

Proof:
$$
\begin{aligned}
a[(a + b) + c] &= a(a + b) + ac && \text{(Post. Vb)} \\
&= a + ac && \text{(Lemma 4.5)} \\
&= a = [(a + b) + c]a
\end{aligned}
$$
$$\text{(Post. IVb, Lemma 4.5)}$$

THEOREM 4.9.   *For any three elements a, b, and c in K, $a + (b + c) = (a + b) + c$ and $a \cdot (bc) = (ab) \cdot c$.*

The reader will recognize Theorem 4.9 as the associative laws from ordinary algebra. Some authors include these laws among the postulates although, as we shall see, this is unnecessary.

Proof: Let

$$Z = [(a + b) + c] \cdot [a + (b + c)]$$
$$= [(a + b) + c]a + [(a + b) + c] \cdot (b + c)$$
$$= a + [(a + b) + c] \cdot (b + c) \qquad \text{(Lemma 4.8)}$$
$$= a + \{[(a + b) + c]b + [(a + b) + c] \cdot c\} \qquad \text{(Post. Vb)}$$
$$= a + \{b + [(a + b) + c] \cdot c\} \qquad \text{(Lemma 4.8, Post. IVb)}$$
$$= a + (b + c) \qquad \text{(Lemma 4.5)}$$

However, we may also write

$$Z = (a + b)[a + (b + c)] + c[a + (b + c)] \qquad \text{(Post. Vb)}$$
$$= (a + b)[a + (b + c)] + c \qquad \text{(Lemma 4.8)}$$
$$= \{a[a + (b + c)] + b[a + (b + c)]\} + c \qquad \text{(Post. Vb)}$$
$$= \{a[a + (b + c)] + b\} + c \qquad \text{(Lemma 4.8)}$$
$$= (a + b) + c \qquad \text{(Lemma 4.5)}$$

Therefore, by transitivity

$$a + (b + c) = (a + b) + c$$

and

$$(a \cdot b)c = a(b \cdot c) \qquad \text{(Duality)}$$

Now that we have established the associative laws, certain expressions may be simplified by omitting parentheses as follows:

$$(a + b) + c = a + b + c \qquad (4.5)$$

and

$$(a \cdot b) \cdot c = abc \qquad (4.6)$$

This format may be extended to Boolean sums and products of any number of variables.

THEOREM 4.10.   *For any pair of elements a and b in K, $a + \bar{a}b = a + b$; $a(\bar{a} + b) = ab$.*

Proof:

$$a + \bar{a}b = (a + \bar{a}\,(a + b) \qquad \text{(Post. Va)}$$
$$a + \bar{a}b = a + b \qquad \text{(Post. VI, Post IIIb)}$$
$$a \cdot (\bar{a} + b) = a \cdot b \qquad \text{(Duality)}$$

THEOREM 4.11. *Foe every pair of elements a and b in K,* $\overline{a + b} = \bar{a} \cdot \bar{b}$ *and* $\overline{a \cdot b} = \bar{a} + \bar{b}$.

The expressions in Theorem 4.11 are the two forms of the very important DeMorgan's law, which we introduced in Chapter 3. The second form is the dual of the first, but, for the sake of instruction, we shall arrive at the second form in another way.

$$(a + b) + \bar{a} \cdot \bar{b} = [(a + b) + \bar{a}] \cdot [(a + b) + \bar{b}] \qquad \text{(Post. Va)}$$

$$(a + b) + \bar{a} \cdot \bar{b} = [\bar{a} + (a + b)] \cdot [\bar{b} + (b + a)] \qquad \text{(Post. IVa)}$$

$$(a + b) + \bar{a} \cdot \bar{b} = 1 \cdot 1 = 1 \qquad \text{(Theorem 4.9, Lemma 4.3)}$$

$$(a + b) \cdot (\bar{a} \cdot \bar{b}) = a(\bar{a} \cdot \bar{b}) + b(\bar{b} \cdot \bar{a}) \qquad \text{(Post. Vb, Post. IVb)}$$

$$(a + b) \cdot (\bar{a} \cdot \bar{b}) = 0 + 0 = 0 \qquad \text{(Theorem 4.9, Lemma 4.3)}$$

Both requirements of Postulate VI have been satisfied, so $a + b$ is the unique complement of $\bar{a} \cdot \bar{b}$. Therefore, we may write

$$a + b = \overline{\bar{a} \cdot \bar{b}} \qquad \text{or} \qquad \overline{a + b} = \bar{a} \cdot \bar{b}$$

The above holds equally well for $\bar{a}$ and $\bar{b}$ in place of a and b, so we can write

$$\overline{\bar{a} + \bar{b}} = \bar{\bar{a}} \cdot \bar{\bar{b}} = a \cdot b$$

or

$$\bar{a} + \bar{b} = \overline{a \cdot b} \qquad \text{(Lemma 4.6, 4.7)}$$

4.5 Set Theory as an Example of Boolean Algebra

We have already discussed truth-functional calculus as one example of Boolean algebra. Since this example satisfies all of Huntington's postulates, it must also satisfy each of the theorems and lemmas proved so far.

Set theory is a second example of a Boolean algebra. A *set* may be regarded as any collection of objects. In order to talk of an algebra of sets, we must limit the objects which make up these sets with some meaningful criteria. We define a *universal set* to include every object which satisfies these criteria. For a reason which will soon be apparent the universal set must necessarily contain at least one object. The universal set may, in fact, contain a finite or an infinite number of objects. The collection of points in a plane or in a finite region of a plane make interesting examples of a universal set. When we say that R is a subset of S, we mean that every object in R is also found in S. This is symbolized by $R \subset S$. Only those sets which are subsets of the universal set are material to the development of an algebra of sets. A second set of

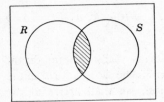

FIGURE 4.2. $R \cap S$.

importance is the *null set*, or the set containing no objects. The universal set and the null set will be symbolized by S_U and S_Z, respectively. The *union* of two sets, R and S, is that set which contains all objects contained in either R or S. This is symbolized by $R \cup S$. The *intersection* of two sets, R and S, is that set containing these elements found in both set R and set S. The intersection is designated by $R \cap S$. The *complement* of a set S, designated $C(S)$, contains those objects found in the universal set but not in set S.

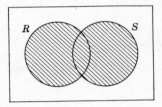

FIGURE 4.3. $R \cup S$.

A helpful illustration of a universal set is the collection of points in a rectangle such as shown in Fig. 4.2. The two sets R and S contain those points of the universal set found within the respective circles.* A representation of this type is known as a *Venn diagram*.

The darkened area in Fig. 4.2 is the intersection of sets R and S. Similarly, $R \cup S$ and $C(S)$ are darkened in Figs. 4.3 and 4.4, respectively.

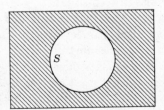

FIGURE 4.4. $C(S)$.

* The student of advanced calculus may be concerned about the status of points on the boundaries of the circles, a problem not critical to this development. One possible resolution of the conflict is to imagine the universal set to include only a large but finite number of points in the rectangle. In this way, boundaries can be deliberately arranged to avoid points in the universal set.

TABLE 4.2

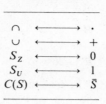

Recalling the postulates in Sec. 4.1, let us choose a set of K objects to be those sets containing no points other than points found in the universal set. Such sets, one of which is the null set, are subsets of the universal set. If equality is taken to relate only identical sets, then Postulate I is satisfied. We shall proceed to show that, in light of the correspondences in Table 4.2, the remaining postulates will also be satisfied.

The sets $A \cap B$ and $A \cup B$ are subsets of S_U, so Postulate II is satisfied. As $R \cup S_Z$ and $R \cap S_U$ contain precisely the same objects as R, Postulate III is satisfied. Satisfaction of Postulate IV, the commutative laws, is evident. Under the set theory interpretation, the distributive laws become

$$A \cup (B \cap C) = (A \cup B) \cap (A \cup C) \tag{4.7}$$

and

$$A \cap (B \cup C) = (A \cap B) \cup (A \cap C) \tag{4.8}$$

The fact that Equations 4.7 and 4.8 are satisfied is best illustrated by using the Venn diagram. Notice in Fig. 4.5a that set $B \cap C$ is shaded with vertical lines while set A is shaded by horizontal lines. The set $A \cup (B \cap C)$ consists of the areas darkened in either manner. In Fig. 4.5b, the set $A \cup B$ is indicated by horizontal lines, and $A \cup C$ is shaded by vertical lines. The intersections of these two sets, $(A \cup B) \cap (A \cup C)$ is, therefore, given by the crisscrossed area in Fig. 4.5b. Notice that the areas representing both sides of Equation 4.7 are identical. The validity of Equation 4.8 can be similarly

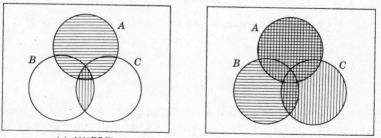

(a) $A \cup (B \cap C)$ (b) $(A \cup B) \cap (A \cup C)$

FIGURE 4.5

demonstrated by using the Venn diagram. By the definition of the complement, it follows immediately that

$$R \cap \bar{R} = S_Z \tag{4.9}$$

and

$$R \cup \bar{R} = S_U \tag{4.10}$$

Thus, Postulate VI is satisfied. The fact that S_U and S_Z are always defined and distinct guarantees the satisfaction of Postulate VII.

4.6 Examples of Boolean Simplification

We are now ready to utilize Boolean algebra in the simplification of some meaningful examples. The first two examples are a problem in set theory and a needlessly complex logic circuit.

Example 4.1

Misters A, B, and C collect old manuscripts. Mr. A is a collector of English political works and of foreign language novels. Mr. B is a collector of all political works, except English novels, and English works which are not novels. Mr. C collects those nonfictional items which are English works or foreign language political works. Determine those books for which there is competition—that is, those desired by two or more collectors

Solution: Let us first define the various sets involved in the problem.

> A set of books collected by Mr. A
> B set of books collected by Mr. B
> C set of books collected by Mr. C
> E set of all English language books
> N set of all novels
> P set of all political works

The set of books collected by two or more persons can be symbolized by

$$Z = (A \cap B) \cup (A \cap C) \cup (B \cap C) \tag{4.11}$$

Translating the problem statements into set theoretic expressions gives

$$A = (E \cap P) \cup (\bar{E} \cap N)$$

$$B = (P \cap \overline{E \cap N}) \cup (E \cap \bar{N})$$

$$C = [E \cup (\bar{E} \cap P)] \cap \bar{N}$$

Notice that it was more convenient to use the Boolean algebra symbol rather than the set theoretic symbol for the complement. In order to utilize Boolean algebra in obtaining a solution, the symbols for intersection and union will also be replaced by the corresponding "." and "+." Before substituting into Equation 4.11, let us write the expression for A, B, and C in Boolean form and make whatever simplifications are possible.

$$A = EP + \bar{E}N$$

$$B = P(\overline{E N}) + E\bar{N} = P(\bar{E} + \bar{N}) + E\bar{N} \qquad \text{(DeMorgan's theorem)}$$
$$= P\bar{E} + P\bar{N}(E + \bar{E}) + E\bar{N}$$
$$= P\bar{E} + P\bar{N}\bar{E} + PE\bar{N} + E\bar{N}$$
$$= P\bar{E}(1 + \bar{N}) + (P + 1)E\bar{N} = P\bar{E} + E\bar{N}$$
$$C = (E + \bar{E}P)\bar{N} = (E + P)\bar{N} \qquad \text{(Theorem 4.10)}$$

Now writing Equation 4.11 as a Boolean expression and substituting we get

$$Z = (EP + \bar{E}N)(P\bar{E} + E\bar{N}) + (EP + \bar{E}N)(E\bar{N} + P\bar{N})$$
$$+ (P\bar{E} + E\bar{N})(E\bar{N} + P\bar{N})$$

or

$$Z = (EP + \bar{E}N)(P\bar{E} + E\bar{N} + E\bar{N} + P\bar{N}) + (P\bar{E} + E\bar{N})(E\bar{N} + P\bar{N})$$
$$\text{(Distributive laws)}$$
$$= (EP + \bar{E}N)(P\bar{E} + E\bar{N} + P\bar{N}) + (P\bar{E} + E\bar{N})(E\bar{N} + P\bar{N})$$
$$= (EP + \bar{E}N)(P\bar{E} + E\bar{N} + P\bar{E}\bar{N} + PE\bar{N}) + (P\bar{E} + E\bar{N})(E\bar{N} + P\bar{N})$$
$$= (EP + \bar{E}N)(P\bar{E}(1 + \bar{N}) + (P + 1)\, E\bar{N}) + (P\bar{E} + E\bar{N})(E\bar{N} + P\bar{N})$$
$$= (EP + \bar{E}N)(P\bar{E} + E\bar{N}) + (P\bar{E} + E\bar{N})(E\bar{N} + P\bar{N})$$
$$= (P\bar{E} + E\bar{N})(EP + \bar{E}N + E\bar{N} + P\bar{N}) \qquad \text{(Distributive law)}$$
$$= E\bar{N} + P\bar{E}(PE + \bar{E}N + P\bar{N}) \qquad \text{(Distributive law)}$$
$$= E\bar{N} + P\bar{E}E + P\bar{E}N + P\bar{E}\bar{N}$$
$$= E\bar{N} + P\bar{E}(N + \bar{N}) = E\bar{N} + P\bar{E}$$

Thus, we conclude that more than one person is interested in English nonfiction and foreign language political works.

An alternative solution of this problem constitutes an interesting application of the Venn diagram.

Notice that sets A, B, and C are represented by the darkened areas in Figs. 4.6a, b, and c, respectively.

By inspection, we see that the area darkened in at least two of these three diagrams is that shown in Fig. 4.7. This can be expressed as

$$(E \cap \bar{N}) \cup (P \cap \bar{E})$$

which is the same result obtained by the algebraic method.   ∎

The reader may have noticed that expressions similar to

$$xy + \bar{x}z + yz \qquad (4.12)$$

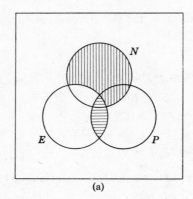

(a)

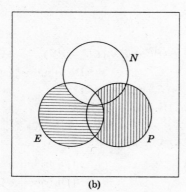

(b)

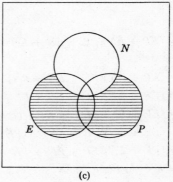

(c)

FIGURE 4.6

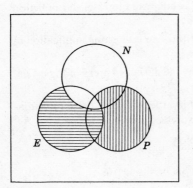

FIGURE 4.7

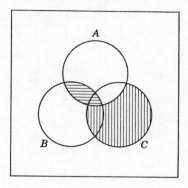

FIGURE 4.8

appeared several times in Example 4.1. Each time, a simplification involving several lines of algebraic manipulation was made. The following theorem makes it possible to short-cut this manipulation in future examples.

THEOREM 4.12. *For any three elements a, b, and c in K,*

$$ab + \bar{a}c + bc = ab + \bar{a}c$$

and

$$(a + b)(\bar{a} + c)(b + c) = (a + b)(\bar{a} + c)$$

Proof:
$$
\begin{aligned}
ab + \bar{a}c + bc &= ab + \bar{a}c + bc(a + \bar{a}) \\
&= ab + abc + \bar{a}c + \bar{a}cb \\
&= ab(1 + c) + \bar{a}c(1 + b) \\
&= ab + \bar{a}c
\end{aligned}
$$

$$(a + b)(\bar{a} + c)(b + c) = (a + b)(\bar{a} + c) \qquad \text{(Duality)}$$

Theorem 4.12 is illustrated by the Venn diagram in Fig. 4.8. Notice that the outlined area bc is partially covered by ab and partially by $\bar{a}c$. The theorem may be applied whenever two products jointly cover a third in this manner.

Example 4.2

The logic circuit shown in Fig. 4.9 was designed by trial-and-error. To simplify the figure, some of the actual connections to the inputs and to the outputs of the inverters are not shown. Let us utilize Boolean algebra to design an equivalent circuit free from unnecessary hardware.

Solution: It is first necessary to determine the function realized by this circuit. To this end, the output functions for the various logic circuits are shown in

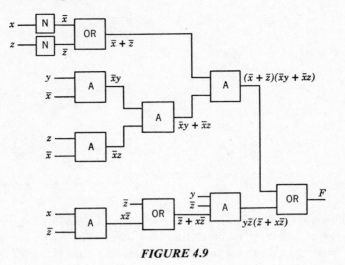

FIGURE 4.9

Fig. 4.9. The output function is given by

$$F(x, y, z) = y\bar{z}(\bar{z} + \bar{z}x) + (\bar{x} + \bar{z})(\bar{x}y + \bar{x}z) \qquad (4.13)$$

$$= y\bar{z} + (\bar{x} + \bar{z})(\bar{x}y + \bar{x}z) \qquad \text{(Lemma 4.5)}$$

$$= y\bar{z} + (\bar{x} + \bar{z})\bar{x}(y + z)$$

$$= \bar{z}y + \bar{x}z + \bar{x}y \qquad \text{(Lemma 4.5)}$$

Now if $\bar{z}$ is identified as a, y as b, and $\bar{x}$ as c in Theorem 4.12, we have

$$F(x, y, z) = \bar{z}y + \bar{x}z \qquad \text{(Theorem 4.12)}$$

We see that the considerably simplified and therefore, cheaper circuit in Fig. 4.10 realizes the same function as the one in Fig. 4.9. ∎

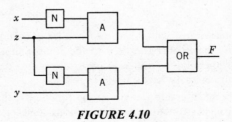

FIGURE 4.10

Example 4.3

Simplify the following Boolean expression, which represents the output of a logical decision circuit.

$$F(w, x, y, z) = x + xyz + yz\bar{x} + wx + \bar{w}x + \bar{x}y$$
$$= x + yz(x + \bar{x}) + \bar{x}y + x(w + \bar{w})$$
$$= (x + \bar{x}y) + yz + x$$
$$= x + y + x + yz$$
$$= x + (y + yz)$$
$$= x + y$$

Another approach to the problem involves the use of Lemma 4.7. That is,

$$\overline{F(w, x, y, z)} = \overline{x + xyz + yz\bar{x} + \bar{x}y + wx + \bar{w}x}$$
$$= \bar{x}\overline{(xyz)}\overline{(\bar{x}yz)}\overline{(\bar{x}y)}\overline{(wx)}\overline{(\bar{w}x)}$$
$$= [\bar{x}(\bar{x} + \bar{y} + \bar{z})](x + \bar{y} + \bar{z})(x + \bar{y})(\bar{w} + \bar{x})(w + \bar{x})$$
$$= [\bar{x} \cdot (\bar{w} + \bar{x})(w + \bar{x})](x + \bar{y})(\bar{y} + \bar{z} + x)$$
$$= [\bar{x}](x + \bar{y})$$
$$\overline{F(w, x, y, z)} = \bar{x} \cdot \bar{y}$$

Therefore, by Lemma 4.7 we have

$$F(w, x, y, z) = \overline{\overline{F(w, x, y, z)}} = \overline{\bar{x} \cdot \bar{y}} = x + y. \quad \blacksquare$$

4.7 Remarks on Switching Functions

The reader may have wondered why the Venn diagram which was so helpful in Example 4.1 was not utilized in the two subsequent examples. It should be recalled that this device has so far been introduced only as an illustration of set theory. The Boolean algebra of sets has many elements rather than only two, as is the case with truth values and logic voltage levels. The Venn diagram can, however, be made useful in the simplification of switching circuits and truth functions if the following definitions and interpretations are kept in mind.

First, it is convenient to distinguish Boolean functions of only two elements by calling them *switching* functions. We then define a *switching variable* as a letter (usually chosen from the beginning or end of the alphabet, excluding *F*) which may take on either of the element values, 0 or 1. In the case of logic circuits, there are a fixed number of variables, say *n*, which serve as inputs to a circuit under consideration. The situation is similar in the case of truth functions.

x_1	x_2	x_3	F
0	0	0	?
0	0	1	?
0	1	0	?
0	1	1	?
1	0	0	?
1	0	1	?
1	1	0	?
1	1	1	?

FIGURE 4.11

There are 2^n possible ways of assigning values to n variables. The three-variable case is illustrated in Fig. 4.11. If the eight question marks are replaced by any combination of zeros and ones, a specific *function* of x_1, x_2, and x_3 is defined. As there are 2^8, or 2^{2^3}, ways of replacing the eight question marks by zeros and ones, there are 2^{2^3} *switching functions* of three variables. The value of F for a particular row of the truth table is known as the functional value for the corresponding combination of input values. Formally, we have the following definition.

Definition 4.1. A *switching function* of n variables is any one particular assignment of functional values (1's or 0's) for all 2^n possible combinations of values of the n variables.

In general, then, there are 2^{2^n} distinct switching functions of n variables.

We have seen that there are many *expressions* for a given switching function or truth table assignment. In the two-element algebra, there are many distinct expressions which are equal to each of the two elements, 1 and 0. For example,

$$1 \cdot (1 + 0) = 1$$

Boolean algebra provides a means of determining whether such expressions are equal to 1 or 0.

Let us consider now the possibility of an extended Boolean algebra whose *elements* are *all possible switching functions* of n variables. Such an algebra will provide us with a means of determining which *function* is represented by a given *expression* in the n variables. For example, we can determine that

$$x + xy + y + yz$$
$$x + \bar{x}y + x(w + \bar{w})$$

and

$$x + y$$

x	y	F
0	0	0
0	1	1
1	0	1
1	1	1

FIGURE 4.12

are all expressions for the same function, specifically that function defined by the truth table of Fig. 4.12.

The theorems which we have proved so far, when specialized for a two-valued algebra, express equality between switching functions or, if you prefer, between elements of the extended algebra. Thus, as in ordinary algebra, two switching functions are identically equal if their evaluations are equal for every combination of variable values. This notion of equality satisfies Postulate I. It remains to show that the proposed algebra of 2^{2^n} elements satisfies the remainder of the postulates and is indeed a Boolean algebra. Functional expressions may certainly be related by the connectives AND, OR, and NOT in the same way variables are related. Therefore, Postulates II, IV, V are satisfied. The zero element is the function which is identically 0, and the unit element is the function which is identically 1. The definition of the negation of F must certainly be $\bar{F}$. If a function h is to satisfy the expression

$$h(x_1, x_2, \ldots, x_n) = \bar{F}(x_1, x_2, \ldots, x_n) \tag{4.14}$$

then h must be 0 whenever F is evaluated as 1 and vice versa. Satisfaction of Postulate VII is immediate.

We shall now see that a Venn diagram provides a convenient illustration of the algebra whose elements are the 2^8 functions of three variables. In Fig. 4.13, areas are assigned to each of the eight combinations of values.

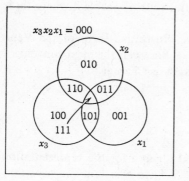

FIGURE 4.13

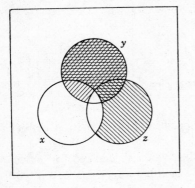

FIGURE 4.14

A function can then be indicated by darkening those areas corresponding to combinations of values for which the function is 1. Every possible area configuration stands for a distinct function. For example, the function

$$F(x, y, z) = \bar{z}y + \bar{x}z + \bar{x}y = \bar{z}y + \bar{x}z \qquad (4.15)$$

resulting from Example 4.2 is depicted in Fig. 4.14. The differently lined regions indicate the simplification resulting from the application of Theorem 4.12.

Almost every problem in Boolean algebra will be found to be some variation of the following statement:

"Given one of the 2^{2^n} functions of n variables, determine from the large number of equivalent expressions of this function one which satisfies some criteria for simplicity."

If the problem concerns an electronic switching circuit, the simplicity criteria will result from the desirability of using the cheapest possible circuit that provides the desired performance.

The Venn diagram is a means of solving this type of problem where three or fewer variables are involved.

For more than three variables, the basic illustrative form of the Venn diagram is inadequate. Extensions are possible, however, the most convenient of which is the Karnaugh map, to be discussed in Chapter 6.

4.8 Summary

In this chapter, we have developed a formal algebra for the representation and manipulation of switching functions. We have also laid the necessary foundation for the development in later chapters of systematic procedures for simplification of algebraic expressions and the corresponding circuits.

Following is a list of the important postulates, lemmas, and theorems for ready reference in problems and proofs.

PIIIa $a + 0 = a$ PIIIb $a \cdot 1 = a$
PIVa $a + b = b + a$ PIVb $ab = ba$
PVa $a + bc = (a + b)(a + c)$ PVb $a(b + c) = ab + ac$
PVIa $a \cdot \bar{a} = 0$ PVIb $a + \bar{a} = 1$

L 2a $a + a = a$ L 2b $a \cdot a = a$
L 3a $a + 1 = 1$ L 3b $a \cdot 0 = 0$
L 4 $\bar{1} = 0$
L 5a $a + ab = a$ L 5b $a(a + b) = a$
L 7 $\bar{\bar{a}} = a$

T 9a $a + (b + c) = (a + b) + c$ T 9b $a(bc) = (ab)c = abc$
$\qquad\qquad = a + b + c$
T10a $a + \bar{a}b = a + b$ T10b $a(\bar{a} + b) = a \cdot b$
T11a $\overline{a + b} = \bar{a} \cdot \bar{b}$ T11b $\overline{a \cdot b} = \bar{a} + \bar{b}$
T12a $ab + \bar{a}c + bc = ab + \bar{a}c$ T12b $(a + b)(\bar{a} + c)(b + c)$
$\qquad\qquad\qquad = (a + b)(\bar{a} + c)$

PROBLEMS

4.1. (a) Verify the distributive law
$$a(b + c) = ab + ac$$
for the two-value Boolean algebra.
(b) Repeat (a) for $a + bc = (a + b)(a + c)$.

4.2. Write out a proof of the second associative law.
$$(a \cdot b) \cdot c = a(b \cdot c)$$

4.3. Verify by a truth-table analysis (using 1 and 0 instead of **T** and **F**) that the following theorems of Boolean algebra hold for the two-valued truth-functional calculus.
(a) Lemma 4.2
(b) Lemma 4.5
(c) Lemma 4.7
(d) Theorem 4.9
(e) Theorem 4.10
(f) Theorem 4.11

4.4. Write in its simplest possible form the Boolean function realized by the logic network of Fig. P4.4.

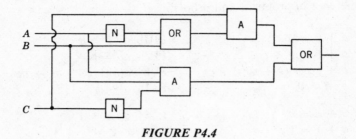

FIGURE P4.4

4.5. Use the Venn diagram to illustrate the validity of expression 4.8.

4.6. Verify the following by Boolean algebraic manipulation. Justify each step with a reference to a postulate or theorem.
(a) $(X + \overline{Y} + XY)(X + \overline{Y})\overline{X}Y = 0$
(b) $(X + \overline{Y} + X\overline{Y})(XY + \overline{X}Z + YZ) = XY + \overline{X}\overline{Y}Z$
(c) $(AB + C + D)(\overline{C} + D)(\overline{C} + D + E) = AB\overline{C} + D$

4.7. Using the postulates and theorems of Boolean algebra, simplify the following to a form with as few occurrences of each variable as possible.
(a) $(X + \overline{Y})[XYZ + \overline{Y}(Z + X)] + XY\overline{Z}(X + \overline{X}Y)$
(b) $(X + \overline{Y}\overline{X})[XZ + X\overline{Z}(Y + \overline{Y})]$

4.8. Simplify the following as far as possible by manipulation of Boolean algebra.
(a) $X\overline{Z}Y + (X\overline{Z}Y + Z\overline{X})[Y(Z + X) + \overline{Y}Z + \overline{Y}X\overline{Z}]$
(b) $[(a \downarrow b)] \uparrow [(a + (\bar{a} + xz)\bar{a})(\bar{z} + \bar{a})]$

4.9. Repeat Problem 4.8, utilizing a Venn diagram.

4.10. Prove that $(a + b)(\bar{a} + c)(b + c) = (a + b)(\bar{a} + c)$.

4.11. By manipulation of Boolean algebra verify that
$$[\overline{X}_1\overline{X}_2(X_3X_1 + \overline{X}_2)] + (X_1 + X_2)\overline{(X_1\overline{X}_2\overline{X}_3)} \downarrow (\overline{X}_1X_2X_3) = X_2\overline{X}_3 + \overline{X}_1X_3.$$

4.12. Let the dual of a Boolean function, f, be designated by f^D. Prove the theorem $f^D(X_1, X_2, \ldots, X_n) = \bar{f}(\overline{X}_1, \overline{X}_2, \ldots, \overline{X}_n)$ for all functions of 1, 2, or 3 variables. Assume that the variables may take on only the values 1 and 0.

4.13. A Boolean function, F, is called self-dual if the dual of the function is equal to the function itself $(F = F^D)$. Show that there are $2^{2^{n-1}}$ self-dual functions of n variables.

4.14. Prove that

$$f(X_1, X_2, \ldots, X_n) = X_1 f(1, X_2, \ldots, X_n) + \bar{X}_1 f(0, X_2, \ldots, X_n)$$

where the variables may take on only the values 1 and 0.

4.15. State and prove the dual of the expression in Problem 4.14.

4.16. Consider the possibility of a Boolean algebra of three elements, say, 0, 1, and 2. Show that a Boolean algebra cannot be defined for three elements.

4.17. Honest John, the used car dealer, has some two-toned, 8-cylinder cars. He also has in stock some air-conditioned cars and some cars with fewer than 8 cylinders, all of which are overpriced. Another dealer, Insane Charlie, has on his lot some cars without air conditioning or 8-cylinders, which are not overpriced. He also has some air-conditioned, 8-cylinder models and some overpriced solid color cars. Write the simplest possible Boolean expression for the categories of cars currently stocked by both dealers.

BIBLIOGRAPHY

1. Huntington, E. V. "Sets of Independent Postulates for the Algebra of Logic," *Trans. American Math. Soc.*, **5**, 288–309 (1904).

2. Sheffer, H. M. "A Set of Five Independent Postulates for Boolean Algebras, with Applications to Logical Constants," *Trans. American Math. Soc.*, **14**, 481–488 (1913).

3. Boole, George. *An Investigation of the Laws of Thought*, Dover Publications, New York, 1954.

4. Shannon, C. E. "Symbolic Analysis of Relay and Switching Circuits," *Transactions AIEE*, **57**, 713–723 (1938).

5. Carroll, Lewis. *The Complete Works*, The Nonesuch Press, London, 1939.

6. Harrison, M. A. *Introduction to Switching and Automata Theory*, McGraw-Hill, New York, 1965.

Chapter 5 *Switching Devices*

5.1 Introduction

The theory discussed in the last two chapters is not dependent on specific hardware for its validity. In the following chapters, we shall generally assume ideal realizations of the various logical operations so that we could theoretically proceed with no knowledge whatever of the actual hardware involved. However, the less-than-ideal characteristics of practical devices impose significant limitations on the designer. It is thus important for the designer to have some knowledge of the characteristics of practical switching devices.

5.2 Switches and Relays

The simplest switching device is the switch itself. A switch is any mechanical device by means of which two (or more) electrical conductors may be conveniently connected or disconnected. The simplest form of switch consists of two strips of spring metal on which are mounted electrical contacts. A lever or push-button controls whether the switch is *open* (contacts separated) or *closed* (contacts touching). The manner in which switches may be used to implement logic functions is illustrated in Fig. 5.1. In the circuit of Fig. 5.1a, the light bulb will be lit only if both switch *A* AND switch *B* are closed. In Fig. 5.1b, the light bulb will be lit if either switch *A* OR switch *B* is closed.

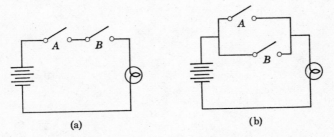

FIGURE 5.1. *Switch realizations of AND and OR.*

A relay is a switch operated by an electromagnet. When an appropriate current flows through the coil, a magnetic force displaces the armature, in turn causing the switch contacts to open or close. The position of the switch contacts when the coil is not energized is known as the *normal* position. Thus, a relay may have both *normally open* (NO) and *normally closed* (NC) contacts (Fig. 5.2a). In Fig. 5.2b, we have a simple relay circuit for controlling a light bulb. The light bulb will be lit if relay *A* OR relay *B* is energized AND relay *C* is NOT energized. Thus, if we let *T* represent the statement, "The bulb is lit," the circuit of Fig. 5.2b could be conveniently described by the equation

$$T = (A + B)\bar{C} \tag{5.1}$$

While relays are far too slow for use in the "main frame" of a digital computer, they are still useful for various auxiliary functions in peripheral equipment. The bilateral nature of relay circuits gives rise to some special problems. These are treated in Appendix B.

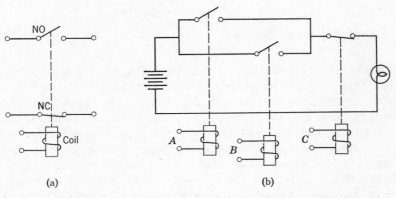

FIGURE 5.2. *Typical relay symbol and relay logic circuit.*

5.3   Logic Circuits

Switches and relays are of great historic importance, but today the vast majority of digital systems are constructed of electronic devices, principally semiconductor diodes and transistors. The schematic symbol for a diode is shown in Fig. 5.3a. Ideally, a diode offers zero resistance to the flow of current from anode to cathode (forward direction) and infinite resistance to current flow from cathode to anode (reverse direction). The ideal current-voltage characteristic would thus be that shown in Fig. 5.3b. A typical characteristic for an actual semiconductor diode is shown in Fig. 5.3c. When the diode is conducting in the forward direction, there is a small, nearly constant positive voltage across the diode. When the voltage across the diode goes negative, a very small reverse current flows. When the voltage goes sufficiently negative, a phenomenon known as the *Zener* breakdown* occurs, and the reverse current increases sharply. In normal switching circuit applications, diodes are never operated in the Zener range.

The circuits for the two basic forms of diode gates are shown in Fig. 5.4. In analyzing them, we first assume that the diodes are ideal and then consider the effects of the nonideal characteristics. In these gate circuits, there are two possible input voltages, V_- and V_+, and two supply voltages, V_L and V_H, such that $V_H > V_+ > V_- > V_L$. In the circuit of Fig. 5.4a, $e_0 = V_+$ only if both inputs are at V_+. If one or more inputs are at V_-, then $e_0 = V_-$. This behavior is summarized in the table of Fig. 5.4b. In the circuit of Fig. 5.4e, $e_0 = V_-$ only if both inputs are at V_-. If one or more inputs are at V_+, then $e_0 = V_+$. This behavior is summarized in the table of Fig. 5.4f.

In order to classify these gates as logic devices, we must specify which voltage levels shall represent the switching values 0 and 1. The specification $V_+ = 1$ and $V_- = 0$ is denoted as *positive logic*. In this case, the voltage

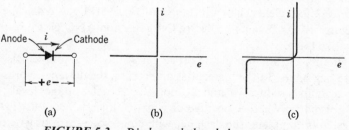

(a)   (b)   (c)

FIGURE 5.3.   *Diode symbol and characteristics.*

---

* There are actually two types of breakdown, *Zener* and *avalanche*, which are due to quite different physical phenomena. From an external point of view, however, there is no important difference.

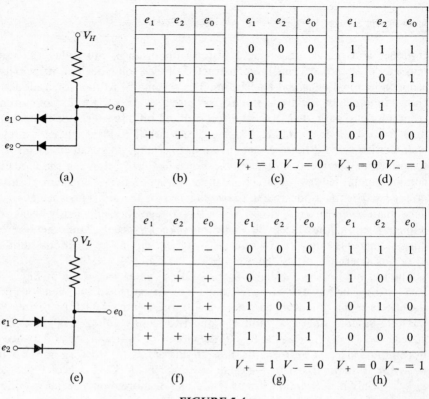

FIGURE 5.4

tables of Figs. 5.4b and 5.4f become the truth tables of Figs. 5.4c and 5.4g. We see that the circuits of Figs. 5.4a and 5.4e are AND gates and OR gates, respectively, for positive logic. The designation $V_- = 1$ and $V_+ = 0$, denoted as *negative logic*, yields the truth tables of Figs. 5.4d and 5.4h. We see that the circuits of Figs. 5.4a and 5.4e are OR gates and AND gates, respectively, for negative logic.

This duality of function of the diode gates may be a source of some confusion to beginners. In the logical design process, the designer works in terms of the switching values, 0 and 1, not the actual voltage levels, so that it does not matter whether positive or negative logic is being used. After the logic design is complete, the designer will choose one form of gate for AND and the other for OR, in accordance with the type of logic desired. The choice of positive or negative logic may be governed by compatibility requirements with connected systems, or power supply limitations, or any of a number of factors beyond the scope of this book.

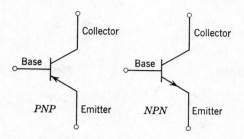

FIGURE 5.5.   *PNP and NPN transistors.*

If the diodes were ideal, there would be no limit on the number of inputs on a single gate (fan-in), or on the number of other gates a single gate could drive (fan-out), or on the number of gates that could be connected in series one after another (levels of gating). However, the voltage across conducting diodes and the finite reverse current of nonconducting diodes result in continual degradation signal level as the circuit complexity increases. For example, even in a single gate, the output voltage will always be slightly different from the input voltages because of the drops across the conducting diodes.

The signal degradation problem associated with diode logic may be alleviated by introducing amplification at appropriate points, most frequently through the use of transistors. The symbols for the two basic types of transistors, *PNP* and *NPN*, are shown in Fig. 5.5.

The transistors are used as switches in logic circuits. In the *PNP* transistor, if the base is driven slightly positive with respect to the emitter, the transistor will be *cut off*, thus acting as an open circuit to current flow from emitter to collector. If the base is driven slightly negative with respect to the emitter, the transitor will be *saturated*, thus acting essentially as a short circuit flow from emitter to collector. There will be a small voltage drop from emitter to collector in saturation, but this drop will be essentially independent of the magnitude of the emitter-to-collector current. The NPN transistor behaves in the same fashion except that the polarities are reversed. A negative voltage from base to emitter will cut off an *NPN* transistor while a positive voltage will saturate it, and the current flow will be from collector to emitter.

Transistors exhibit both voltage and current gain. Figure 5.6 shows an *NPN* transistor connected in the standard "grounded emitter" configuration. When the transistor is cut off, the collector will be at the collector supply level, V_+. When it is saturated, the collector will be typically about 0.1 volt above the emitter supply level, V_-. This collector voltage swing will be in the range of 5 to 50 volts for typical transistors. The base voltage swing necessary to take the transistor from cutoff to saturation will be in the range of 0.1 to 0.5 volt.

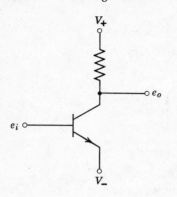

FIGURE 5.6. *Transistor in grounded emitter circuit.*

In addition, the current flowing in the base circuit will typically be on the order of one-hundredth of the current in the collector and emitter circuits.

The basic circuits for *PNP* and *NPN inverters* are shown in Fig. 5.7. As before, $V_H > V_+ > V_- > V_L$. Consider the *NPN* inverter, Fig. 5.7b. When $e_i = V_-$, the voltage at the base will be negative with respect to V_-, so the transistor will be cut off, and $e_o = V_+$. The values of R_b and R_{bb} are so chosen that when $e_i = V_+$, the voltage at the base will be positive with respect to V_-. The transistor will thus be conducting so that $e_o = V_-$, neglecting the small drop across the transistor in saturation. This behavior is summarized in the table of Fig. 5.7c. For positive logic, this will result in the truth table of Fig. 5.7d. For negative logic, the truth table will be the same, except the position

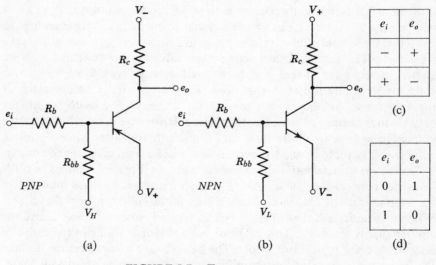

FIGURE 5.7. *Transistor inverters.*

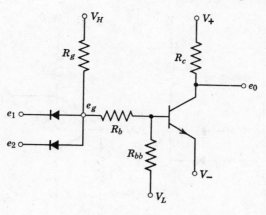

(a)

e_1	e_2	e_0
−	−	+
−	+	+
+	−	+
+	+	−

(b)

e_1	e_2	e_0
0	0	1
0	1	1
1	0	1
1	1	0

$V_+ = 1 \quad V_- = 0$

(c)

e_1	e_2	e_0
1	1	0
1	0	0
0	1	0
0	0	1

$V_+ = 0 \quad V_- = 1$

(d)

FIGURE 5.8.   *Diode-transistor gate, first form.*

of the rows will be reversed. We leave it to the reader to satisfy himself that the *PNP* inverter will function in the same general fashion.

In order to correct the loading deficiencies of the diode gates, we may simply let the output of each gate drive an inverter, as shown in Figs. 5.8 and 5.9. Consider the circuit of Fig. 5.8a, which consists of a first-form diode gate driving a *NPN* inverter. If one or more inputs are at V_-, e_g will also be at V_-, and the voltage at the base will be negative with respect to the emitter. The transistor will be cut off, so $e_0 = V_+$. If both inputs are at V_+, the voltage at e_g, which will depend on the circuit parameters in the base circuit, will be such as to drive the base positive with respect to the emitter. The transistor will conduct, and $e_0 = V_-$. This behavior is summarized in the table of Fig. 5.8b. For positive logic, this results in the truth table of Fig. 5.8c, which is seen to be the table for NOT-AND, or NAND. For negative logic, we obtain the

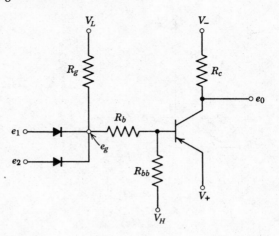

(a)

e_1	e_2	e_0
+	+	−
−	+	−
+	−	−
−	−	+

(b)

e_1	e_2	e_0
1	1	0
0	1	0
1	0	0
0	0	1

$V_+ = 1$ $V_- = 0$

(c)

e_1	e_2	e_0
0	0	1
1	0	1
0	1	1
1	1	0

$V_+ = 0$ $V_- = 1$

(d)

FIGURE 5.9.   *Diode-transistor gate, second form.*

truth table of Fig. 5.8d, the table for NOT-OR, or NOR. By a similar argument, we find that the circuit of Fig. 5.9a implements NOR for positive logic and NAND for negative logic.

The addition of the transistor inverters considerably alleviates the loading problems associated with diode logic, but it does not entirely eliminate them. Consider the situation shown in Fig. 5.10, where one NAND gate drives another. When the output of the first gate is at the low level (transistor in saturation), a load current, I_L, will flow as shown. If other gates are connected in parallel, each will contribute a similar I_L. However, there is a limit, set by base current and current gain, to the amount of current that can flow into the transistor if it is to remain in saturation. If so many gates are connected that this limit is exceeded, the transistor will come out of saturation and the

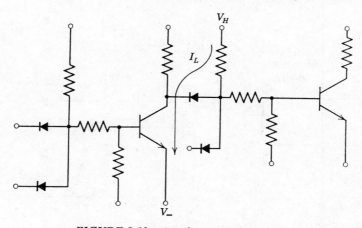

FIGURE 5.10.   *Loading on NAND gate.*

output voltage will rise above the low logic level. I_L is of the same order of magnitude as the base current, so the gain of the transistor nevertheless considerably reduces fan-out restrictions.

The fan-in can also be considerably increased over that possible with diode logic. The finite back resistance of the diodes will still cause deterioration in signal level at the gate output, but now the gate voltage, e_g, need only have enough swing to provide reliable control of the transistor. As long as we observe the fan-in and fan-out restrictions, there will be no deterioration of signal level through successive levels of diode-transistor gating. Because of the small voltage across the transistor in saturation and minor loading effects in cutoff, the signal levels will differ slightly from the supply voltages, V_+ and V_-. However, the power gain of the transistors will eliminate any progressive degradation of signals.

5.4   Speed and Delay in Logic Circuits

Faster operation generally means greater computing power, so there is a continual search for faster and faster logic circuits. The development of faster circuits is the province of the electronic circuit engineer, rather than the logic designer, and the actual speed of the electronics is of only indirect interest to the logic designer. However, the qualitative nature of the transitions and delays in logic circuits plays an important role in the logical theory of sequential circuits.

Figure 5.11 shows a grounded-emitter inverter and the response to a positive pulse at the base, taking the transistor from cutoff to saturation and back.

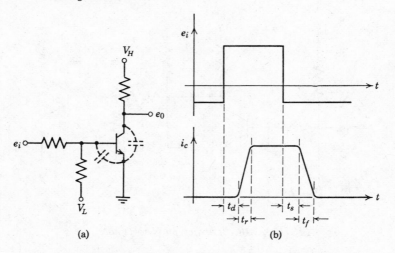

FIGURE 5.11. *Transient response of transistor inverter.*

There are a number of complex physical factors entering into the determination of this response, of which we can consider only a few. First, from the time of the start of the base pulse, there is a delay time, t_d, before the collector current starts to rise. This delay is primarily caused by the effective base-to-emitter capacitance of the transistor (shown dotted in Fig. 5.11a). This capacitance must be charged to a positive level before the transistor starts to turn on. After the transistor starts to conduct, there is a finite rise time, t_r, primarily controlled by the collector capacitance. On the trailing edge of the base pulse, there is again a delay, t_s, before the collector current starts to drop. This delay is due to the base-to-emitter capacitance as well as the storage of charge in the base region during saturation. The delay due to the storage effect t_s is longer than t_d. Finally, there is a finite fall time, t_f, again due to the collector capacitance.

In terms of specifying delay times through gates, the times t_d and t_s shown in Fig. 5.11b are meaningful only if the input rise and fall times are negligible compared to those at the output. This is rarely the case, since logic circuits are usually driven by other logic circuits of the same type, so that rise and fall times are comparable at input and output, and are generally of the same order of magnitude as the delay times. It is thus necessary to specify just what points on the input and output waveforms are to be used in specifying delay times.

In this regard it is important to note that the transistors in logic circuits are operating in the switching mode. Even though the input may vary over the full logic range (0 to 3.6 for TTL, 0 to 5 for DTL), only a very small

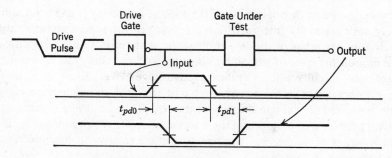

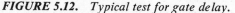

FIGURE 5.12. *Typical test for gate delay.*

change in input voltage is required to cause switching. For both TTL and DTL, the output switches fully from one logic level to the other as the input passes through a narrow range of about 0.1 volt, nominally centered at 1.5 volts. Unless the input change is so slow that it takes longer than the nominal rise time of the circuit to pass through this 0.1 volt range, which is most unlikely, the output rise and fall times will be totally determined by the internal parameters of the logic circuit.

On the basis of this switching at a nominal voltage of 1.5 volts, delay time is normally measured from the time the input waveform passes through 1.5 volts until the corresponding output change passes through 1.5 volts. A typical test setup for measuring delay time is shown in Fig. 5.12, along with the corresponding waveforms. As shown, the gate under test is driven by another gate of the same type, which is in turn driven by a drive pulse having rise and fall times of the same order of magnitude of those of the gates.

As shown, the delay times are measured from the 1.5 volt points of the input transitions to the 1.5 volt points of the corresponding output transitions. Because one transition involves the transistor turning on, the other the transistor turning off, the delays, as noted above, are not equal. The notation for these delays is unfortunately not standardized among manufacturers. The most common notations are t_{pd0} or t_{pd-} for the negative-going output transition and t_{pd1} or t_{pd+} for the positive-going output transition. The average gate delay, t_{pd}, is the arithmetic mean of these two delays. The transition voltages will be different for other logic families. The user should consult the manufacturer's data sheet for details.

5.5 Integrated Circuit Logic

In Section 5.3, we discussed some elementary implementations of logic elements, which could be realized in terms of discrete diodes, transistors, and resistors, interconnected by printed circuits or wires. Such discrete logic was dominant through most of the 1960's, but today it has been almost entirely supplanted by integrated circuit logic. In integrated circuits, entire gate circuits made up of diodes, transistors, and resistors are formed on a single monolithic chip of silicon by chemical and metallurgical techniques.

Integrated circuits offer significant advantages over discrete circuits in three primary areas: size, power, and cost. Several complete gate circuits can be deposited on a single chip about the size of the head of a common pin. Practical problems of connecting these minute circuits require mounting them in much larger packages, but the size advantage relative to discrete circuits is nevertheless substantial.

Closely related to the size advantage is the reduction in power dissipation. Whatever the size of the individual circuit packages, the density with which large numbers of packages can be mounted is largely determined by their power dissipation because of the problems associated with dissipating the heat generated by large numbers of closely-packed circuits. The reduced power requirements also reduce power supply requirements, providing for still further reductions in equipment size and cost. Some families of IC logic require so little power that battery-powered operation of even very complex devices becomes practical, a real advantage where portability is important.

Perhaps most important are the reductions in cost made possible by integrated circuits. The fabrication processes of integrated circuits are basically similar to those of transistors and diodes, and the cost of a complete integrated circuit is generally comparable to the cost of a single transistor. Further, all the costs of interconnecting discrete components are eliminated. As a result, the cost of logic has been reduced dramatically, to the point where the logic cost is a minor part of the cost of most digital systems.

Although integrated circuits are marketed under a variety of brand names, the great majority fall into five basic classes: DTL, RTL, TTL (T^2L), ECL, and MOS. A complete treatment of all of these would properly be the subject of several books, but we shall briefly consider some of the more significant characteristics from the point of view of the system designer.

The circuit of the standard gate in DTL (Diode-Transistor-Logic) is shown in Fig. 5.13. The circuit is basically similar to the discrete NAND gate of Fig. 5.8. The extra transistor provides extra gain and additional speed, and the extra diode provides biasing of the output transistor without a separate negative supply (V_L). The signal levels are the same as the supply levels, 0 and +5. The number of inputs may be anywhere from 2 to 10, with from one to four gates in a single package. The total number of inputs for all gates in a package is approximately constant, controlled by the number of pins on the package. The input X is known as the *expander input* and allows additional diodes to be connected to increase the fan-in to about 20. The fan-out is typically 8 to 10, expressed as the number of similar gates

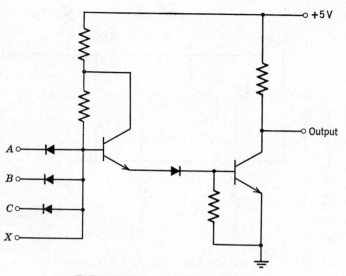

FIGURE 5.13.   *Standard DTL gate.*

that can be driven. The average gate delay (t_{pd}) is typically 30 nsec, and typical power dissipation is 10 mW/gate.

As was the case for the discrete gate of Fig. 5.8, this is a positive logic NAND gate, a negative logic NOR gate. In catalog listings, gates are classified according to their positive logic function, so this gate will be referred to as a NAND gate. In most integrated circuit families, one type of gate is considered standard in the sense that it is least expensive and available in the widest variety of configurations. In DTL, NAND is standard, with AND, OR, and NOR generally available only in 2-input gates. The designer working with integrated circuits will generally use the standard gate almost exclusively. As we shall see in Section 5.6, this does not pose any great problems.

The standard gate circuit for RTL (resistor-transistor logic) is shown in Fig. 5.14. If any of the inputs are at the high level, the corresponding transistor will conduct, resulting in a low output. Only if all inputs are low, cutting off all transistors, will the output be high. The gate is thus a positive-logic NOR. The logic levels are typically 0 and +3.6 volts. The basic fan-in of a single gate is limited to about 5, but expansion, to a maximum fan-in of 20, simply requires connecting more gating transistors in parallel between the output line and ground. The fan-out is less than DTL, typically about 4. The average gate delay is typically 20 nsec, and typical power dissipation is 12 mW/gate.

A standard TTL gate is shown in Fig. 5.15. The multiple-emitter transistor at the input performs the same AND function as the diodes at the input of

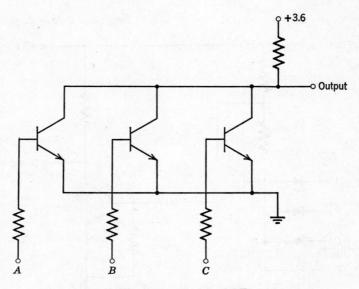

FIGURE 5.14. *Standard RTL gate.*

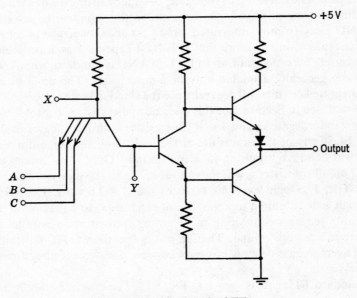

FIGURE 5.15. *Standard TTL gate.*

the DTL gates but with better speed and lower cost. The rather complicated but very fast three-transistor output section provides inversion so that the overall function of the gate is positive logic NAND. Inputs per gate, gate per package, and fan-out are similar to DTL. Expansion, to a maximum fan-in of 20, is accomplished by connecting additional multiple-emitter transistors between terminals X and Y. The power supply levels are the same as DTL, 0 and $+5$ volts, but the logic levels are different, 0 and 3.6 volts, because of the different output stage.

One important advantage of TTL is that the speed and power can be varied over quite a wide range simply by changing the size of the resistors. Increasing the size of the resistors decreases the speed and power; decreasing the size of the resistors increases the speed and power. TTL circuits are available in three types: Standard, with typical gate delays and power dissipation of 10 nsec and 10 mW/gate; low power, with typical gate delays and power dissipation of 33 nsec and 1 mW/gate; and high speed, with typical gate delays and power dissipation of 6 nsec and 25 mW/gate.

Whatever may be done to the resistance values, the speed of TTL is ultimately limited by the time required to pull the output transistors out of saturation. In *Schottky* TTL (STTL), the transistors are kept out of saturation by Schottky barrier diodes connected between base and collector. Standard STTL has typical gate delays and power dissipation of about 3 nsec and 20 mW/gate, and low power STTL has typical gate delays and power dissipation of 10 nsec and 2 mW/gate.

The fastest form of logic currently available is ECL (emitter-coupled logic) in which a totally different circuit configuration is utilized to ensure that all transistors are operated in the *active* region between saturation and cut off. The circuit of the standard ECL gate is shown in Fig. 5.16. The theory of

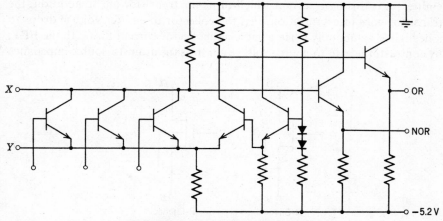

FIGURE 5.16. *Standard ECL gate.*

operation will not be discussed, as it is rather complex, but the reader will note that the gate provides both OR and NOR outputs. This is a distinct advantage, as it virtually eliminates the need for separate inverters. The logic levels are typically −.75 and −1.75 volts. The inputs per gate and gates per packages are somewhat less than for DTL, but the fan-out is considerably increased, typically 20 to 25. Expansion, to a maximum fan-in of 20, is possible by connecting additional input transistors between points X and Y. There are several types of ECL circuits available, but the two most commonly used provide average gate delays of 2 nsec and 1 nsec, with power dissipations of 25 mW/gate and 75 mW/gate, respectively.

The transistors used in all types of circuits discussed so far are known as *bipolar* transistors, with all these types (DTL, RTL, TTL, ECL) being thus classified as bipolar logic. The last type of logic to be discussed, MOS (metal oxide semiconductor), is based on a different type of transistor, the field effect transistor (FET). As with the bipolar transistor, there are two types, known as *N*-channel FET's and *P*-channel FET's, the symbols for which are shown in Fig. 5.17. The three terminals of FET's are known as the Drain, Source, and Gate.

In the *N*-channel FET, if the gate is negative with respect to the source, the FET is an open circuit from drain to source. If the gate is positive with respect to the source, the FET is a short circuit from drain to source. The operation is the same for the *P*-channel FET except that polarities are reversed. The operation of the FET may appear to be the same as that of the bipolar, but there are important differences. In the bipolar, although we characterize the operation in terms of the base-to-emitter voltage, it is the base current which is the controlling factor. For example, in an NPN transistor, when the base goes positive with respect to the emitter, current flows from base to emitter; it is this base current which turns the transistor on. In addition, the transistor goes into saturation, i.e., the collector to emitter voltage drops to a negligible value, only when a sufficient collector current flows. In the FET, by contrast, the gate-to-source voltage controls the drain-to-source impedance

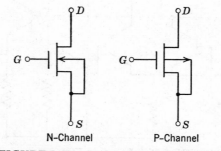

FIGURE 5.17. *N-channel and P-channel FET's.*

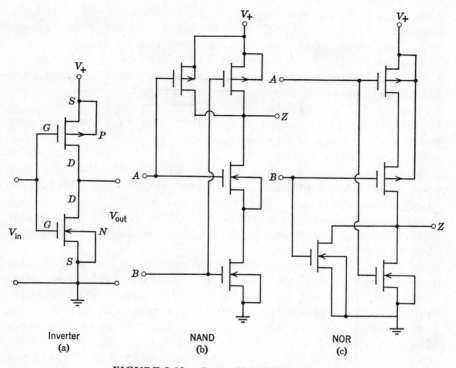

FIGURE 5.18. *Basic CMOS logic circuits.*

by electrostatic field action (hence the name *field effect*) and the gate current is essentially unmeasurable. Further, when the FET is turned on, the drain-to-source voltage is essentially zero even though no current flows in the drain circuit. As we shall see, these special characteristics of the FET make possible logic circuits with extremely low power consumption.

There are three basic types of MOS logic circuits, N-channel (NMOS), P-channel (PMOS), and complementary (CMOS). For reasons which will be discussed briefly in Chapter 16, the NMOS and PMOS technologies are competitive only in the form of large-scale integrated circuits (LSI). Of the the three types, only CMOS is typically utilized in the realization of individual gates and flip-flops.

The circuit of the standard CMOS inverter is shown in Fig. 5.18a. The circuit consists of an N-channel FET and a P-channel FET in series (hence the name *complementary*), normally connected between ground and a positive supply voltage. The signal levels are the same as the supply levels, i.e., 0 and V_+. When the input is at 0 volts, the NFET is cut off and the PFET is turned on. As a result, there is a low impedance path from the output to V_+ and an open circuit to ground, resulting in an output voltage equal to V_+. Similarly, when the input is at V_+ volts, the NFET becomes a low impedance while the PFET becomes an open circuit, resulting in 0 volts at the output.

Note that, in a static condition, one FET or the other is always cut off, and there are no resistive paths. Further, if the circuit drives another CMOS device, there is no output current, since the gate of an FET draws no current. As a result, the only quiescent power dissipation is that due to leakage in the cut off FET. This quiescent dissipation is typically less than 1 μW/gate and in some circuits is in the nanowatt range.

When the inverter switches from one level to another, both gates will be on momentarily, and the change in voltage levels requires the charging of parasitic capacitances within the inverter and the interconnections between circuits. As a result, current will flow during switching and the average power dissipation is directly dependent on the switching rate. At an average switching rate of 1 MHz (one logic change per microsecond), the typical dissipation is 1 mW/gate.

NAND and NOR gates are constructed by using one inverter for each input, with the upper (P) transistors in parallel and the lower (N) transistors in series for NAND, and vice-versa for NOR, as shown in Figs. 5.18b and c. In the NAND gate (Fig. 5.18b), if any of the inputs are at the 0 level, the corresponding upper transistors are biased on and the corresponding lower transistors are biased off, so that the output is at the V_+ level. If all the inputs are at the V_+ level, all the upper transistors are off and all the lower transistors are on, so that the output is at the 0 level. The operation of the NOR gate may be analyzed in an analogous fashion.

Standard CMOS circuits will run properly at supply voltages of 3 to 15 volts, a characteristic which makes it possible to use unregulated power supplies. Special CMOS circuits can be operated at 1.3 volts, making it possible to use single dry cells as the power source. There is no standard gate type in CMOS, with NAND and NOR being equally available. Maximum fan-in is 4. Since FET gates draw no current, fan-out is primarily determined by parasitic wiring capacitances and is, for all practical purposes, unlimited. Average gate delay is dependent on the power supply voltage, being typically 35 nsec at 5 volts, 25 nsec at 10 volts.

5.6   Comparison of IC Logic Families

As we have seen, the logic designer has a wide range of circuit types from which to choose. The main factors which should be considered are summarized in Table 5.1. The first five factors have already been discussed. Noise performance is a complex factor, as there are many types of noise. Two basic factors are noise immunity and noise generation. Noise immunity

TABLE 5.1. Basic Characteristics of Integrated Circuit Logic

Type — Characteristic	DTL	RTL	TTL	STTL	ECL	CMOS
Positive logic function of basic gate	NAND	NOR	NAND	NAND	OR/NOR	NAND or NOR
Maximum fan-in without expansion	10	5	8	8	5	8
Typical fan-out	8–10	4	8–10	8–10	20–25	Unlimited
Typical power dissipation per gate	10 mW	12 mW	1–25 mW	2–20 mW	High	.01 μW static $\approx$ 1 mW at 1 MHz.
Typical gate delay, nsec.	30	20	6–33	3–10	1–2	25–35
Noise performance	Good	Fair	Fair to medium	Fair to medium	Fair	Very good
Cost	Low	Low	Low to medium	Medium to high	High	Medium to high
Availability of complex functions	Fair	Fair to medium	Excellent	Medium	Fair	Medium, growing

generally refers to the ability of circuits to discriminate against spurious signals on the input lines. Such noise usually arises because of stray inductive and capacitive coupling and tends to increase with speed, due to larger di/dt and dv/dt terms. Noise generation refers primarily to noise generated on the power supply lines by the switching action of the gates, and has no necessary relation to noise immunity. For example, TTL has good immunity to noise on the input lines but generates so much noise on the power lines that its overall noise performance is only fair.

Another significant factor is the availability of complex functions. So far, we have considered only the gate functions available. In addition, inverters, which may be considered single-input gates, are available in all families, as are various types of flip-flops. Flip-flops, which are electronic storage elements, are discussed in detail in Chapter 9. Also available in some cases are more complex functions, such as adders, decoders, counters, etc. Such circuits are generally classed as medium-scale integration (MSI), and TTL is the clear leader in this area. Finally, we might note that TTL, STTL, and DTL, which use the same power supply levels, are compatible. CMOS is compatible with DTL and TTL if operated at 5 volts, although care must be exercised because of the widely different current levels. RTL and ECL are not compatible with other types without elaborate interface electronics. As a general rule, mixing of logic families should be avoided as much as possible.

The reader should keep in mind that integrated circuit technology is constantly and rapidly changing so that the data in Table 5.1 must be considered as subject to change. This is particularly true in the cost area, where the situation changes almost daily. For example, RTL was once the dominant type of logic because of its low cost. However, TTL is now cost competitive and technically superior in every way so that RTL is rapidly disappearing and should not be used in new designs. It has been included here primarily for completeness.

Until recently, DTL has been the best choice for good noise performance at low cost. However, CMOS, which has even better noise performance, is rapidly coming down in price, to the point where DTL should not be used for new designs. At present, TTL is the dominant family of logic, but CMOS is improving in speed as well as cost, and STTL is coming down in cost, so that CMOS and STTL, between them, will probably replace TTL in time. There is nothing yet in sight to match the speed of ECL, but its noise and power problems will probably continue to limit its use.

5.7 *Logical Interconnection of NAND and NOR Gates*

In each family of integrated circuits, we have seen that there is one standard gate type which is more economical and more readily available than other types. What constraints, if any, does this situation impose on the designer? We saw in Chapter 3 that both NAND and NOR are sufficient connectives by themselves in the sense that any other binary connective can be realized solely in terms of either NAND or NOR. Any Boolean expression can thus be realized solely with NAND or NOR gates.

Theoretically, then, restricting a design to one standard gate type should cause no problems. But what about the practical problems of finding economical realizations by simple design procedures? As was the case in the example of the overflow detector in Chapter 3, many problems seem to be most naturally formulated in AND-OR terms. Furthermore, as we shall see in the next chapter, the second-order, AND-OR and OR-AND forms of Boolean expressions are the end result of a straightforward design procedure. Nevertheless, it turns out that the use of NAND and NOR circuits causes no particular logical problems.

Consider the simple logical circuit of Fig. 5.19a, which consists of three NAND gates driving another NAND gate. From DeMorgan's law,

$$\overline{XYZ} = \bar{X} + \bar{Y} + \bar{Z}, \qquad \text{(Theorem 4.11)}$$

we see that the final NAND gate can be replaced by an OR gate with inversion on the inputs (Fig. 5.19b). From Lemma 4.7,

$$\bar{\bar{X}} = X \qquad \text{(Lemma 4.7)}$$

so that the two successive inversions on the lines between the input and output gates cancel, giving the circuit of Fig. 5.19c. Algebraically, the same

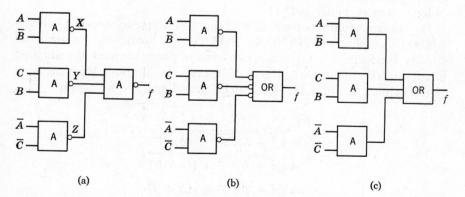

(a) (b) (c)

FIGURE 5.19. *Conversion of NAND-NAND to AND-OR circuit.*

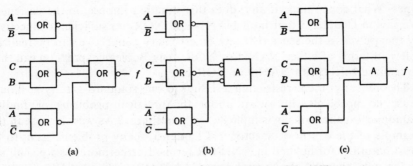

FIGURE 5.20.   *Conversion of NOR-NOR to OR-AND circuit.*

result is obtained as follows:

$$f = \overline{(\overline{A\bar{B}}) \cdot (\overline{BC}) \cdot (\overline{A\bar{C}})}$$
$$= (\overline{\overline{A\bar{B}}}) + (\overline{\overline{BC}}) + (\overline{\overline{A\bar{C}}}) \qquad \text{(Theorem 4.11)}$$
$$= A\bar{B} + BC + \bar{A}\bar{C} \qquad\qquad \text{(Lemma 4.7)}$$

Thus, we see that a two-level NAND circuit is equivalent to a two-level AND-OR circuit. In a similar fashion, we can show that a two-level NOR circuit is equivalent to a two-level OR-AND circuit (Fig. 5.20). These equivalences are of great importance. In the next chapter, it will be shown that *any* switching function can be realized in a two-level AND-OR or OR-AND form. Thus, we may continue to formulate design problems in the natural concepts of AND and OR while realizing the resultant functions in terms of two-level NAND-NAND or NOR-NOR configurations. The use of a single type of logic element contributes greatly to economy in quantity production and also simplifies the stocking of spare parts for servicing. As a result, the majority of computers being constructed today use a single type of logic element, either NAND or NOR.

The reader should be careful to note that this simple relationship between AND-OR circuits and NAND or NOR circuits holds only for two-level circuits. If circuits with more than two levels of gating are used, the situation is considerably more complicated. For example, consider the three-level NAND circuit of Fig. 5.21a. By straightforward analysis, we obtain

$$f = \overline{(\overline{A \cdot B}) \cdot C \cdot D \cdot (\overline{\bar{A} \cdot B}) \cdot D}$$
$$= \overline{(\overline{\bar{A} + \bar{B})CD} \cdot [(\overline{\bar{A}B})D]}$$
$$= \overline{(\overline{\bar{A} + \bar{B})CD}} + \overline{[(\overline{\bar{A}B})D]}$$
$$= (\bar{A} + \bar{B})CD + \overline{\bar{A}B} + \bar{D}$$
$$= (\bar{A} + \bar{B})CD + \bar{A}B + \bar{D}$$

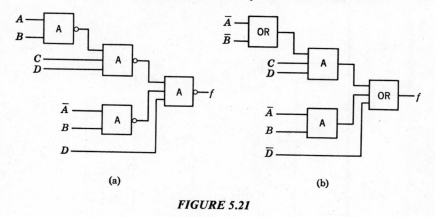

(a)   (b)

FIGURE 5.21

from which we see that the equivalent AND-OR circuit is that shown in Fig. 5.21b.

The similarity in form between the two circuits of Fig. 5.21 suggests the possibility of a direct algorithmic procedure for converting AND-OR circuits to NAND or NOR circuits, and such a procedure will be presented in Chapter 8. However, the circuits obtained by such conversion are sometimes more complex than the AND-OR form and, except for the two-level case, will not generally be the simplest NAND or NOR form for the given function.

Direct design procedures for NAND or NOR circuits have been developed, but they are rather limited in scope and power. By contrast, the design algorithms for two-level forms are completely general, very powerful, and essentially independent of the type of circuit realization chosen. For just this reason, it would seem to be better to stick to two-level design except for special cases.

The standard two-level gate circuit is comprised of a number of input gates of a single type, all driving one output gate. Let us now consider the possibility of eliminating the output gate and just connecting the outputs of all the input gates together. Such a connection is commonly referred to as the *wired-OR* connection, but the actual logical results obtained depend both on the function realized by the gate and also on the form of its output circuit.

Consider first the connection of gates with the common emitter output, as discussed in Section 5.3. Figure 5.22a shows two such gate outputs connected together. Recall that, when the transistor saturates (output low), the output is effectively shorted to ground. Thus, in the circuit of Fig. 5.22a, if either transistor saturates (low output), the resultant output will be low. Only if both transistors are off (high output) will the resultant output be high. Thus,

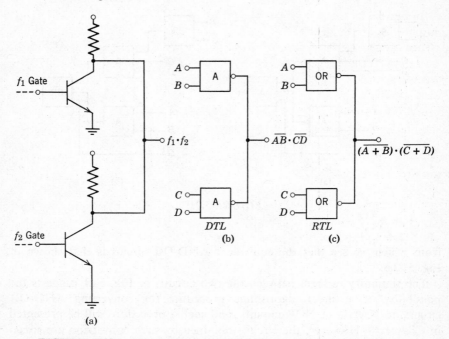

FIGURE 5.22. *Wired-OR connections for common-emitter output gates.*

in terms of positive logic, the wired-OR connection results in the logical ANDing of the functions realized by the gates.

Both DTL and RTL gates have common emitter output. For the interconnection of two DTL NAND gates, as shown in Fig. 5.22b, the resultant function is

$$f = (\overline{AB}) \cdot (\overline{CD})$$

or, by DeMorgan's Law,

$$f = \overline{AB + CD}$$

Note that the function realized is the complement of the function that would be realized if the same gates were connected to a NAND output gate in the standard NAND-NAND connection. For the connection of two RTL NOR gates, as shown in Fig. 5.22c, the resultant function is

$$f = \overline{(A + B)} \cdot \overline{(C + D)}$$

or, by DeMorgan's Law,

$$f = \overline{A + B + C + D}$$

Thus, the wired-OR connection of RTL NOR gates is simply equivalent to input expansion.

In Fig. 5.22a, the reader will note that the two collector resistors are directly in parallel, so the net collector resistance is half that of the individual gates. When large numbers of gates are connected together in the wired-OR connection, the resultant collector resistance may be too small for proper operation. As a result, DTL gates are available in the "open collector" configuration, in which output collector resistor is simply omitted. The wired-OR connection is formed by tying these open collectors together and supplying them through a single external resistor, as shown in Fig. 5.23. The value of this external resistor depends on the number of gates and the loading. The manufacturer's data sheets should be consulted for information on this matter.

ECL gates use a different type of output circuit, the common-collector, or emitter follower, output. With this connection, when the transistor saturates, the output is effectively shorted to the supply voltage, producing a high output. The low output corresponds to the transistor being cut off. When such output circuits are connected together, as in Fig. 5.24a, the output will be high if either transistor is on. Thus, in terms of positive logic, the resultant function is the logical OR of the functions realized by the gates. Standard ECL gates are OR/NOR gates, so there are two possible connections, as shown in Fig. 5.24. For the OR outputs, the result is simply input expansion:

$$f = (A + B) + (C + D)$$

For NOR, the resultant function is given by

$$f = \overline{(A + B)} + \overline{(C + D)}$$

or, by DeMorgan's Law,

$$f = \overline{(A + B) \cdot (C + D)}$$

which corresponds to the complement of the function realized by NOR-NOR connection of the same gates.

Conventional TTL and STTL gates, having the two-transistor output circuit of Fig. 5.15, may not be used in the wired-OR connection. The nature of the circuit is such that connecting outputs together will damage the gates. However, open-collector TTL gates, identical except for input to open-collector DTL gates, are available for wired-OR use. Unfortunately, it is the output circuit that is primarily responsible for the speed of TTL, so that changing the output circuit to that of DTL results in a corresponding degradation in performance. One manufacturer offers a special form of TTL, *Tri-State TTL*, which permits the wired-OR connection without loss of performance, but it is more expensive and complex than standard TTL and ordinarily is used only in busing applications. Finally, the wired-OR connection may not be used with CMOS.

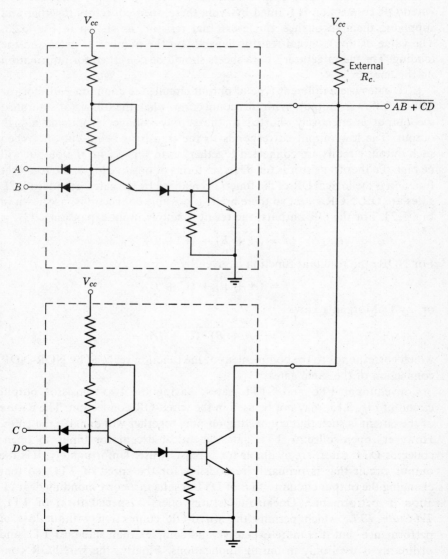

FIGURE 5.23. *Wired-OR connection of open collector gates.*

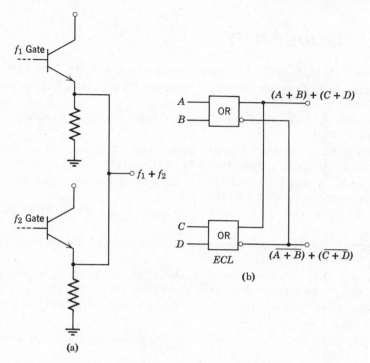

FIGURE 5.24.   *Wired-OR with common-collector output gates.*

5.8   Conclusion

It may seem to the reader that the wide variety of gate types and characteristics will pose a serious problem to the logic designer, possibly requiring different design techniques for each type of logic. Fortunately, this is not the case. In the next chapter, we shall develop powerful design methods which can be readily adapted to any type of logic. There have been some attempts to develop special design methods tailored to some specific type of logic. We feel that this is most unwise, as it will tend to limit the designer to a specific type of logic and prevent him from exploiting the full power of all the many types of logic available. The logic types discussed in this chapter represent only a portion of those currently available, and new types are constantly being developed. An understanding of the basic characteristics and limitations of logic devices will enable the resourceful designer to adapt powerful general design techniques to whatever physical realizations may best suit his purpose.

BIBLIOGRAPHY

1. Caldwell, S. H. *Switching Circuits and Logical Design*, Wiley, New York, 1958.
2. Marcus, M. P. *Switching Circuits for Engineers*, Prentice-Hall, Englewood Cliffs, N.J., 1962.
3. Pressman, A. I. *Design of Transistorized Circuits for Digital Computers*, Rider, New York, 1959.
4. Garrett, L. S. "Integrated-Circuit Digital Logic Families," *IEEE Spectrum*, 46–58 (Oct. 1970), 63–72 (Nov. 1970), 30–42 (Dec. 1970).
5. Maley, G. A. and J. Earle. *The Logic Design of Transistor Digital Computers*, Prentice-Hall, Englewood Cliffs, N.J., 1963.
6. Torero, E. A. "Focus on CMOS," *Electronic Design*, **20**: 8, 54–61 (Apr. 13, 1972).
7. Torero, E. A. "Focus on Fast Logic," *Electronic Design*, **20**: 12, 50–57 (June 8, 1972).
8. Femling, Don. "Enhancement of Modular Design Capability by Use of Tri-State Logic," *Computer Design*, **10**: 6, 59–64 (June 1971).
9. Sheets, John. "Three State Switching Brings Wired-OR to TTL," *Electronics*, **43**: 19, 78–84 (Sept. 14, 1970).
10. Carr, W. N. and J. P. Mize. *MOS/LSI Design and Application*, McGraw-Hill, New York, 1972.
11. Kohonen, T. *Digital Devices and Circuits*, Prentice-Hall, Englewood Cliffs, N.J., 1972.

Chapter 6

Minimization of Boolean Functions

6.1 Introduction

In Chapter 3, we showed how the concept of truth-function analysis could be used in setting up a mathematical model of a logical (or switching) system. In Chapter 4, we stated the formal rules of an algebra governing the manipulation of the mathematical model. In Example 4.2, utilization of the mathematical model resulted in an obvious simplification of the switching circuit. In general, our basic purpose is design; and the reason for using a mathematical model is that it provides a convenient form for exploring possible designs. Further, the *best* design is generally the *simplest* design that will get the job done, from which it follows that our objective in manipulating the mathematical model will usually be simplification or minimization. Due to the myriad possible forms of Boolean expressions, no completely general criteria for the simplest expression have been developed. As we shall see, however, it is possible to define a simplest form of the two-level, or minimum delay-time, circuit as discussed in Chapter 5.

We have seen that the rules of the algebra can be applied formally to manipulation and simplification of Boolean expressions, but such methods are far from easy to apply. The algebraic manipulation of Boolean functions is often quite involved, and finding the right line of attack requires considerable ingenuity and, sometimes, just plain luck. It is obvious that standard, systematic methods of minimization would be very useful. In this chapter and the next, two such methods are set forth—one graphical, the Karnaugh map, and one tabular, the Quine-McCluskey method.

6.2 Standard Forms of Boolean Functions

A standard, or general, method of analysis would seem to imply a standard starting point and a standard objective. In this section, we shall develop two standard forms which may be used as starting points for simplification. The definitions of *literal, product term, sum term,* and *normal term,* which are given in Table 6.1, will be useful in discussing these standard forms. In a sense, the normal form is the "best" form. Since

$$A \cdot A = A = A + A$$

and   (Lemma 4.2)

$$A + \bar{A} = 1 \qquad \text{and} \qquad A\bar{A} = 0$$

a multiple occurrence of a variable in a sum or product term is always either redundant or results in a trivial function.

TABLE 6.1

Term	Definition	Synonym
Literal	A variable or its complement. $(A, \bar{A}, B, \bar{B}, \text{etc.})$	
Product term	A series of literals related by AND, e.g., $A\bar{B}D$, $AC\bar{D}E$, etc.	Conjunction
Sum term	A series of literals related by OR, e.g., $A + C + \bar{D}, A + B + \bar{D} + E$, etc.	Disjunction
Normal term	A product or sum term in which no variable appears more than once.	

As a first step, we shall consider the form known as the *sum of products*. The nature of this form and manner of developing it are illustrated in the following examples.

Example 6.1

Find the sum-of-products form of the function

$$f(A, B, C, D) = (AC + B)(CD + \bar{D})$$

First, let $AC = x$ and $CD = y$. Then

$$f(A, B, C, D) = (x + B)(y + \bar{D})$$

$$= (x + B)y + (x + B)\bar{D} \qquad \text{(Distributive law)}$$

$$= xy + By + x\bar{D} + B\bar{D} \qquad \text{(Distributive law)}$$

$$= ACD + BCD + AC\bar{D} + B\bar{D} \qquad (6.1)$$

■

Example 6.2

Repeat Example 6.1 for the function

$$f(A, B, C, D, E) = \overline{(\overline{AC} + \bar{D})(\overline{B + CE})}$$

$$\begin{aligned}
&= [(\bar{A} + \bar{C}) + \bar{D}](\bar{B} \cdot \overline{CE}) && \text{(DeMorgan's law)}\\
&= [(\bar{A} + \bar{C})\bar{B} + \bar{B}\bar{D}](\bar{C} + \bar{E}) && \text{(DeMorgan's law)}\\
&&& \text{(Distributive law)}\\
&= (\bar{A}\bar{B} + \bar{B}\bar{C} + \bar{B}\bar{D})(\bar{C} + \bar{E}) && \text{(Distributive law)}\\
&= \bar{A}\bar{B}\bar{C} + \bar{B}\bar{C} + \bar{B}\bar{C}\bar{D} + \bar{A}\bar{B}\bar{E} + \bar{B}\bar{C}\bar{E} + \bar{B}\bar{D}\bar{E} && (6.2)
\end{aligned}$$

∎

From these examples, the general method to be followed in converting any Boolean function to the sum-of-products form should be evident. If no terms other than single variables are negated, only repeated application of the second distributive law is required. Where terms other than single variables are negated, DeMorgan's law must also be applied. Recalling the definition of a normal term, we see that Equation 6.1 or 6.2 could be referred to as a *sum of normal products*, but the shorter notation is preferred.

Continuing with the sum of products, Equation 6.1, we can write:

$$\begin{aligned}
f(A, B, C, D) &= ACD + BCD + AC\bar{D} + B\bar{D} = ACD(B + \bar{B})\\
&\quad + BCD(A + \bar{A}) + AC\bar{D}(B + \bar{B}) + B\bar{D}(A + \bar{A})
\end{aligned}$$

$$(a + \bar{a} = 1, a \cdot 1 = a)$$

$$\begin{aligned}
&= ABCD + A\bar{B}CD + ABCD + \bar{A}BCD + ABC\bar{D}\\
&\quad + A\bar{B}C\bar{D} + B\bar{D}A + B\bar{D}\bar{A} && \text{(Distributive law)}\\
&= ABCD + A\bar{B}CD + \bar{A}BCD + ABC\bar{D} + A\bar{B}C\bar{D}\\
&\quad + B\bar{D}A(C + \bar{C}) + B\bar{D}\bar{A}(C + \bar{C})
\end{aligned}$$

$$(a + a = a, a + \bar{a} = 1, a \cdot 1 = a)$$

$$\begin{aligned}
&= ABCD + A\bar{B}CD + \bar{A}BCD + ABC\bar{D} + A\bar{B}C\bar{D}\\
&\quad + ABC\bar{D} + AB\bar{C}\bar{D} + \bar{A}BC\bar{D} + \bar{A}B\bar{C}\bar{D}
\end{aligned}$$

$$\text{(Distributive law)}$$

$$\begin{aligned}
f(A, B, C, D) &= ABCD + A\bar{B}CD + \bar{A}BCD + ABC\bar{D} + A\bar{B}C\bar{D}\\
&\quad + AB\bar{C}\bar{D} + \bar{A}BC\bar{D} + \bar{A}B\bar{C}\bar{D} && (a + a = a)
\end{aligned}$$

$$(6.3)$$

Notice that Equation 6.3 is a sum of normal products with every product containing as many literals as there are variables in the function. Such products are called *canonic products, standard products, or minterms*. The terms *standard* or *canonic sum of products* and *full disjunctive normal form* have been used for expressions of the form of Equation 6.3. *Standard sum of products* will be preferred here.

We are led to the following theorem, the validity of which may already be evident to the reader. A proof of this theorem will be presented in the next section.

THEOREM 6.1.   *Any switching function of n variables* $f(x_1, x_2, \ldots, x_n)$ *may be expressed as a standard sum of products.*

In view of the principle of duality, we expect that a *standard product of sums, full conjunctive normal form,* or *product of maxterms* will exist also.

Example 6.3

Express $f(A, B, C, D) = A + C + \bar{B}\bar{D}$ as a standard product of sums.

$$A + (C + \bar{B}\bar{D}) = A + (C + \bar{B})(C + \bar{D}) \qquad \text{(Distributive law)}$$
$$= (A + C + \bar{B})(A + C + \bar{D})$$

Then, since

$$A \cdot \bar{A} = 0 \qquad \text{(Postulate VI)}$$
$$A + 0 = A \qquad \text{(Postulate IIIa)}$$

and

$$(a + b\bar{b}) = (a + b)(a + \bar{b}) \qquad \text{(Distributive law)}$$

we can write

$$f(A, B, C, D) = (A + \bar{B} + C + D\bar{D})(A + C + \bar{D} + B\bar{B})$$
$$= (A + \bar{B} + C + D)(A + \bar{B} + C + \bar{D})(A + B + C + \bar{D})$$
$$\cdot \overline{(A + \bar{B} + C + D)}$$
$$= (A + \bar{B} + C + D)(A + \bar{B} + C + \bar{D})(A + B + C + \bar{D})$$

$$(6.4)$$

∎

Equation 6.4 is seen to consist of a product of normal sums, with each sum containing as many literals as there are variables in the function. Such sums are known as *canonic sums,* or *standard sums,* or *maxterms.*[*]

Generalizing, we have the dual of Theorem 6.1.

THEOREM 6.2.   *Any switching function of n variables* $f(x_1, x_2, \ldots, x_n)$ *may be expressed as a standard product of sums.*

It may seem to the reader that we are moving in the wrong direction, since the standard forms found in the above examples were more complicated than the expressions we started with. It is often the case that the standard

---

[*] The reasons for the names *minterm* and *maxterm* will become evident in a later section.

sum or product will be a very complex form, but our basic purpose in using them is to provide a common form for starting simplification procedures.

6.3 Minterm and Maxterm Designation of Functions

It may have occurred to the reader, in following the examples of the preceding section, that writing out all the minterms or maxterms of a given function may be rather laborious. A shorthand notation for switching functions would certainly be useful. A switching function is defined by its truth table, a listing of the function values for all possible input combinations. Therefore, a simple and precise means of designating functions is obtained by numbering the rows of the truth table for a function and then listing the rows (input combinations) for which the function has the value 1, or, alternately, the rows for which it has the value 0.

Figure 6.1 shows the truth table of a particular three-variable function, with the rows assigned identifying numbers. The row numbers are simply the decimal equivalents of the input combinations on each row interpreted as binary numbers. For example, the input combination $A = 0, B = 1, C = 1$, interpreted as a binary number, gives us $011_2 = 3_{10}$, so that row is called Row 3. Now we can specify the function by listing the rows for which it has the value 1:

$$f(A, B, C) = \sum m(0, 4, 5, 7) \qquad (6.5)$$

or the rows for which it has the value 0,

$$f(A, B, C) = \prod M(1, 2, 3, 6) \qquad (6.6)$$

The symbols "$\sum m$" and "$\prod M$" in the above lists are not just arbitrarily chosen but rather indicate a direct correspondence between these lists and the standard forms. To show this, we write out the standard sum-of-products form (Equation 6.5a) and the standard product-of-sums form (Equation 6.6a)

Row. No.	A	B	C	f
0	0	0	0	1
1	0	0	1	0
2	0	1	0	0
3	0	1	1	0
4	1	0	0	1
5	1	0	1	1
6	1	1	0	0
7	1	1	1	1

FIGURE 6.1. *Truth table with row numbers assigned.*

for the function of Fig. 6.1:

$$f(A, B, C) = \bar{A}\bar{B}\bar{C} + A\bar{B}\bar{C} + A\bar{B}C + ABC \tag{6.5a}$$

$$f(A, B, C) = (A + B + \bar{C})(A + \bar{B} + C)$$
$$(A + \bar{B} + \bar{C})(\bar{A} + \bar{B} + C) \tag{6.6a}$$

Consider first the sum-of-products form. From Lemma 4.3 $(a + 1 = 1)$, we see that the function will take on the value 1 whenever any one (or more) of the products takes on the value 1. Since these standard products, or minterms, contain all three variables, there is only one combination of inputs for which a given minterm will equal 1. Since each minterm is unique, no two minterms will equal 1 for the same input combination. For example, for input $A = 1$, $B = 0$, $C = 0$ (Row 4), the second minterm in Equation 6.5a equals 1, since

$$A\bar{B}\bar{C} = 1 \cdot 1 \cdot 1 = 1$$

It should be obvious that this is the only input combination for which $A\bar{B}\bar{C}$ equals 1. Thus, there is a one-to-one correspondence between $A\bar{B}\bar{C}$ and the 1 on Row 4 of the truth table. We indicate this by saying that this minterm "produces" the 1 on Row 4, and we designate it as minterm 4, denoted m_4. Similarly, the other minterms in Equation 6.5a are m_0, m_5, and m_7, and Equation 6.5 may then be interpreted as a list of the minterms in the standard sum-of-products form.

Consider next the standard product of sums, Equation 6.6a. From the dual version of Lemma 4.3 $(a \cdot 0 = 0)$, we see that this function will be 0 whenever one (or more) of the products is 0. By the dual of the above argument, a given maxterm can be 0 for only one input combination, and vice versa. For example, for input $A = 0$, $B = 1$, $C = 0$ (Row 2),

$$A + \bar{B} + C = 0 + \bar{1} + 0 = 0 + 0 + 0 = 0$$

Thus, we may say that this maxterm "produces" the 0 in Row 2, and designate it as maxterm 2, denoted M_2. Similarly, the other maxterms in Equation 6.5 are M_1, M_3, and M_6, and Equation 6.6 may be regarded as a list of maxterms in the standard product-of-sums form.

In summary, we associate each minterm with the input combination for which it would produce a 1 in the function and each maxterm with the combination for which it would produce a 0. Figure 6.2 shows the minterms and maxterms associated with each row of a 3-variable truth table. The extension to more variables should be obvious. From this table, we can derive a rule for determining the actual product or sum, given the row number or vice versa. For the minterms, each uncomplemented variable is associated with a 1 in the corresponding position in the binary row number, and each complemented variable is associated with a 0. For the maxterms, the rule is just the opposite.

Row No.	A B C	Minterms	Maxterms*
0	0 0 0	$\bar{A}\bar{B}\bar{C} = m_0$	$A + B + C = M_0$
1	0 0 1	$\bar{A}\bar{B}C = m_1$	$A + B + \bar{C} = M_1$
2	0 1 0	$\bar{A}B\bar{C} = m_2$	$A + \bar{B} + C = M_2$
3	0 1 1	$\bar{A}BC = m_3$	$A + \bar{B} + \bar{C} = M_3$
4	1 0 0	$A\bar{B}\bar{C} = m_4$	$\bar{A} + B + C = M_4$
5	1 0 1	$A\bar{B}C = m_5$	$\bar{A} + B + \bar{C} = M_5$
6	1 1 0	$AB\bar{C} = m_6$	$\bar{A} + \bar{B} + C = M_6$
7	1 1 1	$ABC = m_7$	$\bar{A} + \bar{B} + \bar{C} = M_7$

FIGURE 6.2

Example 6.4

Convert Eq. 6.3 to the minterm list form.

$f(A, B, C, D) =$

$$ABCD + A\bar{B}CD + \bar{A}BCD + ABC\bar{D} + A\bar{B}C\bar{D} + AB\bar{C}\bar{D} + \bar{A}BC\bar{D} + \bar{A}B\bar{C}\bar{D}$$

1111	1011	0111	1110	1010	1100	0110	0100
15	11	7	14	10	12	6	4

$f(A, B, C, D) = \sum m(4, 6, 7, 10, 11, 12, 14, 15)$ ∎

Example 6.5

Convert the following equation to the maxterm list form.

$f(A, B, C, D) =$

$$(A + \bar{B} + C + \bar{D})(\bar{A} + B + C + D)(A + B + \bar{C} + \bar{D})(\bar{A} + B + C + \bar{D})$$

$$0 \quad 1 \quad 0 \quad 1 \quad 1 \quad 0 \quad 0 \quad 0 \quad 0 \quad 0 \quad 1 \quad 1 \quad 1 \quad 0 \quad 0 \quad 1$$

$$5 \qquad\qquad 8 \qquad\qquad 3 \qquad\qquad 9$$

$f(A, B, C, D) = \prod M(3, 5, 8, 9)$ ∎

Since the truth table provides the complete specification of any switching function and we have now demonstrated a procedure for converting any truth table to a sum of minterms or product of maxterms, Theorem 6.1 and 6.2 have now been proven.

6.4   Karnaugh Map Representation of Boolean Functions

The Karnaugh map[1] is one of the most powerful tools in the repertory of the logic designer. The power of the Karnaugh map does not lie in its application

---

* Although the maxterm numbering system of Fig. 6.2 is most common, there are some books which number them in the reverse order, i.e., $\bar{A} + \bar{B} + \bar{C} = M_0, \bar{A} + \bar{B} + C = M_1$, etc.

A	B	A·B
0	0	0
0	1	0
1	1	1
1	0	0

A	B	A + B
0	0	0
0	1	1
1	1	1
1	0	1

FIGURE 6.3. *Truth tables for AND and OR.*

of any marvelous new theorems but rather in its utilization of the remarkable ability of the human mind to perceive patterns in pictorial representations of data. This is not a new idea. Anytime we use a graph instead of a table of numeric data, we are utilizing the human ability to recognize complex patterns and relationships in a graphical representation far more rapidly and surely than in a tabular representation.

A Karnaugh map may be regarded either as a pictorial form of a truth table or as an extension of the Venn diagram. First consider a truth table for two variables. We list all four possible input combinations and the corresponding function values, e.g., the truth tables for AND and OR (Fig. 6.3).

As an alternative approach, let us set up a diagram consisting of four small boxes, one for each combination of variables. Place a "1" in any box representing a combination of variables for which the function has the value 1. There is no logical objection to putting "0's" in the other boxes, but they are usually omitted for clarity. Figure 6.4 shows two forms of Karnaugh maps

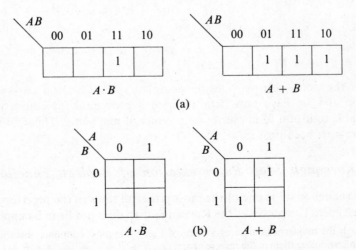

FIGURE 6.4. *Karnaugh maps for AND and OR.*

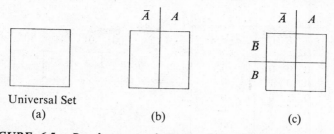

FIGURE 6.5. *Development of Karnaugh maps by Venn diagram approach.*

for AB and $A + B$. The diagrams of Fig. 6.4a are perfectly valid Karnaugh maps, but it is more common to arrange the four boxes in a square, as shown in Fig. 6.4b.

As an alternative approach, recall the interpretation of the Venn diagram discussed in Section 4.7. We interpret the universal set as the set of all 2^n combinations of values of n variables, divide this set into 2^n equal areas, and then darken the areas corresponding to those combinations for which the function has the value 1. We start with the universal set represented by a square (Fig. 6.5a) and divide it in half, corresponding to input combinations in which $A = 1$ and combinations in which $\bar{A} = 1$ (Fig. 6.5b). We then divide it in half again, corresponding to $B = 1$ and $\bar{B} = 1$ (Fig. 6.5c).

With this notation, the interpretations of AND as the intersection of sets and OR as the union of sets make it particularly simple to determine which squares should be darkened (Fig. 6.6).

We note that the shaded areas in Fig. 6.6 correspond to the squares containing 1's in Fig. 6.4. Thus, both interpretations lead to the same result. We might say that the Karnaugh map is essentially a diagrammatic form of truth table and that the Venn diagram concepts of union and intersection of areas aid us in setting up or interpreting a Karnaugh map.

Since there must be one square for each input combination, there must be 2^n squares in a Karnaugh map for n-variables. Whatever the number of variables, we may interpret the map in terms of a graphical form of the

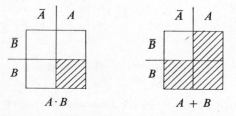

FIGURE 6.6. *Karnaugh maps of AND and OR (Venn form).*

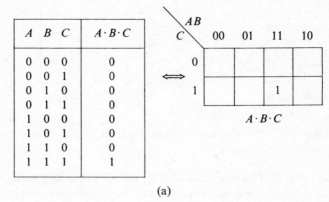

(a)

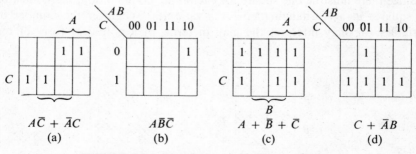

(b)

FIGURE 6.7.   *Karnaugh maps for 3-variable AND and OR.*

truth table (Fig. 6.7a) or in terms of union and intersection of areas (Fig. 6.7b).

In Fig. 6.7b, we have changed the labeling slightly and placed 1's in the squares, rather than darkening them. The two types of map (Figs. 6.7a and b) are thus the same, except for labeling, and both types will be called Karnaugh maps* or, for short, K-maps. The K-maps for some other 3-variable functions are shown in Fig. 6.8 to clarify the concepts involved.

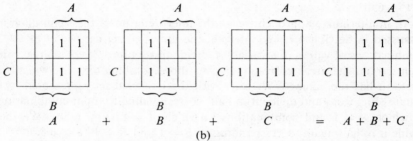

FIGURE 6.8.   *Sample 3-variable Karnaugh maps.*

* Both forms appear in Karnaugh's original paper.[1]

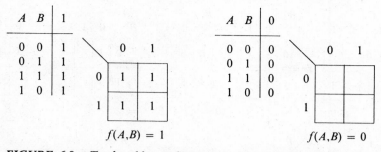

FIGURE 6.9. *Truth tables and K-maps of unit and zero elements.*

Note particularly the functions mapped in Figs. 6.7a and 6.8b. These are both minterms, m_7 and m_4, respectively. Each is represented by one square; obviously, each one of the eight squares corresponds to one of the eight minterms of three variables. This is the origin of the name *minterm*. A minterm is the form of Boolean function corresponding to the minimum possible area, other than 0, on a Karnaugh map. A *maxterm*, on the other hand, is the form of Boolean function corresponding to the maximum possible area, other than 1, on a Karnaugh map. Figures 6.7b and 6.8c are maps of maxterms M_0 and M_3; we see that each map covers the maximum possible area—all the squares but one.

We have just referred to the smallest area other than 0 and the largest area other than 1. In Section 4.7, we remarked that, in the Venn diagram interpretation of switching algebra, the unit element is the *function* that is identically 1 and the zero element, the *function* that is identically 0. In truth table terms, we have the situation shown in Fig. 6.9. Thus, "1" is represented by the *entire* area of the K-map and "0" by *none* of the area of the K-map.

Next, we recall that the negative (complement) of an element is uniquely defined by the relationships

$$A + \bar{A} = 1 \quad \text{and} \quad A \cdot \bar{A} = 0$$

From this, we see that consistent interpretation of the algebra requires that the negation of a function be represented by the squares not covered on the map of the function. For example, consider DeMorgan's laws:

$$\overline{A \cdot B} = \bar{A} + \bar{B} \quad \text{and} \quad \overline{A + B} = \bar{A} \cdot \bar{B}$$

On K-maps, we have the interpretations of Fig. 6.10.

Since each square on a K-map corresponds to a row in a truth table, it is appropriate to number the squares just as we numbered the rows. These standard K-maps (as we shall call them) are shown in Fig. 6.11 for 2 and 3 variables. Now if a function is stated in the form of the minterm list, all we need to do is enter 1's in the corresponding squares to produce the K-map.

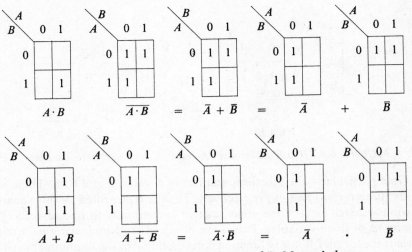

FIGURE 6.10. *K-map interpretation of DeMorgan's laws.*

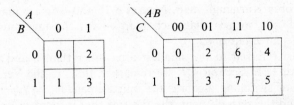

FIGURE 6.11. *Standard K-maps for two and three variables.*

Example 6.6

Develop the K-map of $f(A, B, C) = \sum m(0, 2, 3, 7)$.

Solution:

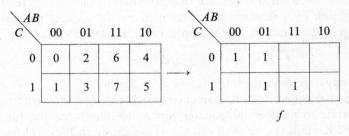

Standard Map

FIGURE 6.12

If a function is stated as a maxterm list, we can enter 0's in the squares listed or 1's in those not listed.

Example 6.7

Develop the K-map of $f(A, B, C) = \prod M(0, 1, 5, 6)$.

Solution:

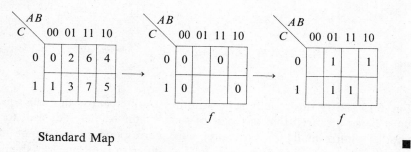

Standard Map

FIGURE 6.13

A map showing the 0's of a function is a perfectly valid K-map, although it is more common to show the 1's.

In developing these basic concepts, we have restricted ourselves to the simple 2- and 3-variable K-maps, but in practical cases we will more often be using maps for functions of more variables. The standard map for 4 variables is shown in Fig. 6.14 in both notations.

We have seen that one requirement for a Karnaugh map is that there must be a square corresponding to each input combination; the maps of

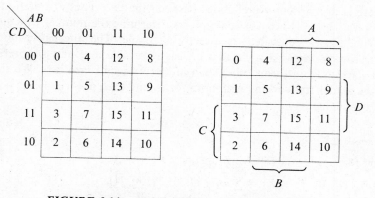

FIGURE 6.14. *Standard K-maps for four variables.*

DE \ BC	$A = 0$ 00	01	11	10
00	0	4	12	8
01	1	5	13	9
11	3	7	15	11
10	2	6	14	10

00	$A = 1$ 01	11	10 BC	DE
16	20	28	24	00
17	21	29	25	01
19	23	31	27	11
18	22	30	26	10

FIGURE 6.15. *Standard K-map for five variables.*

Fig. 6.14 satisfy this requirement. Another requirement is that the squares must be so arranged that any pair of squares immediately adjacent to one another (horizontally or vertically) must correspond to a pair of input conditions that are *logically adjacent*, i.e., differ in only one variable. For example, squares 5 and 13 on the maps in Fig. 6.14 correspond to input combinations $\bar{A}B\bar{C}D$ and $AB\bar{C}D$, identical except in A. Note that squares at the ends of columns or rows are also logically adjacent.

The standard K-map for 5 variables is shown in Fig. 6.15. Here we have two 4-variable maps placed side by side. They are identical in $BCDE$, but one corresponds to $A = 1$, the other to $A = 0$. The standard 4-variable adjacencies apply in each map. In addition, squares in the same relative position on the two maps, e.g., 4 and 20, are also logically adjacent. Any maps which satisfy the requirements of 2^n squares in proper adjacency can be considered K-maps. In Figs. 6.16 and 6.17, we give the most common forms of K-maps, for 2 to 6 variables, in the two alternative notations.

There is no particular preference between the two types of notation. The notation of Fig. 6.16 makes it simple to determine the number of a square, since the binary equivalent is directly available. The alternative notation emphasizes the areas associated with each variable and may be preferable when starting from forms other than minterm or maxterm lists. We suggest that the reader try working with both forms and then use whichever he finds most convenient.

Example 6.8

Find K-maps for the following functions:

(a) $f(V, W, X, Y, Z) = \sum m(9, 20, 21, 29, 30, 31)$.

(b) $f(A, B, C, D, E) = AB + \bar{C}D + DE$.

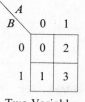

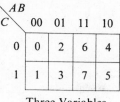

Two Variables

Three Variables

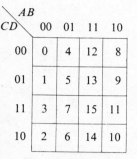

Four Variables

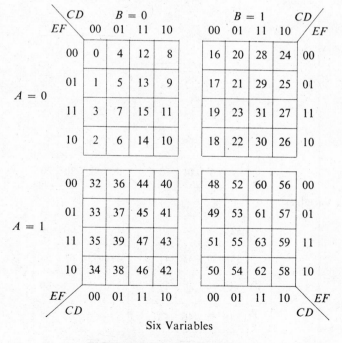

Five Variables

Six Variables

FIGURE 6.16. *Standard K-maps.*

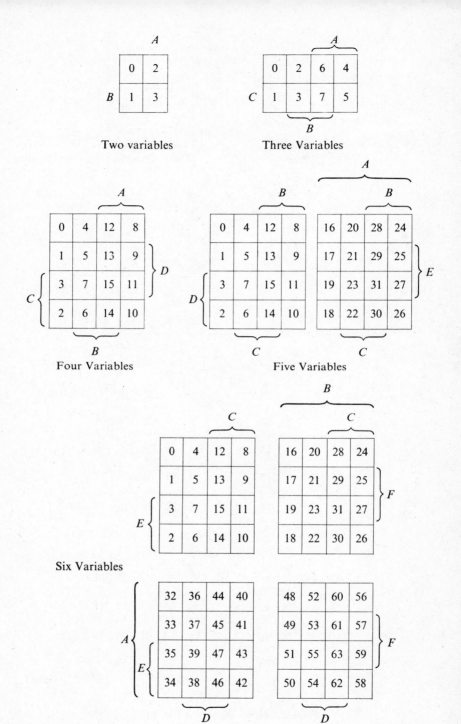

FIGURE 6.17. *Standard K-maps—alternate notation.*

Solution: (a) For the first function, we enter 1's in the listed squares:

WX		V = 0			V = 1			WX
YZ\	00	01	11	10	00	01	11	10	/YZ
00	0	4	12	8	16	20 1	28	24	00
01	1	5	13	9 1	17	21 1	29 1	25	01
11	3	7	15	11	19	23	31 1	27	11
10	2	6	14	10	18	22	30 1	26	10

FIGURE 6.18

(b) Referring to the standard map for 5 variables, we easily identify the maps of the individual product terms, as shown in Fig. 6.19.

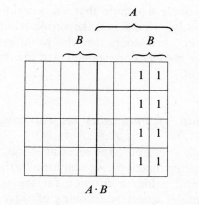

$A \cdot B$

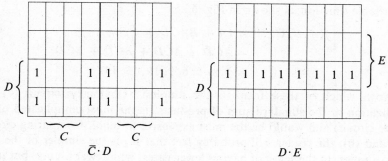

$\bar{C} \cdot D$ $D \cdot E$

FIGURE 6.19

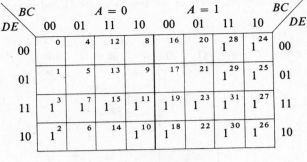

FIGURE 6.20

We then take the union of these to form the final K-map (Fig. 6.20).
The minterm list of this function may now be read directly from the map.

$$f(A, B, C, D, E) = AB + \bar{C}D + DE$$
$$= \sum m(2, 3, 7, 10, 11, 15, 18, 19, 23, 24, 25, 26, 27, 28, 29, 30, 31) \quad \blacksquare$$

In part (a) of Example 6.8, note that the variables do not have to be A, B, C, etc. Obviously, we can call the variables anything we want. The only precaution to be observed is that the variables must appear on the K-map in the proper manner corresponding to the order of their listing in the function statement.

6.5 Simplification of Functions on Karnaugh Maps

It has taken us a while to establish the necessary tools, but we are now ready to use them in minimizing functions. As mentioned earlier, we must have some circuit format in mind before a criterion for the simplest circuit can be defined. Let us consider direct realizations of three different forms of a 4-variable function as shown in Fig. 6.21.

$$f(A, B, C, D) = (A + B)(C\bar{D} + \bar{C}D)$$
$$= AC\bar{D} + A\bar{C}D + BC\bar{D} + B\bar{C}D$$
$$= \sum m(5, 6, 9, 10, 13, 14) \qquad (6.7)$$

Now, which of these forms is the simplest? Obviously, circuit (c), the realization of the standard sum of products, is the most complicated of the three circuits and would be the most expensive to build. Comparing circuits (a) and (b), the reader will probably feel that (a) is the simpler of the two. It is simpler, in the sense of having fewer gates, with fewer inputs; but it has one drawback relative to circuit (b). Note that the signals A and B, entering

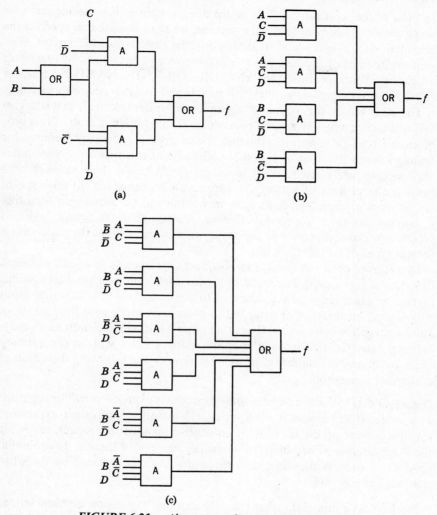

FIGURE 6.21.   *Alternate realizations of a function.*

the left gate, pass through three gates, or three *levels of gating*, before reaching the output. By comparison all of the signals into circuit (b) pass through only two levels of gating.

In Chapter 5, it was pointed out that each level of logic adds to the delay in the development of a signal at the circuit output. In high-speed digital systems, it is desirable that this delay be as small as possible.

The choice between a simpler circuit and a faster circuit is generally a matter of engineering judgment. More speed almost invariably costs more money, and one of the most difficult jobs of the design engineer is to decide just how much speed he can afford. For the present, we shall assume that speed is the dominant factor and we shall design for the fastest possible circuit. Any sum-of-products or product-of-sums expression can be realized by two levels of gating. Whether these be AND-OR, OR-AND, NAND-NAND, or NOR-NOR configurations, they will be referred to as *second-order* circuits.

The other reason for concentrating on second-order circuits is that straight-forward and completely general procedures for the minimization of this type of circuit configuration are available. There are no general procedures for finding designs like that of Fig. 6.21a. Circuits of this type, of higher order than second, are often referred to as *factored* circuits. This name is used because the process of getting the expression implemented by this type of circuit from a second-order form is very similar to the process of factoring a polynomial in conventional algebra. This sort of factoring is based largely on trial-and-error and the ability to remember what the products of certain types of factors look like.

In a sum-of-products form, each product corresponds to a gate and each literal to a gate input. The same holds for each sum in a product-of-sums form. The exact ratio between the cost of a gate and the cost of a gate input will depend on the type of gate, but in practically every case the cost of an additional gate will be several times that of an additional input on an already-existing gate. On this basis, the elimination of gates will be the primary objective of any minimization process, leading to the following definition of a minimal expression.

Definition 6.1. A second-order sum-of-products expression will be regarded as a *minimal* expression if there exists (1) no other equivalent expression involving fewer products, and (2) no other equivalent expression involving the same number of products but a smaller number of literals. The minimal product of sums is the same with the word *products* replaced by the word *sum*, and vice versa.

Notice that *a* minimal rather than *the* minimal expression is characterized by Definition 6.1. As we shall see, there may very well be several distinct but equivalent expressions satisfying this definition and having the same number of both products and literals.

For the remainder of this section, only the sum-of-products form will be considered, for the reason that the K-map as we have defined it represents each minterm by a square. In the next section, adaption of the K-map to the minimization of product-of-sums expressions will be discussed.

Simplification of functions on the K-map is based on the fact that sets of minterms that can be combined into simpler product terms will either be adjacent or will appear in symmetric patterns on the K-map.

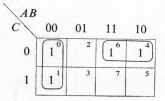

FIGURE 6.22. *K-map for Equation 6.8.*

Consider this function:

$$f(A, B, C) = \sum m(0, 1, 4, 6)$$
$$= \bar{A}\bar{B}\bar{C} + \bar{A}\bar{B}C + A\bar{B}\bar{C} + AB\bar{C} \qquad (6.8)$$

By algebraic manipulation, we can simplify this function as follows:

$$f(A, B, C) = \bar{A}\bar{B}(C + \bar{C}) + A\bar{C}(B + \bar{B})$$
$$= \bar{A}\bar{B} + A\bar{C}$$

On the map, we have the pattern shown in Fig. 6.22. Note that the minterms which have combined into a simpler term are adjacent on the K-map. This is a general principle. *Any pair of n-variable minterms which are adjacent on a K-map may be combined into a single product term of n − 1 literals.* As we noted earlier, K-maps are so arranged that minterms in adjacent squares are identical, except in one variable. This variable appears in the true form in one minterm, in the complemented (false) form in the other. Thus, the value of the function will be independent of the value of this variable. For example, in the function given by Equation 6.8, if $A = 0$ and $B = 0$, then $f = 1$, regardless of the value of C. Similarly, if $A = 1$ and $C = 0$, then $f = 1$, regardless of the value of B.

Now consider

$$f(A, B, C, D) = \sum m(0, 8, 12, 14, 5, 7)$$
$$= \bar{A}\bar{B}\bar{C}\bar{D} + A\bar{B}\bar{C}\bar{D} + AB\bar{C}\bar{D} + ABC\bar{D} + \bar{A}\bar{B}CD + \bar{A}BCD$$
$$= \bar{B}\bar{C}\bar{D}(A + \bar{A}) + AB\bar{D}(C + \bar{C}) + \bar{A}BD(C + \bar{C})$$
$$= \bar{B}\bar{C}\bar{D} + AB\bar{D} + \bar{A}BD \qquad (6.9)$$

which is mapped in Fig. 6.23. Note that here the pairing includes 0 and 8, and 12 and 14, which do not appear to be adjacent. The variables are arranged in "ring" pattern of symmetry, so that these squares would be adjacent if the map were inscribed on a torus (a doughnut-shaped form). If you have difficulty visualizing the map on a torus, just remember that squares in the same row or column, but on opposite edges of the map, may be paired.

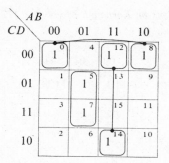

FIGURE 6.23. *Map of Equation 6.9.*

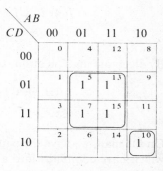

FIGURE 6.24. *Map of Equation 6.10.*

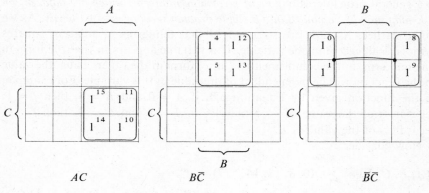

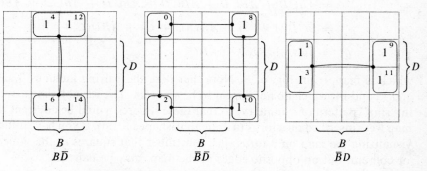

FIGURE 6.25. *Sets of four on the K-map.*

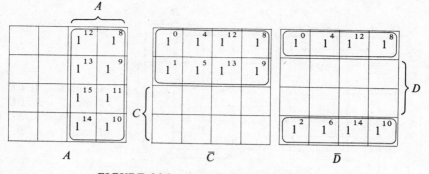

FIGURE 6.26. *Sets of eight on the K-map.*

Now consider the function

$$f(A, B, C, D) = \sum m(5, 7, 10, 13, 15)$$
$$= BD + A\bar{B}C\bar{D} \qquad (6.10)$$

which is displayed on the map in Fig. 6.24. Here we see that a set of four adjacent minterms combines into a single term, with the elimination of two literals. In other words, if $B = 1$ and $D = 1$, then $f = 1$, regardless of the values A and C. Also note that m_{10} does *not* combine with any other minterm of the function. The adjacencies must be row or column adjacencies. For example, the diagonally adjacent m_{10} and m_{15} cannot be combined.

Other possible sets of four are shown in Fig. 6.25. Note the adjacency on the edges and corners.

Since sets of two minterms combine to eliminate one variable, and sets of four combine to eliminate two, it is to be expected that sets of eight will combine to eliminate three variables. Figure 6.26 illustrates some of these sets.

The same general principles apply as we go to the 5- and 6-variable maps, but we also have logical adjacency between squares, or sets of squares, in the same position on different sections of the map.

Some sets on a 5-variable map are shown in Fig. 6.27. Some sets on a 6-variable map are shown in Fig. 6.28. We note that here the "map-to-map" adjacency works both horizontally and vertically but not diagonally. For example, 32 combines vertically with 0 and horizontally with 48; but 0 and 48 do *not* combine, even though they are in the same position in their respective sections of the map. The reader should also keep in mind that the "end-to-end" adjacency applies only to the individual 4-variable section. For example, 48 and 56, at opposite ends of a row in a single section, combine; while 32 and 56, at opposite ends of separate sections, do not combine.

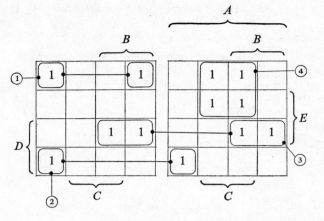

$$
\begin{aligned}
\text{Group } 1 &= \bar{A}\bar{C}D\bar{E} \\
\text{Group } 2 &= \bar{B}\bar{C}D\bar{E} \\
\text{Group } 3 &= BDE \\
\text{Group } 4 &= AC\bar{D}
\end{aligned}
$$

FIGURE 6.27. *Sets on a 5-variable map.*

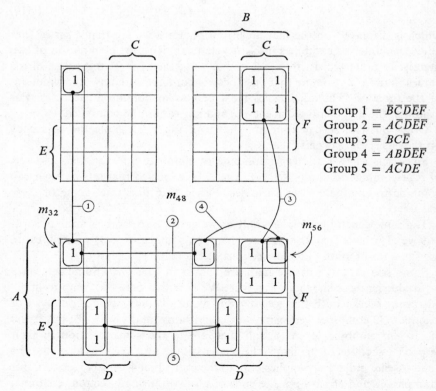

$$
\begin{aligned}
\text{Group } 1 &= \bar{B}\bar{C}D\bar{E}F \\
\text{Group } 2 &= A\bar{C}D\bar{E}F \\
\text{Group } 3 &= BC\bar{E} \\
\text{Group } 4 &= AB\bar{D}\bar{E}F \\
\text{Group } 5 &= A\bar{C}DE
\end{aligned}
$$

FIGURE 6.28. *Sets on a 6-variable map.*

We have seen that a set of two *logically* adjacent minterms eliminates one variable, that a set of four eliminates two variables, that a set of eight eliminates three variables, etc. The way to test whether a set is in fact logically adjacent is to determine whether sufficient variables remain constant over the entire set. On an n-variable map, a pair of minterms is adjacent if $n - 1$ variables remain constant over the pair; a set of four minterms is adjacent if $n - 2$ variables remain constant; a set of eight minterms is adjacent if $n - 3$ variables remain constant over the set, etc. For example, in Fig. 6.28, the fact that 0 and 48 do not combine into a pair is consistent with the fact that they correspond to input combinations differing in the values of two variables, A and B.

The process of simplifying a function on a K-map consists of nothing more than determining the smallest set of adjacencies that covers (contains) all the minterms of the function. Let us illustrate the process by a few examples.

Example 6.9

Simplify $f(A, B, C, D) = \sum m(0, 1, 2, 3, 13, 15)$.

Solution:

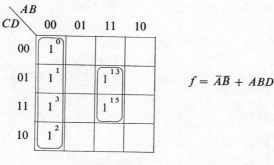

$f = \overline{A}\overline{B} + ABD$

FIGURE 6.29

Example 6.9 presented no difficulty as only one set of adjacencies was possible.

Example 6.10

Simplify $f(A, B, C, D) = \sum m(0, 2, 10, 11, 12, 14)$.

Solution:

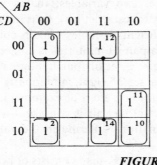

$$f = \bar{A}\bar{B}\bar{D} + AB\bar{D} + A\bar{B}C$$

FIGURE 6.30

■

Here there are choices. We could also combine m_2 and m_{10}, or m_{10} and m_{14}. However, there is no reason to use these combinations since m_2, m_{10}, and m_{14} are already covered by (contained in) the necessary pairings with m_0, m_{11}, m_{12}, respectively, for which there is no choice. *Thus, a rule in using Karnaugh maps is to start by combining those terms for which there is only one possibility.*

Example 6.11
Simplify $f(A, B, C, D) = \sum m(0, 2, 8, 12, 13)$.

Solution:

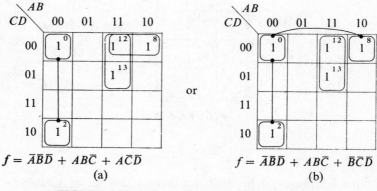

FIGURE 6.31. *Alternative maps for Example 6.11.*

In this case, there are two equally valid choices.   ■

Note that, in both realizations of Example 6.11, there is some redundancy. In the first, m_{12} is covered by the terms $AB\bar{C}$ and $A\bar{C}\bar{D}$; in the second, m_0

is covered by $\bar{A}\bar{B}\bar{D}$ and $\bar{B}\bar{C}\bar{D}$. This procedure of covering a minterm more than once causes no trouble. In terms of AND-OR realization, it simply means that, when the variable values are such that the particular minterm is 1, the output of more than one AND gate will take on the value 1. Since the OR is inclusive, it does not matter how many of the AND gates are at the value 1.

Example 6.12

Simplify $f(A, B, C, D) = \sum m(1, 5, 6, 7, 11, 12, 13, 15)$

Solution:

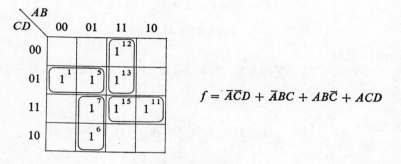

$$f = \bar{A}\bar{C}D + \bar{A}BC + AB\bar{C} + ACD$$

FIGURE 6.32

∎

This example illustrates a possible hazard in using K-maps. The temptation is great to use the set of four in the center; but when we make the *necessary* pairings with the other four minterms, we find that the four in the center have been covered. This emphasizes the importance of determining the essential products first, i.e., the products containing at least one minterm which can be combined in no other way.

We will now present a few more examples, without further comment. It is suggested that the reader study the first two carefully and then try to work the others before looking at the answers. Once mastered, Karnaugh maps will seem almost second nature, but mastery requires practice.

Example 6.13

$$f(A, B, C, D, E) = \sum m(0, 1, 4, 5, 6, 11, 12, 14, 16, 20, 22, 28, 30, 31)$$

Solution:

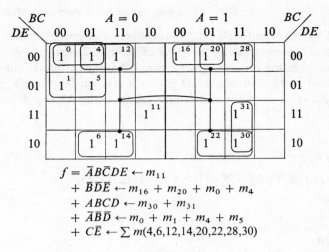

$$f = \bar{A}B\bar{C}DE \leftarrow m_{11}$$
$$+ \ \bar{B}\bar{D}\bar{E} \leftarrow m_{16} + m_{20} + m_0 + m_4$$
$$+ \ ABCD \leftarrow m_{30} + m_{31}$$
$$+ \ \bar{A}\bar{B}\bar{D} \leftarrow m_0 + m_1 + m_4 + m_5$$
$$+ \ C\bar{E} \leftarrow \sum m(4,6,12,14,20,22,28,30)$$

FIGURE 6.33

∎

Example 6.14

$f(A, B, C, D, E, F) = \sum m(2, 3, 6, 7, 10, 14, 18, 19, 22, 23, 27, 37, 42, 43, 45, 46)$

Solution:

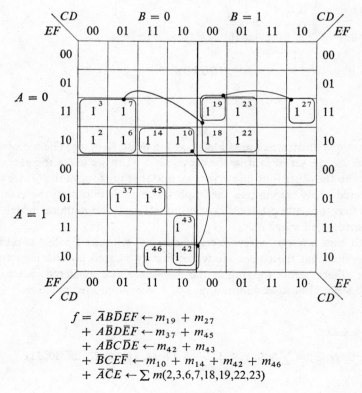

$$f = \bar{A}B\bar{D}EF \leftarrow m_{19} + m_{27}$$
$$+ \ A\bar{B}D\bar{E}F \leftarrow m_{37} + m_{45}$$
$$+ \ A\bar{B}C\bar{D}E \leftarrow m_{42} + m_{43}$$
$$+ \ \bar{B}CE\bar{F} \leftarrow m_{10} + m_{14} + m_{42} + m_{46}$$
$$+ \ \bar{A}\bar{C}E \leftarrow \sum m(2,3,6,7,18,19,22,23)$$

FIGURE 6.34

∎

Example 6.15

$f(A, B, C, D) = \prod M(7, 9, 13)$
Solution:

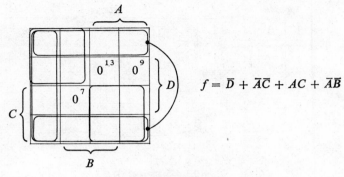

$$f = \bar{D} + \bar{A}\bar{C} + AC + \bar{A}\bar{B}$$

FIGURE 6.35

Example 6.16

$f(A, B, C, D) = \sum m(0, 1, 2, 4, 5, 8, 10)$
Solution:

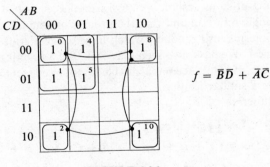

$$f = \bar{B}\bar{D} + \bar{A}\bar{C}$$

FIGURE 6.36

Example 6.17

$f(A, B, C, D, E) = \sum m(0, 1, 3, 4, 5, 7, 8, 9, 10, 12, 13, 21, 24, 25, 26, 28, 29)$

Solution:

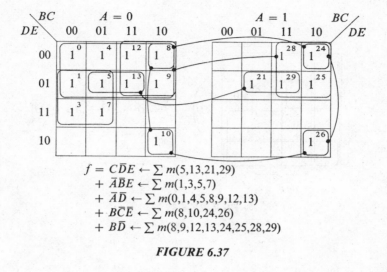

$$f = C\bar{D}E \leftarrow \sum m(5,13,21,29)$$
$$+ \bar{A}\bar{B}E \leftarrow \sum m(1,3,5,7)$$
$$+ \bar{A}\bar{D} \leftarrow \sum m(0,1,4,5,8,9,12,13)$$
$$+ B\bar{C}\bar{E} \leftarrow \sum m(8,10,24,26)$$
$$+ B\bar{D} \leftarrow \sum m(8,9,12,13,24,25,28,29)$$

FIGURE 6.37

6.6 Map Minimizations of Product-of-Sums Expressions

With only minor changes, the procedure described in Section 6.5 can be adapted to product-of-sums design. We have seen that each 1 of a function is produced by a single minterm and that the process of simplification consists of combining minterms into products which have fewer literals and produce more than a single 1. We have also seen that each 0 of a function is produced by a single maxterm. It would seem reasonable, then, to expect that maxterms might combine in a similar fashion.

Consider the function

$$f(A, B, C) = \prod M(3, 6, 7)$$
$$= (A + \bar{B} + \bar{C})(\bar{A} + \bar{B} + C)(\bar{A} + \bar{B} + \bar{C}) \qquad (6.11)$$

which is shown in Fig. 6.38. Maxterm 6 produces a 0 when $A = 1$, $B = 1$,

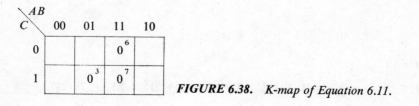

FIGURE 6.38.   *K-map of Equation 6.11.*

$C = 0$, and maxterm 7 produces a 0 when $A = 1$, $B = 1$, $C = 1$. Taking them together, they produce 0's whenever $A = 1$ and $B = 1$, regardless of C. Furthermore, the single sum $(\bar{A} + \bar{B})$ will produce both 0's, since it will be 0 whenever $A = 1$ and $B = 1$, regardless of C. We also note that M_3 and M_7 together produce 0's for $B = 1$ and $C = 1$, regardless of A, as does the single sum $(\bar{B} + \bar{C})$. Algebraically, we have

$$f(A, B, C) = (A + \bar{B} + \bar{C})(A + \bar{B} + C)(\bar{A} + \bar{B} + \bar{C})$$
$$= (A + \bar{B} + \bar{C})(\bar{A} + \bar{B} + \bar{C})(\bar{A} + \bar{B} + C)(\bar{A} + \bar{B} + \bar{C})$$
$$= (A\bar{A} + \bar{B} + \bar{C})(\bar{A} + \bar{B} + C\bar{C})$$
$$= (\bar{B} + \bar{C})(\bar{A} + \bar{B})$$

To summarize, to design product-of-sums (P-of-S) forms, select sets of the 0's of the function. Realize each set as a sum term, with variables being the *complements* of those which would be used if this same set were being realized as a product, to produce 1's. Let us consider some examples.

Example 6.18

Obtain a minimal product-of-sums realization of $f(A, B, C, D) = \sum m(0, 2, 10, 11, 12, 14) = \prod M(1, 3, 4, 5, 6, 7, 8, 9, 13, 15)$

Solution:

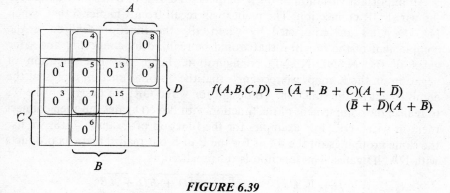

$$f(A, B, C, D) = (\bar{A} + B + C)(A + \bar{D})$$
$$(\bar{B} + \bar{D})(A + \bar{B})$$

FIGURE 6.39

■

Example 6.19

$f(A, B, C, D) = \sum m(0, 2, 8, 12, 13) = \prod M(1, 3, 4, 5, 6, 7, 9, 10, 11, 14, 15)$

Solution:

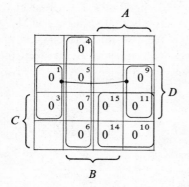

$$f(A,B,C,D) = (\bar{A} + \bar{C})(A + \bar{B})(B + \bar{D})$$

FIGURE 6.40

∎

If we compare Examples 6.18 and 6.10, we see that the sum-of-products (S-of-P) form is preferable to the product-of-sums form for this function, since there are three products in the S-of-P form and four sums in the P-of-S form. Comparing Examples 6.19 and 6.11, we see that the P-of-S form is preferable. The number of products and sums is the same, but there are fewer literals in the sums. It would be useful to have some method of determining in advance which form (S-of-P or P-of-S) will be best for a particular function. Unfortunately, no such method exists. Assuming there is no hardware preference, the designer should try both forms.

An important variation of the K-map method of P-of-S design is design for the wired-OR connection. The reader will recall from Chapter 5 that, when NAND gates are connected by wired-OR, the function realized is the complement of the function that would be realized by connecting the same gates in the NAND-NAND configuration. From our interpretation of negation on the K-map, we recognize that the 0's of a function represent the complement of the function. To design for wired-OR, we design on the 0's to realize the complement of the function with NAND gates but then connect them in wired-OR. For example, for the function of Example 6.18, we use the same groupings on the 0's as for the P-of-S but realize them as products with NAND gates. The function is thus realized as

$$f(A, B, C, D) = \overline{A\bar{B}\bar{C} + \bar{A}D + BD + \bar{A}B}$$

Each product is realized by a NAND gate in the usual manner, but the gates are connected in wired-OR. In a similar manner, the function of Example 6.19 would be realized in wired-OR in the form

$$f(A, B, C, D) = \overline{AC + \bar{A}B + \bar{B}D}$$

6.7 Incompletely Specified Functions

Recall that the basic specification of a switching function is the truth table, i.e., a listing of the values of the function for the 2^n possible combinations of n variables. Our design process basically consists of translating a (generally) verbal description of a logical job to be done into a truth table and then finding a specific function which realizes this truth table and satisfies some criterion of minimum cost. So far, we have assumed that the truth values were strictly specified for all of the 2^n possible input combinations. This is not always the case.

Sometimes the circuit we are designing is a part of a larger system in which certain inputs occur only under circumstances such that the output of the circuit will not influence the overall system. Whenever the output has no effect, we obviously *don't care* whether the output is a 0 or a 1. Another possibility is that certain input combinations never occur due to various external constraints. Note that this does not mean that the circuit would not develop some output if this forbidden input occurred. Any switching circuit will respond in some way to any input. However, since the input will never occur, we don't care whether the final circuit responds with a 0 or a 1 output to this forbidden input combination.

Where such situations occur, we say that the output is *unspecified*. This is indicated on the truth table by entering an "X" as the functional value, rather than 0 or 1.* Such conditions are commonly referred to as *don't-cares* and functions including don't-cares are said to be *incompletely specified*. A realization of an incompletely specified function is any circuit which produces the same outputs for *all input combinations for which output is specified*.

Example 6.20

A digital computer has five output modes as follows:

Mode No.	Description	Code
1	Punched card	001
2	Mag tape	010
3	Typewriter	011
4	Punched tape, numeric	100
5	Punched tape, alphabetic	101

---

* The symbol for the unspecified output condition is unfortunately not standard. Other symbols used include ϕ, d, and $-$.

xyz	f(x,y,z)
000	X
001	0
010	0
011	0
100	1
101	1
110	X
111	X

(a)

xyz	f = x
000	0
001	0
010	0
011	0
100	1
101	1
110	1
111	1

(b)

FIGURE 6.41

The selection of an output device is controlled by signals on three parallel lines, denoted x, y, z, according to the code indicated above. Since both modes 4 and 5 utilize the tape punch, a decoding circuit is required in the tape punch that will have 1 as its output if either 100 or 101 are received. The truth table for this decoding circuit is shown in Fig. 6.41a.

Since the codes 000, 110, and 111 are not used, these are don't-care conditions. A realization of this truth table is *any* circuit which produces the 1's and 0's specified in the truth table, regardless of its output for the three don't-care conditions.

The first step in minimization is to convert the truth table to a K-map. For a sum-of-products design, the 1's and don't-cares of the function map are as shown in Fig. 6.42a. Since the output is optional for the don't-cares, we assign them to be 0's or 1's in whatever manner will result in the simplest realization. On

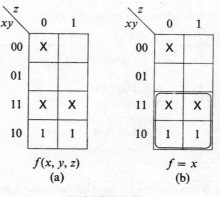

$f(x, y, z)$

(a)

$f = x$

(b)

FIGURE 6.42

the K-map, this means that we group the X's with the 1's whenever this results in a larger set (product with fewer literals), and we ignore the X's if there is no advantage to be gained from using them. In this case, we include the X's in 110 and 111, i.e., we assign them as 1's, and we ignore the X in 000 (Fig. 6.42b). Without the don't-cares, the function would be realized by $f(x, y, z) = xy$, but with them it is $f(x, y, z) = x$. When the truth table of $f(x, y, z) = x$ (Fig. 6.41b) is compared with that of the original incompletely specified function (Fig. 6.41a), we see that it is a correct realization. The output is correct for all input conditions for which the function is specified. ∎

It would be convenient to have some compact algebraic way of stating an incompletely specified function. Since each row in the truth table corresponds to an input combination, we simply add a list of the rows for which the output is unspecified. Thus, the minterm list form for the function of Example 6.20 would be

$$f(x, y, z) = \sum m(4, 5) + d(0, 6, 7)$$

Example 6.21

Obtain a minimal sum-of-products representation for

$$f(w, x, y, z) = \sum m(0, 7, 8, 10, 12) + d(2, 6, 11)$$

Solution:

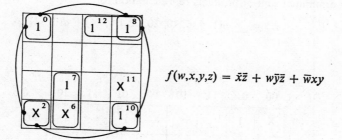

$$f(w,x,y,z) = \bar{x}\bar{z} + w\bar{y}\bar{z} + \bar{w}xy$$

FIGURE 6.43

∎

We use the don't-cares in squares 2 and 6 to obtain larger sets than would otherwise be possible, but we ignore the X in 11, since the only possible combination is with 10, which is already covered.

Example 6.22

Obtain a minimal product-of-sums realization for the function of Example 6.21.

Solution:

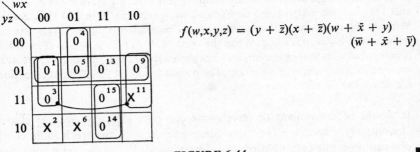

$$f(w,x,y,z) = (y + \bar{z})(x + \bar{z})(w + \bar{x} + y)$$
$$(\bar{w} + \bar{x} + \bar{y})$$

FIGURE 6.44   ∎

Since we wish the P-of-S form in Example 6.22, we design in the zeros, again using the don't-cares if they improve the combination, ignoring them otherwise. Again we start with necessary sets. For example, 4 combines only with 5, and 14 only with 15. We could combine 4 with the don't-care in 6, and also 14 with 6, but this would violate our rule of using don't-cares only if they make *larger* sets possible. The don't-care at 11 is used because it does place 3 in a larger set $(x + \bar{z})$ than could otherwise be obtained.

Example 6.23

Obtain a minimal sum-of-products representation for

$$f(A, B, C, D, E) = \sum m(1, 4, 6, 10, 20, 22, 24, 26) + d(0, 11, 16, 27).$$

Solution:

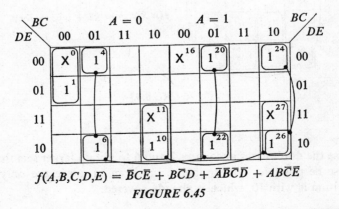

$$f(A,B,C,D,E) = \bar{B}C\bar{E} + B\bar{C}D + \bar{A}B\bar{C}\bar{D} + AB\bar{C}\bar{E}$$

FIGURE 6.45   ∎

Here the don't-care in 16 could be used in several ways, but it would not increase the size of any set.

PROBLEMS

6.1. Convert the following Boolean forms to minterm lists.

(a) $f(w, x, y, z) = wy + x(w + y\bar{z})$

(b) $f(U, V, W, X, Y) = \bar{V}(\bar{W} + \bar{U})(X + \bar{Y}) + \bar{U}\bar{W}\bar{Y}$

(c) $f(V, W, X, Y, Z) = (X + Z)(\overline{Z + WY}) + (VZ + W\bar{X})(\overline{Y + Z})$

6.2. Convert the forms of Problem 6.1 to maxterm lists.

6.3. By using the Karnaugh map, determine minimal sum-of-products realizations of the following functions.

(a) $f(A, B, C, D) = \sum m(0, 4, 6, 10, 11, 13)$

(b) $f(w, x, y, z) = \sum m(3, 4, 5, 7, 11, 12, 14, 15)$

(c) $f(a, b, c, d) = \prod M(3, 5, 7, 11, 13, 15)$

(d) $f(V, W, X, Y, Z) =$
$$\sum m(0, 2, 3, 4, 5, 11, 18, 19, 20, 23, 24, 28, 29, 31)$$

6.4. A logic circuit is to be designed having 4 inputs y_1, y_0, x_1, and x_0. The pairs of bits $y_1 y_0$ and $x_1 x_0$ represent 2-bit binary numbers with y_1 and x_1 as the most significant bits. The only circuit output, z, is to be 1 if and only if the binary number $x_1 x_0$ is greater than or equal to the binary number $y_1 y_0$. Determine a minimal sum of products expression for z.

6.5. Determine a minimal sum-of-products expression equivalent to each of the following Boolean expressions.

(a) $f(A, B, C, D, E) = (\bar{C}\bar{E} + CE)(\bar{A} + B)D + (\overline{A + B})D\bar{C}E$

(b) $f(w, x, y, z) = \overline{(\bar{w} + x) + (\bar{x} + z) + (\bar{y} + \bar{z})}$

6.6. By using Karnaugh maps, determine minimal product-of-sums realizations for the functions of Problems 6.3b and 6.3c.

6.7. (a) Determine the minimal product-of-sums realization for the function of Example 6.15, and compare the cost to that of the minimal sum-of-products realization.

(b) Repeat (a) for Example 6.16.

6.8. A prime number is a number which is only divisible by itself and 1. Suppose the numbers between 0 and 31 are represented in binary in the form of the five bits

$$x_4\, x_3\, x_2\, x_1\, x_0$$

where x_4 is the most significant bit. Design a prime detector. That is, design a combinational logic circuit whose output, Z, will be 1 if and only if the 5 input bits represent a prime number. Do not count zero as a prime. Base your design on obtaining a minimal two-level expression for Z.

6.9. A 2-output, 4-input logic circuit is to be designed which will carry out addition modulo-4. The addition table for modulo-4 addition is given in Fig. P6.9. For example, $(3 + 3)$ mod-4 $= 2$. Therefore, a 2 is entered in row 3 column 3 of the table, and so on. The input numbers are to be coded in straight binary, with one input number given by $x_2 x_1$ and the other by $y_2 y_1$. The output is also to be coded as the binary number $z_2 z_1$. That is, $z_2 z_1 = 00$ if the sum is 0, 01 if the sum is 1, 10 if the sum is 2, and 11 if the sum is 3.

(a) Determine a two-level Boolean expresssion for z_1.

(b) Determine a two-level Boolean expression for z_2.

$\begin{matrix}&X\\Y&\end{matrix}$	0	1	2	3
0	0	1	2	3
1	1	2	3	0
2	2	3	0	1
3	3	0	1	2

$$Z = (X + Y) \text{ Mod-4}$$

FIGURE P6.9

6.10. Determine minimal sum-of-products realizations for the following incompletely specified functions:

(a) $f(A, B, C, D) = \sum m(1, 3, 5, 8, 9, 11, 15) + d(2, 13)$

(b) $f(W, X, Y, Z) = \sum m(4, 5, 7, 12, 14, 15) + d(3, 8, 10)$

(c) $f(A, B, C, D, E) = \sum m(1, 2, 3, 4, 5, 11, 18, 19, 20, 21,$
$\qquad 23, 28, 31) + d(0, 12, 15, 27, 30)$

(d) $f(a, b, c, d, e) = \sum m(7, 8, 9, 12, 13, 14, 19, 23, 24, 27, 29, 30)$
$\qquad + d(1, 10, 17, 26, 28, 31)$

(e) $f(u, v, w, x, y, z) = \sum m(0, 2, 14, 18, 21, 27, 32, 41, 49, 53, 62)$
$\qquad + d(6, 9, 25, 34, 55, 57, 61)$

6.11. Determine minimal realizations of the functions of Problem 6.10 in terms of wired-OR connections of NAND gates.

6.12. The four lines into the combinational logic circuit depicted in Fig. P6.12 carry one binary-coded decimal digit. That is, the binary

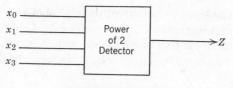

FIGURE P6.12

equivalents of the decimal digits 0–9 may appear on the lines $x_0 x_1 x_2 x_3$. The most significant bit is x_0. The combination of values corresponding to binary equivalents of the decimal numbers 10–15 will never appear on the lines. The single output Z of the circuit is to be 1 if and only if the inputs represent a number which is either 0 or a power of 2. Construct the logic block diagram of a minimal two-level realization of the circuit.

6.13. A shaft-position encoder provides a 4-bit signal indicating the position of a shaft in steps of 30°, using a reflected (Gray) code as listed in Fig. P6.13. It may be assumed that the four possible combinations of

Shaft Position	Encoder Output $E_3 E_2 E_1 E_0$			
0–30°	0	0	1	1
30–60°	0	0	1	0
60–90°	0	1	1	0
90–120°	0	1	1	1
120–150°	0	1	0	1
150–180°	0	1	0	0
180–210°	1	1	0	0
210–240°	1	1	0	1
240–270°	1	1	1	1
270–300°	1	1	1	0
300–330°	1	0	1	0
330–360°	1	0	1	1

FIGURE P6.13

four bits not used above will never occur. Design a minimal S-of-P realization of a circuit to produce an output whenever the shaft is in the first quadrant (0–90°).

6.14. A circuit receives two 3-bit binary numbers, $A = A_2A_1A_0$, and $B = B_2B_1B_0$. Design a minimal S-of-P circuit to produce an output whenever A is greater than B.

6.15. In digital computers, letters of the alphabet are coded in the form of unique combinations of five or more bits. One of the most common codes is the 6-bit Flexowriter code, which is used for punched paper tape. This code is given in Fig. P6.15 in terms of the octal equivalents.

A	30	N	06
B	23	O	03
C	16	P	15
D	22	Q	35
E	20	R	12
F	26	S	24
G	13	T	01
H	05	U	34
I	14	V	17
J	32	W	31
K	36	X	27
L	11	Y	25
M	07	Z	21

FIGURE P6.15

For example,

$$C = 16|_8 = \underbrace{0\,0\,1}_{1}\ \underbrace{1\,1\,0}_{6} \qquad \text{(See Problem 2.15)}$$

Design a minimal S-of-P circuit that will receive this code as an input and produce an output whenever the letter is a vowel. The alphabet uses only 26 of the possible codes and the others are used for numerals and punctuation marks. However, you may assume that the data is pure alphabetic, so that only the codes listed will occur.

6.16. Shown in Fig. P6.16 are the Flexowriter codes for the ten decimal

0	37
1	52
2	74
3	70
4	64
5	62
6	66
7	72
8	60
9	33

FIGURE P6.16

numerals. Assume that a device receives alphanumeric data in this code, i.e., either these codes for numerals or the codes for letters as listed in Problem 6.15. Design a minimal circuit that will develop a 1 out when the data are numeric, a 0 out when they are alphabetic. You may assume that any possible 6-bit codes not used for either letters or numerals will not occur.

6.17. Referring again to the Flexowriter code listed in Problems 6.15 and 6.16, design an "error circuit," i.e., a circuit that will put out a signal if a code other than one of the 36 "legal" alphanumeric codes is received.

6.18. In a certain computer, three separate sections of the computer proceed independently through four phases of operation. For purposes of control, it is necessary to know when two of three sections are in the same phase at the same time. Each section puts on a 2-bit signal (00, 01, 10, 11) in parallel on two lines. Design a circuit to put out a signal whenever it receives the same phase signal from any two or all three sections.

6.19. Prove again the result in Problem 4.14, assuming that the variables may take on values other than 0 and 1. *Hint:* Use the sum of products form:

$$f(x_1, x_2, \ldots, x_n) = x_1 f_1(x_2, x_3, \ldots, x_n) + \bar{x}_1 f_2(x_2, x_3, \ldots, x_n)$$
$$+ f_3(x_2, x_3, \ldots, x_n)$$

BIBLIOGRAPHY

1. Karnaugh, M. "The Map Method for Synthesis of Combinational Logic Circuits." *Trans. AIEE*, **72,** Pt. I, 593–598 (1953).

2. Veitch, E. W. "A Chart Method for Simplifying Truth Functions," *Proc. ACM*, Pittsburgh, Pa., 127–133 (May 2, 3, 1952).

3. Tanana, E. J. "The Map Method," in E. J. McCluskey and T. G. Bartee (Eds.), *A Survey of Switching Circuit Theory*, McGraw-Hill, New York, 1962.

4. Caldwell, S. H. *Switching Circuits and Logical Design*, Wiley, New York, 1958.

5. McCluskey, E. J. *Introduction to the Theory of Switching Circuits*, McGraw-Hill, New York, 1965.

6. Torng, H. C. *Introduction to the Logical Design of Switching Systems*, Addison-Wesley, Reading, Mass., 1964.

7. Marcus, M. P. *Switching Circuits for Engineers*, Prentice-Hall, Englewood Cliffs, N.J., 1962.

7

Tabular Minimization and Multiple-Output Circuits

7.1 Cubical Representation of Boolean Functions

At this point, it is convenient to introduce a new representation for Boolean functions, which will provide convenient terminology for further work. We are all familiar with the notion of a geometric representation of a continuous variable as a distance along a straight line (Fig. 7.1a). In a similar fashion, a switching variable, which can take on only two values, can be represented by two points at the ends of a single line (Fig. 7.1b). We extend this to represent 2 variables by points in a plane (Fig. 7.2a). Similarly, the four possible values of two switching variables can be represented by the four vertices of a square (Fig. 7.2b). The extension to three variables, as shown in Fig. 7.3, should be obvious. The extension to more than three variables, requiring figures of more than three dimensions, is geometrically difficult but conceptually simple enough.

In general, we say that we represent the various possible combinations of n variables as points in *n-space*, and the collection of all 2^n possible points will be said to form the *vertices of an n-cube*, or a *Boolean hypercube*.

To represent functions on the n-cube, we set up a one-to-one correspondence between the minterms of *n*-variables and the vertices of the

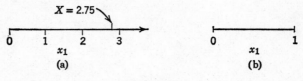

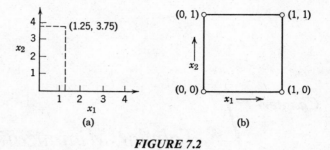

FIGURE 7.1

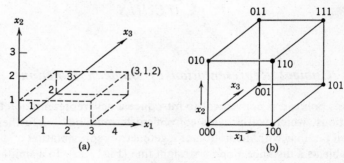

FIGURE 7.2

FIGURE 7.3

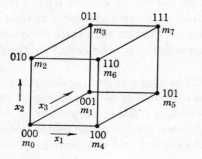

FIGURE 7.4

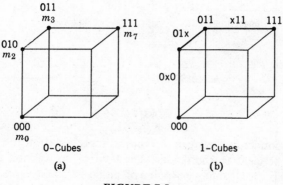

0-Cubes   1-Cubes

(a)   (b)

FIGURE 7.5

n-cube. Thus, in the 3-cube, the vertex (000) corresponds to m_0, the vertex (001) to m_1, etc., as shown in Fig. 7.4. The cubical representation of a function of *n*-variables then consists of the set of vertices of the *n*-cube corresponding to the minterms of the function. For example, the function

$$f(A, B, C) = \sum m(0, 2, 3, 7) \tag{7.1}$$

would be represented on the 3-cube as shown in Fig. 7.5a, where the vertices corresponding to m_0, m_2, m_3, and m_7 are indicated by heavy black dots. These vertices, corresponding to the minterms, may also be referred to as the *0-cubes* of the function.*

Two 0-cubes of a function are said to form a 1-cube if they differ in only one coordinate. In the function of Fig. 7.5, we have three 1-cubes, consisting of the pairs of 0-cubes 000 and 010, 010 and 011 and 011 and 111. The 1-cubes may be denoted by placing an x in the coordinate having different values and darkening the line between the pair of 0-cubes (Fig. 7.5b). In a similar fashion, a set of four 0-cubes, whose coordinate values are the same in all but two variables, are said to form a 2-cube of the function. Pictorially, a 2-cube may be represented as a shaded plane. Figure 7.6 shows the cubical representation of a function exhibiting five 0-cubes, five 1-cubes, and one 2-cube.

When all the vertices (0-cubes) of a *k*-cube are in the set of vertices making up a larger *k*-cube, we will say that the smaller cube is *contained in*, or *covered by*, the larger cube. In Fig. 7.6, for example, the 0-cube 100 is contained in the 1-cubes x00, 10x, and 1x0 and in the 2-cube 1xx. Similarly, the 1-cubes 1x0, 10x, 11x, and 1x1 are all contained in the 2-cube 1xx.

---

* The 0-cubes, 1-cubes, 2-cubes, etc., are more formally known as *k-dimensional product-subcubes* (*k* = 0, 1, 2, etc.) but this lengthy expression is generally shortened to *k-cube*. See Reference [1], pp. 100–101.

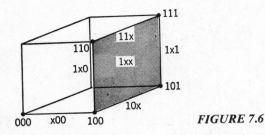

FIGURE 7.6

The correspondences between the cubical representation and the Karnaugh map should be clear to the reader. The 0-cubes correspond to the squares on the K-map, the 1-cubes to pairs of adjacent squares, etc. The cubical *nomenclature* can be applied directly to the K-map without reference to the cubical *representation*, but the cubical representation makes the origin and significance of these terms much more obvious. The cubical representation will also be very helpful in discussing coding in Chapter 8.

7.2   Determination of Prime Implicants

The Karnaugh map is a very powerful design tool, but it does have certain drawbacks. First, it is essentially a trial-and-error method which does not offer any guarantee of producing the best realization. Second, its dependence on the somewhat "intuitive" human ability to recognize patterns makes it unsuitable for any form of mechanization, such as programming for a digital computer. For functions of 6 or more variables, it is difficult for the designer to be sure that he has selected the smallest possible set of products from a Karnaugh map. You will recall that no specific straightforward procedure has been developed by which this proposition might be checked.

A tabular method based principally on the work of W. V. Quine[2] and E. J. McCluskey[3] corrects these deficiencies. It guarantees the best second-order realization (best in the sense stated in Definition 6.1, p. 116); and it can be described in an algorithmic form, i.e., as a series of exactly specified steps, suitable for computer programming.

The Quine-McCluskey method is basically a very cleverly organized procedure for making an exhaustive search for all possible combinations of 0-cubes (minterms) into larger (higher-dimension) cubes and then selecting the minimal combination of cubes required to realize the function.

The starting point for the Quine-McCluskey method is the minterm list of the function. If the function is not in this form, it must be converted to this form by the methods discussed earlier. Let us assume the following function:

$$f(A, B, C, D) = \sum m(0, 2, 3, 6, 7, 8, 9, 10, 13) \qquad (7.2)$$

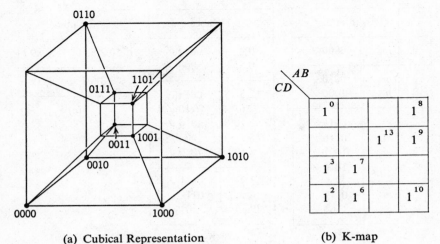

$$f(A,B,C,D) = \sum m(0,2,3,6,7,8,9,10,13)$$

FIGURE 7.7

The cubical representation and K-map for this function are shown in Fig. 7.7.

The first step is to find all possible 1-cubes of the function. By definition, a 1-cube is found by combining two 0-cubes, which are identical except in one variable. First, convert the minterms to binary form, and then count (and list) the number of 1's in the binary representation (Fig. 7.8). Now reorder the minterm list in accordance with the number of 1's in the binary representations (Fig. 7.9a). Separate the minterms into groups with the same number of 1's by horizontal lines. This grouping of minterms is made to reduce the number of comparisons that must be made in determining the 1-cubes. If two minterms are to combine, their binary representation must be

Minterm	Binary Form	No. of 1's
m_0	0000	0
m_2	0010	1
m_3	0011	2
m_6	0110	2
m_7	0111	3
m_8	1000	1
m_9	1001	2
m_{10}	1010	2
m_{13}	1101	3

FIGURE 7.8. *Conversion of minterm list to binary form.*

1's		Minterms
0	m_0	0000 ✓
1	m_2	0010 ✓
	m_8	1000 ✓
2	m_3	0011 ✓
	m_6	0110 ✓
	m_9	1001 ✓
	m_{10}	1010 ✓
3	m_7	0111 ✓
	m_{13}	1101 ✓

(a)

1-Cubes	
0,2	00x0 ✓
0,8	x000 ✓
2,3	001x ✓
2,6	0x10 ✓
2,10	x010 ✓
*8,9	100x
8,10	10x0 ✓
3,7	0x11 ✓
6,7	011x ✓
*9,13	1x01

(b)

2-Cubes	
*0,2,8,10	x0x0
*2,3,6,7	0x1x

(c)

FIGURE 7.9. *Determination of prime implicants by Quine-McCluskey.*

identical except in one position, in which one minterm will have a 0, the other a 1. Thus, the latter has one more 1 than the former, so that the two minterms must be in adjacent groups in the list of Fig. 7.9a.

With the minterms grouped, the procedure for finding the 1-cubes is quite simple. Compare each minterm in the top group with each minterm in the next lower group. If two minterms are the same in every position but one, place a check (✓) to the right of both minterms to show that they have been covered and enter the 1-cube in the next column (Fig. 7.9b). List the cubical form of the 1-cube with an x in the position which does not agree, and also list the decimal numbers of the combining minterms.

In this case, m_0 (0000) combines with m_2 (0010) to form (00x0). The combination of two vertices into a 1-cube represents an application of the distributive law just as does combining two squares on a Karnaugh map. For example, this combination of m_0 and m_2 is equivalent to the algebraic operation

$$\bar{A}\bar{B}\bar{C}\bar{D} + \bar{A}\bar{B}C\bar{D} = \bar{A}\bar{B}\bar{D}(C + \bar{C}) = \bar{A}\bar{B}\bar{D} \tag{7.3}$$

Minterm m_0 also combines with m_8 to form (x000). This completes the comparison between minterms in the first two groups, so we draw a line beneath the resultant 1-cubes. Next, the second and third groups of minterms are compared in the same manner. This comparison results in five more 1-cubes, formed from m_2 and m_3, m_8 and m_9, etc. A line is drawn beneath these five 1-cubes to indicate completion of comparisons between the second and third groups. Note that each minterm in a group must be compared with *every* minterm in the other group, even if either or both have already been checked ($\sqrt{}$) as having formed a 1-cube. *Every* 1-cube must be found, although there is no need to check off a minterm more than once.

This comparison process is repeated between successive groups until the minterm list is exhausted. In this case, all the minterms have been checked, indicating that all combine at least into 1-cubes. Thus, none of the minterms will appear explicitly in the final sum-of-products form.

The next step is a search of Fig. 7.9b for possible combinations of pairs of 1-cubes into 2-cubes. Again, cubes in each group need to be compared only with cubes in the next group down. In addition, cubes need be compared only if they have the same digit replaced by an x. In this case, 1-cube 0,2 (00x0) need be compared only with 8,10 (10x0). They differ in only one of the specified (non-x) positions and, therefore, combine to form the 2-cube 0,2,8,10 (x0x0), which is entered in the next column (Fig. 7.9c). The two 1-cubes are also checked off to show that they have combined into a 2-cube. Next, 1-cube 0,8 (x000) is found to combine with 2,10 (x010) to form 0,8,2,10 (x0x0), which is the same 2-cube as that already formed. Therefore, the two 1-cubes are checked off; but no new entry is made in the 2-cube column.

To clarify the above process, consider the algebraic interpretation. First, we have

$$
\begin{aligned}
m_0 + m_2 + m_8 + m_{10} &= \bar{A}\bar{B}\bar{C}\bar{D} + \bar{A}\bar{B}C\bar{D} + A\bar{B}\bar{C}\bar{D} + A\bar{B}C\bar{D} \\
&= \bar{A}\bar{B}\bar{D}(C + \bar{C}) + A\bar{B}\bar{D}(C + \bar{C}) \qquad (7.4) \\
&= \bar{A}\bar{B}\bar{D} + A\bar{B}\bar{D} \\
&= \bar{B}\bar{D}(A + \bar{A}) \qquad (7.5) \\
&= \bar{B}\bar{D}
\end{aligned}
$$

Here we see the reason for requiring the x to be in the same position in both 1-cubes. The x represents the variable eliminated. If the same variable (C) had not been eliminated from both 1-cubes in Equation 7.4, it is obvious that the second elimination (of A) could not have been made in Equation 7.5. The second combination of 1-cubes into the same 2-cubes is seen in the following.

$$
\begin{aligned}
m_0 + m_8 + m_2 + m_{10} &= \bar{A}\bar{B}\bar{C}\bar{D} + A\bar{B}\bar{C}\bar{D} + \bar{A}\bar{B}C\bar{D} + A\bar{B}C\bar{D} \\
&= \bar{B}\bar{C}\bar{D}(A + \bar{A}) + \bar{B}C\bar{D}(A + \bar{A}) \qquad (7.6) \\
&= \bar{B}\bar{C}\bar{D} + \bar{B}C\bar{D} \\
&= \bar{B}\bar{D}(C + \bar{C}) \qquad (7.7) \\
&= \bar{B}\bar{D}
\end{aligned}
$$

Equations 7.6 and 7.7 represent the same elimination of variables as Equations 7.4 and 7.5, except in the reverse order.

No further combinations can be made between the first and second groups of Fig. 9.7b, so we draw a line beneath the 2-cube formed. The second and third groups are then compared in the same fashion, resulting in another 2-cube. This completes the comparison of 1-cubes, and any unchecked entries—8,9 and 9,13 in this case—are marked with an asterisk (*), to indicate that they are *prime* implicants.

Definition 7.1. A *prime implicant* is any cube of a function which is not totally contained in some larger cube of the function.

If any minterms had failed to combine in the first step, they would also have been marked as prime implicants. The importance of prime implicants will become evidently shortly.

Finally, the 2-cubes are checked for possible combination into 3-cubes. Since the x's are in different positions, these cubes do not combine and are thus prime implicants. For this example, the determination of prime implicants is now complete. In the general case, the same procedure continues as long as larger cubes can be formed.

This procedure of comparing binary representations is quite suitable for computer mechanization, particularly for computers which have some sort of "binary compare" command. For manual processing, this method has the drawback that errors will inevitably occur in trying to compare 0's and 1's over long lists of cubes. This difficulty is considerably reduced by the following alternative method, which utilizes only decimal notation. We shall use the same example as above.

Again, the first step is to list the minterms, grouped according to the number of 1's in their binary representation. However, only the decimal minterm numbers are listed (Fig. 7.10a). To make the comparison, we draw on the fact that the numbers of minterms that combine must differ by a power of two. Consider the combination of m_{13} (1101) and m_9 (1001):

$$13 = 1 \times 2^3 + \left(1 \times 2^2\right) + 0 \times 2^1 + 1 \times 2^0$$
$$9 = 1 \times 2^3 + \left(0 \times 2^2\right) + 0 \times 2^1 + 1 \times 2^0$$

Each 1 in a binary number has the numeric weight of a power of two. The two combining minterms have the same 0's and 1's, except in one position. Thus, the number of the minterm with the extra 1 must be larger than the number of the other minterm by a power of two. The procedure, then, is as follows: compare each minterm number with each *larger* minterm number in the next group down. If they differ by a power of two, they combine to form a 1-cube.

To represent the 1-cube, list the two minterm numbers, followed by their numeric difference in parentheses. For the above example, the listing would be 9,13(4). The number in parentheses indicates the position in which the x

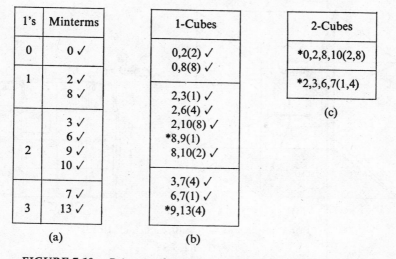

1's	Minterms
0	0 ✓
1	2 ✓
	8 ✓
	3 ✓
	6 ✓
2	9 ✓
	10 ✓
	7 ✓
3	13 ✓

(a)

1-Cubes
0,2(2) ✓
0,8(8) ✓
2,3(1) ✓
2,6(4) ✓
2,10(8) ✓
*8,9(1)
8,10(2) ✓
3,7(4) ✓
6,7(1) ✓
*9,13(4)

(b)

2-Cubes
*0,2,8,10(2,8)
*2,3,6,7(1,4)

(c)

FIGURE 7.10. *Prime implicant determination, decimal notation.*

would appear in the binary representation. Since the binary digit with the weight of 4 is the third from the right, the (4) in parentheses corresponds to the x in the third position in the binary notation 9,13 (1x01). Except for this change in notation, the process of comparing minterms and listing 1-cubes is the same as before. The resultant 1-cube column is shown in Fig. 7.10b.

In the binary-character method, it was seen that 1-cubes could combine only if the x's were in the same position. Since the number in parentheses indicates the position of the x, it follows that only 1-cubes with the same number in parentheses can combine. Therefore, compare each 1-cube with all 1-cubes in the next lower group which have the same number in parentheses. If the lowest minterm number of the 1-cube in the lower group is greater by a power of two than the corresponding number in the other cube, they combine. List the 2-cube by listing all four minterm numbers, followed by both powers of two, in parentheses. For example, in Fig. 7.10b, 0,2(2) and 8,10(2) have the same number in parentheses and 8 is greater than 0 by 8. The resultant 2-cube is thus 0,2,8,10(2,8).

Figure 7.10c shows the complete 2-cube column in decimal notation. To compare 2-cubes, the same procedure applies. Just as both x's, in both 2-cubes, have to be in the same position for combination, so *both* numbers in parentheses must be the same for both 2-cubes. A comparison of Figs. 7.9 and 7.10 will show that the resultant charts are identical. Furthermore, the procedures are identical in principle, differing only in the substitution of mental subtraction of decimal numbers for visual comparison of binary numbers.

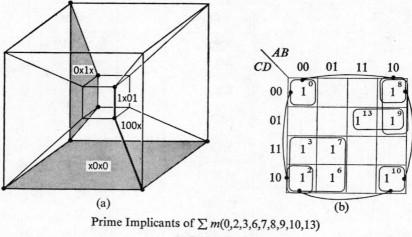

(a)   (b)

Prime Implicants of $\sum m(0,2,3,6,7,8,9,10,13)$

FIGURE 7.11

7.3   Selection of an Optimum Set of Prime Implicants

All of the prime implicants determined for the example of the previous section are indicated on the 4-cube and on the map of Fig. 7.11.

Interpreting the k-cubes as sets of 1's on the K-map, we see that each k-cube will be represented in a sum-of-products realization by a product term with $n - k$ literals. Thus, any set of k-cubes containing all the 0-cubes (minterms) of the function will provide an S-of-P realization. We now have a list of all the k-cubes of the function. To complete the design, we must select a set of k-cubes which contains all the 0-cubes and, furthermore, will lead to a *minimal* S-of-P realization. Fortunately, we do not have to consider all the k-cubes of the function, as shown by the following theorem, due to Quine.[2]

THEOREM 7.1.   *Any sum-of-products realization that is minimal in the sense of Definition 6.1 must consist of a sum of products representing prime implicants.*

Proof: Consider any set of k-cubes containing all the 0-cubes of the function. Any cube which is not a prime implicant is, by definition, contained in some prime implicant. Therefore, we can replace this cube in the set by any prime implicant containing it. Since the prime implicant is of higher dimension, its product representation will have fewer literals than that of the cube it replaces. The replacement will therefore lower the cost of the S-of-P realization.

Theorem 7.1 tells us that we can restrict our attention to the prime implicants in selecting an S-of-P realization. (In future we shall use the term *prime*

		0	2	3	6	7	8	9	10	13
	0,2,8,10(2,8)	✓	✓				✓		✓	
	2,3,6,7(1,4)		✓	✓	✓	✓				
	8,9(1)						✓	✓		
	9,13(4)							✓		✓

FIGURE 7.12. *Prime implicant table—first step.*

implicant to mean either the cubes themselves, or the corresponding product.) However, the theorem does not say that *any* sum of prime implicants covering all the minterms is minimal. To find a minimal sum, we construct a *prime implicant table* (Fig. 7.12).

Each column corresponds to a minterm. At the left of each row are listed the prime implicants of the function, arranged in groups according to cost (number of literals in the product). Also included are an extra column on the left and an extra row at the bottom, for reasons that will be explained shortly. In each row (except the bottom), checks are placed in the columns corresponding to the minterms contained in the prime implicant listed on that row. For example, the first prime implicant listed in Fig. 7.12 contains the 0-cubes 0,2,8, and 10, so checks are placed in the first row in columns 0, 2, 8, and 10.

The completed prime implicant table is then inspected for columns containing only a single check mark. In this example, we find only one check in the first column, on the first row. This indicates that the corresponding prime implicant, 0,2,8,10(2,8), is the only one containing m_0 and must be included in the S-of-P realization. We therefore say that it is an *essential prime implicant* and mark it with an asterisk in the leftmost column. We also place checks in the bottom row in columns 0, 2, 8, and 10 to indicate that these are included in an already-selected prime implicant (Fig. 7.13).

Continuing the search, we find a single check in column 3, indicating that the prime implicant 2,3,6,7(1,4) is the only one covering m_3. We mark this prime implicant as essential and check off the minterms it covers, which were not covered by the first one selected. Finally, we find that 9,13(4) is also essential (Fig. 7.14).

When the search for essential prime implicants is complete, we inspect the bottom row to see if all the columns have been checked. If so, then all

		0	2	3	6	7	8	9	10	13
*	0,2,8,10(2,8)	✓	✓				✓		✓	
	2,3,6,7(1,4)		✓	✓	✓	✓				
	8,9(1)						✓	✓		
	9,13(4)							✓		✓
		✓	✓				✓		✓	

FIGURE 7.13.   *Prime implicant table—second step.*

the 0-cubes are contained in the essential prime implicants and the sum of the essential prime implicants is the minimal S-of-P realization. This is the case for our example.

The final step is to determine the actual products. To do this, we recall that difference numbers for each prime implicant indicate the variables which have been eliminated. Write out the binary equivalent of any one of the minterms in the prime implicant, cross out the positions having the binary weights in parentheses, and convert the remaining digits to the appropriate literals.

For the above example, we have

$$0,2,8,10(2,8) = \cancel{0}0\cancel{0}0 \rightarrow \bar{B}\bar{D}$$
$$2,3,6,7(1,4) \;\;= 0\cancel{0}1\cancel{0} \rightarrow \bar{A}C$$
$$9,13(4) \;\;\;\;\;\;\;\;= 1\cancel{0}01 \rightarrow A\bar{C}D$$

		0	2	3	6	7	8	9	10	13
*	0,2,8,10(2,8)	✓	✓				✓		✓	
*	2,3,6,7(1,4)		✓	✓	✓	✓				
	8,9(1)						✓	✓		
*	9,13(4)							✓		✓
		✓	✓	✓	✓	✓	✓	✓	✓	✓

FIGURE 7.14.   *Prime implicant table—final.*

Thus, the minimal sum-of-products is

$$f(A, B, C, D) = \sum m(0, 2, 3, 6, 7, 8, 9, 10, 13) = \bar{B}\bar{D} + \bar{A}C + A\bar{C}D \quad (7.8)$$

This very simple example was chosen to clarify the exact significance of the various steps rather than as a practical illustration of the value of the technique. Example 7.1 illustrates the application of the method to a problem of more practical scope.

Example 7.1

Determine the minimal S-of-P form for

$$f(A, B, C, D, E) = \sum m(0, 6, 8, 10, 12, 14, 17, 19, 20, 22, 25, 27, 28, 30),$$

for which the prime implicant chart and table are shown in Fig. 7.15.

Solution: Note that we have assigned a letter to represent each prime implicant, in order to simplify the prime implicant table. Thus,

$$\mathbf{a} = 8, 10, 12, 14(2, 4) = 01\cancel{0}\cancel{0}0 = \bar{A}B\bar{E}, \text{ etc.}$$

Translating all essential prime implicants in this fashion, we obtain the desired minimal form

$$f(A, B, C, D, E) = \bar{A}B\bar{E} + CD\bar{E} + A\bar{C}E + AC\bar{E} + \bar{A}\bar{C}\bar{D}\bar{E} \quad (7.9)$$

∎

In both of the previous examples, the essential prime implicants covered all the minterms. This is not always the case. Figure 7.16a shows the prime implicant table for the function

$$f(A, B, C, D, E) = \sum m(1, 2, 3, 5, 9, 10, 11, 18, 19, 20, 21, 23, 25, 26, 27).$$

When we select the essential prime implicants on this table, we find that they do not cover all the minterms of the function. We must now apply some cost criteria in making a selection from the remaining nonessential terms.

To make it easier to find the appropriate terms, a *reduced* prime implicant table is set up, listing only the 0-cubes not contained in the essential prime implicants as columns and those nonessential prime implicants containing any of the uncovered 0-cubes as rows (Fig. 7.16b). In the reduced table, we note that there could be no possible advantage to using $\mathbf{e}$, since it covers only m_5, while $\mathbf{d}$, which has the same cost, covers both m_5 and m_1. Thus, removing $\mathbf{e}$ from the table cannot prevent us from finding a minimal sum. We next note that $\mathbf{g}$ and $\mathbf{h}$ cover the same minterm at the same cost. Thus, removing $\mathbf{h}$ from the table cannot prevent our finding a minimal sum.

Removing rows $\mathbf{e}$ and $\mathbf{h}$ from the table, we obtain a still further reduced table, as shown in Fig. 7.16c. On this table, it may be seen that $\mathbf{d}$ is the only prime implicant left to cover m_5 and $\mathbf{g}$ is the only left to cover m_{23}. We therefore mark them with double asterisks (**) to indicate that they are *secondary*

0 ✓	*0,8(8) **f**	*8,10,12,14(2,4) **a**
8 ✓	8,10(2) ✓ 8,12(4) ✓	*6,14,22,30(8,16) **b**
6 ✓ 10 ✓ 12 ✓ 17 ✓ 20 ✓	6,14(8) ✓ 6,22(16) ✓ 10,14(4) ✓ 12,14(2) ✓ 12,28(16) ✓	*12,14,28,30(2,16) **c** *17,19,25,27(2,8) **d** *20,22,28,30(2,8) **e**
14 ✓ 19 ✓ 22 ✓ 25 ✓ 28 ✓	17,19(2) ✓ 17,25(8) ✓ 20,22(2) ✓ 20,28(8) ✓	
27 ✓ 30 ✓	14,30(16) ✓ 19,27(8) ✓ 22,30(8) ✓ 25,27(2) ✓ 28,30(2) ✓	

	0	6	8	10	12	14	17	19	20	22	25	27	28	30
*a			✓	✓	✓	✓								
*b		✓				✓				✓				✓
c					✓	✓							✓	✓
*d							✓	✓			✓	✓		
*e									✓	✓			✓	✓
*f	✓		✓											
	✓	✓	✓	✓	✓	✓	✓	✓	✓	✓	✓	✓	✓	✓

FIGURE 7.15

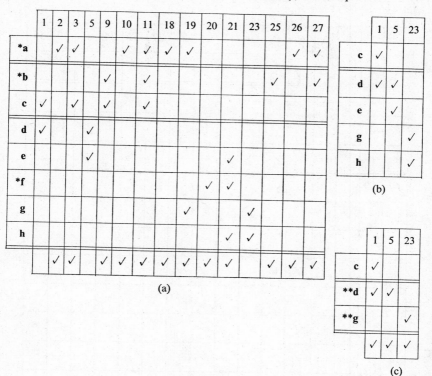

FIGURE 7.16

essential prime implicants. We also note that they cover all of the remaining minterms, so the selection process is now completed and *a* minimal sum is given by

$$f(A, B, C, D, E) = \underbrace{\mathbf{a} + \mathbf{b} + \mathbf{f}}_{\text{essential}} + \underbrace{\mathbf{d} + \mathbf{g}}_{\substack{\text{secondary} \\ \text{essential}}} \tag{7.10}$$

$$= (2, 3, 10, 11, 18, 19, 26, 27) + (9, 11, 25, 27)$$
$$+ (20, 21) + (1, 5) + (19, 23) \tag{7.11}$$
$$= \bar{C}D + B\bar{C}E + A\bar{B}C\bar{D} + \bar{A}\bar{B}\bar{D}E + A\bar{B}DE \tag{7.12}$$

To formalize the above procedure, we state the following definitions and theorem.

Definition 7.2. Two rows, **a** and **b**, of a reduced prime implicant table, which cover the same minterms, i.e., have checks in exactly the same columns, are said to be *interchangeable*.

	1	2	3	8	9	10	17	24	25	27	33	34	36	37	59
a											✓		✓	✓	
c	✓						✓								
d	✓										✓			✓	
e	✓				✓		✓		✓						
f	✓				✓						✓				
g		✓				✓						✓			
i													✓	✓	
j				✓	✓			✓	✓						
k				✓	✓										
l				✓		✓									
m				✓				✓							
n												✓			
q	✓		✓												
r		✓	✓												
s									✓	✓					
t										✓					✓
u															✓

FIGURE 7.17. Reduced prime implicant table.

Definition 7.3. Given two rows, **a** and **b**, in a reduced prime implicant table. Row **a** is said to *dominate* row **b** if row **a** has checks in all the columns in which row **b** has checks and also has a check in at least one column in which row **b** does not have a check.

THEOREM 7.2: *Let* **a** *and* **b** *be rows of a reduced prime implicant table, such that the cost of* **a** *is less than or equal to the cost of* **b**. *Then if* **a** *dominates* **b** *or* **a** *and* **b** *are interchangeable, there exists a minimal sum of products which does not include* **b**.

Example 7.2

Given the function

$$f(A, B, C, D, E, F) = \sum m(1, 2, 3, 4, 5, 8, 9, 10, 17, 20, 21, 24, 25, 27, 32,$$
$$33, 34, 36, 37, 40, 41, 42, 43, 44, 45, 46, 47, 48,$$
$$56, 59, 62).$$

By the procedures already covered, the essential prime implicants have been determined. The reduced prime implicant table left after this removal is shown in Fig. 7.17. Complete the selection of prime implicants.

Solution: We first see that prime implicant **n** covers only m_{34} while **g** covers m_{34}, m_{10}, and m_3. So **g** dominates **n**, and **n** can be removed. Similarly, **e** dominates **c**, **a** dominates **i**, **j** dominates **k** and **m**, and **t** dominates **u**. Fig. 7.18 is a new reduced prime implicant table, with the dominated prime implicants removed.

We next check the new table for secondary essential prime implicants by searching for columns with only one check, as before. For example, with the dominated terms eliminated, **e** is the only prime implicant left that covers m_{17}, so **e** must be used. We mark it (******) as secondary essential and check off the minterms it covers, m_1, m_9, m_{17}, and m_{25}, as before. Similarly, **j** is secondary essential by virtue of m_{24}, **g** because of 34, **a** because of 36, and **t** because of 59. Having completed the selection of secondary essential implicants, we find that only one minterm, m_3, is left. We can see by inspection that **q** and **r** are equal with respect to the one remaining minterm. We arbitrarily select **q**, and a minimal form will be given by the sum of **a** + **e** + **g** + **j** + **t** + **q** + the essential prime implicants. ■

In Example 7.2, one application of row dominance was sufficient to complete the design. In some cases, the first selection of secondary essential prime implicants may leave a large group of minterms still uncovered. In this case, simply repeat the process, i.e., reduce the table by removing the secondary essential rows and apply row dominance again. This process may be repeated as many times as necessary.

Consider the reduced prime implicant table shown in Fig. 7.19. This table cannot be further reduced, since no column has a single check and no row dominates any other row. A technique sometimes known as Petrick's method[5] may be used to make the final selection of prime implicants as follows.

	1	2	3	8	9	10	17	24	25	27	33	34	36	37	59
**a											✓		✓	✓	
d	✓										✓			✓	
**e	✓				✓		✓		✓						
f	✓				✓						✓				
**g		✓				✓						✓			
**j				✓	✓			✓	✓						
l				✓		✓									
q	✓		✓												
r		✓	✓												
s									✓	✓					
**t											✓				✓
	✓	✓		✓	✓	✓	✓	✓	✓	✓	✓	✓	✓	✓	✓

FIGURE 7.18. Prime implicant table—second reduction.

	6	7	15	38	46	47
a	✓	✓				
b			✓			✓
c		✓	✓			
d					✓	✓
e				✓	✓	
f	✓			✓		

FIGURE 7.19. Reduced prime implicant table.

A list of all possible irredundant sets of prime implicants which will cover the remaining minterms may be obtained in the following manner. The letters representing the remaining prime implicants are interpreted as Boolean variables, which will take on the value 1 if that prime implicant is selected, the value 0 if it is not. From the first column, either **a** or **f** must be selected to cover m_6, so we write

$$(\mathbf{a} + \mathbf{f}) = 1 \tag{7.13}$$

Similarly, either **a** or **c** must be selected to cover m_7; so

$$(\mathbf{a} + \mathbf{f})(\mathbf{a} + \mathbf{c}) = 1 \tag{7.14}$$

In this fashion, we build up a product of sums, each sum listing the prime implicants which contain a particular minterm. In this case, the complete expression is

$$(\mathbf{a} + \mathbf{f})(\mathbf{a} + \mathbf{c})(\mathbf{b} + \mathbf{c})(\mathbf{e} + \mathbf{f})(\mathbf{d} + \mathbf{e})(\mathbf{b} + \mathbf{d}) = 1 \tag{7.15}$$

Applying the distributive laws (multiplying out), we obtain

$$\mathbf{abe} + \mathbf{abdf} + \mathbf{acde} + \mathbf{bcef} + \mathbf{cdf} = 1 \tag{7.16}$$

To satisfy Equation 7.16, at least one of the five products must be 1. Of these, **abe** and **cdf** represent a covering of the table by three prime implicants. The total cost is least for **a**, **b**, and **e** (smallest total number of inputs), so these three are included with any essential prime implicants in a minimal sum of products.

The Quine-McCluskey method can be modified to handle don't-care terms very simply. In determining the prime implicants, include the don't-cares in the minterm list. In this manner, you insure that you have included groupings which can be formed by using the don't-cares. Then, in setting up the prime implicant table, list only the minterms across the top. This insures that only prime implicants necessary to cover minterms will be selected.

Example 7.3

Find the minimal S-of-P realization of

$$f(A, B, C, D, E) = \sum m(1, 4, 7, 14, 17, 20, 21, 22, 23) + d(0, 3, 6, 19, 30).$$

On the chart in Fig. 7.20a, we list both minterms and don't-cares in column I.

Solution: On the prime implicant table (Fig. 7.20b), we list only the minterms. From the table, we have

$$f(A, B, C, D, E) = \underset{\uparrow\;\text{essential}}{\mathbf{e}} + \underset{\underset{\text{secondary}}{\uparrow}\;\;\text{essential}}{\underbrace{\mathbf{a} + \mathbf{b} + \mathbf{d}}} + \underset{\uparrow\;\text{optional}}{\begin{cases} \mathbf{f} \\ \text{or } \mathbf{g} \end{cases}}$$

or

$$f(A, B, C, D, E) = CD\bar{E} + \bar{B}\bar{C}E + \bar{B}C\bar{E} + \bar{B}CD + \begin{cases} A\bar{B}E \\ \text{or } A\bar{B}C \end{cases} \tag{7.17}$$

■

I

0 ✓
1 ✓
4 ✓
3 ✓
6 ✓
17 ✓
20 ✓
7 ✓
14 ✓
19 ✓
21 ✓
22 ✓
23 ✓
30 ✓

II

*0,1(1) **h**
*0,4(4) **i**

1,3(2) ✓
1,17(16) ✓
4,6(2) ✓
4,20(16) ✓

3,7(4) ✓
3,19(16) ✓
6,7(1) ✓
6,14(8) ✓
6,22(16) ✓
17,19(2) ✓
17,21(4) ✓
20,21(1) ✓
20,22(2) ✓

7,23(16) ✓
14,30(16) ✓
19,23(4) ✓
21,23(2) ✓
22,23(1) ✓
22,30(8) ✓

III

*1,3,17,19(2,16) **a**
*4,6,20,22(2,16) **b**

*3,7,19,23(4,16) **c**
*6,7,22,23(1,16) **d**
*6,14,22,30(8,16) **e**
*17,19,21,23(2,4) **f**
*20,21,22,23(1,2) **g**

(a)

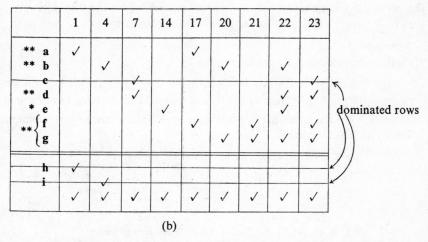

(b)

FIGURE 7.20. *Chart and prime implicant table, Example 7.3.*

Like the K-map, the Quine-McCluskey method can be adapted to P-of-S design by designing on the 0's instead of the 1's of the function. Only two changes are necessary. In column I of the Quine-McCluskey chart, list the 0's (maxterms) of the function plus the don't-cares (if any). Then proceed exactly as before until the last step, the conversion of the prime implicants, at which point you convert to sums instead of products, using complemented variables, just as with the K-map method.

It should be evident that the Quine-McCluskey method does not qualify as a labor-saving technique. The map method is always quicker, but for more than 5 variables, the reliability in determining a minimal realization is questionable. The map method can generally be relied on to come fairly close to a minimal design. If you are designing a circuit that will only be built once, the savings resulting from eliminating one or two gates or literals won't justify the extra time required for the tabular method. On the other hand, if you are designing a circuit to be produced in large quantity, the elimination of a single literal may warrant many extra hours of engineering time.

The real importance of the Quine-McCluskey method lies in the fact that it is a formal procedure, in no way dependent on human intuition. It is, therefore, suitable for computer mechanization, and a number of programs based on this method have been written.[6,7] A thorough understanding of the basic principles of this method is thus vitally important to any logic designer even though he may seldom use it personally.

7.4 Multiple-Output Circuits

So far we have considered only the implementation of a single function of a given set of variables. In the design of complete systems, we frequently wish to implement a number of different functions of the same set of variables. We can implement each function completely independently by the techniques already developed, but considerable savings can often be achieved by the sharing of hardware between various functions.

Example 7.4

The circuit **BB** is to serve as an interface between the two computers of Fig. 7.21. The first four letters of the alphabet must intermittently be transmitted from

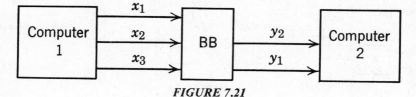

FIGURE 7.21

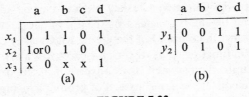

	a	b	c	d
x_1	0	1	1	0	1
x_2	1 or 0	1	0	0
x_3	x	0	x	x	1

(a)

	a	b	c	d
y_1	0	0	1	1
y_2	0	1	0	1

(b)

FIGURE 7.22

computer 1 to computer 2. In computer 1, these letters are coded on three lines, x_1, x_2, x_3, as shown in Fig. 7.22a. In computer 2, they are coded on two lines, y_2 and y_1, as shown in Fig. 7.22b.

The translating functions $y_1(x_1, x_2, x_3)$ and $y_2(x_1, x_2, x_3)$ are easily compiled directly on Karnaugh maps, as in Fig. 7.23. To accomplish this, we determine from Fig. 7.22b which letters of the alphabet require 1's for either y_1 or y_2. We then determine from Fig. 7.22a which input combinations correspond to these

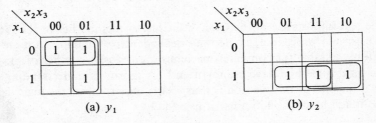

(a) y_1 (b) y_2

FIGURE 7.23

letters. For example, y_1 is to be 1 for c or d (Fig. 7.22b). From Fig. 7.22a, we see that c or d are represented by $x_1 x_2 x_3 = 000, 001, 101$, so we enter 1's in the corresponding squares of Fig. 7.23a.

From these maps, the minimal realizations of y_1 and y_2, when considered individually, are easily seen to be those of Fig. 7.24. Alternatively, we note that y_1 and y_2 could be expressed as

$$y_1 = \bar{x}_1 \bar{x}_2 + x_1 \bar{x}_2 x_3 \quad \text{and} \quad y_2 = x_1 x_2 + x_1 \bar{x}_2 x_3 \quad (7.18)$$

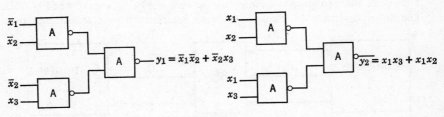

FIGURE 7.24

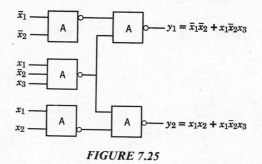

FIGURE 7.25

Taking advantage of the common term, $x_1\bar{x}_2 x_3$, permits the implementation found in Fig. 7.25. Notice that this configuration requires only five NAND gates as opposed to 6 in Fig. 7.24. ■

7.5   *Map Minimization of Multiple-Output Circuits*

The value of a systematic procedure which would identify common products and guarantee an optimum realization (minimum total number of gates for all outputs) of multiple-output circuits is clear. The general convenience of Karnaugh maps leads us to explore this approach first.

Let us assume that we wish to find a minimal S-of-P realization for the three functions:

$$f_\alpha(A, B, C, D) = \sum m(2, 4, 10, 11, 12, 13)$$

$$f_\beta(A, B, C, D) = \sum m(4, 5, 10, 11, 13)$$

$$f_\gamma(A, B, C, D) = \sum m(1, 2, 3, 10, 11, 12).$$

We first draw the Karnaugh maps of all three functions (Fig. 7.26). We also draw maps of the *intersections*, or products, of the three functions, by pairs and all together. Recall that, in terms of K-maps, the intersection of two functions consists of their common squares.

We now locate the prime implicants on the maps, starting with the product of all the functions. The only prime implicant of $f_\alpha \cdot f_\beta \cdot f_\gamma$ is 10,11(1). We now look for prime implicants of the function-pairs, but we do not mark a grouping as a prime implicant if it is a prime implicant of a higher-order function product. For example, we do not mark 10,11(1) on $f_\beta \cdot f_\gamma$ since it is a prime implicant of $f_\alpha \cdot f_\beta \cdot f_\gamma$. On $f_\alpha \cdot f_\gamma$, we mark only 2,10(8) and 12, since the other group is also in $f_\alpha \cdot f_\beta \cdot f_\gamma$. Similarly, we mark only 4 and 13 on $f_\alpha \cdot f_\beta$. Now we go to the maps of the functions themselves, marking only those prime implicants which do not also appear in the function products.

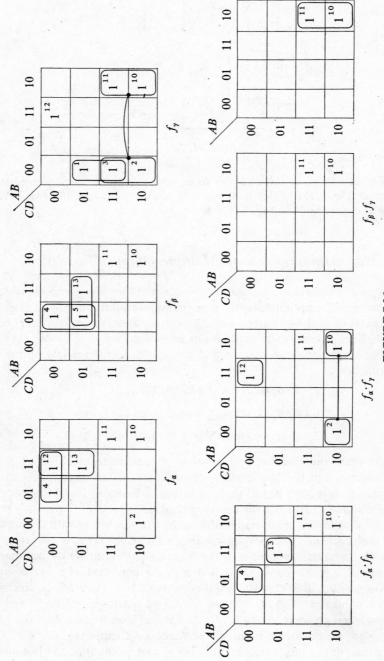

FIGURE 7.26

The next task is to select the sets of prime implicants leading to the realization with the lowest total cost. For a problem of this size, the selection could be made directly from the maps. A more systematic approach uses a modified version of the prime implicant table (Fig. 7.27). The columns correspond to all the minterms for each function. The rows correspond to the prime implicants grouped by cost. Also indicated are the function, or function products, in which they appear. On each row, we check the minterm columns *for the function(s) in which the prime implicant appears.* For example, in Fig. 7.27 we check columns 2, 3, 10, 11 under f_γ, *only*, since the prime implicant $\mathbf{a} = 2,3,10,11(1,8)$ appears only in f_γ, although some of the minterms also appear in other functions. For prime implicant $\mathbf{g} = 2,10(8)$, we check columns 2 and 10 under both f_α and f_γ, since this 1-cube appears in both functions.

We next check for essential prime implicants, just as in the single-output Quine-McCluskey method, by looking for columns with only one check. In this case, the essential prime implicants do not cover all the minterms of all three functions, so a reduced prime implicant table is formed (Fig. 7.28).

The next step is to apply dominance to eliminate rows in exactly the same manner as for the single-output case. Note that the entire table must be considered in determining dominance, not just that part for a single function. For example, in the table of Fig. 7.28, row $\mathbf{d}$ would dominate row $\mathbf{i}$ if only function f_β were considered. But if the whole table is considered, it is seen that $\mathbf{i}$ has a check in a column in which $\mathbf{d}$ does not. Therefore, $\mathbf{d}$ does not satisfy the conditions for dominance of $\mathbf{i}$.

In this example, we cannot eliminate any rows through dominance, so there are no secondary essential prime implicants, and we proceed to apply Petrick's method. As before, we develop a product-of-sums form, each sum listing the prime implicants available to cover one of the minterms not covered by the essential implicants *for all functions.* In this case, we have

$$(\mathbf{b} + \mathbf{i})(\mathbf{c} + \mathbf{k})(\mathbf{d} + \mathbf{i})(\mathbf{d} + \mathbf{e})(\mathbf{e} + \mathbf{k}) = 1 \qquad (7.19)$$

By successive multiplications, we convert this to the sum-of-products form:

$$(\mathbf{i} + \mathbf{bd})(\mathbf{c} + \mathbf{k})(\mathbf{e} + \mathbf{dk}) = 1$$

$$(\mathbf{ci} + \mathbf{bcd} + \mathbf{ik} + \mathbf{bdk})(\mathbf{e} + \mathbf{dk}) = 1$$

$$\mathbf{cei} + \mathbf{cdik} + \mathbf{bcde} + \mathbf{bdek} + \mathbf{eik} + \mathbf{dik} + \mathbf{bcdk} + \mathbf{bdk} = 1$$

and then eliminate redundant products by application of Lemma 4.5 $(\mathbf{a} + \mathbf{ab} = \mathbf{a})$:

$$\mathbf{cei} + \mathbf{bcde} + \mathbf{eik} + \mathbf{dik} + \mathbf{bdk} = 1 \qquad (7.20)$$

Each product represents a sufficient set of prime implicants to cover the remaining minterms, and we wish to select the set having the lowest cost.

Figure 7.27 — Prime implicant chart for multiple-output functions f_α, f_β, f_γ.

Fcn.	Pr. Imp.		f_α						f_β					f_γ					
			2	4	10	11	12	13	4	5	10	11	13	1	2	3	10	11	12
γ	a														✓	✓	✓	✓	
α	b			✓			✓												
α	c						✓	✓											
β	d								✓	✓									
β	e									✓			✓						
γ	f	*												✓		✓			
$\alpha\gamma$	g	*	✓		✓										✓		✓		
$\alpha\beta\gamma$	h	*			✓	✓					✓	✓					✓	✓	
$\alpha\beta$	i			✓					✓										
$\alpha\gamma$	j	*					✓												✓
$\alpha\beta$	k							✓					✓						

a = 2,3,10,11(1,8)   b = 4,12(8)   c = 12,13(1)   d = 4,5(1)   e = 5,13(8)

f = 1,3(2)   g = 2,10(8)   h = 10,11(1)   i = 4   j = 12   k = 13

FIGURE 7.27

Fcn.	P. Imp.	f_α		f_β		
		4	13	4	5	13
α	**b**	✓				
α	**c**		✓			
β	**d**			✓	✓	
β	**e**				✓	✓
$\alpha\beta$	**i**	✓		✓		
$\alpha\beta$	**k**		✓			✓

FIGURE 7.28.   *Reduced multiple-output prime implicant table.*

Since each literal represents a prime implicant, we consider only the products with the fewest literals, **cei**, **bdk**, **dik**, and **eik**. The first two represent two 1-cubes and one 0-cube; the latter two, one 1-cube and two 0-cubes. Thus, **cei** and **bdk** represent the lower-cost sets, and we arbitrarily select **cei**.

From the total set of selected prime implicants, **c**, **e**, **f**, **g**, **h**, **i**, and **j**, we now determine those required for each function. If a prime implicant appears in only one function, it obviously must be used in the realization of that function. For prime implicants of function-products, we must be more careful. Consider prime implicant $\mathbf{j} = m_{12}$, which appears in f_α and f_γ. It is essential to f_γ and must be used in its realization. However, m_{12} is covered by prime implicant $\mathbf{c} = 12,13(1)$ in f_α, so **j** would be redundant in the realization of f_α. On the basis of such considerations, we make the following selection of prime implicants for the three functions:

$$f_\alpha = \mathbf{c} + \mathbf{g} + \mathbf{h} + \mathbf{i}$$
$$f_\beta = \mathbf{e} + \mathbf{h} + \mathbf{i}$$
$$f_\gamma = \mathbf{f} + \mathbf{g} + \mathbf{h} + \mathbf{j}$$

Either by locating the cubes on the maps or by the method given earlier, we convert these to algebraic notation:

$$f_\alpha = AB\bar{C} + \bar{B}C\bar{D} + A\bar{B}C + \bar{A}BC\bar{D} \tag{7.21}$$
$$f_\beta = A\bar{B}C + B\bar{C}D + \bar{A}B\bar{C}\bar{D} \tag{7.22}$$
$$f_\gamma = \bar{A}\bar{B}D + \bar{B}C\bar{D} + A\bar{B}C + AB\bar{C}\bar{D} \tag{7.23}$$

The realization of these equations is shown in Fig. 7.29.

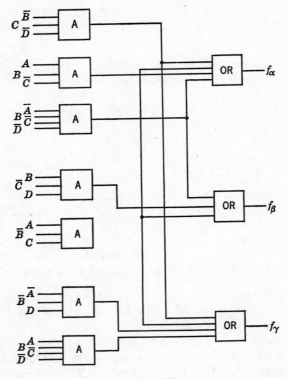

FIGURE 7.29

7.6   *Tabular Determination of Multiple-Output Prime Implicants*

We have already adapted the prime implicant table to multiple output in the last section,* so all that remains to complete the tabular method is to adapt the chart method to the determination of multiple output prime implicants. Let us consider the same set of functions used in the last section:

$$f_\alpha(A, B, C, D) = \sum m(2, 4, 10, 11, 12, 13)$$

$$f_\beta(A, B, C, D) = \sum m(4, 5, 10, 11, 13)$$

$$f_\gamma(A, B, C, D) = \sum m(1, 2, 3, 10, 11, 12)$$

---

* Such combinations of map and tabular methods are, obviously, equally valid for single functions and are sometimes the most efficient method.

Minterms			1-cubes			2-cubes		
1	γ	✓	*1,3(2)	γ	**f**	*2,3,10,11(1,8)	γ	**a**
2	$\alpha\gamma$	✓	2,3(1)	γ	✓			
* 4	$\alpha\beta$	**i**	*2,10(8)	$\alpha\gamma$	**g**			
			*4,5(1)	β	**d**			
3	γ	✓	*4,12(8)	α	**b**			
5	β	✓						
10	$\alpha\beta\gamma$	✓	3,11(8)	γ	✓			
*12	$\alpha\gamma$	**j**	*5,13(8)	β	**e**			
			*10,11(1)	$\alpha\beta\gamma$	**h**			
11	$\alpha\beta\gamma$	✓	*12,13(1)	α	**c**			
*13	$\alpha\beta$	**k**						

FIGURE 7.30. *Tabular determination of multiple-output prime implicants.*

In the first column of Fig. 7.30 are listed all the minterms of the functions, mixed together without regard for which function they appear in, and grouped by number of 1's in their binary representations as before. To each minterm we append a *tag* to indicate which functions it appears in. We then proceed to form 1-cubes in essentially the same fashion as for the single function but with some important differences.

First, we cannot combine two cubes unless they have at least one letter in common in their tags. That is, we cannot combine a cube in one function with a cube in another. For example, in Fig. 7.30, m_1 and m_5 meet the conditions for combination, except that m_1 is in f_γ and m_5 is in f_β.

Second, when two cubes combine, the tag of the resultant larger cube consists only of those letters common to the tags of both cubes. Again referring to Fig. 7.30, when we form 5,13(8), the tag is β, since this cube is found only in f_β, as shown by the fact that β is the only letter common to the tags of 5 and 13.

Third, we check off a cube as having combined only if its entire tag appears in the tag of the larger cube. Again referring to the combination of 5 and 13 into 5,13(8), we check off 5 as having combined; but we do *not* check 13. Although 13 has combined to form a 1-cube of f_β, it has not combined in f_α and is therefore a prime implicant of the product $f_\alpha \cdot f_\beta$.

Subject to these special rules, we complete the determinations of prime implicants as described in Section 7.2. The reader will note that the prime implicants thus formed are precisely those found on the maps in the previous section. We now proceed to the prime implicant table, Fig. 7.27, and complete the design as before.

PROBLEMS

7.1. Minimize the following functions by Quine-McCluskey (S-of-P).

(a) $f(W, V, X, Y, Z) = \sum m(0, 1, 3, 8, 9, 11, 15, 16, 17, 19,$
$24, 25, 29, 30, 31)$

(b) $f(A, B, C, D) = \sum m(0, 1, 4, 5, 6, 7, 8, 9, 10, 12, 14)$

(c) $f(a, b, c, d) = \sum m(0, 2, 3, 7, 8, 10, 11, 12, 14)$

(d) $f(A, B, C, D, E) = \sum m(0, 1, 2, 3, 4, 6, 9, 10, 15, 16, 17, 18, 19,$
$20, 23, 25, 26, 31)$

(e) $f(a, b, c, d, e, f) = \sum m(0, 2, 4, 5, 7, 8, 16, 18, 24, 32, 36,$
$40, 48, 56)$

7.2. Determine minimal P-of-S realizations of the functions of Problem 7.1 by Quine-McCluskey.

7.3. Repeat Problem 6.3 by Quine-McCluskey.

7.4. Repeat Problem 6.4 by Quine-McCluskey.

7.5. Determine minimal S-of-P realizations of the following incompletely specified functions by Quine-McCluskey.

(a) $f(A, B, C, D, E) = \sum m(4, 5, 10, 11, 15, 18, 20, 24, 26, 30, 31)$
$+ d(9, 12, 14, 16, 19, 21, 25)$

(b) $f(A, B, C, D, E) = \sum m(0, 2, 3, 6, 9, 15, 16, 18, 20, 23, 26)$
$+ d(1, 4, 10, 17, 19, 25, 31)$

(c) $f(a, b, c, d, e, f) = \sum m(0, 2, 4, 7, 8, 16, 24, 32, 36, 40, 48)$
$+ d(5, 18, 22, 23, 54, 56)$

(d) $f(A, B, C, D) = \sum m(0, 1, 4, 6, 8, 9, 10, 12) + d(5, 7, 14)$

(e) $f(a, b, c, d) = \sum m(2, 4, 8, 11, 15) + d(1, 10, 12, 13)$

(f) $f(W, X, Y, Z) = \sum m(1, 4, 8, 9, 13, 14, 15) + d(2, 3, 11, 12)$

7.6. Repeat Problem 6.8 by Quine-McCluskey.

7.7. Determine minimal P-of-S realizations of the functions of Problem 7.5 by Quine-McCluskey.

7.8. A circuit receives two 2-bit binary numbers $Y = y_1 y_0$ and $X = x_1 x_0$. The 2-bit output $Z = z_1 z_0$ should equal 11 if $Y = X$, 10 if $Y > X$ and 01 if $Y < X$. Design a minimal sum-of-products realization.

7.9. Five men judge a certain competition. The vote of each is indicated by 1 (Pass) or 0 (Fail) on a signal line. The five signal lines form the input to a logic circuit. The rules of the competition allow only 1 dissenting vote. If the vote is 2–3 or 3–2, the competition must continue. The logic circuit is to have two outputs, xy. If the vote is 4–1 or 5–0 to pass, $xy = 11$. If the vote is 4–1 or 5–0 to fail, $xy = 00$. If the vote is 3–2 or 2–3, $xy = 10$. Design a minimal S-of-P circuit.

7.10. A 5-bit binary number $N = x_4x_3x_2x_1x_0$ appears at the inputs to a combinational logic circuit. The circuit has two outputs.
z_1 indicates the number is evenly divisible by 6.
z_2 indicates the number is evenly divisible by 9.
Design a minimal S-of-P realization.

7.11. A logic circuit has five inputs, x_4, x_3, x_2, x_1, and x_0. Output z_0 is to be 1 when a majority of the inputs are 1. Output z_1 is to be 1 when less than four of the inputs are 1, provided that at least one input is 1. Output z_2 is to be 1 when two, three, or four of the inputs are 1. Design a minimal S-of-P circuit.

7.12. Determine minimal S-of-P forms for the following multiple-output systems.

(a) $f_1(a, b, c) = \sum m(0, 1, 3, 5)$
$f_2(a, b, c) = \sum m(2, 3, 5, 6)$
$f_3(a, b, c) = \sum m(0, 1, 6)$

(b) $f_1(A, B, C, D) = \sum m(0, 1, 2, 3, 6, 7)$
$f_2(A, B, C, D) = \sum m(0, 1, 6, 7, 14, 15)$
$f_3(A, B, C, D) = \sum m(0, 1, 2, 3, 8, 9)$

(c) $f_1(A, B, C, D) = \sum m(4, 5, 10, 11, 12)$
$f_2(A, B, C, D) = \sum m(0, 1, 3, 4, 8, 11)$
$f_3(A, B, C, D) = \sum m(0, 4, 10, 12, 14)$

(d) $f_1(A, B, C, D, E) = \sum m(0, 1, 2, 3, 6, 7, 20, 21, 26, 27, 28)$
$f_2(A, B, C, D, E) = \sum m(0, 1, 6, 7, 14, 15, 16, 17, 19, 20, 24, 27)$
$f_3(A, B, C, D, E) = \sum m(0, 1, 2, 3, 8, 9, 16, 20, 26, 28, 30)$

(e) $f_1(A, B, C, D, E) = \sum m(0, 1, 2, 8, 9, 10, 13, 16, 17, 18, 19, 24, 25)$
$f_2(A, B, C, D, E) = \sum m(0, 1, 3, 5, 7, 9, 13, 16, 17, 22, 23, 30, 31)$
$f_3(A, B, C, D, E) = \sum m(2, 3, 8, 9, 10, 11, 13, 15, 16, 17, 18, 19, 22, 23)$

7.13. In the yard tower of a railroad yard, a controller must select the route of freight cars entering a section of the yard from point A as shown on his control panel, Fig. P7.13. Depending on the positions of the switches, a car can arrive at any one of four destinations. Other cars may enter from points B or C. Design a circuit which will receive as inputs signals S_1 to S_5, indicating the positions of the corresponding switches, and will light a lamp, D_0 to D_4, showing which destination the car from A will reach. For the cases when cars can enter from B or C (S_2 or S_3 in the 0 position), all output lamps should light, indicating that a car from A cannot reach its destination safely.

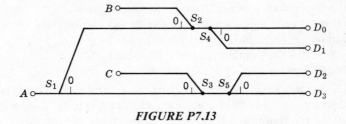

FIGURE P7.13

7.14. Determine minimal S-of-P forms for the following multiple output systems.

(a) $f_1(a, b, c, d) = \sum m(0, 2, 9, 10) + d(1, 8, 13)$
   $f_2(a, b, c, d) = \sum m(1, 3, 5, 13) + d(0, 7, 9)$
   $f_3(a, b, c, d) = \sum m(2, 8, 10, 11, 13) + d(3, 9, 15)$

(b) $f_1(A, B, C, D) = \sum m(2, 3, 6, 10) + d(8)$
   $f_2(A, B, C, D) = \sum m(2, 10, 12, 14) + d(6, 8)$
   $f_3(A, B, C, D) = \sum m(2, 8, 10, 12) + d(0, 14)$

(e) $f_a(W, X, Y, Z) = \sum m(0, 5, 7, 14, 15) + d(1, 6, 9)$
   $f_b(W, X, Y, Z) = \sum m(13, 14, 15) + d(1, 6, 9)$
   $f_c(W, X, Y, Z) = \sum m(0, 1, 5, 7) + d(9, 13, 14)$

(d) $f_1(A, B, C, D, E) = \sum m(2, 8, 10, 12, 18, 26, 28, 30)$
   $\qquad\qquad\qquad\quad + d(0, 14, 22, 24)$
   $f_2(A, B, C, D, E) = \sum m(2, 3, 6, 10, 18, 24, 26, 27, 29)$
   $\qquad\qquad\qquad\quad + d(8, 19, 25, 31)$
   $f_3(A, B, C, D, E) = \sum m(1, 3, 5, 13, 16, 18, 25, 26)$
   $\qquad\qquad\qquad\quad + d(0, 7, 9, 17, 24, 29)$

7.15. Design a minimal second-order circuit to convert binary-coded decimal input to bi-quinary (or 2-out-of-7) output. This will have four inputs and seven outputs, as shown in Fig. P7.15a. The codes for

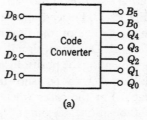

(a)

FIGURE P7.15a

Digit	BCD				Bi-Quinary						
	D_8	D_4	D_2	D_1	B_5	B_0	Q_4	Q_3	Q_2	Q_1	Q_0
0	0	0	0	0	0	1	0	0	0	0	1
1	0	0	0	1	0	1	0	0	0	1	0
2	0	0	1	0	0	1	0	0	1	0	0
3	0	0	1	1	0	1	0	1	0	0	0
4	0	1	0	0	0	1	1	0	0	0	0
5	0	1	0	1	1	0	0	0	0	0	1
6	0	1	1	0	1	0	0	0	0	1	0
7	0	1	1	1	1	0	0	0	1	0	0
8	1	0	0	0	1	0	0	1	0	0	0
9	1	0	0	1	1	0	1	0	0	0	0

(b)

FIGURE P7.15b

input and output for the ten decimal digits are listed in Fig. P7.15b. It may be assumed that the six possible input combinations not listed (corresponding to 10–15) will never occur.

7.16. Repeat Problem 7.15 to convert from the Excess-3 code to the 2-out-of-5 code, as listed in Fig. 8.12, p. 183.

7.17. A certain digital computer uses 6-bit "OP codes," that is, a unique combination of 6 bits is used to specify each of the possible operations of the machine. The OP codes are received by a decoder, which must decide what is to be done and issue control signals accordingly. This decoding is generally done in several steps, with the codes first being classified into basic types. Thus, some operations will require a word to be read from memory, in which case the decoder must issue the signal RO (Read Operand). Other operations require that a word be stored in memory, in which case the decoder will issue the signal WO (Write Operand). Other operation categories are Jump (J), Zero Address (ZA), Iterative (I), and Input/Output (IO). These categories are not mutually exclusive, i.e., an operation may fall into more than one category.

The operations falling into each category are listed in Fig. P7.17. The operations are designated by the octal equivalents of the binary codes. Thus, for example, Operation 17 has the code 001 111. Design

$$\underbrace{001}_{1}\ \underbrace{111}_{7}$$

a minimal S-of-P circuit to produce the control signals, RO, WO, J, ZA, I, IO in response to the 6-bit Op code. Designate the 6 bits of

Category	Signal	Operations
Read Operand	RO	12–17, 24–33, 36–46, 52, 53, 70–73
Write Operand	WO	20–23, 27, 47, 56–61, 70–73, 75, 76
Jump	J	22, 23, 75, 76
Zero Address	ZA	01–11, 34, 35, 50, 51, 54, 55
Iterative	I	24–33
Input/Output	I/O	62–67, 74

FIGURE P7.17

the OP code as $C_5C_4C_3C_2C_1C_0$. You may assume that 00 and 77 are not used.

7.18. In *floating point* operations, numbers are represented in a computer in the binary equivalent of the familiar "scientific notation" of a *mantissa* (or coefficient) multiplied by the radix raised to a power denoted by the *exponent*, e.g., 0.314×10^6, where 0.314 is the mantissa and 6 the exponent. If two floating-point numbers are to be added or subtracted, their exponents must be equal. Thus, if we wish to add 0.6124×10^8 to 0.4042×10^7, we first shift the mantissa of the number with the smaller exponent to the right to equalize the exponent. Here we form $0.6124 \times 10^8 + 0.0404 \times 10^8 = 0.6528 \times 10^8$. Thus, the first step in adding two floating-point numbers is to inspect the exponents to determine which number should be shifted. The number of digits in the mantissa is fixed by the size of the computer registers, so that the least significant digits of the mantissa shifted will be lost. In the example just given, the 2 in the smaller number was lost. If the difference between the exponents is greater than the number of digits in the mantissa, all the digits of the smaller number will be lost and the sum will be equal to the larger number. For example, to four places of accuracy, $0.6124 \times 10^8 + 0.1048 \times 10^3 = 0.6124 \times 10^8$. Thus, if we can detect this condition, we can skip the actual addition of the mantissa and simply transfer the larger operand to the result register, with a resultant saving in time.

The comparison is carried out by subtracting one exponent from the other and inspecting the difference for magnitude and sign. The magnitude is determined by a decoder which puts out a signal, G, if the magnitude of the difference is greater than N, the number of digits in the mantissa. Let the exponents of the operands be A and D, let $A > D$ denote that A is larger than D by an amount no greater than

C	A_s	D_s	E_s	G	$A > D$	$A \gg D$	$D > A$	$D \gg A$
0	0	0	0	0	1	0	0	0
0	0	0	0	1	0	1	0	0
0	0	0	1	0	0	0	1	0
0	0	0	1	1	0	0	0	1
0	0	1	0	0	1	0	0	0
0	0	1	0	1	0	1	0	0
0	0	1	1	0	1	0	0	0
0	0	1	1	1	0	1	0	0
0	1	0	0	0	0	0	0	1
0	1	0	0	1	0	0	0	1
0	1	0	1	0	0	0	0	1
0	1	0	1	1	0	0	0	1
0	1	1	0	0	1	0	0	0
0	1	1	0	1	0	1	0	0
0	1	1	1	0	0	0	1	0
0	1	1	1	1	0	0	0	1
1	0	0	0	0	1	0	0	0
1	0	0	0	1	0	1	0	0
1	0	0	1	0	0	0	1	0
1	0	0	1	1	0	0	0	1
1	0	1	0	0	0	1	0	0
1	0	1	0	1	0	1	0	0
1	0	1	1	0	0	1	0	0
1	0	1	1	1	0	1	0	0
1	1	0	0	0	0	0	1	0
1	1	0	0	1	0	0	0	1
1	1	0	1	0	0	0	1	0
1	1	0	1	1	0	0	0	1
1	1	1	0	0	1	0	0	0
1	1	1	0	1	0	1	0	0
1	1	1	1	0	0	0	1	0
1	1	1	1	1	0	0	0	1

FIGURE P7.18

N, and $A \gg D$ denote that A is larger than D by an amount greater than N. $D > A$ and $D \gg A$ will have similar meaning for D greater than A. Let A_s be the sign bit of A, D_s the sign bit of D, and E_s the sign bit of their difference, and let C be the carry from the most significant bit when the subtraction is carried out. Now it can be shown[8] that the truth table in Fig. P7.18 applies for two's complement arithmetic. Develop a minimal S-of-P circuit to produce two outputs, AG and MG, corresponding to the following conditions:

AG: A is equal to or greater than D, by any amount.

MG: A is larger than D by an amount greater than N, or D is larger than A by an amount greater than N.

Note that $A > D$ also includes $A = D$, so AG includes $A = D$.

BIBLIOGRAPHY

1. Miller, R. E. *Switching Theory*, Vol. I, Wiley, New York, 1965.

2. Quine, W. V. "The Problem of Simplifying Truth Functions," *Am. Math. Monthly*, **59**, No. 8, 521–531 (Oct., 1952).

3. McCluskey, E. J. "Minimization of Boolean Functions," *Bell System Tech. J.*, **35**, No. 5, 1417–1444 (Nov., 1956).

4. ———. *Introduction to the Theory of Switching Circuits*. McGraw-Hill, New York, 1965.

5. Petrick, S. R. "On the Minimization of Boolean Functions," *Proc. Symp. on Switching Theory*, ICIP, Paris, France, June, 1959.

6. Bartee, T. C. "Automatic Design of Logical Networks," *Proc. Western Joint Computer Conf.*, 1959; pp. 103–107.

7. Rhyne, V. T., *Fundamentals of Digital Systems Design*, Prentice-Hall, Englewood Cliffs, N. J., 1973.

8. Flores, I. *The Logic of Computer Arithmetic*, Prentice-Hall, Englewood Cliffs, N.J., 1963.

Chapter 8 Special Realizations and Codes

8.1 Introduction

The material in this chapter will serve as a bridge between the combinational and sequential sections of the book. So far we have approached combinational logic design on the assumption that two-level AND-OR-NOT forms will always provide the best realization. In Sections 8.2–8.4, we consider some important practical types of circuits for which this is often not true. All of the theory and methods so far considered were developed at a time when the only logic elements available were individual gates constructed of discrete components. The development of integrated circuits has made entirely new types of realizations practical and has radically altered the economic constraints on design. Some of the problems and implications of these developments will be treated briefly in Sections 8.5 and 8.6 and extensively in Chapters 14 and 16.

The material in the remaining sections is an introduction to certain techniques which are best realized in sequential circuits. It will thus, hopefully, provide motivation for the study of sequential circuits and will also be a source of examples in succeeding chapters.

8.2   The Binary Adder

In Chapters 6 and 7, only two-level realizations of Boolean functions were considered. This approach always results in the fastest possible circuits and, for typical functions of a few variables, often is the most economical. There are, however, some very important functions which do not lend themselves to two-level realizations.

Let us consider the design of a circuit to accomplish the addition of two binary mumbers. The addition of two n-bit binary numbers is depicted in Fig. 8.1. A two-level realization of this process would require a circuit with $n + 1$ outputs and $2n$ inputs. In a typical computer, n might be 32. Thus, an attempt at two-level design would seem a staggering problem, even for a computer implementation of the tabular method of minimization.

Fortunately; a more natural approach to the problem is possible. Each pair of input digits X_i and Y_i may be treated alike. In each case, the carry, C_i, from the previous position is added to X_i and Y_i to form the sum bit, S_i, and the carry, C_{i+1}, to the next higher-order position. If exactly one or three of the bits X_i, Y_i, and C_i are 1, then $S_i = 1$. If not, then $S_i = 0$. The carry, C_{i+1}, will be 1 if two or more of X_i, Y_i, and C_i are 1. The carry, C_i, into the first digit position, generally 0,* is available as an input at the same time X and Y become available. The addition process may thus be implemented one digit at a time, starting with the least significant digit. As soon as X, Y, and C_1 are available, C_2 may be generated, in turn making possible the generating of C_3, and so on.

A circuit which accepts one bit of each operand and an input carry and produces a sum bit and an output carry is known as a *full adder*. From the above discussion, we can set down the equations and K-maps for the full adder as shown in Fig. 8.2. Two-level realizations of these two functions require a total of nine gates with 25 inputs plus one inverter to generate $\bar{C}_i$. A NAND circuit for this form is shown in Fig. 8.3.

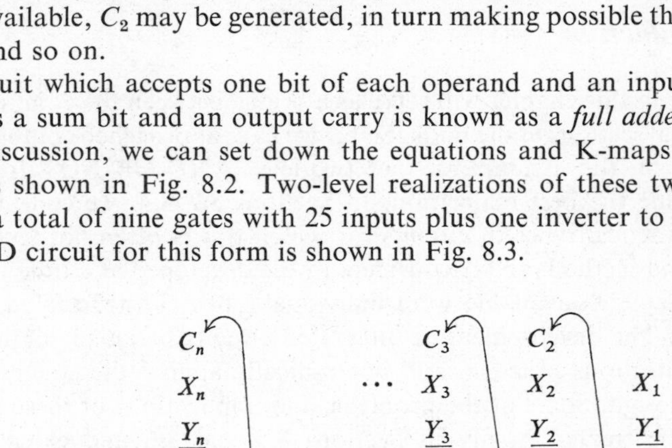

$$
\begin{array}{cccccc}
C_n & & C_3 & C_2 & & \\
X_n & \cdots & X_3 & X_2 & X_1 & \\
Y_n & & Y_3 & Y_2 & Y_1 & \\
S_{n+1} = C_{n+1} \quad S_n & S_{n-1} & S_3 & S_2 & S_1 &
\end{array}
$$

FIGURE 8.1

* In some cases involving complement arithmetic, the input carry is a 1. See, for example, Flores,[1] p. 40.

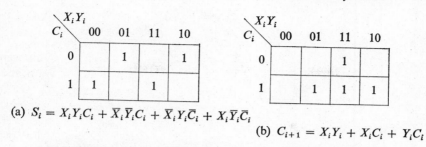

(a) $S_i = X_i Y_i C_i + \overline{X}_i \overline{Y}_i C_i + \overline{X}_i Y_i \overline{C}_i + X_i \overline{Y}_i \overline{C}_i$

(b) $C_{i+1} = X_i Y_i + X_i C_i + Y_i C_i$

FIGURE 8.2.   *Sum and carry functions for the binary full adder.*

Since the full adder is of basic importance in digital computers, a great deal of effort has gone into the problem of producing the most economical realization. One interesting form can be seen on the K-maps as shown in Fig. 8.4. We generate C_{i+1} as before and use it in the generation of S_i. Notice that the map of $X_i + Y_i + C_i$ in Fig. 8.4a agrees with S_i in five of the eight squares. In Fig. 8.4b, we see the intersection of this map and a map

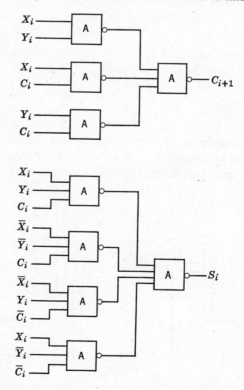

FIGURE 8.3.   *Basic two-level realization of full adder.*

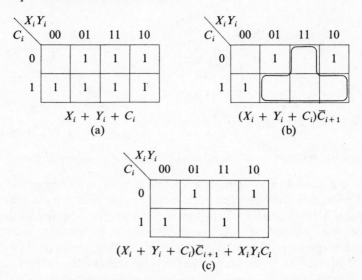

FIGURE 8.4.   *K-maps for alternate realization of sum bit in full adder.*

of $\bar{C}_{i+1}$ (complement of Fig. 8.2b). It is only necessary to include one more term, $X_i Y_i C_i$, to form the sum

$$S_i = \bar{C}_{i+1}(X_i + Y_i + C_i) + X_i Y_i C_i \tag{8.1}$$

as indicated in Fig. 8.4c. A realization of Equation 8.1 is shown in Fig. 8.5a, requiring eight gates with 19 inputs. Application of DeMorgan's laws to Equation 8.1 leads to Equation 8.2 and to a circuit of the same basic form but using different gates (Fig. 8.5b). The choice between the two circuits would

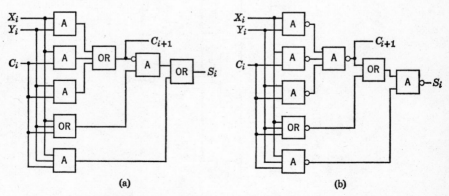

FIGURE 8.5.   *Circuits for alternate realizations of Figure 8.4.*

depend on the exact electronics chosen, particularly on where the necessary inversions could be best implemented.

$$
\begin{aligned}
S_i &= \overline{\overline{C}_{i+1}(X_i + Y_i + C_i)} + X_iY_iC_i \\
&= \overline{C_{i+1} + \overline{(X_i + Y_i + C_i)}} + X_iY_iC_i \\
&= \overline{\overline{C_{i+1} + \overline{(X_i + Y_i + C_i)}}} + X_iY_iC_i \\
&= \overline{[C_{i+1} + \overline{(X_i + Y_i + C_i)}](\overline{X_iY_iC_i})}
\end{aligned}
\tag{8.2}
$$

Still another version of the full adder is shown in Fig. 8.6. This circuit requires nine gates with 18 inputs, but it has the advantage of using identical 2-input gates and not requiring the inverted values of X_i and Y_i.

In the preceding designs, we see examples of the usual compromise between speed and cost. The more economical circuits have more levels of gating and therefore more delay from input to output. However, before concluding that the two-level realization should be chosen if speed is the dominant consideration, the reader should note that carry delay is the limiting factor in a cascade of n full adders forming an n-bit parallel adder (Fig. 8.7). For some combinations of input bits, the presence or absence of the first carry, C_1, will determine the final two bits, S_n and C_{n+1}. In such cases, C_1 must propagate through at least two levels of logic in each adder before C_{n+1} and S_n can assume their final values. The adders of Figs. 8.5 and 8.6 have four and six levels, respectively, of logic from input to sum but only two levels from carry-out. Thus, the carry propagation delay in a parallel adder using either

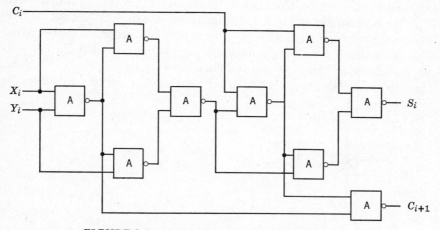

FIGURE 8.6. *Full adder using nine identical gates.*

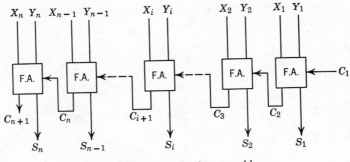

FIGURE 8.7.   n-*bit binary adder.*

of these designs would be the same as for one using the two-level design of Fig. 8.3. The only extra delay would be in the development of the sum bits in the last one or two stages. This would seldom be important, and if it were, the two-level design could be used in the last one or two stages and one of the more economical designs in the other stages.

This case provides a good example of obtaining economy without sacrificing speed by taking advantage of characteristics of the entire system. There are, unfortunately, no simple rules or completely specified procedures for finding such designs. Even with computer mechanization, completely determinate design procedures, such as the Quine-McCluskey method, will always be limited to relatively simple devices or subsystems. The efficient integration of such subsystems into complete large-scale systems is dependent on the ingenuity and resourcefulness of the designer.

As a means of increasing speed, we note that it is not necessary to consider only one digit at a time. Consider the possibility of generating two sum-bits in a step, as illustrated in Fig. 8.8. The intermediate carry, C_{i+1}, is not actually generated but is implicit in the logic of the adder. The process is similar to the addition of three 2-bit binary numbers except that one of those numbers may not exceed 1. Thus, the sum will not exceed 7_{10}, as indicated in Fig. 8.8c. The function S_i may be generated as before, and C_{i+2} will be 1 whenever the

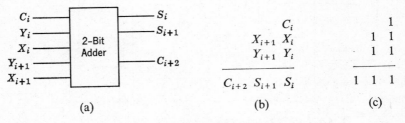

(a)   (b)   (c)

FIGURE 8.8

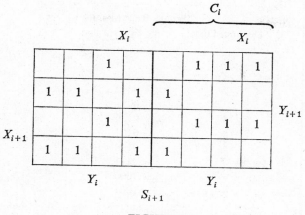

FIGURE 8.9

sum is 4_{10} or more. Therefore,

$$C_{i+2} = X_{i+1}Y_{i+1} + (X_{i+1} + Y_{i+1})(C_iX_i + C_iY_i + X_iY_i) \qquad (8.3)$$

The bit S_{i+1} will be 1 whenever the sum is 2, 3, 6, or 7_{10}, leading to the map in Fig. 8.9. From this map, we determine the minimal second-order form for S_{i+1}, Equation 8.4. Thus, thirteen gates are required for a two-level realization of S_{i+1} compared to five gates for the two-level realization of S_i.

$$\begin{aligned} S_{i+1} = {} & \bar{X}_{i+1}Y_{i+1}\bar{X}_i\bar{C}_i + X_{i+1}\bar{Y}_{i+1}\bar{X}_i\bar{C}_i + \bar{X}_{i+1}\bar{Y}_{i+1}X_iY_i \\ & + X_{i+1}Y_{i+1}X_iY_i + \bar{X}_{i+1}Y_{i+1}\bar{Y}_i\bar{C}_i + X_{i+1}\bar{Y}_{i+1}\bar{Y}_i\bar{C}_i \\ & + \bar{X}_{i+1}Y_{i+1}\bar{X}_i\bar{Y}_i + X_{i+1}\bar{Y}_{i+1}\bar{X}_i\bar{Y}_i + \bar{X}_{i+1}Y_{i+1}X_iC_i \\ & + X_{i+1}Y_{i+1}X_iC_i + \bar{X}_{i+1}\bar{Y}_{i+1}Y_iC_i + X_{i+1}Y_{i+1}Y_iC_i \end{aligned}$$

$$(8.4)$$

For groups of three bits, a similar increase in functional complexity could be expected, and so on. Thus, as conjectured earlier, the cost of a two-level realization of the entire n-bit addition would be prohibitive.

8.3 Coding of Numbers

As discussed briefly in Section 2.5, in some computers, decimal numbers are not converted directly into the corresponding binary numbers but are instead converted on a digit-by-digit basis, with each decimal digit being represented by some binary code. The most commonly used code is the BCD code, in which each digit is represented by the corresponding 4-bit binary number. This code is repeated in Fig. 8.10.

Decimal Digit	Binary Representation $X_3\,X_2\,X_1\,X_0$
0	0000
1	0001
2	0010
3	0011
4	0100
5	0101
6	0110
7	0111
8	1000
9	1001

FIGURE 8.10. *Binary-coded decimal digits.*

Addition in a binary-coded decimal machine is usually accomplished in a digit-by-digit fashion. The basic configuration of a binary-coded decimal adder is shown in Fig. 8.11. The two decimal digits to be added are represented by $X_{i3}X_{i2}X_{i1}X_{i0}$ and $Y_{i3}Y_{i2}Y_{i1}Y_{i0}$. The carry from the previous stage, C_i, may take on only the values 0 or 1 as only two decimal digits are being added. The adder must be a 5-output function of nine variables in order to generate the binary representation of the sum $S_{i3}S_{i2}S_{i1}S_{i0}$, and the carry to the next stage C_{i+1}.

The discussion of the previous section would indicate that a two-level realization of the circuit in Fig. 8.11 would not be simple. A computer implementation of tabular minimization would seem to be a reasonable approach.

It turns out that the sum bits differ only slightly from the output bits of a four-stage binary adder. In the usual realization, the 4-bit binary addition is accomplished first, and the sum and carry bits are then adjusted to their proper values in an additional two levels of logic. For details, the reader is referred to Marcus[2] or Torng.[3]

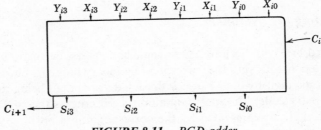

FIGURE 8.11. *BCD adder.*

Decimal Digit	Excess-3 $X_3\ X_2\ X_1\ X_0$				2-out-of-5 $X_4\ X_3\ X_2\ X_1\ X_0$					Gray Code $X_3\ X_2\ X_1\ X_0$			
0	0	0	1	1	0	0	0	1	1	0	0	1	0
1	0	1	0	0	0	0	1	0	1	0	1	1	0
2	0	1	0	1	0	0	1	1	0	0	1	1	1
3	0	1	1	0	0	1	0	0	1	0	1	0	1
4	0	1	1	1	0	1	0	1	0	0	1	0	0
5	1	0	0	0	0	1	1	0	0	1	1	0	0
6	1	0	0	1	1	0	0	0	1	1	1	0	1
7	1	0	1	0	1	0	0	1	0	1	1	1	1
8	1	0	1	1	1	0	1	0	0	1	1	1	0
9	1	1	0	0	1	1	0	0	0	1	0	1	0

FIGURE 8.12

The code of Fig. 8.10 is not the only scheme used for representing decimal digits as sets of binary bits. In Fig. 8.12 is a listing of three other possible codes. The Excess-3 code is particularly useful where it is desired to perform arithmetic by the method of complements. The 9's complement of a decimal digit a is defined as $9-a$. See Ware,[4] for a discussion of 9's complement arithmetic. The 9's complement of a decimal digit expressed in Excess-3 code may be obtained by complementing each bit individually. This fact is easily verified by the reader.

Not all codes of decimal digits use only 4 bits. The 2-out-of-5 code represents each decimal digit by one of the ten possible combinations of two 1's and three 0's. The third code in Fig. 8.12 utilizes the center ten characters of a 4-bit Gray code. The distinguishing feature of the Gray code is that successive coded characters never differ in more than one bit. The basic Gray code configuration is shown in Fig. 8.13a. A 3-bit Gray code may be obtained by merely reflecting the 2-bit code about an axis at the end of the code and assigning a third bit as 0 above the axis and as 1 below the axis. This is illustrated in Fig. 8.13b. By reflecting the 3-bit code, a 4-bit code may be obtained as in Fig. 8.13c, etc.

We have already employed the Gray code in the designation of rows and columns on the Karnaugh map. This code is particularly useful in minimizing errors in analog-digital conversion systems. Consider a shaft-position encoder, such as discussed in Chapter 1, which puts out a digital signal to indicate which of ten equal segments the shaft is in. When the shaft moves from segment seven to segment eight, for example, the code must change from that for seven to that for eight. As the shaft moves across the segment boundary, if more than one bit has to change, it is possible, due to slight

```
0 0        0 0 0      0 0 0 0
0 1        0 0 1      0 0 0 1
1 1        0 1 1      0 0 1 1
1 0        0 1 0      0 0 1 0
--------
           1 1 0      0 1 1 0
           1 1 1      0 1 1 1
           1 0 1      0 1 0 1
           1 0 0      0 1 0 0
           ---------------
                      1 1 0 0
                      1 1 0 1
                      1 1 1 1
                      1 1 1 0
                      1 0 1 0
                      1 0 1 1
                      1 0 0 1
                      1 0 0 0
   (a)       (b)          (c)
```

FIGURE 8.13

mechanical inaccuracies, that not all will change at exactly the same time. If the BCD code were used, the code would have to change from 0111 to 1000, a change of all 4 bits. If the most significant digit were to change from 0 to 1 before any of the other bits changed, we would then momentarily have the code 1111, (15_{10}), and thus a very large error. With the Gray code, since only one bit is to change at a time this sort of error cannot occur. This code may also bring about a reduction in the amount of logic required in certain types of registers and counters.

Codes are also encountered in connection with the control of output devices, or indicators. In recent years, light-emitting diodes (LED's) and liquid crystals have become increasingly popular for use with logic circuits. Their main advantage is that they are directly compatible with logic circuits in contrast to incandescent or neon indicators, which usually require special drivers.

For display of decimal digits, the standard form is the seven-segment display, shown in Fig. 8.14a. Each of the seven segments is a separate LED or crystal which can be turned on or off individually. It should be apparent to the reader that each decimal digit can be formed by lighting some subset of the seven segments. To control this display, we must generate a 7-bit code to indicate whether each segment should be on or off. If we let 0 correspond to OFF and 1 to ON, the *seven-segment code* for the decimal digits will be as shown in Fig. 8.14b. Since the seven-segment code is highly redundant, it

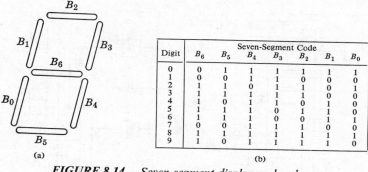

Digit	Seven-Segment Code						
	B_6	B_5	B_4	B_3	B_2	B_1	B_0
0	0	1	1	1	1	1	1
1	0	0	1	1	0	0	0
2	1	1	0	1	1	0	1
3	1	1	1	1	1	0	0
4	1	0	1	1	0	1	0
5	1	1	1	0	1	1	0
6	1	1	1	0	0	1	1
7	0	0	1	1	1	0	0
8	1	1	1	1	1	1	1
9	1	0	1	1	1	1	0

(a)   (b)

FIGURE 8.14. *Seven-segment display and code.*

would be unsuitable for internal use in a computer. Thus, the use of this form of display will require logic to convert from some internal code, such as BCD, to the seven-segment code. Before considering the design of a circuit for this particular application, let us consider the design of another, more general, type of circuit associated with coding and decoding problems.

8.4 The Decoder

The multiple output configuration most often found in digital computers is one of the simplest of such circuits. In addressing core memory, for example, it is necessary to use a circuit with a distinct output line corresponding to each possible combination of values which may be assumed by a set of inputs. This output line will be 1 if and only if the inputs take on the corresponding set of values. The input values will be the address of a word in memory. A 1 on the corresponding output line will make it possible to read the addressed word from memory. Circuits of this type are called *n-to-2^n line decoders* and are often referred to merely as *decoders*. Decoders are also found in many applications not directly involving memory.

A 3-to-2^3 decoder is illustrated in Fig. 8.15a. Notice that the decoding function consists of generating all eight minterms and connecting each to an output line. Often the circuit technology will dictate the use of NOR or NAND gates to implement the decoder. A simple application of DeMorgan's theorem will show that the NOR gate decoder of Fig. 8.15b is equivalent to Fig. 8.15a.*

In theory, the networks of Fig. 8.15 could be extended directly to any number, n, inputs using 2^n gates with n inputs each. In practice, a fan-in limitation will be reached in any technology. Thus, some type of

---

* Many of the 3-2^3 decoders available as single MSI parts replace the AND gates of Fig. 8.15a with NAND gates. In that case, the "addressed" OUTPUT will be O while all other outputs are 1.

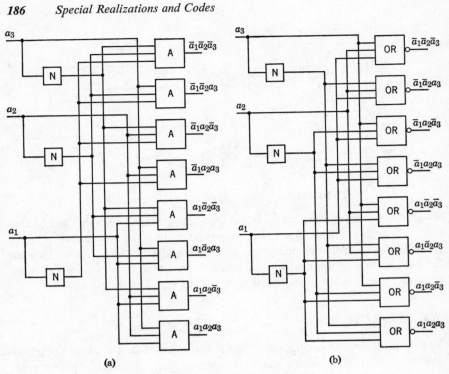

FIGURE 8.15.   *3-to-2^3 line decoders.*

multilevel network must be employed in large decoders. It is impossible to make an unqualified statement as to the most economical decoder design, as cost factors vary with different technologies. If we assume that decoders of the form of Fig. 8.15 are available as single-package integrated circuits, the following procedure will generally lead to the most economical design. Given m, the number of inputs of the largest integrated circuit decoder available, determine the largest integer r satisfying Equation 8.5.

$$r \leq \frac{n}{k} \leq m \qquad k = \text{a power of two} \tag{8.5}$$

If $r = n/k$, we use k r-input decoders to form the minterms of disjoint subsets of r variables. These minterms will then be combined in as many levels of 2-input AND gates as required. For example, suppose $n = 12$ and $m = 5$; then the largest r satisfying Equation 8.5 is given by

$$r = 3 = \tfrac{12}{4} \leq 5 \tag{8.6}$$

So we use four 3-input decoders, as shown in Fig. 8.16.

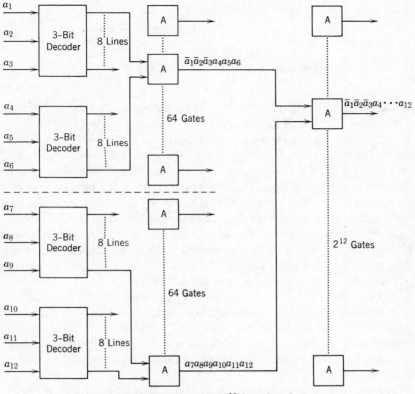

FIGURE 8.16. *12-to-2^{12} line decoders.*

To keep the diagram readable, only a few connections are actually shown. The three-bit decoders may be considered to be copies of either circuit of Fig. 8.15 in integrated circuit form. There are 64 pairs of output lines, one from each of the upper two 3-bit decoders. These pairs form the inputs to the upper 64 second-level AND gates. The outputs of these gates are the 64 possible minterms of the variables a_1, a_2, a_3, a_4, a_5, a_6. The lower 64 second-level gate outputs are the minterms of a_7, a_8, a_9, a_{10}, a_{11}, and a_{12}. The 2^{12} 12-bit minterms are formed by using all possible pairs of outputs of the second level gates (one from the upper 64 and one from the lower 64) to form inputs to the final 2^{12} AND gates.

Assuming the fan-in level permits, it might seem that it would be better to combine the outputs of all four decoders in 4-input AND gates, thus eliminating the second level of gates. Let us assume the cost is proportional to the number of gate inputs. The number of gate inputs, exclusive of the 3-bit decoders, is given by Equation 8.7. Clearly the dominant cost is the output gates. No matter what the form of the decoder, the

$$\text{number of inputs} = \overbrace{2 \cdot 2 \cdot 64}^{\text{2nd level}} + \overbrace{2 \cdot 2^{12}}^{\text{output gates}}$$

$$= 2^8 + 2^{13} \approx 2^{13} \tag{8.7}$$

cost will always be dominated by the 2^n output gates, so these should have the minimum number of inputs—two.

If the 2^n output gates each have two inputs, there will have to be $2 \times 2^{n/2}$ gates at the next level of gating; and by the same reasoning, these should have only two inputs. If we continue this same general reasoning, it would appear that the r-bit decoders should be multi-level, using only 2-input gates. This would be true if the decoder were constructed of discrete gates. However input count is a significant measure of cost primarily because of the cost of wiring to the inputs. Thus, if the decoder of Fig. 8.15 is implemented as an integrated circuit, with all gate connections internal, the effective input count is only three. This is a first example of the manner in which the availability of integrated circuits may influence cost considerations.

If r in Equation 8.5 is not an integer, i.e., if n does not divide evenly by a power of two, there are several possible circuits, the choice depending on the relative cost of the gates themselves compared to the costs of interconnections. In most cases, the best design results from using as many r-bit decoders as necessary, followed by as many levels of 2-input AND gates as required. (See Problem 8.7.)

While fan-in problems limit the size of decoders of the form of Fig. 8.5, fan-out limits can also be a problem in very large decoders. In Fig. 8.16, for example, each of the 128 second-level gates provides an input to 64 of the output gates. Fan-out problems can be relieved by the use of a tree decoder, in which new input variables are introduced at each level of the tree. Figure 8.17 shows a tree decoder for three variables. There are still fan-out problems in that the variables introduced at the later levels must drive many gates. However, it is easier to provide heavy drive capability for a few input variables than for $2 \times 2^{n/2}$ gates at the second level in a circuit of the form of Fig. 8.16. Tree decoders, having more levels than the form of Fig. 8.16, are slower and contain almost twice as many gates, but for very large decoders they may be the only choice. Obviously, various combinations of the two basic forms could be devised.

8.5   Code Conversion and Read-Only Memories

At the close of Section 8.3, we pointed out the need for circuits to convert from one code to another. The decoder may be regarded as a special form of

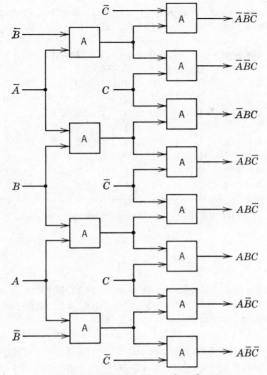

FIGURE 8.17.   *Tree decoder.*

code converter, converting from some *n*-bit code to a one-out-of-2^n code. Further, the decoder can be used as the basis for more general code converters.

Example 8.1

Consider the problem of converting from a BCD code to a seven-segment code, as discussed in Section 8.3. The corresponding codes are tabulated in Fig. 8.18. We require a circuit simultaneously producing seven functions of four inputs. We could apply the formal multiple-output techniques of Chapter 7; but with seven functions to be realized, it is obvious that this would be rather tedious. (There turn out to be 127 maps of functions and function products required.)

A more practical design approach is found by noting that a full decoder produces all minterms of the input variables and that any function can be obtained by an ORing of minterms. In this case, since we are using BCD code, the decimal digits in Fig. 8.18 are the minterm numbers. So we can construct our code converter from a partial decoder, producing the ten required minterms as shown in Fig. 8.19. To complete the design, we note that the table of Fig. 8.18 effectively lists the minterms making up each output function. For example,

$$B_6 = \sum m(2, 3, 4, 5, 6, 8, 9) \tag{8.8}$$

Dec.	BCD Input				Seven-Segment Output						
Digit	A_3	A_2	A_1	A_0	B_6	B_5	B_4	B_3	B_2	B_1	B_0
0	0	0	0	0	0	1	1	1	1	1	1
1	0	0	0	1	0	0	1	1	0	0	0
2	0	0	1	0	1	1	0	1	1	0	1
3	0	0	1	1	1	1	1	1	1	0	0
4	0	1	0	0	1	0	1	1	0	1	0
5	0	1	0	1	1	1	1	0	1	1	0
6	0	1	1	0	1	1	1	0	0	1	1
7	0	1	1	1	0	0	1	1	1	0	0
8	1	0	0	0	1	1	1	1	1	1	1
9	1	0	0	1	1	0	1	1	1	1	0

FIGURE 8.18.   *BCD to seven-segment code.*

and

$$B_0 = \sum m(0, 2, 6, 8) \tag{8.9}$$

so we OR these minterms together as shown. The OR gates developing the other output functions would be similarly connected, although we have not shown the connections in the interest of clarity. ■

It can be seen that we have here the basis of a completely general design technique, not just for code converters but for logic circuits of any type. The resultant designs will seldom be even close to minimal in the classic sense of Chapters 6 and 7, but the simplicity of the design technique makes it quite attractive. The larger the number of output functions, the greater the savings in design effort will be and the closer to minimal the circuits will be. And as the integrated circuit decoders become ever less expensive, this approach becomes economically practical for an ever wider class of circuits.

We can carry this approach further by noting that the circuit form of Fig. 8.19 could be generalized to a collection of 4-input AND gates connected to a set of output OR gates through an interconnection matrix. If the interconnection matrix can be easily altered, we have the basis for a sort of "general purpose" integrated circuit that can be custom-tailored to individual applications. Such a circuit is commonly known as a *Read-Only Memory* (ROM).

There are many types of ROM's, but the type best suited to applications requiring logic level outputs as a function of level inputs is the transistor-coupled, or semiconductor, ROM. The basic structure of a transistor-coupled ROM is shown in Fig. 8.20. The lines making up the interconnection matrix are known as *word lines* and *bit lines*. The word lines are driven by an

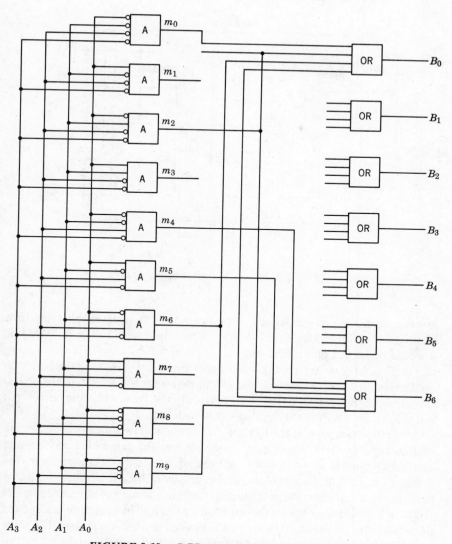

FIGURE 8.19. *BCD to seven-segment converter.*

n-to-2^n decoder, and the outputs are taken from the bit lines. Wherever a connection is to be made between a word line and bit line, a transistor is connected, emitter to the bit line and base to the word line. All collectors are tied to a common supply voltage.

In the absence of any input, all word lines are held at a sufficiently negative level to cut off the transistors so that the bit lines are at 0 volts. When an input appears, the corresponding word line is raised to a sufficiently positive

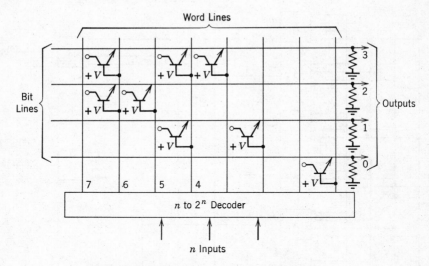

FIGURE 8.20.   *Transistor-coupled ROM.*

level to turn on the transistors, thus raising the connected bit lines to a positive level. For example, in Fig. 8.20, if the input were 100 (m_4), word line 4 would be turned on and the output would be $b_3b_2b_1b_0 = 1010$. Comparing this to Fig. 8.19, we see that the decoder is the array of AND gates, and the bit lines with their associated transistors make up the array of OR gates. (Please note that we have described here only the basic principles involved; the precise details of the technology vary widely.)

As mentioned above, if the ROM is to be practical as a means of providing custom logic circuits, an economical procedure for generating any desired connection matrix is a necessity. A number of such procedures exist, the details of which are quite complex and generally of little interest to the ROM user. In the usual case, the manufacturer offers a variety of standard sizes of ROM's. The customer selects the appropriate size and indicates, on a standard form provided by the manufacturer, the desired connection matrix. If a large number of copies is required, the manufacturer will usually prepare a metalization mask from this data, for which service there will be a fixed charge plus a small per-unit charge for each device delivered. If only a few copies are needed, an electrically alterable ROM may be used. In this type, the device is initially manufactured with transistors connected at every intersection. Where 0's are to be stored, the connections are burned out with large currents. Field alterable ROM's, which can be programmed and sometime reprogrammed by the user (usually with some effort), are becoming increasingly available.

Example 8.2

The BCD-to-Seven-Segment code converter is to be realized using a ROM. The closest standard size is 16 × 8. Specify the connection matrix.

Solution: Figure 8.21 shows a typical form of the type provided by ROM

CUSTOMER: _____	THIS PORTION TO BE COMPLETED BY XY MFG.
P.O. NO.: _____	PART NO.: _____
YOUR PART NO.: _____	S.D. NO.: _____
DATE: _____	DATE RECEIVED: _____

WORD	INPUTS				OUTPUTS							
	A_3	A_2	A_1	A_0	B_7	B_6	B_5	B_4	B_3	B_2	B_1	B_0
0	0	0	0	0	×	○	/	/	/	/	/	/
1	0	0	0	1	×	○	○	/	/	○	○	○
2	0	0	1	0	×	/	/	○	/	/	○	/
3	0	0	1	1	×	/	/	/	/	/	○	○
4	0	1	0	0	×	/	○	/	/	○	/	○
5	0	1	0	1	×	/	/	/	○	/	/	○
6	0	1	1	0	×	/	/	/	○	○	/	/
7	0	1	1	1	×	○	○	/	/	/	○	○
8	1	0	0	0	×	/	/	/	/	/	/	/
9	1	0	0	1	×	/	○	/	/	/	/	○
10	1	0	1	0	×	×	×	×	×	×	×	×
11	1	0	1	1	×	×	×	×	×	×	×	×
12	1	1	0	0	×	×	×	×	×	×	×	×
13	1	1	0	1	×	×	×	×	×	×	×	×
14	1	1	1	0	×	×	×	×	×	×	×	×
15	1	1	1	1	×	×	×	×	×	×	×	×

FIGURE 8.21. *ROM truth table for BCD-to-seven-segment converter.*

manufacturers for the customer to specify the desired connection matrix. All that is necessary is to fill in the truth tables of the desired functions in the output section. In this case, we simply enter functions B_0 to B_6 from Fig. 8.18 as outputs B_0 to B_6, as shown in Fig. 8.21. Since output B_7 and words 10–15 are not used in this case, ×'s (for don't-care) are entered in these squares. (The reader should note that BCD-to-Seven-Segment conversion is so common that standard circuits are available, but this basic approach can be used for any arbitrary codes.) ■

In view of the application of the ROM discussed above as a realization of custom logic in a standard package, the very name, *Read-Only Memory*, may seem inappropriate. However, if we regard the signals appearing on the m output lines not as distinct functions of the n input variables but rather as the bits making up an *m*-bit binary word, then the *memory* interpretation becomes more obvious. The patterns of 0's and 1's stored on each word line by connections to the bit lines now represent individual units, or *words*, of binary information; and the input bits represent the *address* of the word to be read out on the bit lines. The term *read-only* represents the fact that, once the pattern of connections has been established, it either cannot be altered at all or else can be altered only by mechanical means. This is in contrast to a *read-write* memory, such as a core memory, in which the information stored can be altered by electronic processes basically the same as those used to read information out.

Used as a means of information storage, ROM's find wide application in the control units of computers, in which application they store coded instructions representing the sequences of internal operations which the computer must carry out in executing program instructions. The execution of a single instruction, such as ADD, will involve a whole sequence of register transfers, memory accesses, and logical operations. When the appropriate sequences are stored in an ROM, the computer is often said to be *micro-programmed*. ROM's used for this application are generally much larger than those used for implementing logic, typically containing several thousand words of 16–64 bits in length. The semiconductor type used for logic may be prohibitively expensive in large sizes, so many other types, such as capacitor-coupled, diode-coupled, and transformer-coupled ROM's, have been developed. These types are usually slower and less costly than the transistor-coupled ROM. The interested reader will find more information on ROM's and micro-programming in Chapters 3 and 8 of Reference 6.

8.6 NAND and NOR Implementation

While the ROM offers an attractive means of realizing complex logic functions, there will still be many situations where technical or economic constraints dictate the use of discrete gates. As we have seen, logic circuits are usually laid out initially in terms of AND, OR, and NOT gates, regardless of how many levels may be required. However, economic and technical limitations applying to a given logic family may restrict us to inverting gates, i.e., NAND or NOR.

We saw in Chapter 3 that any switching function can be realized exclusively in terms of NAND or NOR; and we further saw, in Chapter 5, that, for two-level forms, the translation from AND-OR to NAND-NAND or OR-AND to NOR-NOR simply requires a direct one-for-one replacement of gates. For multi-level networks, the situation is more complex; and in this section we wish to consider some methods for converting multilevel networks

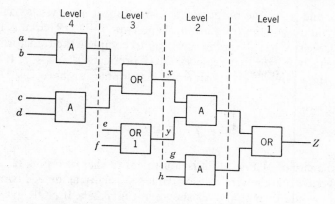

FIGURE 8.22. *Typical multilevel network.*

to NAND and NOR forms. The number of possible design constraints and the number of possible network forms preclude an exhaustive treatment. The best we can do is offer a sampling of methods for handling commonly encountered problems, with the hope that the reader can adapt these methods to fit specific situations.

For purposes of discussion, the levels of gating in a k-level network will be numbered from 1 to k, starting with the output gate, as shown in Fig. 8.22. If the multi-level network consists of alternate levels of AND and OR gates, with variables entering the network only at even levels, then the network is simply a cascade of 2-level circuits, in which all gates may be replaced by NAND for AND-OR, or NOR for OR-AND. Thus, the circuit of Fig. 8.22, except for gate OR-1, is a cascade of two AND-OR circuits so that all gates except OR-1 may be directly replaced by NAND gates. The variables entering OR-1 at the third level pose a special problem, which is dealt with by the following rule.

Rule 8.1. Arrange the network in the form of alternate levels of AND and OR gates with an output OR (AND) gate. Replace all AND and OR gates by NAND (NOR) gates. Complement all inputs which enter the network at odd levels.

The validity of this rule may be justified by reference to the equation for the circuit of Fig. 8.22,

$$z = ((a \cdot b + c \cdot d) \cdot (e + f)) + g \cdot h \qquad (8.10)$$

or

$$z = x \cdot y + g \cdot h \qquad (8.11)$$

where x and y are the output of the two third-level gates, as shown. If we assume validity of the translation of the two AND-OR circuits to NAND-NAND, we see that all that is necessary to complete the transformation is to ensure that the gate that replaces OR-1 produces $y = e + f$. If we replace OR-1 with a NAND gate with $\bar{e}$ and $\bar{f}$ as inputs, we have, by DeMorgan's Law,

$$\overline{\bar{e} \cdot \bar{f}} = \bar{\bar{e}} + \bar{\bar{f}} = e + f$$

Thus, the circuit of Fig. 8.23a is equivalent to that of Fig. 8.22.

If inverted literals were not available as usually assumed, two additional inverters would be required, as shown in Fig. 8.23b. If NOR gates as well as NAND gates were available, an inverter could be saved by using the NOR-NOT combination of Fig. 8.23c.

Not all multilevel networks are conveniently alternating levels of AND's and OR's. Sometimes optimal realizations will include inverters at intermediate points in the network. Sometimes, for reasons of fan-out limitation, a gate will feed into another gate of the same type. The basic procedure for handling such situations is to break the circuit up into subcircuits satisfying Rule 8.1, convert these to NAND or NOR form, and then combine them, inserting inverters where the output of a subcircuit enters the complete circuit at the odd level.

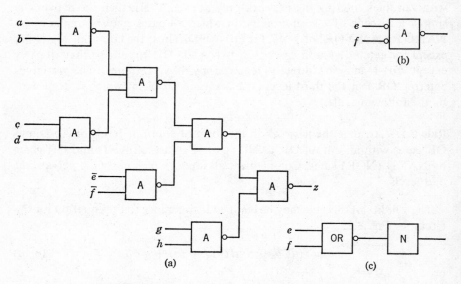

FIGURE 8.23.   *NAND realization of Figure 8.22.*

Example 8.3

Convert the circuit of Fig. 8.24a to NAND form.

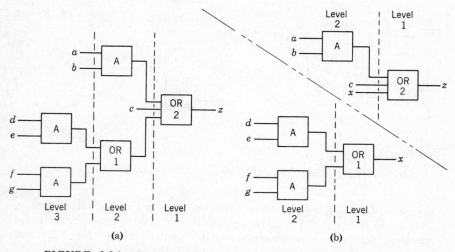

FIGURE 8.24. *Division of circuit into subcircuits, Example 8.3.*

Solution: The original circuit does not conform to Rule 8.1 because of the two successive levels of OR gating, so we divide the circuit into two subcircuits of two-level AND-OR form, as shown in Fig. 8.24b. We then translate these individually to NAND form according to Rule 8.1, as shown in Fig. 8.25a. The

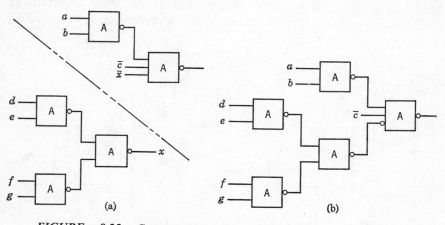

FIGURE 8.25. *Conversion and combination of subcircuits, Example 8.3.*

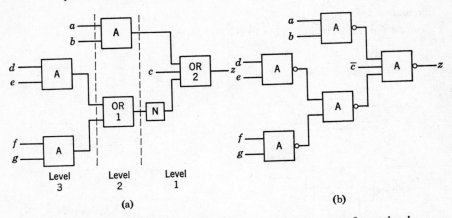

FIGURE 8.26. *Transformation with inversion at output of even level.*

input to OR-2 which will come from OR-1 we consider as an odd-level input, x, which must be inverted in accordance with Rule 8.1. When the subcircuits are combined, an inverter is therefore placed at this input, as shown in Fig. 8.25b. ■

The same general method can sometimes result in the elimination of an inverter. The circuit of Fig. 8.26a is the same as that of Fig. 8.24a, except that now the output of OR-1 is inverted. Breaking the circuit into subcircuits as before, the required inversion at the input to OR-2 cancels the existing inversion, resulting in the NAND equivalent of Fig. 8.26b.

It is important to note that this insertion or elimination of inverters applies only where the subcircuit outputs enter the complete circuit at the odd level. For example, inversion at the input to an even level, as in Fig. 8.27a,

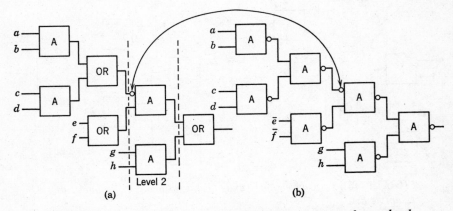

FIGURE 8.27. *Transformation with inversion at input of even level.*

does not affect the validity of the two-level transformation and should be left in place, as in Fig. 8.27b.

The reader should note carefully that we have not considered the question of whether the multilevel circuits to be transformed were minimal. The problem of finding minimal multilevel forms is very difficult and, except for a few very special functions, multilevel forms are seldom significantly simpler than their two-level equivalents. The most common reason for resorting to multilevel forms is to satisfy fan-in or fan-out constraints. Assuming that such constraints are a characteristic of a logic family rather than a particular type of gate, it is reasonable to concentrate, as we have done, on isomorphic transformations, i.e., transformations producing one-to-one. "gate-for-gate," equivalents.

8.7 Parity

We noted above that the multilevel form produces significant cost improvement relative to the two-level form for only a few special functions. In this section, we discuss one of these, the parity function. As we shall see, the parity function is the worst possible function for two-level realization. It consists of 2^{n-1} minterms, none of which combine to form simpler products. Because it does not lend itself to second-order realization, it is most frequently realized either as a sequential circuit or as a cascade of identical combinational circuits. Sequential circuits are treated in the next four chapters. Arrays of identical combinational logic circuits, known as *iterative* circuits, are treated in Chapter 15.

The avoidance of error is of major interest in any data-handling system. Among the major causes of error are component failure and intermittent signal deviation resulting from additive noise. Errors of the first type may be reduced in frequency by using duplicate or redundant circuits. Random errors due to noise occur with greater frequency in some parts of a digital system than in others. Errors are particularly likely where transmission of information from one system or subsystem to another is involved. The longer and noisier the transmission path, the greater the likelihood of error. Errors may also occur in computer memories where information is generally read out in the form of low-level pulses. Noise will generally have less effect in an integral subsystem, such as an arithmetic unit, where information is primarily in the form of levels rather than pulses.

If it were possible to determine that data received from a transmission line or sampled from a memory were erroneous, it might be possible to effect a retransmission or a resampling of the data. It is possible to reduce the probability of an error going undetected. The simplest approach is called a *parity check.* Suppose, for example, that information is to be stored on a magnetic tape in characters of seven binary bits each. Let us add an eight bit to each character in such a way that the number of 1-bits in the character will always be even. We say then that the character is coded with even parity. After a character has been read from the magnetic tape, a parity check is made to see that the number of 1-bits is still even.

Example 8.4

Establish even parity by adding a bit to each of the 7-bit characters 1101001 and 0101111. These even-parity characters are to be transmitted through a medium in which an error occurs once in every 10^4 bits (one out of 10^4 bits is received in complemented form). Determine the probability that an erroneous character will be developed which will go undetected by a parity check at the receiver.

Solution: Determination of the parity bits is immediate.

Data Character							Parity Bit Coded Character							
X_6	X_5	X_4	X_3	X_2	X_1	X_0	X_7	X_6	X_5	X_4	X_3	X_2	X_1	X_0
1	1	0	1	0	0	1	0	1	1	0	1	0	0	1
0	1	0	1	1	1	1	1	0	1	0	1	1	1	1

For the above, or any even-parity character, complementing two bits will result in a character which still has even parity, as will the complementation of any even number of bits. The parity checker at the receiver would be unable to distinguish such characters from valid coded characters. There are

$$_8C_2 = \frac{8!}{6! \cdot 2!}$$

combinations of 2 of the 8 bits. For each of these combinations, the probability that those 2 bits are in error and the remaining 6 are correct is $10^{-8}(1 - 10^{-4})^6$. Proceeding similarly for 4, 6, and 8 bits, we arrive at

$$P(\text{even number of bits in error}) = \frac{8!}{6! \cdot 2!} 10^{-8}(1 - 10^{-4})^6$$

$$+ \frac{8!}{4! \cdot 4!} 10^{-16}(1 - 10^{-4})^4 + \frac{8!}{6! \cdot 2!} 10^{-24}(1 - 10^{-4})^2 + 10^{-32} \quad (8.12)$$

as the expression for the probability of an undetectable erroneous character. Neglecting the obviously insignificant terms yields the close approximation.

$$P \text{ (even number of bits in error)} = 28(10^{-8}) \quad (8.13)$$

This is very significantly less than $7 \cdot 10^{-4}$, which is the approximate probability of a single error in a 7-bit character transmitted without a parity bit. ∎

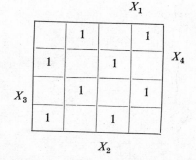

FIGURE 8.28. *4-variable odd parity check function.*

The Boolean function for checking odd parity is the classic example of a function which is not efficiently realized in two levels. The function $f(X_1, X_2, X_3, X_4)$ which is 1 when an odd number of the variables $X_1, X_2, X_3,$ or X_4 are 1 is depicted in Fig. 8.28. An odd-parity check over any number of variables will appear as a "checkerboard" pattern on a Karnaugh map. A two-level parity check over eight variables would require 129 NAND gates.

Let us consider an alternative arrangement. The circuit for detecting odd parity over two bits is the familiar exclusive-OR circuit depicted in Fig. 8.29.

If exactly one of the four variables $X_1, X_2, X_3,$ and X_4 is 1, then exactly one of the outputs, Z_1 or Z_2, must be 1. This must also be true if three of the four X's are 1. If all of the X's are 0 or if all are 1, clearly $Z_1 = Z_2 = 0$. If the two inputs to one gate are 1 while the two inputs to the other gate are 0, both Z_1 and Z_2 must be 0. If one input to each gate is 1, then $Z_1 = Z_2 = 1$. This discussion is summarized in the table of Fig. 8.30b.

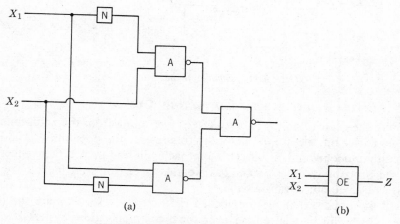

FIGURE 8.29. *Exclusive-OR network.*

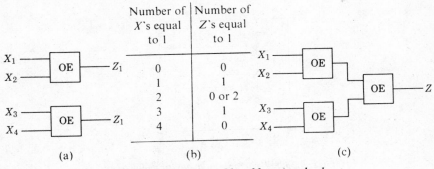

FIGURE 8.30. *4-variable odd parity checker.*

From the table, the number of outputs equal to 1 is odd if and only if an odd number of inputs are 1. Thus, a check of odd parity may be accomplished by adding a third exclusive OR gate, as shown in Fig. 8.30c. Extending the argument to eight variables leads to the circuit of Fig. 8.31, which may be realized by using 21 NAND gates and 14 inverters. We could have arrived at this same result by verifying that $X_1 \oplus X_2$ is an associative operation. This task is left to the reader.

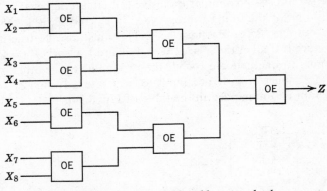

FIGURE 8.31. *8-variable odd parity checker.*

8.8 Error-Detecting-and-Correcting Codes

Noting the decrease in probability of error resulting from the addition of a single parity bit, we are led naturally to consider the possibility of adding more than one redundant bit. This approach will lead to a further reduction in the probability of error and, as we shall see, can even facilitate error correction without retransmission.

Let us suppose that the four coded characters in Fig. 8.32 are the only ones which will be transmitted over some communications system. Notice that each of the four coded characters differs from the other coded characters

```
A   0   0   0   0   0
B   1   1   1   0   0
C   0   0   1   1   1
D   1   1   0   1   1
```

FIGURE 8.32

in at least 3 of the 5 bits. We say that the *minimum distance* of the code is 3. In general, the minimum distance, M, of a code is defined as the minimum number of bits in which any two characters of a code differ.

As there are four distinct 2-bit coded characters, we see that $M = 3$ has been achieved at the price of adding 3 redundant bits. Suppose that character D is transmitted as shown but is received as 11000. Although the last 2 bits are in error, 11000 will not be confused with any of the bits A, B, and C. Indeed, 2 erroneous bits in any character of the code will not distort it into any other character of the code. It should be apparent to the reader that we may generalize our observation to the statement: Errors in 2 or fewer bits may be detected in any minimum-distance-3 code.

Errors in 3 or more bits cannot always be detected in a minimum-distance-3 code. Notice, for example, that errors in the first 3 bits of B in Fig. 8.32 will distort this character into the code for A.

The probability of an error in 2 bits may be sufficiently small in some applications to be disregarded. In this case, it is possible to employ a minimum-distance-3 code as an error-correcting code. Suppose that the character 11000 is received in a communications system employing the code of Fig. 8.32. If it is assumed that no more than 1 bit will be in error, we may deduce that the transmitted character was 11100. Thus, circuitry at the receiving end may be provided to convert 11000 directly to 11100. In general, we may say that a received character with a single-bit error will differ in only 1 bit from the transmitted character but in at least 2 bits from any other character in the code. Thus, circuitry may be provided to correct all single-bit errors without retransmission. The above is true for any minimum-distance-3 code.

Suppose a 2-bit error actually occurs in a character of a minimum-distance-3 code set up to correct single-bit errors. The correction process is still carried out on the erroneous character, but it will be corrected to the wrong coded character. Thus, the 2-bit error will in effect go undetected.

We have discussed above the special case of the general relation

$$M - 1 = C + D, \qquad C \leq D \tag{8.14}$$

where C is the number of erroneous bits which will be corrected and D is the number of errors which will be detected. For a specified minimum distance, M, various combinations of values of C and D will satisfy Equation 8.14. A partial list is given in Fig. 8.33. The single parity-bit code and the 2-out-of-5 code are examples of minimum-distance-2 codes for detecting single-bit errors. As discussed above, a minimum-distance-3 code may be used either for

M	1	2	3	4	5	6
D	0	1	2 1	3 2	4 3 2	5 4 3
C	0	0	0 1	0 1	0 1 2	0 1 2

FIGURE 8.33

detecting 2-bit errors or for detecting and correcting single-bit errors. It is necessary to choose between rejecting all erroneous characters and requiring retransmission or correcting single-bit errors on the spot and passing 2-bit errors. We cannot have the benefit of both modes of operation.

A minimum-distance-4 code may be used for detecting triple errors. Alternatively, it may be assumed that triple errors are sufficiently improbable to be ignored. A 2-bit error will result in a character differing in 2 bits from the correct and in only 2 bits from at least one other character in the code. Thus, 2-bit errors cannot be corrected. Single-bit errors, however, may be corrected. This correction process will alter characters with 3-bit errors to the wrong coded character. Two-bit errors, however, will not be mistaken for single-bit errors. Thus, it is again necessary to choose between the two modes of operation listed in Fig. 8.33.

The solutions of Equation 8.14 for still larger M may be similarly justified.

8.9 Hamming Codes

One of the most convenient minimum-distance-3 codes was devised by R. W. Hamming.[5]

Let us number the bit positions in sequence from left to right. Those positions numbered as a power of 2 are reserved for parity check bits. The remaining bits are information bits.

The 7-bit code is shown in Fig. 8.34, where P_1, P_2, and P_4 indicate parity bits. The bits X_3, X_5, X_6, and X_7 make up the information character to be transmitted. From these bits, the parity bits are determined as follows:

P_1 is selected so as to establish even parity over bits 1, 3, 5, and 7.
P_2 is selected so as to establish even parity over bits 2, 3, 6, and 7.
P_4 is selected so as to establish even parity over bits 4, 5, 6, and 7.

$$
\begin{array}{ccccccc}
1 & 2 & 3 & 4 & 5 & 6 & 7 \\
P_1 & P_2 & X_3 & P_4 & X_5 & X_6 & X_7
\end{array}
$$

FIGURE 8.34

Example 8.5

Determine the Hamming-coded character corresponding to the information character $X_3 X_5 X_6 X_7 = 1010$.

Solution: We may write

$$P_1 = X_3 \oplus X_5 \oplus X_7 \tag{8.15}$$

The above extension of the exclusive-OR notation of Chapter 3 is defined to mean that P_1 will be 1 if there are an odd number of 1's among X_3, X_5, and X_7 and that P_1 will be zero otherwise. Thus

$$P_1 \oplus X_3 \oplus X_5 \oplus X_7 = 0 \tag{8.16}$$

establishing even parity over bits 1, 3, 5, and 7. In this case,

$$P_1 = 1 \oplus (0 \oplus 0) = 1 \oplus 0 = 1 \tag{8.17}$$

Similarly,

$$P_2 = X_3 \oplus X_6 \oplus X_7 = 1 \oplus (1 \oplus 0) = 0$$

and

$$P_4 = X_5 \oplus X_6 \oplus X_7 = 0 \oplus (1 \oplus 0) = 1 \tag{8.18}$$

The final coded character is 1011010. ∎

The correction process at the receiving end is very convenient in the case of a Hamming code. Since we must assume that only 1 bit is in error, it is only necessary to locate that bit. This may be accomplished by checking for odd parity over the same three combinations of bits for which even parity was established at the transmitting end as follows:

$$C_1 = P_1 \oplus X_3 \oplus X_5 \oplus X_7$$
$$C_2 = P_2 \oplus X_3 \oplus X_6 \oplus X_7 \tag{8.19}$$
$$C_4 = P_4 \oplus X_5 \oplus X_6 \oplus X_7$$

If $C_1 = 1$, there must have been an error in one of the 4 bits, 1, 3, 5, and 7, and so on.

The erroneous bit may be determined directly from Fig. 8.35. If all 3 check bits indicate 0 or even parity, then no single bit is in error. This was the situation upon transmission, and we again remind ourselves that we have assumed that there are no 2-bit errors.

C_4	(odd parity over 4 5 6 7)	0	0	0	0	1	1	1	1
C_2	(odd parity over 2 3 6 7)	0	0	1	1	0	0	1	1
C_1	(odd parity over 1 3 5 7)	0	1	0	1	0	1	0	1
Erroneous bit		none	1	2	3	4	5	6	7

FIGURE 8.35

If $C_1 = 1$ but $C_2 = C_4 = 0$, we may conclude that one of the bits 1, 3, 5, and 7 is in error but that 4, 5, 6, and 7 as well as 2, 3, (6, 7) are all correct. Thus, bit 1 must be in error as indicated in the second column of Fig. 8.35. Should $C_4 = C_2 = C_1 = 1$, we may conclude that bit 7 is in error as this is the only bit influencing all three parity checks. The entries at the bottom of remaining columns of Fig. 8.35 may be similarly determined.

Example 8.6

Suppose the character $C_1 C_2 X_3 C_4 X_5 X_6 X_7 = 1101101$ is received in a communications system employing the Hamming code. In this case,

$$C_1 = 1 \oplus 0 \oplus 1 \oplus 1 = 1$$
$$C_2 = 1 \oplus 0 \oplus 0 \oplus 1 = 0$$
$$C_3 = 1 \oplus 1 \oplus 0 \oplus 1 = 1$$

Thus, bit 5 is in error and the correct character is 1101001. ∎

If the 7-bit Hamming code is employed for single-bit detection and correction, the probability of an undetected error is the same as the original probability of error in 2 or more of the 7 bits. This is approximately*

$$P_2 = \frac{7!}{5!\,2!}\,(P)^2(1 - P)^5 \tag{8.20}$$

where P is the probability of an error in any one bit.

By rejecting and calling for the retransmission of any impossible characters, the 7-bit Hamming code may be used for double-error detection. In this case, the probability of an undetected error would be the probability of a 3-or-more-bit error, which is approximately

$$P_3 = \frac{7!}{4!\,3!} = P^3(1 - P)^4 \tag{8.21}$$

If $P = 10^{-4}$, then $P_2 = 20.9(10^{-8})$ and $P_3 = 34.9(10^{-12})$.

Minimum-distance-3 Hamming codes are possible for any $2^n - 1$ bits, where $2^n - n - 1$ of these are information bits. In these cases, all bits numbered as a power of 2 will be parity-check bits. The groups of bits checked for parity continue in the same binary pattern. For example, in the 15-bit code, even parity is achieved over the last 8 bits through P_8. The bit P_4 falls in the group consisting of every other series of 4 bits, beginning at the left, and so on. Each of these codes may be extended to minimum-distance-4 by adding one more bit to provide even parity over the entire character.

---

* This expression neglects the probability that more than 3 bits may be in error, which is considerably smaller still.

Example 8.7

Suppose information is to be transmitted in blocks of several coded characters. Each row of Fig. 8.36, for example, is one character of a five-character block.

$$
\begin{array}{ccccc}
X_{11} & X_{12} & X_{13} & X_{14} & P_{15} \\
X_{21} & X_{22} & X_{23} & X_{24} & P_{25} \\
X_{31} & X_{32} & X_{33} & X_{34} & P_{35} \\
X_{41} & X_{42} & X_{43} & X_{44} & P_{45} \\
P_{51} & P_{52} & P_{53} & P_{54} & P_{55}
\end{array}
$$

FIGURE 8.36

Establishing even parity over the corresponding bits of each character, as well as over the individual characters, results in a minimum distance of two between valid blocks.

Even parity is established over the corresponding bits of the several characters of Fig. 8.36 by $P_{51} \ldots P_{55}$, respectively. The single parity bits of each character are $P_{15} \ldots P_{55}$. Neglecting 2-bit errors, if odd parity is determined for row i and column j of a received block, the bit ij is in error. ■

PROBLEMS

8.1. Design a combinational circuit which will accomplish the multiplication of the 2-bit binary number $X_1 X_0$ by the 2-bit binary number $Y_1 Y_0$. Is a two-level circuit the most economical?

8.2. Verify that the circuit of Fig. 8.6 is a full adder.

8.3. The bits of two binary-coded decimal digits $X_3 X_2 X_1 X_0$ and $Y_3 Y_2 Y_1 Y_0$ serve as the input to a 4-bit binary adder. The output bits of the adder are $Z_4 Z_3 Z_2 Z_1 Z_0$. Let the bits $S_3 S_2 S_1 S_0$ be the binary-coded decimal representation of the least significant bit of the decimal sum. Let $S_4 = 1$ if the most significant bit of the sum is 1, and let $S_4 = 0$ otherwise. Determine a minimal multiple-output realization of S_4, S_3, S_2, S_1, and S_0 as functions of Z_4, Z_3, Z_2, Z_1, and Z_0.

8.4. Design a circuit to convert from the 2-out-of-5 code of Fig. 8.12 to the seven-segment code, using discrete gates.
(a) Assume only the codes shown in Fig. 8.12 will occur.
(b) Include provision for detection of illegal codes.

8.5. Develop an expression for the number of gate inputs required in a tree type decoder. Assume two input gates and assume one variable is introduced at each succeeding level of the tree.

8.6. Assume a 4-input decoder is required. There is a choice between realizing it from discrete gates in the general form of Fig. 8.15 or using a complete decoder in a single integrated circuit package. The cost of the decoder is $5.32. Dual 4-input AND gates (two 4-input gates/package) are available at $0.48 per package. The mounting and wiring costs are $0.50 per package plus $0.05 per input or output connected. Determine the costs of the two methods of realizing the decoder.

8.7. A 12-input decoder is to be realized with a structure similar to that of Fig. 8.16, except that the only decoders available are 4-input decoders at $5.32 each. These decoders could be used in the circuit of Fig. 8.16, leaving one input unused; or three decoders could be used, requiring a different, and possibly more complex, arrangement of AND gates. Available gates that might be useful are Dual 4-input AND at $0.48/package, triple 3-input AND, and quad 2-input AND, both at $0.63/package. Mounting and wiring costs are as given in Problem 8.6. Determine the most economical design.

8.8. Repeat Problem 8.4 using a 32×8 ROM. Use a form similar to Fig. 8.21.

8.9. The network of Fig. P8.9 has been designed for comparing the magnitudes of two 2-bit binary numbers, a_2, a_1 and b_2, b_1. If $z = 1$ and $y = 0$, a_2, a_1 is larger. If $z = 0$ and $y = 1$, b_2, b_1 is larger. If

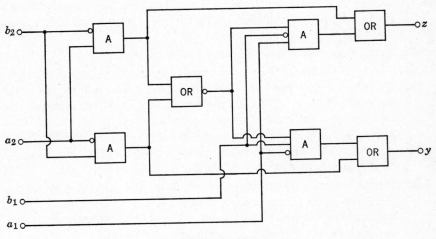

FIGURE P8.9

$z_2 = z_1 = 0$, the two numbers are equal. Translate the network into a form which uses only NAND gates. Do not attempt to minimize the network.

8.10. Determine NAND equivalents of the circuits of Fig. P8.10.

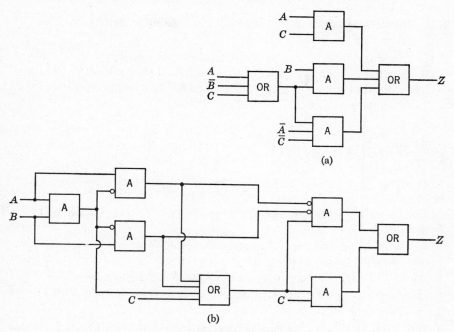

(a)

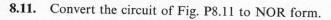

(b)

FIGURE P8.10

8.11. Convert the circuit of Fig. P8.11 to NOR form.

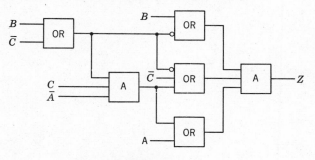

FIGURE P8.11

8.12. Determine a circuit which checks for odd parity over 7 bits using exclusive-OR ($\oplus$) gates.

8.13. Show that $X_1 \equiv X_2$ and $X_1 \oplus X_2$ are associative operations.

8.14. Encode the information characters 0 1 1 0 1 1 1 0 1 0 1 according to the 15-bit Hamming code.

8.15. Determine which bit if any, is in error in the Hamming-coded character 1 1 0 0 1 1 1.

8.16. Determine the exact versions of the approximate expressions given by Equations 8.20 and 8.21.

8.17. Suppose the 5×5 cross-parity configuration of Fig. 8.36 is used for single-bit error detection and correction. If the probability that any 1 bit will be in error is 10^{-8}, determine the approximate probability of an undetected error.

BIBLIOGRAPHY

1. Flores, I. *The Logic of Computer Arithmetic*, Prentice-Hall, Englewood Cliffs, N.J., 1963.

2. Marcus, M. P. *Switching Circuits for Engineer*, Prentice-Hall, Englewood Cliffs, N.J., 1962.

3. Torng, H. C. *Introduction to the Logical Design of Switching Systems*, Addison-Wesley, Reading, Mass., 1964.

4. Ware, W. H. *Digital Computer Technology and Design*, Wiley, New York, 1963.

5. Hamming, R. W. "Error Detecting and Error Correcting Codes," *BSTJ*, **29**: 2, 147–160 (April, 1950).

6. Hill, F. J. and G. R. Peterson. *Digital Systems: Hardware Organization and Design*, Wiley, New York, 1973.

Chapter 9 *Introduction to Sequential Circuits*

9.1 General Characteristics

Our attention up to this point has been directed to the design of *combinational* logic circuits. Consider the general model of a switching system in Fig. 9.1. Here we have n input, or excitation, variables, $x_k(t)$, $k = 1, 2, \ldots, n$ and p output, or response, variables, $Z_i(t)$, $i = 1, 2, \ldots, p$, all assumed to be functions of time. If at any particular time the present value of the outputs is determined solely by the present value of the inputs, we say that the system is *combinational*. Such a system can be totally described by a set of equations of the form

$$Z_i = F_i(x_1, x_2, \ldots, x_n) \tag{9.1}$$

where the dependence on time need not be explicitly indicated, since it is understood that the values of all the variables will be those at some single time. If, on the other hand, the present value of the outputs is dependent not only on the present value of the inputs but also on the past history of the system, then we say that it is a *sequential system*.

As an example to clarify the distinction, consider two types of combination*

---

* The word *combination* as used to identify the type of lock has no connection with the technical word *combinational* under discussion here.

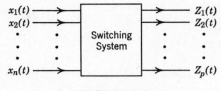

FIGURE 9.1

locks (Fig. 9.2). One type, the conventional combination padlock (Fig. 9.2a), has only one dial. The second type, sometimes used in luggage, has several dials (Fig. 9.2b). The inputs are the dial settings, the outputs the condition of the locks, open or closed. For the first type, the present condition of the lock (open or closed) depends not only on the *present* setting of the dial but also on the manner in which the dial has previously been manipulated. The second type will open whenever the three dials are set to the correct numbers; it does not matter what the previous settings were or in what order the dials were set. Obviously, then, the single-dial lock is a sequential device, the multiple-dial lock a combinational device.

The telephone system is another example of a sequential system. Suppose you have dialed the first six digits of a telephone number and are now dialing the last digit. This seventh digit is the present input to the system, and the output is a signal causing you to be connected to the desired party. Obviously, the present input is not the only factor determining which number you reach, since the other six digits are equally important.

Digital computers are very important examples of sequential systems. There are many levels of sequential operation in a digital computer involving

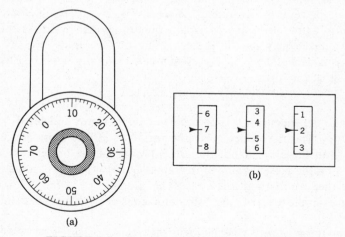

FIGURE 9.2.   *Two types of combination lock.*

sequential use of subsystems which are themselves sequential. Consider the addition of a number into the accumulator of a digital computer; the resultant sum depends both on the number and on what has been put into the accumulator previously.

9.2 Flip-Flops

Clearly evident in the sequential devices and systems discussed above is the requirement to store information about previous events. In the combination lock, information about the previous dial settings is stored in the mechanical position of certain parts. In the dial telephone system, each number is stored as it is dialed in relays or electronic devices. In a computer, there are many different functions which involve the storage of information.

There are many types of storage, or memory, devices. In electronic sequential circuits or systems, the most common memory device is the flip-flop.* Figure 9.3 shows the circuit for a flip-flop constructed from two NOR gates and the timing diagram for a typical operation sequence. We have also repeated the truth table for NOR for convenience in explaining the operation.

At the start, both inputs are at 0, the Y output is at 0, and the X output at 1. Since the outputs are fed back to the inputs of the gates, we must check to

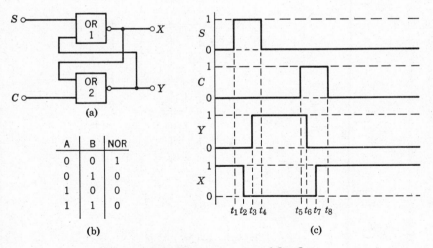

FIGURE 9.3. *Operation of flip-flop.*

---

* The more formal name is *bistable multivibrator*. The origin of the name flip-flop is uncertain, but it has become standard.

see that the assumed conditions are consistent. Gate 2 has inputs of $C = 0$ and $X = 1$, giving an output $Y = 0$, which checks. Similarly, at gate 1, we have $S = 0$ and $Y = 0$, giving $X = 1$. At time t_1, input S goes to 1. The inputs of gate 1 are thus changed from 00 to 01. After a delay (as discussed in Chapter 5), X changes from 1 to 0 at time t_2. This changes the inputs of gate 2 from 01 to 00, so Y changes from 0 to 1 at t_3. This changes the inputs of gate 1 from 01 to 11, but this has no effect on the outputs. Similarly, the change of S from 0 at t_4 has no effect. When C goes to 1, Y goes to 0, driving X to 1, thus "locking-in" Y so that the return of C to 0 has no further effect.

A flip-flop operating in the above fashion is known as a Set-Clear or S-C flip-flop* and has the block diagram symbol in Fig. 9.4a.

The S and C inputs correspond to those in Fig. 9.3, and the Q and $\bar{Q}$ outputs correspond to the Y and X outputs, respectively. A pulse on the S input will "Set" the flip-flop—that is, drive the Q output to the 1 level and the $\bar{Q}$ output to the 0 level. A pulse on the C line will "Clear" (Reset) the flip-flop—that is, drive the Q output to the 0 level and the $\bar{Q}$ output to the 1 level. This behavior is summarized in tabular form in Fig. 9.4c. S^v, C^v, and Q^v indicate the values

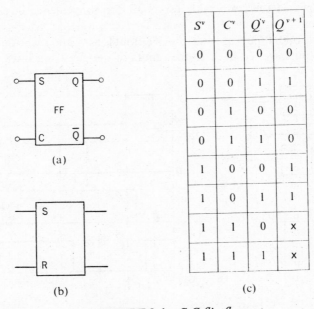

S^v	C^v	Q^v	Q^{v+1}
0	0	0	0
0	0	1	1
0	1	0	0
0	1	1	0
1	0	0	1
1	0	1	1
1	1	0	x
1	1	1	x

(a)

(b)

(c)

FIGURE 9.4. *S-C flip-flop.*

* The symbol shown in Fig. 9.4a is that specified by the 1962 standards (see p. 32). In the 1973 standards the name has been changed to RS (Reset-Set) and the symbol to that shown in Fig. 9.4b. The older standard is followed in this book.

of the inputs and output at some arbitrary time t_v, and Q^{v+1} indicates the value to which the output will go as a result of an input at t_v. The reader should satisfy himself that this table can be represented by the equation

$$Q^{v+1} = S^v + Q^v \bar{C}^v \qquad \text{(where } S^v \cdot C^v = 0) \qquad (9.1)$$

The don't-care entries for Q^{v+1} in the last two rows reflect the fact that in normal operation both inputs should not be permitted to be 1 at the same time. There are two reasons for this restriction. First, if both inputs are 1, both outputs will be driven to 0, which violates the basic definition of flip-flop operation, which requires that the outputs should always be the complements of each other. Second, if both inputs are 1 and then go to 0 at the same time, both NOR gates will see 00 at the inputs and both outputs will try to go to 1. However, because of the feedback, it is impossible for both outputs to be 1 at the same time (the reader should convince himself of this), with the result that the flip-flop will switch unpredictably and may even go into oscillation. With this restriction on the inputs, the S-C flip-flop functions reliably as a memory device in which the state of the outputs indicates which of the inputs was last at the 1 level.

A Set or Clear input must remain 1 for some minimum duration if a change in flip-flop state is to be accomplished reliably. To illustrate, consider Fig. 9.3 with Y initially 0 and X initially 1. If S returned to 0 before time t_3, both inputs to OR-1 would again be 0 at least momentarily, tending to cause X to return to 1. In this situation, the operation of the circuit would be unpredictable. The minimum input pulse duration is the *first of three timing restrictions* which we shall place on sequential circuits containing flip-flops.

As we shall see in Chapter 10, the design of a sequential circuit will be greatly simplified if state changes can take place only at periodically spaced points in time. This can be assured if we synchronize all state changes with pulses from an electronic clock. A *pulse* is simply a signal which normally remains at one level (usually 0) and goes to the other level only for a very short duration. By contrast, a *level* signal is one which may remain at either the 1 or 0 level for indefinite periods of time and which changes values only at intervals long compared to the pulse duration. Just what is meant by "very short" depends on the speed of the circuits but is normally about the same as the delay time of the flip-flops. A typical clock waveform is illustrated in Fig. 9.5a. *A sequential circuit synchronized with this clock will change state only on the occurrence of a clock pulse and will change state no more than once for each clock pulse.* Under these conditions, the circuit is said to be operating in the *clock mode*. Figure 9.5b illustrates how a conventional (nonclocked) S-C flip-flop may be converted to a clocked S-C flip-flop. As shown, the S and C inputs are now AND-gated with the clock (trigger) pulse before being applied to the inputs of the flip-flop proper. The only

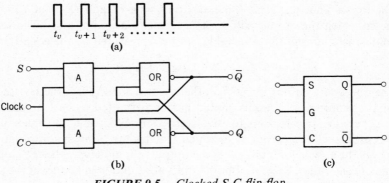

FIGURE 9.5. *Clocked S-C flip-flop.*

restriction on the timing of S and C signals is that they must not change during the duration of clock pulse. As we shall see, this restriction is not as difficult to meet as it might seem. When a clock pulse arrives, it is gated through to the appropriate input of the flip-flop section in accordance with the current values of S and C. The truth table and equation are the same as for the unlocked flip-flop, except that now S^v, C^v, and Q^v refer to the values during the clock pulse duration and Q^{v+1} refers to the value of the output after the clock pulse has ended. Although the AND gating may be external, the flip-flop and clock gating are generally included in the same integrated circuit package. The complete network is symbolized as shown in Fig. 9.5c, with the clock, or trigger, input denoted by G.*

While the clocked S-C is perfectly practical in a theoretical sense, it is rarely encountered in practice, having been almost totally supplanted by the clocked

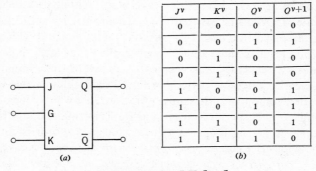

J^v	K^v	Q^v	Q^{v+1}
0	0	0	0
0	0	1	1
0	1	0	0
0	1	1	0
1	0	0	1
1	0	1	1
1	1	0	1
1	1	1	0

(a)   *(b)*

FIGURE 9.6. *J-K flip-flop.*

* For all flip-flops other than S-C, the input notation is consistent with the 1973 standards, but the older output notation (Q and $\bar{Q}$) is more descriptive and is, therefore, retained. Also note that notation for the clock input may be inconsistent, with some manufacturers using C, T, CP, *Clock*, and *Pulse*.

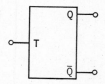

FIGURE 9.7. *T flip-flop.*

J-K flip-flop. In the J-K flip-flop, the restriction that both inputs may not be 1 at the same time, which may be inconvenient for the designer, is removed. The block diagram symbol and truth table are shown in Fig. 9.6. The reader will note that the *J-K* is the same as the S-C, with *J* corresponding to *S* and *K* to *C*, except that both inputs may be 1 at the time of the clock pulse, in which case the flip-flop changes state. The equation for this flip-flop is

$$Q^{v+1} = \bar{K}^v Q^v + J^v \bar{Q}^v \tag{9.2}$$

Note that *J-K* flip-flops are always clocked.

Another type is the *T* flip-flop (Fig. 9.7), in which the trigger (clock) input is the only input. Whenever the *T* input is pulsed, the flip-flop changes state, so that the equation is simply

$$Q^{v+1} = \bar{Q}^v \tag{9.3}$$

Note that a *J-K* flip-flop becomes a *T* flip-flop if both the *J* and *K* inputs are permanently connected to the 1-level.

A fourth type of flip-flop is the *D* flip-flop (Fig. 9.8). For this flip-flop, the output after the clock pulse is equal to the *D* input at clock time, so that the equation is

$$Q^{v+1} = D^v \tag{9.4}$$

Note that a *J-K* flip-flop will function as a *D* flip-flop when connected as shown in Fig. 9.8b.

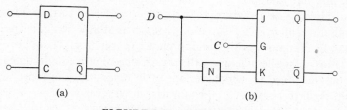

(a)   (b)

FIGURE 9.8. *D flip-flop.*

9.3   *Why Sequential Circuits?*

We have seen some examples of systems that are sequential, and it is obvious that many existing systems behave in a similar manner. As designers, however, our job is not merely to analyze but rather to choose the best possible approach to a synthesis. Given a certain logical task to perform, we are faced with a specific decision with regard to implementation: Should we use a

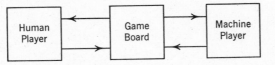

FIGURE 9.9.   *Model of tic-tac-toe game.*

combinational or sequential circuit? Now that we know how information can be stored in sequential circuit, let us consider some of the factors that should be considered in making this decision.

Let us consider the design of a tic-tac-toe machine. The overall "game system" may be represented as shown in Fig. 9.9. The game board might consist of switches by which the human player could indicate his moves and lights to indicate the present state of the game. The machine player will consist of some sort of logic system to decide what moves should be made in response to the moves of the human player.

Clearly, the overall game progresses sequentially, but this does not mean that the machine player has to be a sequential device. It could be a combinational circuit, receiving from the game board information as to the state of the board at any given time and deciding on a move solely on the basis of that current information. There are nine squares on the board and three possible conditions (blank, 0, X) for each square, so there are $3^9 = 19,683$ possible game situations. Thus, the combinational approach involves a circuit to implement the following logical requirement: Given any one of 3^9 possible input combinations, choose the proper one of nine possible responses. It is fairly evident that satisfaction of this requirement is going to involve a massive combinational logic circuit.

As an alternative approach, let us provide the machine player with some means of altering its structure as the game progresses to take advantage of the fact that not all of the 3^9 situations can occur at any particular point in the game. Let us assume that the human player moves first. There are only nine possible moves, so the machine will start in a configuration designed to select the proper response to whichever move is made. Furthermore, when the machine selects its response, it then knows which seven responses are available to the human on the next move, and it can alter its circuitry to take this into account. A similar sequence of events is repeated at each successive move.

In this sequential approach, we are simply breaking a large problem into a number of smaller problems. Rather than building one large circuit capable of dealing with all the situations which might occur at any point in the game, we build a number of smaller circuits,* each capable of dealing with the

---

* As will become evident in succeeding chapters, these circuits will usually not be completely independent.

limited number of situations possible at some particular point in the game. The machine player then sequences from one smaller circuit to another as the game progresses.

In this case, the problem is inherently sequential, since it involves a continuous interchange of information between the two players. Interchanges closely resembling a question-and-answer format are common between individual units of a digital system. Such inherently sequential processes lead naturally to a sequential design approach, which is generally more economical than a purely combinational approach. The methods to be developed in the following chapters will enable the reader, should he be so inclined, to design his own tic-tac-toe machine and check the statement that a sequential design is more economical.

As another example, consider an 8-bit odd-parity checker. If all 8 bits are available simultaneously, the desired function can be realized by a six-level combinational circuit described by the equation

$$Z = [(x_1 \oplus x_2) \oplus (x_3 \oplus x_4)] \oplus [(x_5 \oplus x_6) \oplus (x_7 \oplus x_8)] \qquad (9.5)$$

requiring 21 gates in a NAND realization. It will be recalled that in designing the similar even-parity checker of Fig. 8.15 we broke it down into parts, considering each pair of digits, then each pair of pairs, etc.

Next, let us assume that the 8-bit word is a unit of data which must be transmitted over a long distance—in an airline reservation system, for example. We could use eight lines, one for each bit, but this would be expensive. Instead, let us use a single line and send the 8 bits in sequence, x_1 at time t_1, x_2 at t_2, etc. On a second line, we send *clock*, or synchronizing, pulses to mark these times t_1, t_2, etc.* These two lines will provide the inputs to a circuit which is to put out a signal if the number of 1's in each group of 8 bits is not even (Fig. 9.10).

Each bit could be stored as it arrived until all 8 were present. They could then be applied to the six-level combinational circuit; but this would be very

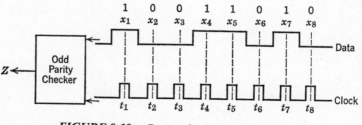

FIGURE 9.10. *Parity checker for serial data.*

---

* The synchronizing line would also be needed with eight-line approach to indicate when a data word was present.

expensive, requiring eight flip-flops for storage plus the 21-gate network. Instead, we note that the exclusive-OR is associative, and rewrite Equation 9.5 in the form

$$Z = \left\{ \left[\left[\left\{ \left[\left[(x_1 \oplus x_2) \oplus x_3 \right] \oplus x_4 \right\} \oplus x_5 \right] \oplus x_6 \right] \oplus x_7 \right] \oplus x_8 \right\} \quad (9.6)$$

Functionally, this means that first 2 bits are checked for parity, then the third bit is checked against the results of the first check, etc. (The reader should try this method on a specific example to convince himself that it works.)

We do not suggest a combinational realization of Equation 9.6, which would be very inefficient. Instead, the first 2 bits are checked in an exclusive-OR network and the results stored until the third bit arrives. Then the third bit is compared to the stored results *in the same exclusive-OR* network. This result is stored in turn until the fourth bit arrives, etc. Figure 9.11 shows a simple circuit for realizing this method.

Initially, the flip-flop must be cleared to $Z = 0$ before x_1 arrives so that the first operation is

$$x_1 \oplus Z = x_1 \oplus 0 = x_1$$

This is stored in the flip-flop when a clock pulse arrives at time t_1. That is, if $x_1 + z = 1$ when the clock pulse arrives, a pulse will be gated to the S input setting the flip-flop to 1. Similarly, the flip-flop will be set to 0 if $x_1 + z = 0$ when the clock pulse arrives. Then x_2 arrives, and the exclusive-OR forms

$$x_2 \oplus Z = x_2 \oplus x_1$$

which is stored in the flip-flop by the clock pulse at t_2. This procedure continues until, after the clock pulse at t_8, the output Z is the desired parity

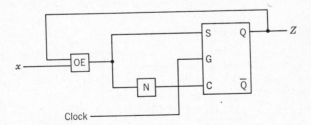

FIGURE 9.11.   *Sequential parity checker.**

---

* An apparently even simpler sequential parity checker may be designed by using a trigger flip-flop. In effect, the exclusive-OR gate is inherent in the electronics of that type of flip-flop.

check, i.e., $Z = 1$ only if there has been an odd number of 1's in the 8 data bits. In fact, the output Z provides an odd-parity check on whatever number of data bits have been received since the flip-flop was last cleared. Thus, Fig. 9.11 is a general n-bit odd-parity checker.

The tic-tac-toe machine and the serial parity checker are examples of the type of situation in which the sequential approach is most often preferred. In neither case is there any speed advantage to using a combinational approach. In the tic-tac-toe machine, the speed of the game is dictated by the speed with which the human player makes his moves. The sequential parity checker is slower than the combinational circuit if we measure the speed of the combinational circuit *from the time that all 8 bits are available*; but if the bits are going to arrive in sequence anyway, there is no difference in speed. The sequential circuit, however, is simpler, and therefore cheaper, than the combinational circuit.

Data always arrive sequentially at some level or another, usually for reasons of cost or hardware limitations rather than logical necessity. This is particularly true for data modulated on a carrier and transmitted over a long distance. Disc and drum memories generally put out data 1 bit at a time, while data from magnetic or paper tape generally arrives one *character* at a time (i.e., 6 or 8 bits at a time). Data from punched cards may arrive one character at a time or eighty characters at a time. The output of a core memory will arrive in groups (words) of anywhere from 12 to 100 bits at a time. Whatever the size of the data groupings, there will certainly be no speed disadvantage to processing the data as it arrives, and there may be a definite cost advantage.

9.4 *Shift Registers and Counters*

In Fig. 9.11, we saw a sequential circuit employing a single flip-flop. As we might expect, most sequential circuit applications require the storage of more than one binary bit of information. In general, the information storage capability of a sequential circuit is reflected by the number of flip-flops in the circuit. Thus, a typical sequential circuit will contain several flip-flops.

In the next four chapters, we shall consider in detail the design of arbitrary sequential circuits. At this point, however, let us introduce in general terms two particular configurations of flip-flops which are used in a significant percentage of sequential-circuit applications. In shift registers and counters, the one-to-one correspondence between number of flip-flops and information storage capability is particularly clear.

In a digital computer, instructions are commonly stored in sequential locations in memory. Thus, if the first instruction were in location 501, for

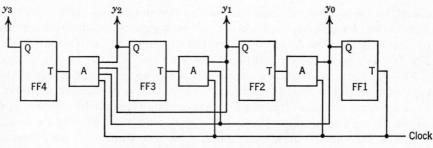

FIGURE 9.12. *Modulo-16 counter.*

example, the second instruction would be in 502, the third in 503, etc. To control the sequencing of the program, the computer uses an *instruction counter*. At the start of the program, the contents of the counter are set to the address of the first instruction. As each instruction is executed, the counter is incremented by one, thus changing the address to that of the next instruction. This is just one example of the use of counters in digital systems.

Although counters may be designed by using any kind of flip-flop, the functioning of a trigger flip-flop counter is most quickly understood. A modulo-16 counter is shown in Fig. 9.12. By modulo-16 we mean that the counter counts from 0 through 15 and then back to 0. In Fig. 9.12, y_0 represents the least significant bit.

Suppose that initially the count is 0, that is, $y_3 y_2 y_1 y_0 = 0000$. With the arrival of the first input pulse, flip-flop 1 will change state. As no pulses will appear at the output of the AND gates, the other three flip-flops will remain in the 0-state. Thus, the count will have advanced to $y_3 y_2 y_1 y_0 = 0001$. With y_0 now 1, the second input pulse will propagate through the first AND gate, triggering y_1 to 1 while at the same time triggering y_0 to 0. Thus, the count will advance to $y_3 y_2 y_1 y_0 = 0010$. In general, a flip-flop will change state at the time of an input pulse if all less significant flip-flops are in the 1 state, so that the count progresses as shown in Fig. 9.13.

Most counters will sometimes have to be reset to zero or set to some specified starting value. Circuitry for this purpose will generally be superimposed on that shown in Fig. 9.12.

In addition to counting, a counter also *stores* the current count in the intervals between input pulses. In general, any set of flip-flops used to store related bits of information is known as a *register*. A counter may thus be referred to as a *counting register*. Often a register is used to assemble and store information arriving from a serial source. The circuit shown in Fig. 9.14, called a *shift register*, is set up to accomplish this task. It is assumed that a new bit of information may be sampled from line X with the arrival of each clock pulse. If the first bit is 1, flip-flop 1 will be set to 1 by the

Count Before Clock Pulse $y_3 y_2 y_1 y_0$				Count After Clock Pulse $y_3 y_2 y_1 y_0$			
0	0	0	0	0	0	0	1
0	0	0	1	0	0	1	0
0	0	1	0	0	0	1	1
0	0	1	1	0	1	0	0
0	1	0	0	0	1	0	1
0	1	0	1	0	1	1	0
0	1	1	0	0	1	1	1
0	1	1	1	1	0	0	0
1	0	0	0	1	0	0	1
1	0	0	1	1	0	1	0
1	0	1	0	1	0	1	1
1	0	1	1	1	1	0	0
1	1	0	0	1	1	0	1
1	1	0	1	1	1	1	0
1	1	1	0	1	1	1	1
1	1	1	1	0	0	0	0

FIGURE 9.13

clock pulse. If the first bit is 0, flip-flop 1 will be reset to 0. The second clock pulse will similarly shift the second bit into flip-flop 1 while simultaneously shifting the first bit into flip-flop 2 and so on as each successive bit is sampled.

9.5   *Speed-Versus-Cost Trade-Off*

In situations such as those described earlier, in which the information to be processed arrives in time sequence, sequential circuits are generally the natural choice. Since sequential circuits often use the same equipment repetitively on successive pieces of information, they generally provide the most economical means of accomplishing a specified task. If all the information required for a certain task is available simultaneously, it is obviously possible to process it all simultaneously. However, if the nature of the task is

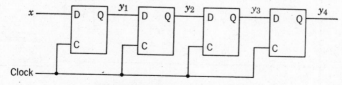

FIGURE 9.14.   *Shift register.*

such that it may be broken up into identical (or at least similar) segments, then it may be advantageous to serialize the process deliberately in order to gain the economy of repetitive application of a single unit of equipment. In such situations, a "trade-off" between speed and cost may exist.

There are a number of complex factors involved in deciding whether serialization is desirable, some of which are illustrated by the following example. Let us assume that a 32-bit word is to be read out of core memory into a 32-bit flip-flop register and checked for even parity. Since the bits are all present at once, a combinational circuit might seem logical. If we follow the same approach as used in the 8-bit checker of Fig. 8.15, we find that a rather expensive 10-level, 93-gate circuit is required. We recall that the sequential checker, Fig. 9.11, can be used for any number of bits, so perhaps it would be cheaper to provide circuitry to sequence the 32-bits into this checker one at a time. The system of Fig. 9.15 provides a straightforward realization of this approach.

The clock-pulses drive a 5-bit (modulo-32) counter, which in turn drives a decoder. The decoder produces gating pulses one at a time, in sequence, on 32 output lines. These gating pulses are applied to AND gates to sequence

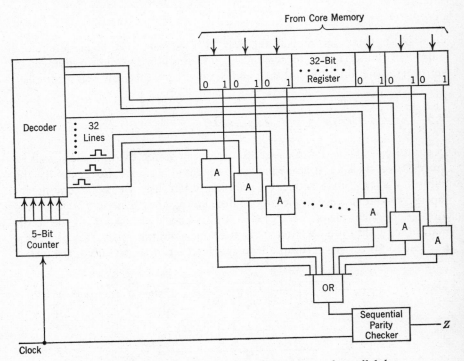

FIGURE 9.15. *Parity check by serialization of parallel data.*

the 32 bits into the parity checker. Using the simplest possible design for the counter and decoder, this circuit requires 6 flip-flops and 72 gates so that the cost of the two methods would be roughly comparable. However, in a computer, the counter and decoder might be used for other purposes, in which case only a part of their cost should be charged to the parity checker.

We have a general and perhaps obvious principle that the evaluation of any circuit must consider its relation to the overall system. A cost evaluation certainly must take into account any possible sharing of circuitry with other parts of the system, as above. Similarly, an evaluation of speed advantage must consider the value of speed to the overall system. The combinational circuit in the above example would be about 32 times as fast as the sequential, but perhaps the parity check can be carried out simultaneously with some other operation in such a manner that the speed advantage is of little value.

In the general case, we have a device, or system, which is presented with n units of information all at the same time. The units may be bits, words, characters, or any other convenient grouping. In a fully *parallel* system, there is a separate circuit, or subsystem, provided for each of the n units of information, and the n units of information are supplied to all circuits simultaneously. Because of differences between the various circuits or logical dependence of one on another, not all circuits may complete their operations at the same time. But they all start together, working on all the available information at once.

In a fully *serial* system, all n units of information are applied to a single circuit, one at a time, in sequence. The parity checker of Fig. 9.11 is thus fully serial. Assuming circuit elements of the same speed and cost, the fully parallel system will be the fastest and most expensive approach, the fully serial the slowest and least expensive. Generally, there are alternatives available between these two extremes. We might divide the n units of information into two equal groups and process the first half in one time period, followed by the second half in a second time period. This approach would generally be less expensive than the fully parallel and about half as fast. Similarly, we could divide the information into three groups, four groups, etc., until we reach the fully serial approach.

If we were to plot cost versus speed for all these possible alternatives, we would get a plot looking something like that shown in Fig. 9.16. Starting with the fully parallel system, the cost goes down as the speed goes down. The cost may continue to decrease until the fully serial system is reached. More often, however, the curve will level off before reaching the limit of fully serial operation. This occurs because the cost of sequencing the data tends to cancel out the savings resulting from the reduction in the number of circuits required for the actual processing. We saw this to be the case in the parity checker of Fig. 9.15. Just where this leveling off occurs will depend

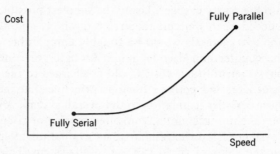

FIGURE 9.16. *Typical speed-cost curve.*

on many factors, as indicated in the parity checker example. Once a cost-speed relationship has been determined, the designer must then decide how much he can afford.

9.6 A General Model for Sequential Circuits

The general model for sequential circuits, shown in Fig. 9.17, consists of two parts, a combinational logic section and a memory section. The memory stores information about past events required for proper functioning of the circuit. This information is represented in the form of r binary outputs, $y_1, y_2, \ldots, y_r$, known as the *state variables*. Each of the 2^r combinations of state variables defines a *state* of the memory; and in general, each state corresponds to a particular combination of past events.

The combinational logic section receives as inputs the state variables and the *n circuit inputs*, $x_1, x_2, \ldots, x_n$. It generates *m outputs*, $z_1, z_2, \ldots, z_m$ and *p excitation variables*, $Y_1, Y_2, \ldots, Y_P$, which specify the next state to be assumed by the memory. The number of excitations is often equal to the

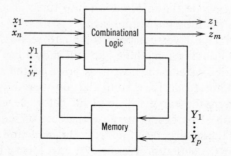

FIGURE 9.17. *General model of sequential circuits.*

number of state variables, but this depends on the type of memory. The combinational logic network may be defined by Boolean equations of the standard forms, i.e.,

$$z_i = f_i(x_1, x_2, \ldots, x_n, y_1, y_2, \ldots, y_r) \quad (i = 1, 2, \ldots, m) \qquad (9.7)$$

$$Y_j = f_j(x_1, x_2, \ldots, x_2, y_1, y_2, \ldots, y_r) \quad (j = 1, 2, \ldots, p) \qquad (9.8)$$

These are time-independent equations, i.e., they are valid at any time all the inputs and outputs are stable, as for any combinational circuit.

We have not yet specified the relationship between the inputs and outputs of the memory portion of Fig. 9.17. This relationship will depend on whether the memory is realized in terms of an array of memory elements or flip-flops or is merely a set of feedback loops with time delay. These two points of view will be considered in Sections 9.7 and 9.9.

9.7   Clock Mode and Pulse Mode Sequential Circuits

A model for the *clock mode*, or simply *clocked*, sequential circuit is shown in Fig. 9.18. The memory consists of r clocked flip-flops, which may be any type, although we have shown *J-K*. The excitations thus consist of r *J*-signals and r *K*-signals. We have shown the state variables coming from the center

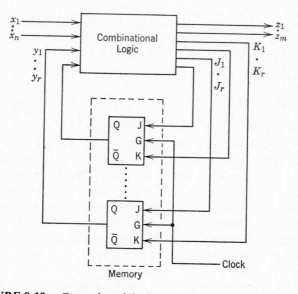

FIGURE 9.18.   *General model of clock mode sequential circuit.*

of the flip-flops since either the true or complemented outputs, or both, may be used.

All inputs and outputs of the combinational logic section are *level* signals, i.e., signals which may take on either the 1 or 0 values for arbitrary periods of time. As discussed above, the combinational logic is time-independent, so that the outputs and excitations may change at any time in response to changes in the circuit inputs. However, the state variables will change only when the flip-flops are triggered by a clock pulse. The J and K excitation variables may change at any time in response to changes in the x input variables, but no change in state will occur until a clock pulse arrives. The clock pulse contains no information in the sense of determining *what* state change will take place, it simply times, or synchronizes,* the state change. The values of the J and K signals at the time of the clock pulse determine *what* transition takes place. The times of arrival of the clock pulses are denoted by $t_v(v = 1, 2, 3, \ldots)$, and the values of the variables at these times are denoted by the superscript v. At time t_v, the excitation variables, $J_1, K_1, J_2, K_2, \ldots, J_r, K_r$, must be at stable levels determined by the current values of the inputs and state variables. The outputs are similarly stable at t_v and may be used as the inputs to other circuits synchronized by the same clock. The clock pulse arriving at t_v will cause a transition of the state variables from y_i^v to y_i^{v+1}, thus placing the circuit in a new state. The excitation and output variables will in turn adjust to this state change, as well as any input changes, reaching new stable levels before the next clock pulse, at t_{v+1}.

This circuit behavior may be described by the equations

$$z_i^v = f_{zi}(x_1^v \cdots x_n^v, y_1^v \cdots y_r^v) \qquad (i = 1, 2, \ldots, p) \qquad (9.9)$$

$$J_j^v = f_{J_j}(x_1^v \cdots x_n^v, y_1^v \cdots y_r^v) \qquad (j = 1, 2, \ldots, r) \qquad (9.10a)$$

$$K_j^v = f_{K_j}(x_1^v \cdots x_n^v, y_1^v \cdots y_r^v) \qquad (j = 1, 2, \ldots, r) \qquad (9.10b)$$

$$y_j^{v+1} = \bar{K}^v y^v + J^v \bar{y}^v \qquad (j = 1, 2, \ldots, r) \qquad (9.11)$$

Equation 9.11 characterizes the behavior of a *J-K* flip-flop, as discussed in Section 9.2. In that discussion, we used the notation Q^v and Q^{v+1} to refer to the "present" and "next" values of the flip-flop output. Here y^v refers to the value of y at the time of the clock pulse at clock time t_v and y^{v+1} to the value at clock time t_{v+1}. This does not imply that y does not change to y^{v+1} until t_{v+1}; it generally changes immediately in the conclusion of the clock pulse, t_v. However, this value (y^{v+1}) is the value it will have at t_{v+1}, and that is the value that is significant in terms of the next transition.

---

* Systems of this type are sometimes referred to as *synchronous* systems, but the word has been used so many different ways as to be virtually meaningless.

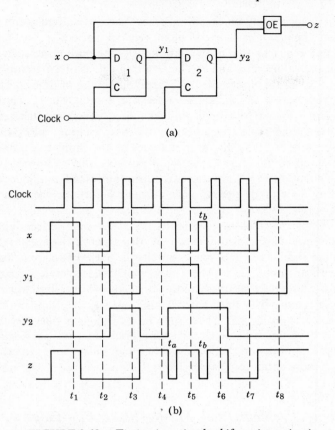

FIGURE 9.19.   *Timing in a simple shift register circuit.*

To illustrate the fact that the inputs, outputs, and state variables in a clock mode circuit are only of interest at the time of a clock pulse, let us consider the circuit of Fig. 9.19a. This 2-bit shift register circuit will have an output of 1 if and only if the present input value is *not* the same as the input two clock periods earlier, which is stored in flip-flop 2. Notice first in Fig. 9.19b that y_1 is a near replica of the input x delayed by one clock period. At time t_1, for example, $x = 1$, so that the pulse at time t_1 causes y_1 to take on this value and remain 1 until after the clock pulse at time t_2. Succeeding values of x are similarly shifted into flip-flop 1 by succeeding clock pulses. Notice next the brief pulse appearing on input line x at time t_b. This pulse does not overlap a clock pulse so it has no effect on y_1.

At time t_2, $y_1 = 1$, so that the corresponding clock pulse triggers this value into flip-flop 2. The value of y_2 remains 1 until after the pulse at time

t_3. With each succeeding clock pulse, a new value is shifted from y_1 to y_2. Thus, y_2 is a replica of y_1 delayed one more clock period.

The output z may be found by continuously computing the exclusive-OR function of the values of y_2 and x. The resulting values of z are plotted as the last line of Fig. 9.19b. Notice the brief zero-going pulse at time t_a. This pulse occurs because both x and y change values after time t_1, but y_2 changes first. Also notice that the narrow input pulse at time t_b causes another zero-going output pulse at the same time. Neither of these output pulses overlap a clock pulse. Suppose that this output serves as the input to another clock mode sequential circuit. Just as the input pulse at time t_b did not affect the state of the circuit of Fig. 9.19a, the two output pulses will not affect the state of succeeding clock mode circuits since they end before the occurrence of the next clock pulse.

Closely related to clock mode circuits are *pulse mode* sequential circuits. Pulse mode circuits are distinguished by the absence of a separate clock line. They are similar to clock mode circuits in that every state change must coincide with an input pulse. The pulses which trigger state changes must appear on the circuit input lines, and unclocked S-C or T flip-flops must be used. These input pulses must satisfy timing restrictions similar to any set forth for a clock pulse line. They must be of sufficient duration, and the circuit must have time to stabilize from an input pulse on one line before a second pulse can occur on that or any other line. Pulse mode circuits will be discussed in more detail in Chapter 11.

9.8 Timing Problems and Master-Slave Flip-Flops

We noted earlier a first timing restriction on flip-flops—that the clock or trigger pulses must have a certain minimum duration. In this section, we wish to investigate some additional restrictions, some applying to any type of circuit, some peculiar to circuits employing flip-flops. One fairly obvious restriction, which applies to all types of logic circuits, combinational as well as sequential, is that the interval between successive changes of a single variable must not be less than the basic delay time of the logic circuits. If the basic delay time of a gate is 10 ηsec, for example, it certainly cannot respond reliably to signal changes only 5 ηsec apart. In simpler terms, the designer must select a family of logic circuits fast enough to keep up with expected signal rates.

In discussing flip-flops, we introduced clocking as a means of minimizing timing problems in flip-flops, but this should not be taken to mean that clocked flip-flops are free of timing problems. They are not, but it is often simpler to enforce timing restrictions on a single source of control pulses than on a variety of level signals. In discussing the S-C flip-flop with reference to the timing diagram of Fig. 9.3, we noted that, if S changed back to 0 before the resultant change of Y at t_3, the flip-flop would not trigger reliably. This is simply a specific example of the restriction noted above, that the interval between two successive changes in must not be too short. In terms of flip-flops, this means there is a minimum duration for trigger pulses, primarily determined by the basic gate delay but also affected by other complex electronic factors. This minimum trigger duration will be a basic specification provided by the flip-flop manufacturer.

Next, if one pulse is followed immediately by another, the leading edge comprises a third successive change in the same variable and, by the same argument as above, this cannot occur too soon. Thus, there is a minimum interval between trigger pulses, most frequently stated in terms of a maximum clock rate. For example, the minimum interval between trigger pulses for a 10 MegaHertz flip-flop is 100 ηsec. Because of electronic factors too complex to discuss here, the minimum trigger interval is generally much longer than the minimum trigger duration. For example, a typical 10 MHz flip-flop would have a minimum pulse duration of 20 ηsec, one fifth the minimum interval.

The above restrictions on trigger duration and interval apply to all flip-flops regardless of how they are used. When flip-flops are used in clock mode or pulse mode circuits, still other problems arise. In a clock mode circuit, the arrival of a clock pulse triggers state changes in the flip-flops in accordance with the values of the excitation variables which are functions of the flip-flop outputs. Assume that changes in these outputs propagate through the combinational logic, causing new values of the excitation to appear at the flip-flops *while the clock pulse is still present*. The flip-flops may then make additional state changes and may continue to do so as long as the clock pulse is present. Such operation would be in contradiction to a basic premise of clock mode operation—that each clock pulse may cause only a single change of state. Similar comments would apply to excessively long input pulses in pulse mode circuits.

As a specific example of this type of problem, consider the simple frequency division circuit shown in Fig. 9.20a. It is intended that the input x be a square wave of half the clock frequency, as shown in Fig. 9.20b. It is also intended that the circuit output z change value each time the input x is 1 at clock time. In this way, the output should be a square wave of $1/2$ the input frequency or $1/4$ the clock frequency.

Consider the response of the circuit to clock pulse 3. At the beginning of this pulse, x and z have been stable for a sufficient period of time for d to have stabilized at 1. The clock pulse is therefore gated onto the S line, causing z to go to 1 after some delay. At this time, x is still 1 so d goes briefly to 0 until x goes to 0. We thus have a brief 0-going pulse on d, but this

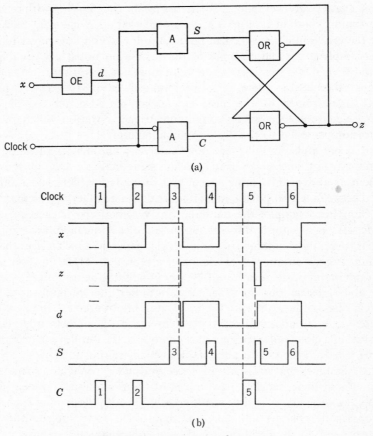

(a)

(b)

FIGURE 9.20.   *Frequency division circuit.*

causes no problems because the clock pulse has ended before the change it caused can propagate around to cause a change in *d*.

Consider next the situation of a clock pulse 5, which is assumed to be considerably longer than other clock pulses. At the start of this pulse, *d* is 0, so the clock pulse is gated onto the C line, causing *z* to go to 0 and, since *x* is still 1, this in turn causes *d* to go to 1. But this time the clock pulse is still present, so the change in *d* causes C to go to 0 and S to 1, causing *z* to go back to 1.

Because pulse 5 was too wide, the zero output lasted only briefly rather than two clock periods as desired. In effect, pulse 5 caused two state changes from 1 to 0 and then from 0 to 1. If the width of pulse 5 had been within the design tolerance, circuit delays would have prevented the appearance of a new value of *d* at the flip-flop input until after the clock had returned to 0.

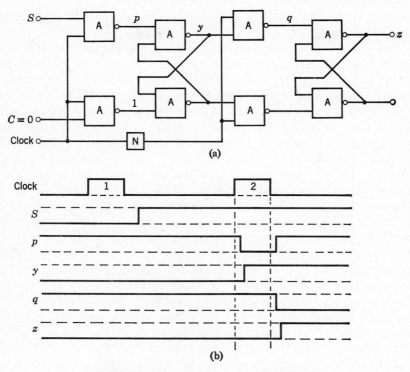

FIGURE 9.21.   *Clocked master-slave flip-flop.*

In all sequential circuits using elementary clocked flip-flops, it is necessary to enforce both a minimum and a maximum clock pulse width. Unfortunately it is difficult to do this in most integrated circuit technologies where gate delays exhibit a substantial statistical variance.

The need to limit the maximum clock pulse width can be avoided by using a more sophisticated flip-flop called a *master-slave flip-flop*. One form of a clocked master-slave S-C flip-flop is depicted in Fig. 9.21a together with a timing diagram in Fig. 9.21b. Initially, the flip-flop output $z = 0$, as shown. In the absence of set or clear inputs or a clock pulse, the output y of the first pair of cross-coupled gates, the *master* element, is the same as z. Initially, then, $y = 0$. The set input goes to 1 prior to the onset of clock pulse 2. Therefore, the value of p goes to 0 one gate delay following the leading edge of pulse 2. After one more gate delay, the output y of the master element goes to 1. Notice that there is no further circuit activity until the clock returns to 0. In particular, the output z of the second pair of cross-coupled gates, the *slave* element, remains 0. When the clock returns to 0, q goes to 0; and after another gate delay, z goes to 1. The symmetry of the circuit is apparent, so that the dual two-step process would occur following a clear input.

Now consider a clocked sequential circuit containing only master-slave memory elements. Assume also that the clock inputs to all flip-flops are perfectly synchronized. Therefore, at the onset of each clock pulse, some of the master elements will change state, but the flip-flop outputs will remain at the previous values. After the clock pulse returns to 0, no further set or clear signals can be propagated into the master elements. At this same time, some of the flip-flop outputs, or slave elements, will change state. Clearly none of these new state values will have an effect on any of the master elements until the next clock pulse. *Thus, no more than one state change can take place for each clock pulse regardless of the clock pulse width.*

The state changes in a sequential circuit composed of master-slave flip-flops coincide with the trailing edge of clock pulses. Another commonly used type of flip-flop synchronizes the state change with the leading edge of the clock pulse. Such devices are usually called *edge-triggered* flip-flops. Flip-flops commonly available in integrated circuit form are of one of the above two varieties, and most clocked systems are constructed using one or the other of these two types. Under certain circumstances in which the clock input to one flip-flop is out of phase with the clock input to other flip-flops, the clock mode hypothesis of only one state change per clock pulse may still be violated. This phenomenon is known as *clock skew*. Clock skew is most likely to appear in very high-speed systems in which transmission delays in the interconnections between gates are significant. Where clock skew has been a problem, the designers have resorted to a variety of special techniques to synchronize state changes effectively. We shall not pursue this topic further at this point in our discussion.

It is possible to construct unclocked master-slave flip-flops which can be used in pulse-mode circuits, but unfortunately this type of flip-flop is not readily available in most integrated circuit logic families. One form of an unclocked master-slave flip-flop is given in Fig. 9.22. A timing analysis of this circuit will be left as a problem for the reader.

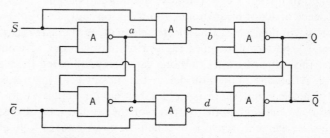

FIGURE 9.22.   *Unclocked master-slave flip-flop.*

9.9 *Level Mode Sequential Circuits*

Clock mode and pulse mode circuits both require pulses to control state transitions, and the very concept of a pulse implies a considerable degree of control over signal timing. We now relax these timing restrictions even further and consider circuits for which all inputs are levels, signals which may remain at either value (0 or 1) for arbitrary periods and may change at arbitrary times. Such circuits will be called *level mode* sequential circuits.

If we place a restriction on the inputs that only one may change at a time and that no further changes may occur until all outputs and state variables have stabilized, we will say that the level mode circuit is operating in the *fundamental mode*. Circuit design is considerably simplified if these restrictions can be enforced on circuit inputs. The reason for this will be discussed in detail in Chapter 13, but we can see intuitively that simultaneous changes of inputs may cause problems. If we use fine enough division of time, no events are ever exactly simultaneous. Thus, response to presumably simultaneous events might depend on the order in which they actually occur, an order which may be determined by unpredictable circuit delays.

The basic mode for the level-mode circuit is shown in Fig. 9.23. We see that the flip-flop memory of the clock mode and pulse mode circuits has been replaced by delay in the feedback loops. As a rule, this is not a physical delay deliberately inserted in the feedback loops but rather represents a "lumping" of the inherent distributed delay in the combinational logic developing the excitation variables. With this type of memory, the excitations are themselves the next values of the state variables. The excitations will be labeled with capital Y's, while the theoretically delayed values of the excitations

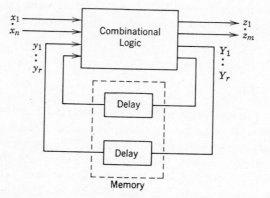

FIGURE 9.23. *Basic model of level mode sequential circuits.*

which serve as inputs to the combinational logic network will be called secondaries and be labeled with lower case y's. Where the distinction is not important, the excitations and secondaries will be called state variables.

The equations for the z and Y variables are the same as for the general model (Equations 9.6 and 9.7). The equation describing the state change is simply

$$y_j(t + \Delta t) = Y_j(t) \qquad (j = 1, 2, \ldots, r) \tag{9.11}$$

The meaning of these expressions will become clear as we consider the design of level mode circuits in Chapter 13.

It should be noted that the level mode circuit is the most general type of sequential circuit in the sense that it can respond to any type of signal, pulse or level. As pointed out earlier, the distinction between a pulse and a level is a matter of definition. Physically, a pulse may be considered as two successive changes in a level signal occurring in a short period of time. The most important distinction lies in the manner in which the signal is interpreted. In a clock mode or pulse mode circuit, a pulse is considered as a single event which can trigger only a single change of state. In a level mode circuit, a pulse is considered as two events, two level changes, each of which may cause a change in state.

Recognition of this dual nature of pulses enables us to apply level mode circuits to pulse situations in which pulse mode circuits cannot be used. Any pulse may cause a change in state, and a second pulse arriving while a circuit is still in transition as a result of a first pulse clearly may cause unreliable operation. Since, as noted earlier, no two events are ever exactly simultaneous, we must therefore limit pulse mode circuits to situations where only one pulse can occur at a time. But there are many practical situations where we must deal with pulses from unrelated sources, so that such a restriction cannot be enforced. In such cases, we must apply level mode techniques, with each pulse edge considered a separate event. Simultaneous changes can cause problems in level mode circuits, as noted, but they can be resolved by level mode techniques.

9.10   Conclusion

In this chapter, we have attempted to give the reader an overview of the general characteristics of sequential circuits, of some of the factors governing their use, and of some of their special problems. We have treated the timing of clock mode circuits in more detail to prepare the reader for Chapter 10. He should now be able to consider the logical analysis and synthesis of such sequential circuits with confidence that physical realizations will work as postulated.

We should note that nomenclature regarding sequential circuits is not standardized. As we noted, clocked circuits are often referred to as *synchronous* circuits, leaving all other types lumped in the category of *asynchronous* circuits. But this division groups pulse mode and level mode circuits, which are very dissimilar, and separates pulse mode and clock mode circuits, which are quite similar. McCluskey[1] first proposed the division into *pulse mode* and *fundamental mode* circuits, with clock mode as subdivision of pulse mode circuits. Unfortunately, this leaves the important group of circuits which are neither pulse mode nor fundamental mode in limbo, with no specific name. We have therefore adopted the name *level mode* for all nonpulsed circuits, with fundamental mode as a subdivision of this category. But names are not that important; the important thing is that you fully understand the characteristics of and the differences between the various types of circuits. With such understanding, you can always figure out what the other guy is talking about, whatever names he may use.

PROBLEMS

9.1. Consider the circuit realization of an edge-triggered D flip-flop given in Fig. P9.1. Suppose that, prior to the occurrence of a clock pulse, the

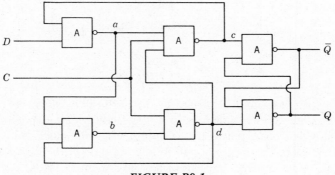

FIGURE P9.1

circuit output Q and the circuit input D are stable at the values 0 and 1, respectively.

(a) Under the above conditions, what must be the logical values at points a, b, c, and d?

(b) Start with the initial conditions determined above and let the clock go to 1 and remain at that value for 200 ηsec. Construct as accurately as possible a timing diagram of D, a, b, c, d, and Q beginning

50 ηsec prior to the clock pulse and extending 100 ηsec after the trailing edge of the pulse. Assume that the delay associated with each gate is 20 ηsec. Be sure to verify that changes at the D input while the clock is at the 1 level do not affect the output.

9.2. Assume that $\bar{S} = \bar{C} = Q = 1$ in the circuit of Fig. 9.22 and that the circuit is stable.

(a) Determine the values of a, b, c, and d.

(b) Beginning with the initial conditions listed above let $\bar{C}$ go to 0 and remain at this value for 200 ηsec. Construct as accurately as possible a timing diagram for $\bar{C}$, a, b, c, d, and Q beginning 50 ηsec prior to the pulse on $\bar{C}$ and continuing 100 ηsec after the trailing edge of this pulse.

9.3. By means of a timing diagram indicating signals levels at relevant points, show that the circuit in Fig. P9.3 functions as a *J-K* flip-flop.

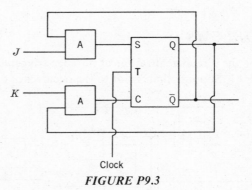

Clock

FIGURE P9.3

9.4. Figure P9.4 shows a variant of the D flip-flop known as a *latch*. The behavior of this circuit is as follows. Information present at the D input is transferred to the Q output when the clock is high (at the 1 level), and the Q output will follow the state of the D input as long as the clock remains high. Information present at the Q output will be

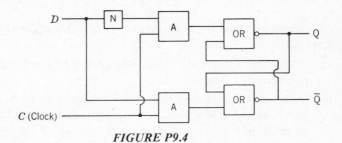

FIGURE P9.4

retained as the clock goes low until such time as the clock again goes high. Verify this operation by means of a timing diagram showing signal levels at relevant points.

BIBLIOGRAPHY

1. McCluskey, E. J. "Fundamental Mode and Pulse Mode Sequential Circuits," *Proc, IFIP Congress 1962*, North-Holland Publ. Co., Amsterdam, 1962.

2. Mealy, G. H. "A Method for Synthesizing Sequential Circuits," *Bell System Tech J.*, **34**: 5, 1045–1080 (Sept. 1955).

3. Moore, E. F. *Sequential Machines: Selected Papers*, Addison-Wesley, Reading, Mass., 1964.

Chapter 10 *Synthesis of Clock Mode Sequential Circuits*

10.1 *Analysis of a Sequential Circuit*

In this chapter, we shall develop a complete design procedure for clock mode sequential circuits. In Chapters 11 and 13, we shall develop similar procedures for pulse mode and fundamental mode (pulseless) sequential circuits. The clock mode will be considered first because the clock mode design process is the most straightforward throughout. A disadvantage of the clock mode as the initial topic might be the difficulty in constructing practical examples. Most real life systems which include a clock contain a large number of states and are often better described by a system design language,[15] rather than a state diagram. Sequential circuits which are both interesting and limited in size tend to be pulse or fundamental mode. We have chosen several clock mode examples which we hope the reader will not regard as overly artificial.

As is the case with electrical networks, the analysis of a sequential circuit is easier than the synthesis. To provide the reader with insight into the goal of the synthesis process, we shall first analyze an already designed sequential circuit. Consider the circuit given in Fig. 10.1. Notice that this is an example of the general clock mode model presented in Chapter 9.

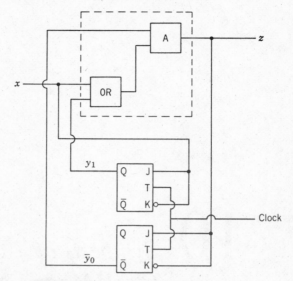

FIGURE 10.1.   *Sequential circuit example.*

The *memory element input equations*, 10.1 and 10.2, can be determined by tracing connections in the simple combinational logic block. From these expressions

$$J_{y_1} = x \qquad K_{y_1} = \bar{x} \tag{10.1}$$

$$J_{y_0} = z = (x + y_1)\bar{y}_0 \qquad K_{y_0} = \overline{(x + y_1)\bar{y}_0} \tag{10.2}$$

we can use the general next state equation for the J-K flip-flop, $y^{v+1} = J \cdot \bar{y}^v + \bar{K}^v \cdot y^v$, to determine the *next state equations* for the memory elements y_0 and y_1 as given by Equations 10.3a and 10.3b. In both

$$y_1^{v+1} = x^v \cdot \bar{y}_1^v + x^v \cdot y_1^v = x^v \tag{10.3a}$$

$$y_0^{v+1} = [(x^v + y_1^v)\bar{y}_0^v] \cdot \bar{y}_0^v + [\overline{(x^v + y_1^v)\bar{y}_0^v}]y_0^v$$

$$= (x^v + y_1^v)\bar{y}_0^v \tag{10.3b}$$

of these expressions $K = J$, leading to a considerable simplification. In effect, the circuit could have been implemented equally well using D flip-flops rather than J-K flip-flops. Although it will usually not be possible to simplify the expressions so readily, the use of $J \cdot \bar{y}^v + \bar{K} \cdot y^v = y^{v+1}$ to determine next state equations is a completely general procedure where J-K flip-flops are used.

From Equations 10.3a and 10.3b, we can tabulate what we shall define as the *transition table* for the sequential circuit. This is merely an evaluation of y_1^{v+1}, y_0^{v+1}, and z^v for all possible combinations of values of x^v, y_0^v, and y_1^v. The transition table for the circuit under discussion is given in Fig. 10.2.

y_1^v	y_0^v		y_1^{v+1}, y_0^{v+1}				z^v	
		x^v	0		1		0	1
0	0		0	0	1	1	0	1
0	1		0	0	1	0	0	0
1	1		0	0	1	0	0	0
1	0		0	1	1	1	1	1

FIGURE 10.2. *Transition table for circuit of Fig. 10.1.*

For example, if $x^v = y_0^v = 0$ and $y_1^v = 1$, then y_1^{v+1} and y_0^{v+1}, as given by Equations 10.3a and b, are 0 and 1 respectively. These values are entered in the square enclosed by the dashed line in Fig. 10.2. The reader can easily determine the remaining values in the table. The values for z^v are similarly determined. In this case, $z^v = y_0^{v+1}$.

Transition tables of the form of Fig. 10.2 will tend to become rather bulky as the number of variables increases. To simplify the tables, we will assign a shorthand notation, as follows. First, we represent the states by the letter q, with arbitrarily assigned decimal subscripts. For example, we shall represent $y_1 y_0 = 00$ by q_1, $y_1 y_0 = 01$ by q_2, $y_1 y_0 = 11$ by q_3, and $y_1 y_0 = 10$ by q_4. In this example, there is only one input and one output. If there are multiple inputs and outputs, these values may be coded also, in which case we represent the set of input variables $(x_1, x_2, \ldots, x_n)$ by X and the set of output variables $(z_1, z_2, \ldots, z_q)$ by Z. Particular combinations of values will be represented by the decimal equivalents to the binary interpretations of the combinations. Thus, $x_2 x_1 = 00$ will be represented by $X = 0$, $x_2 x_1 = 01$ by $X = 1$, $z_2 z_1 = 11$ by $Z = 3$, etc. We also combine the output and next-state portions of the transition table. This notation results in the *state table*, as shown in Fig. 10.3a. Now the entry in the dashed square, for example, may be interpreted as follows: If the circuit is in state q_4 and receives input 0, the output at clock time will be 1 and the next state will be q_2.

The same information contained in the state table may be displayed more graphically by a *state diagram*. The state diagram corresponding to the state table of Fig. 10.3a is shown in Fig. 10.3b. There is a circle for each state and an arrow leaving each state for each input. The arrows terminate at the appropriate next states. Each is labeled with the number of the input that causes that particular transition. The inputs are indicated on the left of each slash

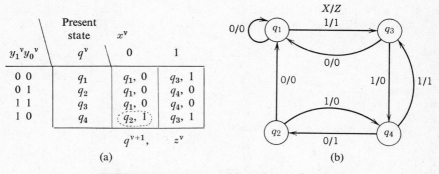

$y_1^v y_0^v$	Present state q^v	x^v 0	1
0 0	q_1	$q_1, 0$	$q_3, 1$
0 1	q_2	$q_1, 0$	$q_4, 0$
1 1	q_3	$q_1, 0$	$q_4, 0$
1 0	q_4	$q_2, 1$	$q_3, 1$
		$q^{v+1},$	z^v

(a)   (b)

FIGURE 10.3.   *State table and state diagram.*

and the output corresponding to that particular input present state combination on the right, i.e., X/Z. For example, the square enclosed by the dashed circle in Fig. 10.3a corresponds to the arrow from q_4 to q_2. The square is in the column corresponding to $X = 0$, and the output entry is 1. Hence, the arrow is denoted $0/1$. While state table and state diagram contain the same information, both are useful at different steps in the synthesis process.

10.2   Design Procedure

The analysis process of the previous section followed a sequence of steps which successively transformed the circuit description as follows:

$$\text{Circuit} \rightarrow \begin{matrix} \text{Memory} \\ \text{element} \\ \text{input} \\ \text{equations} \end{matrix} \rightarrow \begin{matrix} \text{Next} \\ \text{state} \\ \text{equations} \end{matrix} \rightarrow \begin{matrix} \text{Transition} \\ \text{table} \end{matrix} \rightarrow \begin{matrix} \text{State} \\ \text{table} \end{matrix} \rightarrow \begin{matrix} \text{State} \\ \text{diagram} \end{matrix}$$

The synthesis process resembles an analysis in reverse. The principal distinction is that synthesis begins prior to the existence of a state diagram. This diagram must be obtained from some other description of the problem. Usually this is merely an unambiguous (hopefully) natural language description of the intended circuit function. The synthesis procedure is depicted in Fig. 10.4.

We shall consider the complete process in this chapter. However, for the reader impatient to get back to the circuit, we must remark that first considerable attention must be paid to steps 1 and 2. Sections 10.3 and 10.4 are concerned with various approaches to obtaining a state table. The next two sections are devoted to finding an equivalent state table with a minimum number of states. A circuit can be obtained from the minimal state table without great difficulty. Obtaining the most economical circuit is not so easy.

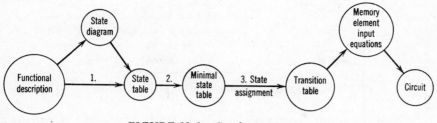

FIGURE 10.4.   *Synthesis process.*

The complexity of the memory element input equations will vary with the assignment of combinations of flip-flop output values to states in the state table. The attempt to optimize this correspondence is called the *state assignment problem.*

Minimizing the state table is a programmable procedure. Determining the state table in the first place will require insight and ingenuity on the part of the designer. An algorithm for state table determination is possible if the behavior of the circuit is first described in the form of a *regular expression.* We shall leave a discussion of regular expressions to books on automata theory and for our purposes let the state diagram be the first formal description of the sequential circuit.

10.3   Synthesis of State Diagrams

As depicted in Fig. 10.4, one approach to obtaining the state table involves the use of a state diagram. This method is particularly convenient for certain types of circuits which possess a readily identifiable state called a *reset state.* Where such a state exists, there is a mechanism by which the circuit may be returned to the reset state from any other state in a single operation. This operation may be specified as a special input column in the state table, or it may function independently of the state table. For example, a special line may be connected to all flip-flops which will clear them to zero. This line may be connected to separate input points rather than those used in the implementation of the state table. Example 10.1 involves a simple circuit with a reset state.

Example 10.1

The beginning of a message in a particular communications system is denoted by the occurrence of three consecutive 1's on a line x. Data on this line have been synchronized with a source of clock pulses. A clock mode sequential circuit is to be designed which will have an output of 1 only at the clock time coinciding with the third of a sequence of three 1's on line x. The circuit will serve to warn the

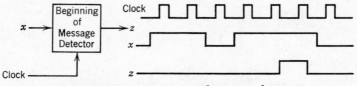

FIGURE 10.5. *Beginning-of-message detector.*

receiving system of the beginning of a message. It will be provided with a separate reset mechanism to place it in state q_0 following the end of a message. Typical behavior of the circuit is depicted in Fig. 10.5.

Solution: The synthesis procedure begins by designating the reset state as q_0, as shown in Fig. 10.6a. Essentially the circuit must keep track of the number of consecutive 1's it has received. It does this by moving to a different state with

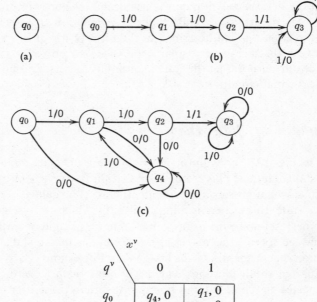

(a)

(b)

(c)

	x^ν	
q^ν	0	1
q_0	$q_4, 0$	$q_1, 0$
q_1	$q_4, 0$	$q_2, 0$
q_2	$q_4, 0$	$q_3, 1$
q_3	$q_3, 0$	$q_3, 0$
q_4	$q_4, 0$	$q_1, 0$

$q^{\nu+1}, z$

(d) State Table

FIGURE 10.6. *State diagram for beginning-of-message detector.*

each input of 1. As shown in Fig. 10.6b, the circuit goes to state q_1 with the first 1 input and to state q_2 with second. The outputs associated with both of these inputs are 0, as shown. The third consecutive input of 1 generates a 1 output and causes the circuit to go to q_3. Once in q_3, the circuit will remain in this state emitting 0 outputs until it is externally reset to q_0. This reset mechanism is not shown in the state table.

It is possible for an input of 0 to appear during any clock period, possibly interrupting a sequence of 1's. A state q_4 is provided for this eventuality. Every input of 0 (unless the circuit is already in q_3) generates a 0 output and causes the circuit to go to q_4, as shown in Fig. 10.6c. A second 0 merely causes the circuit to remain in state q_4. An input of 1 while the circuit is in q_4 may be the first of a sequence of three 1's. Thus, a 1 will cause the circuit to change state from q_4 to q_1.

All possible input sequences which can occur have been considered. The reader may verify this by noting that arrows for both 0 and 1 inputs have been defined leaving each of the five states. Thus, Fig. 10.6c is the completed state diagram. This translates directly to the state table of Fig. 10.6d. ■

The perceptive reader may have noticed that the circuit could be returned to q_0 for inputs of 0. This would eliminate the need for state q_4. As a general rule, already existing states should be used as next states wherever possible. At some point, the designer must stop defining new states. However, since state diagram construction is an intuitive procedure, there may be doubt as to whether an existing state will serve. *When in doubt, define a new state*, thus avoiding serious error.

Example 10.2

Numbers between 0 and 7 expressed in binary form are transmitted serially on line x from a digital measuring device. The bits appear in descending order of significance, synchronized with a clock. Design a circuit which will have an output $z = 1$ at the clock time of the third bit of a binary number representing an extreme value (0 or 7) on line x. Otherwise, $z = 0$. The circuit will have two inputs, x and d, in addition to the clock. If $d = 1$ at a given clock time, then the most significant bit of a measurement will appear on line x at the next clock time.

Assume that $d = 1$ will cause the circuit to assume a reset state a. Construct a state table regarding line d as an external reset mechanism. Then revise the state table to include input d as part of the state table.

Solution: Beginning with state a, we construct a tree of input arrows as shown in Fig. 10.7a. Each arrow leads to a new state from which two new arrows (in general one for each possible input) emerge. In this problem, it is only necessary to generate two levels of the tree. The numbers 0 and 7 are detected at the time of the third input. The binary number is zero if an input of 0 occurs when the circuit is in state d. An input of 1 for state g indicates a seven. Outputs of 1 are shown for these cases.

The third input must take the circuit to a state for which subsequent outputs of 1 are impossible. Notice that states e and f satisfy this requirement. Once the circuit is in either of these states, it remains there permanently; and the outputs

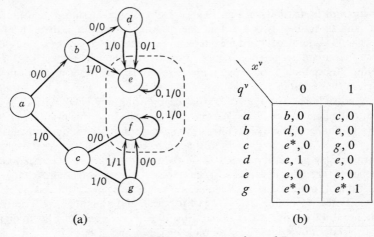

FIGURE 10.7. *Detector of 0 and 7.*

are always 0. In fact, the circuit will behave in the same manner whether it is in state e or state f. Therefore, only one of these states is necessary, as emphasized by the dashed closed curve in Fig. 10.7a. In translating the state diagram to the state table in Fig. 10.7b, state f is omitted. Where f would be a next state, it is replaced by e, as indicated by asterisks.

The state table with the effect of the reset input d included is given in Fig. 10.8b. The four combinations of the inputs x and d are coded as shown in Fig. 10.8a. The two columns for $X = 0$ and $X = 1$ correspond to $d = 0$. These two columns are a direct copy of Fig. 10.7b. Anytime $d = 1$, the circuit is reset to state a, as shown. If the sequential circuit is implemented as given by Fig. 10.8b, no external reset mechanism is required.

dx	X
00	0
01	1
10	2
11	3

(a)

q^ν \ X	0	1	2	3
a	$b, 0$	$c, 0$	$a, 0$	$a, 0$
b	$d, 0$	$e, 0$	$a, 0$	$a, 0$
c	$e, 0$	$g, 0$	$a, 0$	$a, 0$
d	$e, 1$	$e, 0$	$a, 0$	$a, 0$
e	$e, 0$	$e, 0$	$a, 0$	$a, 0$
g	$e, 0$	$e, 1$	$a, 0$	$a, 0$

(b) $q^{\nu+1}, z$

FIGURE 10.8. *Including input d in the state table.*

The response tree of Fig. 10.7 illustrates a general procedure which can be used in the absence of other insight. Where no specific reset state exists, an arbitrary state may be used as a starting point. Preferably this state will have some past history to distinguish it clearly from other states. In the absence of a well-defined starting state, the tree may grow very large before the branches are terminated. In general, the designer should prune the tree as he goes by routing arrows to existing states which lead to the same future behavior as might be defined for new states. If the designer is slow in applying insight, the size of a response tree can get out of hand rapidly.

10.4   Finite Memory Circuits

An equivalent version of the circuit of Fig. 10.1 is shown in Fig. 10.9a. The J-K flip-flops with inverters on the clear input are replaced with equivalent D flip-flops, with the clock input omitted for clarity.

Recalling that the output of a D flip-flop is equal to the value of the input at the immediately previous clock time, $t^{\nu-1}$, we see that in this case the present output is a function only of the present input and the immediately previous

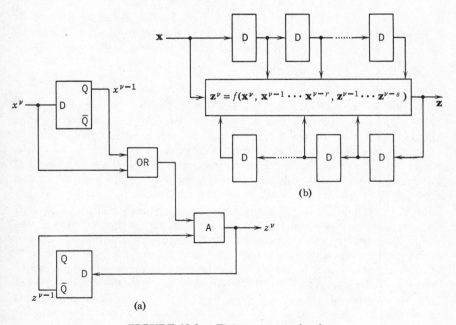

FIGURE 10.9.   *Finite memory circuits.*

input and output, i.e.,

$$z^v = f(x^v, x^{v-1}, z^{v-1}). \tag{10.4}$$

This circuit is an example of a *finite memory* circuit, *finite* in the sense that it need "remember" only a finite number of past inputs and outputs in order to determine the appropriate present output and next state. Note that *finite* here has nothing to do with the memory capacity in bits which is, of couse, always finite.

A generalized version of a finite memory circuit is shown in Fig. 10.9b. Notice that r previous inputs and s previous outputs are stored. The current output is a function of this stored information and the current input. The symbols X and Z are vectors which may represent several input and output lines. The blocks labeled D are not individual flip-flops but represent one-dimensional arrays of flip-flops, one for each input or output line.

Not all sequential circuits are finite *memory*. In Examples 10.1 and 10.2, the circuits were to issue an output on the occurrence of some particular input sequences. They were then to issue no further output after being reset, regardless of what additional inputs (other than RESET) might occur. Thus, these circuits must remember a particular input or output sequence which may have occurred arbitrarily far back in time.

It is often possible to recognize a finite memory circuit from the initial specifications, and this knowledge can often simplify the synthesis procedure. If the designer knows the circuit is finite memory, he can generally specify an upper bound on the memory size, i.e., the maximum number of internal states. He can then set up either the transition table or state table directly, as shown in the following example.

Example 10.3

Information bits are encoded on a single line x so as to be synchronized with a clock. Bits are encoded so that two or more consecutive 1's or four or more consecutive 0's should never appear on line x. An error-indicating sequential circuit is to be designed which will indicate an error by generating a 1 on output line z coinciding with the fourth of every sequence of four 0's or the second of every sequence of two 1's. If, for example, three consecutive 1's should appear, the output would remain 1 for the last two clock periods.

Solution: The circuit to be designed may be immediately recognized as finite memory. In this case, we may call it *finite input memory* in that only inputs need be remembered. In the worst case, the circuit will have to remember the value of three previous inputs. If these stored inputs are all 0 and the present input is 0, then the circuit output is 1. A circuit of the form shown in Fig. 10.10a can be made to do the job. As the input bits arrive, they are shifted through the string of flip-flops so that all times $y_2 = x^{v-1}$, $y_1 = x^{v-2}$, $y_0 = x^{v-3}$. The combinational logic then determines the output on the basis of these stored inputs and the present input.

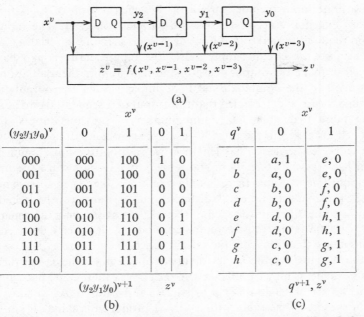

(a)

$(y_2y_1y_0)^v$	x^v				q^v	x^v	
	0	1	0	1		0	1
000	000	100	1	0	a	$a, 1$	$e, 0$
001	000	100	0	0	b	$a, 0$	$e, 0$
011	001	101	0	0	c	$b, 0$	$f, 0$
010	001	101	0	0	d	$b, 0$	$f, 0$
100	010	110	0	1	e	$d, 0$	$h, 1$
101	010	110	0	1	f	$d, 0$	$h, 1$
111	011	111	0	1	g	$c, 0$	$g, 1$
110	011	111	0	1	h	$c, 0$	$g, 1$

$(y_2y_1y_0)^{v+1}$ $\qquad$ z^v $\qquad\qquad$ q^{v+1}, z^v

(b) $\qquad\qquad\qquad\qquad\qquad$ (c)

FIGURE 10.10.   *Synthesis of circuit with finite input memory of length three.*

The transition table for this circuit can be set up directly, as shown in Fig. 10.10b. First list all eight combinations of state variables as the present state entries. The next state entries, $(y_2y_1y_0)^{v+1}$, follow directly from noting that the state variables shift to the right each clock time. For example, if the present values are 000 and the input is 1, the next values are 100. To complete the transition table, the appropriate outputs should be entered. The output is to be 1 if the three previous inputs and the present input are all 0 or if the immediately previous and present inputs are both 1. If the designer wishes to use the circuit form shown in Fig. 10.10a, the only step left is design of the combinational logic to generate the output, z. For this case,

$$z = xy_2 + \bar{x}\bar{y}_2\bar{y}_1\bar{y}_0 \tag{10.5}$$

■

Although the circuit form of Fig. 10.10a will work, it may not be minimal. In only one case is it necessary for the circuit to "remember" three inputs back. If $x^{v-1} = 1$, it doesn't matter what x^{v-2} and x^{v-3} were. Thus, it appears that less than eight internal states may be sufficient. To check on this, the designer may convert the transition table to a state table simply by assigning an arbitrary state designation to each combination of state variable in the table of Fig. 10.10b. The resultant state table, Fig. 10.10c, may then be minimized by the formal techniques to be developed in later sections.

In a finite input memory circuit of the general form of Fig. 10.10a, the memory essentially acts as a serial-to-parallel converter. The input information required to make an output decision is arriving in serial, i.e., in time sequence, so the circuit simply stores the inputs until enough have been received to make the output decision and then applies them all at once, i.e., in parallel, to the combinational logic implementing the output function. Any time it can be determined that the output of a sequential circuit depends on no more than some maximum number of previous inputs, then the approach discussed in the previous example can be used.

If the longest sequence has length n, we say that the finite input memory circuit has *memory of length n*, and n memory elements will be required for each input line. Assume there is only one input line. Then the memory will have 2^n possible states, i.e., 2^n distinct input messages can be stored. If the output is to be different for each of these 2^n messages, then the general form of Fig. 10.10a is the most economical, and the designer's task reduces to nothing more than realizing the output function.

If the outputs are the same for many of the input messages, then fewer memory elements may be sufficient. In the above example, we have a memory of length 3 so that eight distinct messages can be stored. As we have already seen, however, all messages with $x^{v-1} = 1$ produce the same output, so that the memory need not distinguish among the four messages with this characteristic. Thus, by processing, or classifying, the messages as they arrive, we may be able to reduce the memory requirements.

If we reduce the memory in this manner, the circuit form will not be the same as Fig. 10.10a, in that the outputs of the memory flip-flops will not be equal to previous inputs but will be functions of them. Nevertheless, we may still use the state table derived from this basic circuit form as a starting point in the design process. The situation is somewhat analogous to first specifying a combinational function as a standard or expanded sum of products. This may not be minimal, but it allows us to express the function in Karnaugh map form, from which a minimal expression can be determined.

Once we have any state table describing the appropriate behavior, we can derive from it any number of equivalent tables, including a unique minimal table. For an assumed circuit form such as Fig. 10.10a, the transition table, with regard to next-state entries, is unique for a given memory length and number of inputs since the next state entries are simply derived by shifting the inputs through the memory. We may thus set up standard transition tables for finite input memory circuits. These tables are shown in Fig. 10.11 for single-input circuits of length 2, 3, and 4. To form the complete transition tables for any specific circuit, the designer need only fill in the appropriate outputs. If the number of distinct outputs is more than half the number of states, the number of memory devices cannot be reduced, so the designer

$x^{\nu-1}x^{\nu-2}$	q^ν	x^ν 0	1
00	a	a	d
01	b	a	d
11	c	b	c
10	d	b	c

$q^{\nu+1}$

(a) $n = 2$

$x^{\nu-1}x^{\nu-2}x^{\nu-3}x^{\nu-4}$	q^ν	x^ν 0	1
0000	a	a	i
0001	b	a	i
0011	c	b	j
0010	d	b	j
0100	e	d	l
0101	f	d	l
0111	g	c	k
0110	h	c	k
1000	i	e	m
1001	j	e	m
1011	k	f	n
1010	l	f	n
1100	m	h	q
1101	n	h	q
1111	p	g	p
1110	q	g	p

$q^{\nu+1}$

(c) $n = 4$

$x^{\nu-1}x^{\nu-2}x^{\nu-3}$	q^ν	x^ν 0	1
000	a	a	e
001	b	a	e
011	c	b	f
010	d	b	f
100	e	d	h
101	f	d	h
111	g	c	g
110	h	c	g

$q^{\nu+1}$

(b) $n = 3$

FIGURE 10.11. *Standard state tables for single input finite input memory circuits of length n.*

should use the standard circuit form and proceed directly to the design of the output logic. If the number of distinct outputs is less than half the number of states, the designer should generally minimize the corresponding state table before deciding on a circuit form.

Although *finite input memory* circuits are possibly the easiest to recognize and design, general finite memory circuits (of the form of Fig. 10.9b) are also important, and recognizing their finite memory character can often simplify the design process. This is illustrated in the next example.

Example 10.4

Design a 3-bit (modulo-8) up-down counter. The count is to appear as a binary number on three output lines, z_2, z_1, and z_0. The most significant bit is z_2. The count will change with each clock pulse. If the input $x = 1$, the count will increase. If $x = 0$, the count will decrease. Modulo-8 implies that increasing the count from 7 results in a count of 0, and decreasing it from 0 gives a count of 7.

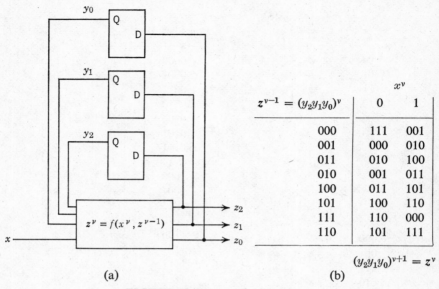

(a)   (b)

FIGURE 10.12.   *Up-down counter.*

Solution: One can immediately recognize the up-down counter as a *finite output memory* machine of memory length 1. The value of the output z is a function of only the immediately previous output, z^{v-1}, and the present input, x. z, however, is a 3-bit vector with components z_2, z_1, and z_0, so that three flip-flops are required to represent one past output.

A circuit configuration and the corresponding transition table are shown in Fig. 10.12. The transition table is constructed directly by noting that the desired present values of the outputs are the next values of the state variables. For example, for the circled entry in Fig. 10.13b, the previous count (present state) was 011, so if $x = 1$, the output (next state) should be 100.

	x^v	
q^v	0	1
q_0	q_7, 7	q_1, 1
q_1	q_0, 0	q_2, 2
q_2	q_1, 1	q_3, 3
q_3	q_2, 2	q_4, 4
q_4	q_3, 3	q_5, 5
q_5	q_4, 4	q_6, 6
q_6	q_5, 5	q_7, 7
q_7	q_6, 6	q_0, 0

q^{v+1}, z

FIGURE 10.13.   *State table for up-down counter.*

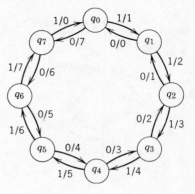

FIGURE 10.14.   *State diagram for up-down counter.*

Clearly, eight states are needed for a modulo-8 counter, so the table of Fig. 10.12b is minimal. If the circuit form of Fig. 10.12a is satisfactory, the designer can proceed directly to the design of the combinational logic circuits to realize the required outputs. However, even though the table is minimal, the form of Fig. 10.12 is not the only possible circuit form. Different types of flip-flops might be used, in which case the state variables need not be equal to the most recent outputs. For completeness, we assign state numbers to the transition table to obtain the state table of Fig. 10.13. The corresponding state diagram is given in Fig. 10.14.   ■

Two distinct approaches to state table synthesis have been described, both of which should be mastered by the student. Sometimes the most difficult task for the novice designer is deciding which method to use in a specific case. If the problem can be identified as finite memory, the direct synthesis of the state table or transition table is generally preferable. The state diagram approach works best when a reset or starting state can be identified; and, as it happens, machines with reset states are often non-finite memory machines. A non-finite memory machine will usually be driven to some holding state or caused to cycle within a subset of its states after it has made the desired response to some input sequence. Examples 10.1 and 10.2 illustrate machines with holding states. If such a machine is to be reusable, there must be an input sequence which will return it to its initial state. A fixed sequence which always will return a sequential machine to its intial state is called a *synchronizing sequence*. An initial state which can be reentered in this manner will be called a *reset state*. Often the synchronizing sequence will be a single input on a special line called the reset line.

Example 10.5

Repeat the design of the 0, 7 detector of Example 10.2, except that now the reset line d is to be omitted. Instead, a second 7 (three 1's in a row), coming any time after a 3-bit group has been tested, will signal that a new character will start arriving at the next clock time.

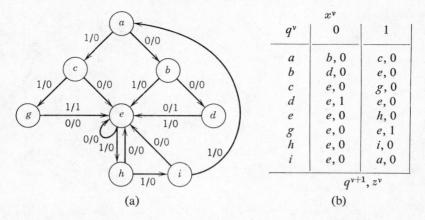

The state table is shown at right.

$$x^v$$

q^v	0	1
a	$b, 0$	$c, 0$
b	$d, 0$	$e, 0$
c	$e, 0$	$g, 0$
d	$e, 1$	$e, 0$
e	$e, 0$	$h, 0$
g	$e, 0$	$e, 1$
h	$e, 0$	$i, 0$
i	$e, 0$	$a, 0$

$$q^{v+1}, z^v$$

(a)   (b)

FIGURE 10.15. *State diagram and state table, Example 10.6.*

Solution: The state diagram (Fig. 10.15a) through state g is the same as Fig. 10.7a with e and f combined, as discussed in Example 10.2. We now recognize e as the holding state characteristic of non-finite memory circuits. The reset is now provided by a sequence of three 1's taking the circuit from e back to a via h and i. The resultant state table is shown in Fig. 10.15b. Since the sequence of three 1's must be applied after the circuit has reached state e, it is not strictly speaking a synchronizing sequence.

In some problems, it might not be possible to identify either a reset state or a bound on the memory. In such cases, one must arbitrarily choose some state whose past history can be specified as closely as possible to use as an initial state. From here, special care should be applied in the generation of a state table. Occasionally the reader may encounter the dilemma of a finite memory machine with a reset state. In that case, either approach may be used.

10.5   Equivalence Relations

The procedures discussed in Section 10.4 provide means of obtaining a state table for a given specification, but the resultant table is not unique and may not be optimum in terms of the ultimate realization. Since the number of flip-flops in a circuit increases as the number of states increases, we may say that a minimal state table is one with a minimum number of states. In some cases, different states in the original state table will be found to perform the same function. Or more specifically, it may not be possible, by means of output measurements, to distinguish between two or more states.

In such cases, the states are said to be *equivalent* and may be replaced by a single state. In order to develop a more concrete definition of the equivalence of states, let us first pursue the concept of an *equivalence relation*.

When an ordered pair of elements, (x, y), possesses some relating property we may speak of x as being *R-related* to y. This notion is symbolized by xRy. The relation, R, is defined as the set of all ordered pairs which possess this same particular property, whatever it may be. It is perhaps worth remarking that the elements x and y might be statement variables, real numbers, integers, or most anything. For our purposes, it can be assumed that R is a relation on a set of elements such that either x or y may represent any element in the set.

If xRx for every x in the set of interest, then the relation R is said to be *reflexive*. The relation "is greater than or equal to" is reflexive, while the relation "is greater than" is not reflexive.

If yRx whenever xRy, then the relation R is *symmetric*. On sets of usual interest, a symmetric relation will be reflexive as well. An exception is a relation such as "is a cousin of," which is symmetric. In the usual sense, one is not his own cousin, so this relation is not reflexive. Neither of the relations in the previous paragraph ($x \geq y$ or $x > y$) is symmetric.

Finally, if xRy and yRz imply xRz, then R is a *transitive* relation. A relation satisfying all three of the above criteria is called an *equivalence* relation. An example of an equivalence relation might be "is a member of the same team as."

An equivalence relation on a set will always partition the set into disjoint subsets, known as *equivalence classes*. For example, the set

$$\{A, a, B, b, C, c, D, d, E, e\}$$

is partitioned into the subsets

$$\{A, B, C, D, E\} \quad \text{and} \quad \{a, b, c, d, e\}$$

if xRy is interpreted to mean that x and y are the same case. A different group of disjoint subsets is obtained if xRy is taken to mean that x and y are the same letter.

Our intuitive concept of equality is an example of an equivalence relation since

$$x = x \qquad \text{(Reflexive)}$$

and

$$y = x \quad \text{if} \quad x = y \qquad \text{(Symmetric)}$$

then

$$\left. \begin{array}{l} x = y \quad \text{and} \quad y = z \\[2mm] x = z \end{array} \right\} \qquad \text{(Transitive)}$$

10.6   Equivalent States and Circuits

In order to define the notion of equivalent states and equivalent sequential circuits, we must introduce some new notation. State tables comprise tabular representations of two functions, the output function and the next-state function. If the outputs and next states are specified for every combination of inputs and present states, the circuits are called *completely specified* circuits. Let us denote the next-state function as δ and the output function as λ. For example, in the circuit of Fig. 10.16,

$$\lambda(q_2, 3) = 0 \tag{10.6}$$

and

$$\delta(q_2, 3) = q_4 \tag{10.7}$$

If a circuit is in some initial state, it will respond to a sequence of inputs with a specific sequence of outputs. Consider the case where the circuit of Fig. 10.16 is initially in state q_1 and is subjected to the series of inputs 0 2 3 0 0 1. In order to determine the output sequence in response to this input sequence it is necessary to determine also the next-state sequence.

$$\lambda(q_1, 0) = 0 \quad \text{and} \quad \delta(q_1, 0) = q_3$$

$$\lambda(q_3, 2) = 2 \quad \text{and} \quad \delta(q_3, 2) = q_1$$

$$\lambda(q_1, 3) = 0 \quad \text{and} \quad \delta(q_1, 3) = q_2$$

$$\lambda(q_2, 0) = 0 \quad \text{and} \quad \delta(q_2, 0) = q_3$$

$$\lambda(q_3, 0) = 0 \quad \text{and} \quad \delta(q_3, 0) = q_3$$

and, finally,

$$\lambda(q_3, 1) = 1 \quad \text{and} \quad \delta(q_3, 1) = q_1$$

The output sequence and final state may now be summarized as a function

x^ν Present State	0	1	3	2
q_1	$q_3,0$	$q_1,0$	$q_2,0$	$q_2,0$
q_2	$q_3,0$	$q_3,0$	$q_4,0$	$q_4,0$
q_3	$q_3,0$	$q_1,1$	$q_1,3$	$q_1,2$
q_4	$q_4,0$	$q_4,0$	$q_2,0$	$q_2,0$

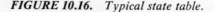

$$q^{\nu+1}, z$$

FIGURE 10.16.   *Typical state table.*

of the initial state and the input sequence, as follows:

$$\lambda(q_1, 023001) = 020001 \qquad (10.8)$$

$$\delta(q_1, 023001) = q_1$$

Now let us take a different point of view. Suppose that the output sequence for a given initial state and input sequence is known. Is the sequence of internal states then of any real interest? The answer is "No." To the "outside world," only the output response of a circuit to a sequence of inputs is significant. In a sense, the function of a sequential circuit may be regarded as the translation of one sequence of signals to a second sequence.

We are thus led toward a meaningful definition of the equivalence of two sequential circuits. First, let us define the equivalence of two states within these circuits.

Definition 10.1.   Let S and T be completely specified circuits, subject to the same possible input sequences. Let $(X_1, X_2, \ldots, X_n)$ represent a sequence of possible values of the input set X, of arbitrary length. States $p \in T$ and $q \in S$ are *indistinguishable* (equivalent), written $p \equiv q$, iff (if and only if) $\lambda_T(p, X_1, X_2, \ldots, X_n) = \lambda_s(q, X_1, X_2, \ldots, X_n)$ for every possible input sequence.

This definition may be applied equally well to different states of a single circuit. Less formally, p and q are equivalent if there is no way of distinguishing between them on the basis of any output sequences starting from these states. We shall use the terms indistinguishable and equivalent almost synonymously. Indistinguishable will be used where necessary to emphasize that indistinguishability is only one of many equivalence relations which might be defined on the states of a state table.

The equivalence of states in a single circuit is an equivalence relation. This can be easily verified as follows. Since the machine is completely specified, the output sequence resulting from the application of a given input sequence to a particular initial state q will always be the same. Therefore, $q \equiv q$, satisfying the reflexive law. Clearly, no distinction can be made between $p \equiv q$ and $q \equiv p$, so the symmetric law is satisfied. Furthermore, if p and q lead to identical output sequences for every input sequence and q and r lead to identical output sequences for every input sequence, certainly p and r must lead to identical output sequences for every input. That is, if $p \equiv q$ and $q \equiv r$, then $p \equiv r$; and the transitive law is satisfied.

Definition 10.2.   The sequential circuits S and T are said to be equivalent, written $S \equiv T$ iff for each state p in T there is a state q in S such that $p \equiv q$ and, conversely, for each state q in S there is a state p in T such that $q \equiv p$.

S	$X = 0$	$X = 1$
q_1	$q_3,0$	$q_2,1$
q_2	$q_1,1$	$q_2,0$
q_3	$q_1,0$	$q_2,1$

T	$X = 0$	$X = 1$
p_1	$p_1,0$	$p_2,1$
p_2	$p_1,1$	$p_2,0$

FIGURE 10.17

Less formally, two circuits are equivalent if there is no way to distinguish between them simply by observing the response to input sequences.

As an example, consider the simple circuits described by the state tables of Fig. 10.17. We first examine circuit S. Suppose that this circuit is initially in state q_1. An input of 0 will take the circuit to state q_3 with an output of 0. As long as 0 inputs continue, the circuit will cycle back and forth between q_3 and q_1 with 0 outputs. Should a 1 input occur, the circuit will go to q_2, and the output will be 1. A 0 input would then send the circuit back to q_1.

Notice that if the circuit were initially in state q_3, exactly the same output sequences as outlined above would occur. This is illustrated in Fig. 10.18 for two input sequences. These two input sequences are, in fact, sufficient to demonstrate that states q_1 and q_3 are equivalent. If the reader remains in doubt, he should test the circuit with input sequences of his own design until he becomes convinced.

Notice now that these or any other input sequences will also result in the same outputs for initial state p_1 in circuit T. Thus, p_1 is equivalent to both q_1 and q_3. Similarly, p_2 is equivalent to q_2. Thus, corresponding to every state in T, there is an equivalent state in S, and vice versa. Therefore, by Definition 10.2, S and T are equivalent.

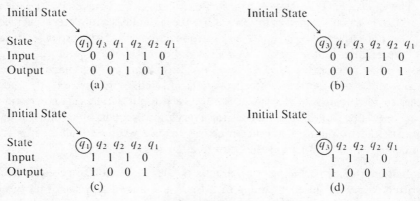

FIGURE 10.18

Our basic problem is to find an optimum state table. In view of the above definitions, the problem can now be stated in more specific terms, as follows: Given a state table, find an equivalent state table with as few states as possible. A table with fewer states may require less memory, thus generally leading to a more economic realization. For example, circuit S of Fig. 10.17 has three states and would therefore require two flip-flops for realization. Circuit T, having only two states, would require only a single flip-flop.

10.7   Determination of Classes of Indistinguishable States

Our basic approach to finding an optimum state table will be to partition the state table into the smallest possible number of *equivalence classes* of indistinguishable states. We shall then show that an equivalent sequential circuit can be formed by defining one state corresponding to each class of indistinguishable states. There are many ways to partition the states of a sequential circuit into disjoint classes. Not all such partitions result in equivalence classes of indistinguishable states. The following theorem provides a test for this condition.

THEOREM 10.1.   *Let the states of a sequential circuit be partitioned into disjoint classes. $p \overset{\Delta}{=} q$ denotes that states p and q fall into the same class in the partition. The partition is composed of equivalence classes of indistinguishable states (two indistinguishable states must be in the same class) if and only if the following two conditions are satisfied for every pair of states p and q in the same class $(p \overset{\Delta}{=} q)$ and every single input X.*

(1) $\lambda(p, X) = \lambda(q, X)$

(2) $\delta(p, X) \overset{\Delta}{=} \delta(q, X)$

Proof: A. (Necessity.) Assume that the partition consists of equivalence classes of indistinguishable states. Let p and q be any pair of states such that $p \overset{\Delta}{=} q$. Therefore, $p \equiv q$. By Definition 10.1, we can write

$$\lambda(p, XX_1 X_2 \cdots X_k) = \lambda(q, XX_1 X_2 \cdots X_k) \tag{10.9}$$

for any sequence $X_1 X_2 \cdots X_k$ and any single input X. Both sides of Equation 10.9 may equally well be written as the concatenation of a single output and the remaining sequence of outputs.

$$\lambda(p, X)\lambda[\delta(p, X), X_1X_2 \cdots X_k] = \lambda(q, X)\lambda[\delta(q, X), X_1X_2 \cdots X_k] \tag{10.10}$$

Therefore,

$$\lambda(p, X) = \lambda(q, X)$$

and
$$\lambda[\delta(p, X), X_1 X_2 \cdots X_k] = \lambda[\delta(q, X), X_1 X_2 \cdots X_k] \tag{10.11}$$
so that
$$\delta(p, X) \equiv \delta(q, X) \quad \text{and} \quad \delta(p, X) \overset{\Delta}{=} \delta(q, X) \quad \text{Q.E.D.}$$

B. (Sufficiency.) Assume that the states of a sequential circuit are partitioned into a set of equivalence classes such that two states fall into the same class if and only if conditions (1) and (2) are satisfied. Let p and q be any two states in the same equivalence class, i.e., $p \overset{\Delta}{=} q$. Suppose that $\delta(p, X_1 X_2 \cdots X_i) \overset{\Delta}{=} \delta(q, X_1 X_2 \cdots X_i)$. Therefore, by condition (2),
$$\delta(p, X_1 X_2 \cdots X_{i+1}) \overset{\Delta}{=} \delta(q, X_1 X_2 \cdots X_{i+1})$$
so that, by induction,
$$\delta(p, X_1 X_2 \cdots X_R) \overset{\Delta}{=} \delta(q, X_1 X_2 \cdots X_R) \tag{10.12}$$
for all R. Since
$$\lambda(p, X_1 \cdots X_R) = \lambda(p, X_1)\lambda(\delta(p, X_1), X_2) \cdots \lambda(\delta(p, X_1 \cdots X_{R-1}), X_R)$$
and
$$\lambda(q, X_1 \cdots X_R) = \lambda(q, X_1)\lambda(\delta(q, X_1), X_2) \cdots \lambda(\delta(q, X_1 \cdots X_{R-1}), X_R \tag{10.13}$$
and the sequences of symbols on each side of these expressions are identical, we have
$$\lambda(p, X_1 \cdots X_R) = \lambda(q, X_1 \cdots X_R)$$
for any k. Therefore, by Definition 10.1, $p \equiv q$, and the partition is composed of equivalence classes of indistinguishable states. Q.E.D.

We are now ready to use Theorem 10.1 to partition example sequential circuits into equivalence classes of indistinguishable states. As a first example, we shall use the partitioning technique to reduce the number of states in the error-detecting sequential circuit of Example 10.3. A technique usually referred to as the Huffman-Mealy method will be illustrated in Example 10.6.

Example 10.6

Partition the states of the sequential circuit in Fig. 10.19 into equivalence classes of indistinguishable states.

The method consists of two steps. The first step is to partition the states into the smallest possible number of equivalence classes so that states in the same class have the same outputs, i.e., satisfy condition (1) of Theorem 10.1. This is done by inspection of the state table. For the state table in Fig. 10.19, we see that three classes are required. Class **a** will include states whose output is 1 for input 0 and 0 for input 1. Class **b** will include states whose output is 0 for either input. Class **c**

$$x^v$$

q^v	0	1
q_0	$q_0, 1$	$q_4, 0$
q_1	$q_0, 0$	$q_4, 0$
q_2	$q_1, 0$	$q_5, 0$
q_3	$q_1, 0$	$q_5, 0$
q_4	$q_2, 0$	$q_6, 1$
q_5	$q_2, 0$	$q_6, 1$
q_6	$q_3, 0$	$q_7, 1$
q_7	$q_3, 0$	$q_7, 1$

$$q^{v+1}, z^v$$

FIGURE 10.19.   *State table for error detector.*

will include states whose output is 0 for input 0 and whose output is 1 for input 1.

$$
\begin{array}{ccc}
\mathbf{a} & \mathbf{b} & \mathbf{c} \\
(q_0) & (q_1, q_2, q_3) & (q_4, q_5, q_6, q_7)
\end{array}
$$

The classes of indistinguishable states will always be composed of subclasses of classes **a**, **b**, and **c**. These classes are the largest classes which will satisfy condition (1). Insisting on the satisfaction of condition (2) will only further subdivide the classes.

This partition satisfies the first condition of Theorem 10.1. If the next states for each state in class **b** (and for the states in class **c**) are in the same class for every possible single input, the second condition of Theorem 10.1 would be satisfied. In such a case, classes **a**, **b**, and **c** would be the desired equivalence classes. Let us determine for each state whether the next states are in classes **a**, **b**, or **c**. Below, the next-state class for a 0 input is written to the left of each state; and the next-state class for a 1 input is written to the right.

Class	**a**	**b**			**c**			
States Next Class	0 a c	1 a c	2 b c	3 b c	4 b c	5 b c	6 b c	7 b c

Clearly class **b** contains pairs of states which do not satisfy condition (2) and are, therefore, not indistinguishable. We must separate class **b** into two classes **b** and **d**. The new partition may or may not consist of equivalence classes of indistinguishable states. To find out, the process of listing the classes containing the next states is repeated.

Class	a	b		c				d
States	0	2	3	4	5	6	7	1
Next Class	a c	d c	d c	b c	b c	b c	b c	a c

In this case, the four classes, **a**, **b**, **c**, and **d** are equivalence classes of indistinguishable states. Notice that all states in a class have next states in the same class for inputs or 0 and 1. Thus, condition (2) is finally satisfied. ∎

The partitioning of states into classes whose outputs agree and whose next states are in the same class is graphically illustrated in Fig. 10.20 for the above example. It should be clear to the reader that the same inputs applied to two states in the same class will result in the same output and send the circuit to next states in the same class. Any input applied to both of these states will yield the same output and again send the circuit to next states in the same class, and so on indefinitely. Therefore, the class containing the two initial states is an equivalence class. Similar arguments apply to the other classes.

Having arrived at a set of equivalence classes of states within a circuit, S, as in the previous example, it remains to determine from them a minimal state circuit equivalent to S.

Equivalence Class Circuit. Corresponding to a completely specified sequential circuit, S, form a circuit, T, with one state, p_j, corresponding to each

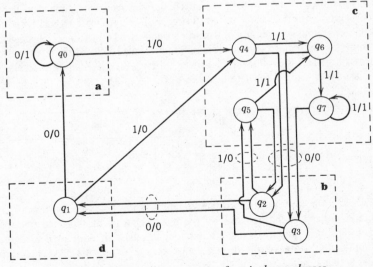

FIGURE 10.20. *Illustration of equivalence classes.*

equivalence class, C_j, of the states of S. As each state in C_j has the same output, $\lambda(C_j, X)$, for any input X, let $\lambda(p_j, X) = \lambda(C_j, X)$. Similarly, each state of C_j has, for any input X, a next state in the same class, i.e., $\delta(C_j, X) \equiv q_k$ is any member of the equivalence class C_k. Therefore, we let $\delta(p_j, X) \equiv p_k$. Given this construction it can be shown by an argument similar to the proof of Theorem 10. 1 that for every class C_k, $p_k \equiv q_{ki}$ for every $q_{ki} \in C_k$. A circuit T, formed in this manner, is known as an equivalence class circuit.

THEOREM 10.2. *The circuit, T, as defined above is equivalent to S. Furthermore, no other circuit equivalent to S has fewer states than T, and any circuit equivalent to S having the same number of states as T must be T.*

Proof: Let q_j be any state in S. Since it falls in some equivalence class C_j, there is a corresponding state, p_j, in T such that, according to the above definition, $q_j = p_j$. Conversely, for any state p_j in T, there is a corresponding equivalence class C_j such that, by the above definition, p_j is equivalent to any state in the class. Thus, by Definition 10.2, the circuits are equivalent.

No single state of any circuit can be equivalent to two states, q_1 and q_2 of circuit S, which are in different equivalence classes. By definition, these states must have different output sequences for some input sequence. Thus, the minimal state equivalent of S can have no fewer states than T, and one state must be assigned to each class. Q.E.D.

Example 10.6 (*continued*)

Determine the minimal state equivalent of the sequential circuit in Fig. 10.19.

Solution: Corresponding to each of the classes **a**, **b**, **c**, and **d**, we define states p_1, p_2, p_3, p_4 in a circuit T. The outputs and next states, defined according to Theorem 10.2, are listed in the state table of Fig. 10.21a. The corresponding state diagram is given in Fig. 10.21b. It is interesting to observe that an 8-state circuit, apparently requiring three flip-flops in the direct realization of Fig. 10.10, has been reduced to a four-state machine. Conceivably one might have arrived at this four-state circuit directly by synthesizing the state diagram. Fig. 10.21b

p^ν \ x^ν	0	1	Class	Equivalent States of S
p_1	$p_1, 1$	$p_3, 0$	a	q_0
p_2	$p_4, 0$	$p_3, 0$	b	q_2, q_3
p_3	$p_2, 0$	$p_3, 1$	c	q_4, q_5, q_6, q_7
p_4	$p_1, 0$	$p_3, 0$	d	q_1

$p^{\nu+1}, z^\nu$

FIGURE 10.21. (a)

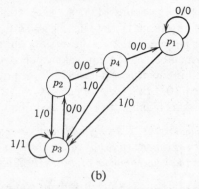

(b)

FIGURE 10.21.   *Minimal state equivalent of Fig. 10.19.*

has been deliberately arranged to illustrate this possibility. The pair of states p_2 and p_3 may have been set down as a starting point. These will be the states of the machine following a single 0 or a single 1, respectively. It is not always possible to be clever, and arriving at Fig. 10.21 after a minimization process would probably be the normal circumstance. ■

The state minimization process resulted in the saving of a flip-flop in the previous example and will do so again in Example 10.7. This will not always happen. It may be possible to reduce the number of states without reducing the number of flip-flops. Assume that the original number of states, N_0, which satisfies Equation 10.15 is reduced to N_1. The number of states, N_1, in the minimal circuit must be less than or equal to 2^{r-1} if a saving in memory elements is to be achieved. If the designer is certain that sufficient states cannot be eliminated, he may decide to skip the state minimization process.

$$2^r \geq N_0 > 2^{r-1} \tag{10.15}$$

Example 10.7

The state table for the "beginning of message" detector is repeated as Fig. 10.22a. Obtain a minimal state equivalent of this table.

q^v \\ x^v	0	1
q_0	$q_4, 0$	$q_1, 0$
q_1	$q_4, 0$	$q_2, 0$
q_2	$q_4, 0$	$q_3, 1$
q_3	$q_3, 0$	$q_3, 0$
q_4	$q_4, 0$	$q_1, 0$

q^{v+1}, z^v

(a)

q^v \\ x^v	0	1
a	$a, 0$	$c, 0$
b	$a, 0$	$d, 1$
c	$a, 0$	$b, 0$
d	$d, 0$	$d, 0$

q^{v+1}, z^v

(b)

FIGURE 10.22.   *State tables for beginning-of-message detector.*

Solution: We first partition the states into classes satisfying condition (1) and then find the classes containing the next states. We notice that state q_1 must

Class	a				b
States	q_0	q_1	q_3	q_4	q_2
Next Class	a a	a b	a a	a a	a a

be removed from class **a**. We thus define class **c** and repeat the process of classifying next states. This time we find that state q_3 must be removed from class **a**,

Class	a			b	c
States	q_0	q_3	q_4	q_2	q_1
Next Class	a c	a a	a c	a a	a b

so we define class **d**. After another pass, we find that states q_0 and q_4 are equivalent, and the process terminates. States a, b, c, and d are defined corresponding to the respective classes. Using the already tabulated information, the minimal state

Class	a		b	c	d
States	q_0	q_4	q_2	q_1	q_3
Next Class	a c	a c	a d	a b	d d

table of Fig. 10.22b is easily generated according to Theorem 10.2. Once again, one might have recognized at the outset that q_0 and q_4 are equivalent.

10.8 Simplification by Implication Tables

An alternative method for determining the equivalence classes involves the use of the *implication table*, based on the work of M. C. Paull and S. H. Unger.[2] This method will probably seem more time-consuming than the partitioning method. However, unlike the partitioning method, it can be extended to the incompletely specified case, as discussed in Chapter 12.

Definition 10.3. A set of states **P** is *implied* by a set of states **R** if, for some specific input X_j, **P** is the set of all next states $\delta(r, X_j)$ for all present states r in **R**.

Present State q^v	Input X^v			
	0	1	2	3
1	3	4	2	4
2	4	4	3	4
3	1	1	3	4
4	1	2	1	4
	q^{v+1}			

FIGURE 10.23.   *Next-state table.*

Consider the next-state table shown in Fig. 10.23. Assume that $\mathbf{R}_1 = (q_1, q_2)$. Then the sets *implied* by $\mathbf{R}_1$ are $\mathbf{R}_2 = (q_3, q_4)$ and $\mathbf{R}_3 = (q_2, q_3)$, since $\delta(q_1, 0) = q_3$ and $\delta(q_2, 0) = q_4$; and $\delta(q_1, 2) = q_2$ and $\delta(q_2, 2) = q_3$. Similarly, set $\mathbf{R}_2 = (q_3, q_4)$ *implies* $\mathbf{R}_1 = (q_1, q_2)$ and $\mathbf{R}_4 = (q_1, q_3)$. Note that a given set, e.g., $\mathbf{R}_1$, can both *imply* and *be implied*. By Theorem 10.1, the states in a set $\mathbf{R}$ are equivalent only if all the states in any set $\mathbf{P}$ implied by $\mathbf{R}$ are equivalent. This follows directly from the definition of implication and condition (2) of Theorem 10.1. Again, referring to Fig. 10.23, it is apparent that the states in $\mathbf{R}_1 = (q_1, q_2)$ cannot be equivalent unless the states in $\mathbf{R}_2 = (q_3, q_4)$ and $\mathbf{R}_3 = (q_2, q_3)$ are also equivalent. The implication table provides a systematic procedure for determining the equivalence classes of the states of the circuit.

Example 10.8
Use an implication table to find a minimal state equivalent to the state table given in Fig. 10.24.

q^v	$x^v = 0$	$x^v = 1$	q^v	$x^v = 0$	$x^v = 1$
1	2, 0	3, 0	7	10, 0	12, 0
2	4, 0	5, 0	8	8, 0	1, 0
3	6, 0	7, 0	9	10, 1	1, 0
4	8, 0	9, 0	10	4, 0	1, 0
5	10, 0	11, 0	11	2, 0	1, 0
6	4, 0	12, 0	12	2, 0	1, 0
	q^{v+1}, z			q^{v+1}, z	

FIGURE 10.24.   *State table, Example 10.8.*

Solution: *For completely specified* state tables, it is sometimes possible to reduce the number of states informally before using a formal minimization technique. This is worthwhile since considerable labor is involved in both formal methods. Notice in Fig. 10.24 that outputs and next states of states 11 and 12 are identical. Thus, these two states are indistinguishable; and state 12 may be eliminated from the table. Since state 12 is a next state of states 6 and 7, we replace these entries with 11. The state table thus modified is given in Fig. 10.25a. Notice that only next state information is provided in this table. State 9, which has different outputs than the other states, is listed separately at the bottom of the table.

In Fig. 10.25a states 5 and 7 now seem to be indistinguishable. Eliminating state 7 and replacing all next state entries of 7 with 5 yields the state table of Fig. 10.25b. No further reductions are apparent. This version may be further simplified using an implication table.*

The implication table is shown in Fig. 10.25. Vertically down the left side are listed all the states of the reduced state table except the first, and horizontally across the bottom are listed all but the last. The table thus contains a square for each pair of states. We start by placing an X in any square corresponding to a pair of states which have different outputs and therefore cannot be equivalent. In this case, q_9 has a different output than any other state, so we place an X in every square in row 9. The other states all have the same output, so no more squares are marked X.

We now enter in each square the pairs of states implied by the pair of states corresponding to that square. We start in the top square of the first column, corresponding to the pair of sets (q_1, q_2). From the state table, we see that this pair implies the sets (q_2, q_4) and (q_3, q_5), so we enter 2–4 and 3–5 in this square. Proceeding to the next lower square, we see that (q_1, q_3) implies (q_2, q_6) and (q_3, q_5). We proceed in this manner to complete the first column. Note that there is only one entry in the square for (q_1, q_{11}). Since one of the implied sets is the single set, (q_2), only one equivalence pair, (q_3, q_1), is implied. We next proceed to the second column, and then the third, etc., until the table is complete (Fig. 10.25c).

The next step is to make a second "pass" through the table to see if any of the implied sets are ruled out as possible equivalence pairs by the squares which have been marked X. In general, if there is an X in a square with "coordinates" *j-k*, then any square containing *j-k* as an implied set may be X'd out. In this case, q_9 cannot be equivalent to any other state, so any implied sets containing 9 are also ruled out. So we place an X in every square containing a 9 (Fig. 10.25d). When this has been done, every square with coordinate 4 has been crossed out, indicating that state q_4 cannot be equivalent to any other state. We therefore proceed to cross out every square containing a 4 entry (Fig. 10.25d).

No other state is ruled out completely, so we now make a systematic check of all squares still not crossed out, moving from right to left. The first such square encountered is 8–11, which contains the entry 8–2. Square 8–2 has

---

* This reduced table could equally well form the starting point for the partitioning method.

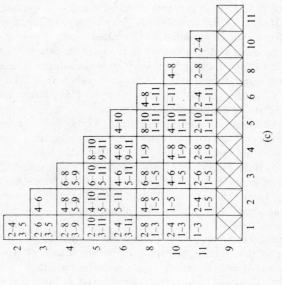
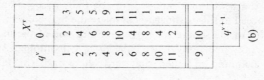

FIGURE 10.25. *Simplification by inspection and implication.*

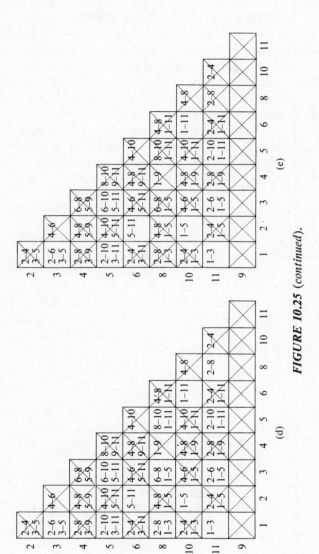

FIGURE 10.25 (continued).

already been crossed out, so we must cross out square 8–11 (Fig. 10.25e). Next, we find entry 1–11 in square 6–10. Square 1–11 is not crossed out, so we leave square 6–10 as is. In a similar fashion, we find that squares 5–8, 3–8, and 1–8 must be crossed out (Fig. 10.25e). When we have completed this "pass," we must make still another check of all remaining squares, since, for example, the elimination of square 1–8 might require the elimination of one of the remaining squares to the right. In general, this process must be repeated until a "pass" is completed without the elimination of any more squares. In this case, the final form is that shown in Fig. 10.25e. We have recopied the implication table twice for clarity in explaining the method, but in practice the complete procedure can be carried out on a single table.

In the completed implication table, each square not crossed out represents a pair of equivalent states. The equivalence classes may now be determined as follows. We first list the states corresponding to the columns of the implication table in reverse order (Fig. 10.26). We then check each column of the final implication table for any squares not crossed out, working from right to left. In Fig. 10.25e, there are no such squares in columns 11, 10, and 8, so dashes (—) are placed opposite 11, 10, and 8 in Fig. 10.26.

In column 6, we find square 6–10 not crossed out, so we enter the equivalence pair (6, 10) opposite 6 in Fig. 10.26. In column 5, we find the square 5–11 not crossed out, so we enter (5, 11) opposite 5 in Fig. 10.25 and recopy the previously determined pair (6, 10). In column 4, there are no squares not crossed out, so we simply recopy the already selected pairs opposite 4. In column 3, we find that q_3 is equivalent to both q_5 and q_{11}, and since q_5 and q_{11} are already equivalent, we add 3 to this set opposite 3 in Fig. 10.26. Similarly, column 2 adds q_2 to the set (q_6, q_{10}) and column 1 adds q_1 to the set (q_3, q_5, q_{11}). The

11	—
10	—
8	—
6	(6,10)
5	(5,11) (6,10)
4	(5,11) (6,10)
3	(3,5,11) (6,10)
2	(3,5,11) (2,6,10)
1	(1,3,5,11) (2,6,10)
Equivalence Classes	(1,3,5,11) (2,6,10) (4) (8) (9)

FIGURE 10.26. *Determination of equivalence classes from the implication table.*

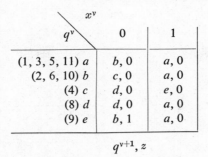

q^v	x^v 0	1
$(1, 3, 5, 11)\ a$	$b, 0$	$a, 0$
$(2, 6, 10)\ b$	$c, 0$	$a, 0$
$(4)\ c$	$d, 0$	$e, 0$
$(8)\ d$	$d, 0$	$a, 0$
$(9)\ e$	$b, 1$	$a, 0$

$$q^{v+1}, z$$

FIGURE 10.27. *Minimal state table for Example 10.8.*

resultant list opposite 1 in Fig. 10.26 comprises all equivalence classes with more than one member. Any states in the state table but not already on this list are not equivalent to any other state and must be included as single-state equivalence classes. These are added in the last line of Fig. 10.26.

Since there are five equivalence classes, the original state table of Fig. 10.24 reduces to the equivalent five-state table of Fig. 10.27.

Example 10.9

Find the minimal table equivalent to the state table of Fig. 10.28.

Solution: First we recopy the next-state portion of the state table, partitioning the states into groups with like output, as shown in Fig. 10.29a.

Present State q^v	Input x^v 0	1	2	3	0	1	2	3
1	6	2	1	1	0	0	0	0
2	6	3	1	1	0	0	0	0
3	6	9	4	1	0	0	1	0
4	5	6	7	8	1	0	1	0
5	5	9	7	1	1	0	1	0
6	6	6	1	1	0	0	0	0
7	5	10	7	1	1	0	1	0
8	6	2	1	8	0	0	0	0
9	9	9	1	1	0	0	0	0
10	6	11	1	1	0	0	0	0
11	6	9	4	1	0	0	1	0
	q^{v+1}				z^v			

FIGURE 10.28. *State table, Example 10.9.*

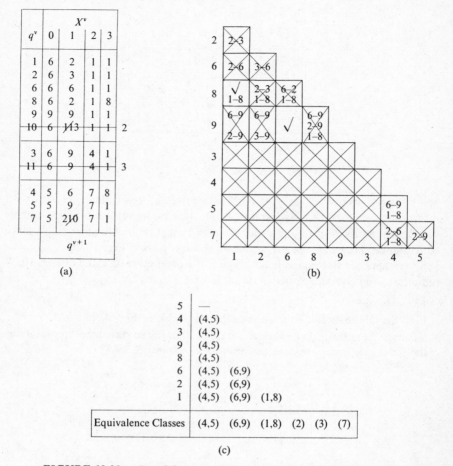

FIGURE 10.29.   *Simplification of Example 10.9 by implication table.*

We next check for states identical by inspection. States 3 and 11, which are in the same output group, have identical next state entries, so 11 may be replaced by 3. State 10 has 11 as a next state entry, so we replace it with 3, after which 10 is seen to be identical to state 2. We then replace the next state entry of 10 with 2 at state 7, but no further identities are found, so we proceed to the implication table, Fig. 10.29b.

Note the ($\sqrt{}$) in squares 1–8 and 6–9 of Fig. 10.29b. These indicate sets that imply only themselves and are therefore equivalent by inspection. Consider set (q_1, q_8), which has identical next state entries except for $X = 3$, for which the implied set is also (q_1, q_8). Thus, (q_1, q_8) does not imply any

x^v				
q^v	0	1	2	3
(4, 5) a	$a, 1$	$b, 0$	$f, 1$	$c, 0$
(6, 9) b	$b, 0$	$b, 0$	$c, 0$	$c, 0$
(1, 8) c	$b, 0$	$d, 0$	$c, 0$	$c, 0$
(2) d	$b, 0$	$e, 0$	$c, 0$	$c, 0$
(3) e	$b, 0$	$b, 0$	$a, 1$	$c, 0$
(7) f	$a, 1$	$d, 0$	$f, 1$	$c, 0$

$$q^{v+1}, z^v$$

FIGURE 10.30. *Minimal state table, Example 10.9.*

other sets; and since they have the same output, they are equivalent. Squares marked with a check will never be crossed out in the processing of the implication table. If any pairs with identical next-state entries and outputs should be overlooked in the preliminary reduction by inspection, they should also be marked with a ($\sqrt{}$) in the implication table. The minimal equivalent table is shown in Fig. 10.30.

10.9 State Assignment and Memory Element Input Equations

Once a minimal state table has been obtained, it remains to design a circuit realizing that table. It must be recognized that the question of economics has not been exhausted with the minimization of the number of states. It is true that the number of memory elements required by such a state table is minimum, but we also require combinational logic to develop the output and memory element input equations. It may be that some state table other than a minimal table would require significantly less combinational logic and, therefore, less overall hardware than the latter. However, we shall ignore this problem for now and concentrate our attention on the problem of realizing a given state table as economically as possible.

If the original state table synthesis was done by the finite memory approach and the number of states has not reduced sufficiently to eliminate any flip-flops, then the basic finite memory form of Fig. 10.9b should be used. In this circuit form, since the inputs and outputs are simply stored, the next-state equations require no combinational logic for realization. Another form might result in simpler output logic, but it is unlikely that this saving would outweigh the cost of memory element input logic.

If the finite memory form is not appropriate, the problem of choosing the best circuit realization is unfortunately a difficult problem in its own right. We recall that each of the states will correspond to one of the 2^r combinations of the r state variables. The first problem in designing a circuit to realize a given state table is to decide which of the 2^r combinations shall be assigned to each state. If the number of states m satisfies

$$2^{r-1} < m \le 2^r$$

then r state variables will be required. There will be

$$\frac{2^r!}{(2^r - m)!} \tag{10.16}$$

ways to assign the 2^r combinations of state variables to the m states.

The problem of assigning state-variable combinations to four states, say a, b, c, d, can be visualized as the assignment of the four states to the four vertices of the 2-cube. Three distinct assignments are shown in this fashion in Fig. 10.31. Any other possible assignments would amount to rotations or reversals of these three assignments and would thus correspond either to reversing the order of the variables or complementing one or both variables.

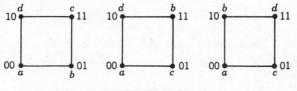

FIGURE 10.31

Such changes will not affect the form of any Boolean function and are thus irrelevant with respect to cost. The three distinct assignments for three rows can be obtained in a similar manner. The distinct assignments for three and four rows are listed in Table 10.1. The number of distinct assignments for various numbers of states is tabulated in Table 10.2.(2).

TABLE 10.1

States	3-state assignments			4-state assignments		
	1	2	3	1	2	3
a	00	00	01	00	00	00
b	01	11	11	01	11	10
c	11	01	00	11	01	01
d	—	—	—	10	10	11

TABLE 10.2

No. of states *m*	No. of state variables *r*	No. of distinct assignments
2	1	1
3	2	3
4	2	3
5	3	140
6	3	420
7	3	840
8	3	840
9	4	10,810,800

For circuits with only two states, there is no choice, and therefore no assignment problem. For three and four states, the simplest thing is to try all three possible assignments and see which produces the most economical circuit. For more than four states, complete enumeration is obviously impossible, so we will need to develop methods of picking a good assignment, preferably on the basis of the final state table. Let us start by taking a four-state example and trying all three possible assignments to see if we can gain some insight into the characteristics of a good assignment.

Figure 10.32a shows the minimal state table for the beginning-of-message detector, Examples 10.1 and 10.7. Figures 10.32b,c,d show the resultant next-state portion of the transition tables for assignments 1, 2, 3 of Table 10.2. Since there is only one output, the assignment will not affect the cost of realizing z, so the output portions of the transition tables have been omitted for clarity. Note that only the order of the states is changed; the state variables are always listed in K-map order, as required for transition tables. This is important, as the next step is to translate the transition tables into K-maps, known as the excitation maps, representing the memory element input equations.

In Chapter 9, we developed K-maps representing the next states as a function of the present state and inputs for the common types of flip-flops. As we synthesize, we shall use this information in a slightly different form. In a J-K flip-flop, for example, we must have at our fingertips the values of J and K which will cause any particular state transition to take place. The values of J and K required to effect each of the four input transitions are given as a *transition list* in Fig. 10.33b. This table can be derived immediately from the K-map of Fig. 10.33a. Transition lists for SC, T, and D flip-flops are given in Fig. 10.33c,d, and e, respectively.

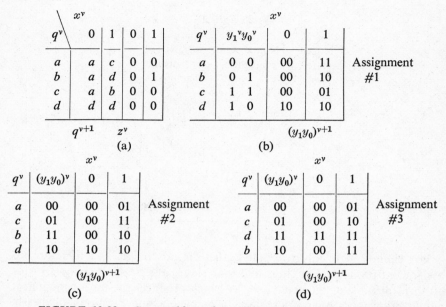

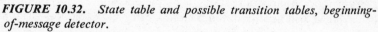

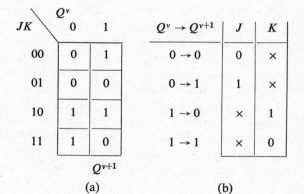

FIGURE 10.32. *State table and possible transition tables, beginning-of-message detector.*

FIGURE 10.33. *Transition lists.*

The left column in each diagram lists the four possible combinations of present value and desired next-value for the output, Q. The entries give the necessary values at the inputs to achieve these combinations. Consider for example, the diagram for the J-K flip-flop. If the present value is 0, i.e., the flip-flop is cleared, and it is to remain that way, then J must equal 0 and K can be either 0 or 1 (don't care). If Q is now 0 and is to be 1, i.e., the flip-flop is to be set, then $J = 1$ and $K = 0$ or 1. The other entries follow by similar reasoning.

To develop the excitation maps for assignment #1, we start with partial transition tables, each showing the transition in only a single state variable (Figs. 10.34a, 10.34d). From these, we generate the excitation maps for the flip-flops controlling each state variable by reference to the transition diagram for the type of flip-flop to be used. (*JK*, Fig. 10.33b.)

Consider the entries on the second row. For $(y_1 y_0 x)^v = 010$, we see from Fig. 10.34a that $y_1^v = 0$, $y_1^{v+1} = 0$, so $J_{y_1} = 0$ and $K_{y_1} = \times$. For $(y_1 y_0 x)^v = 011$, $y_1^v y_1^{v+1} = 01$, so $J_{y_1} = 1$ and $K_{y_1} = \times$. Similarly, from Fig. 10.34d, for both $(y_1 y_0 x)^v = 000$ and $(y_1 y_0 x)^v = 001$, $y_0^v = 1 \rightarrow y_0^{v+1} = 0$, so $J_{y_0} = \times$ and $K_{y_0} = 1$. This process can be speeded up by filling in all the don't cares first. Note that if the present value of the state variable is 0, $K = \times$ regardless of the next value, while $J = \times$ if the present value is 1.

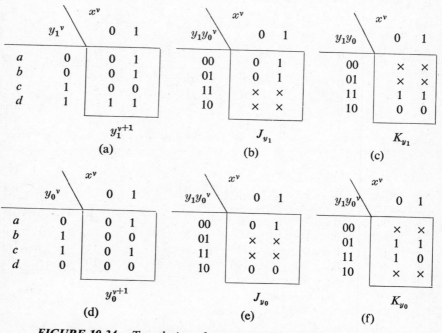

FIGURE 10.34.   *Translation of transition table to excitation maps.*

From these maps (Figs. 10.34b,c,e,f), we read off the minimal realizations, noting that the problem is multiple output, so products should be shared where possible. The minimal groupings are as shown. The equations for the assignment are:

$$J_{y_1} = x \qquad J_{y_0} = x\bar{y}_1 \qquad K_{y_1} = y_0$$
$$K_{y_0} = \bar{x} + \bar{y} \qquad z = x\bar{y}_1y_0 \tag{10.17}$$

Figure 10.35 shows the maps for Assignments 2 and 3. The resultant equations are:

Assignment #2

$$J_{y_1} = xy_0 \qquad K_{y_1} = \bar{x}y_0 \qquad z = xy_1y_0$$
$$J_{y_0} = x\bar{y}_1 \qquad K_{y_0} = \bar{x} + y_1 \tag{10.18}$$

Assignment #3

$$J_{y_1} = xy_0 \qquad K_{y_1} = \bar{x}\bar{y}_0 \qquad z = xy_1\bar{y}_0$$
$$J_{y_0} = x \qquad K_{y_0} = \bar{y}_1 \tag{10.19}$$

Assignments #1 and #3 are seen to have the same basic gate cost, but #3 is preferable because it uses only AND gates while #1 requires one OR gate. The circuit for #3 is shown in Fig. 10.36.

Before analyzing the final assignment in detail, let us consider another example to be sure the basic process is clear.

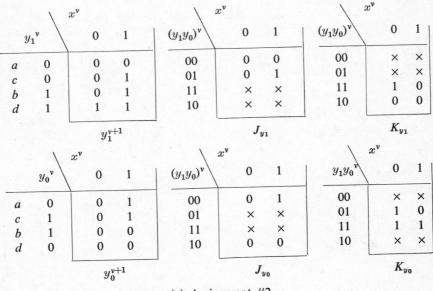

(a) Assignment #2

FIGURE 10.35. *Excitation maps for alternate assignments.*

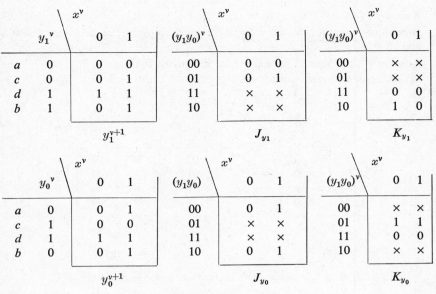

y_1^v		x^v 0	1
a	0	0	0
c	0	0	1
d	1	1	1
b	1	0	1

y_1^{v+1}

$(y_1y_0)^v$	x^v 0	1
00	0	0
01	0	1
11	×	×
10	×	×

J_{y_1}

$(y_1y_0)^v$	x^v 0	1
00	×	×
01	×	×
11	0	0
10	1	0

K_{y_1}

y_0^v		x^v 0	1
a	0	0	1
c	1	0	0
d	1	1	1
b	0	0	1

y_0^{v+1}

(y_1y_0)	x^v 0	1
00	0	1
01	×	×
11	×	×
10	0	1

J_{y_0}

$(y_1y_0)^v$	x^v 0	1
00	×	×
01	1	1
11	0	0
10	×	×

K_{y_0}

(b) Assignment #3

FIGURE 10.35 (*continued*).

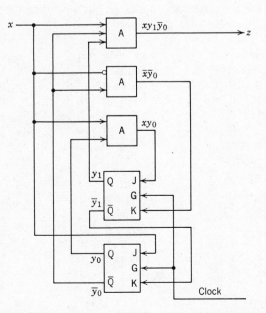

FIGURE 10.36. *Final circuit, beginning-of-message detector.*

Example 10.10

Determine a circuit realization for the error detector circuit of Examples 10.3 and 10.6 using S-C flip-flops.

Solution: The minimal state table for the circuit is shown in Fig. 10.37a, and the resultant transition table for Assignment #2 is shown in Fig. 10.37b.

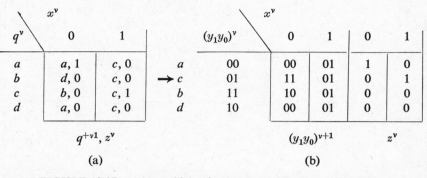

q^v \ x^v	0	1
a	$a, 1$	$c, 0$
b	$d, 0$	$c, 0$
c	$b, 0$	$c, 1$
d	$a, 0$	$c, 0$

q^{+v1}, z^v

(a)

$(y_1 y_0)^v$ \ x^v	0	1	0	1
a	00	00	01	1	0
c	01	11	01	0	1
b	11	10	01	0	0
d	10	00	01	0	0

$(y_1 y_0)^{v+1}$ z^v

(b)

FIGURE 10.37.   *State table and transition table, Example 10.10.*

The translation of the partial transition tables into the excitation maps for S-C flip-flops is shown in Fig. 10.38. The breaking up of the transition tables into partial tables for each variable is not absolutely necessary. One could work directly from the complete transition table, but separating the variables reduces

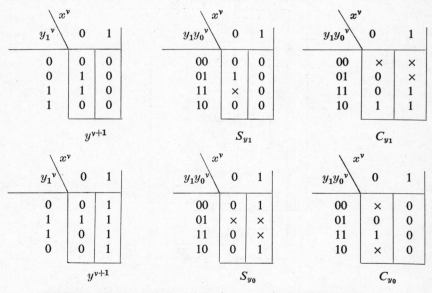

y_1^v \ x^v	0	1
0	0	0
0	1	0
1	1	0
1	0	0

y^{v+1}

$y_1 y_0^v$ \ x^v	0	1
00	0	0
01	1	0
11	×	0
10	0	0

S_{y1}

$y_1 y_0^v$ \ x^v	0	1
00	×	×
01	0	×
11	0	1
10	1	1

C_{y1}

y_1^v \ x^v	0	1
0	0	1
1	1	1
1	0	1
0	0	1

y^{v+1}

$y_1 y_0^v$ \ x^v	0	1
00	0	1
01	×	×
11	0	×
10	0	1

S_{y0}

$y_1 y_0^v$ \ x^v	0	1
00	×	0
01	0	0
11	1	0
10	×	0

C_{y0}

FIGURE 10.38.   *Derivation of excitation maps, Example 10.10.*

the likelihood of error. The equations for this assignment are:

Assignment #2

$$S_{y_1} = \bar{x}y_0 \qquad C_{y_1} = x + \bar{y}_0 \qquad S_{y_0} = x \qquad C_{y_0} = \bar{x}y_1$$
$$z = \bar{x}\bar{y}_1\bar{y}_0 + x\bar{y}_1y_0 \tag{10.20}$$

We will leave it to the reader to verify that the equations for assignment #1 and #3 are as follows:

Assignment #1

$$S_{y_1} = x + \bar{y}_1y_0 \qquad C_{y_1} = \bar{x}y_1 \qquad S_{y_0} = x \qquad C_{y_0} = \bar{x}\bar{y}_1$$
$$z = \bar{x}\bar{y}_1\bar{y}_0 + xy_1y_0 \tag{10.21}$$

Assignment #3

$$S_{y_1} = \bar{x}\bar{y}_1y_0 \qquad C_{y_1} = x + y_1y_0 \qquad S_{y_0} = x + y_1\bar{y}_0 \qquad C_{y_0} = \bar{x}y_0$$
$$z = \bar{x}\bar{y}_1\bar{y}_0 + xy_1y_0 \tag{10.22}$$

Assignment #2 is seen to lead to the simplest equations. We will leave it to the reader to draw the final circuit.

When there are only three or four states, all three possible assignments can be tried, as in the previous examples. The process may seem very complex and time consuming to the reader at this point, but it becomes surprisingly efficient as one gains experience. As we see from Table 10.2, however, if there are five or more states, trying all possible assignments is clearly out of the question. We need some rules or guidelines that will enable us to find assignments that are at least close to minimal with a reasonable amount of effort.

In an effort to find such rules, let us look more closely at the assignments in the previous examples to see if there are any characteristics which might have enabled us to predict in advance which assignment would lead to the minimal realization. In Fig. 10.39 are shown the state tables for the three assignments in Examples 10.7 and 10.10, with the states arranged in the K-map orders corresponding to each assignments. The minimal assignments for each case are enclosed by dashed lines.

A comparison of these maps leads to the observation that the selected assignments tend to group like next-state entries. For example, in Fig. 10.39a, we note that d is the next state for $x^v = 1$ for both states b and d, i.e.,

$$d = \delta(b, 1) = \delta(d, 1)$$

As a result, the next-state values of the state variables will, of course, be identical. If we give b and d adjacent assignments (i.e., place them on adjacent rows on the K-map), the present values of one of the state variables will be identical. Thus, the transitions in that state variable will be the same for both

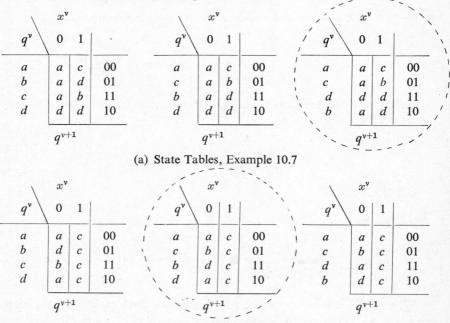

(a) State Tables, Example 10.7

(b) State Tables, Example 10.10

FIGURE 10.39.   *State tables for possible assignments.*

entries, and the corresponding entries on the excitation maps will be the same.

These identical transitions are important because the excitation maps will lead to the simplest equations if 1's, 0's, and don't-care's are grouped as much as possible. In this case, the selected assignments assign like values of y_1 to b and d, so the corresponding transitions in y_1 are identical. This effect is seen on the excitation maps (Fig. 10.35b), where the corresponding entries are also identical—don't cares for J_{y_1}, 0's for K_{y_1}.

Our basic rule, then, will be to group the next-state entries as much as possible. But there may be several ways of grouping like entries and this basic rule can be expanded and made more specific. Note that in the state table for Example 10.10 (Fig. 10.39b), states a and d have identical next-state entries in both columns, which is likely to produce a simpler map than identical entries in only one column, as was the case in Example 10.7. Further, if these two next-states, a and c, are given adjacent assignments, then the next-state entries for one of the variables will be the same in both columns of a given row, probably resulting in even further simplification. This can be seen in the excitation map of Fig. 10.38, where we have four 0's in the corresponding entries for S_{y_1} and a group of four 1's and don't cares for C_{y_1}.

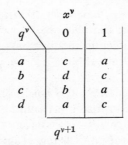

FIGURE 10.40. *Example state table.*

Next consider the state table shown in Fig. 10.40, in which a and c are again the next-states of a and d but this time not in the same columns. Nevertheless, if we make a and c adjacent, one state variable will have identical next-state entries in both columns of a given row, and if we make a and d adjacent, these rows will then pair off. A check of the three possible assignments will show that Assignment #2, which makes a,c and a,d adjacent, does lead to the best circuit for this table. Our discussion so far has been in justification of the three parts of Rule I given below.

In some tables, it will be possible to simplify one particular row of each excitation map by giving the various next states on a given row of the state table adjacent state variable assignments. This observation is formalized as Rule II. If further flexibility remains in the assignment after applying Rules I and II, the assignment may be completed in a way which will simplify the output map as stated in Rule III.

Rule I*a. Check for rows of the state table which have identical next-state entries in every column. Such rows should be given adjacent assignments.
If possible, the next-state entries in these rows should be given adjacent assignments according to Rule II.

b. Check for rows of the state table which have the same next-state entries but in different column-order. Such rows should be given adjacent assignments if the next-state entries can be given adjacent assignments.

c. Rows with identical next-state entries in some, but not all, columns should be given adjacent assignments, with rows having more identical columns having the higher priority.

Rule II.* Next-state entries of a given row should be given adjacent assignments.

Rule III. Assignments should be made so as to simplify the output maps.

---

* Although our justification and our emphasis are different in this edition, Rules I and II are the same rules presented in the first edition.

It is important to note that these rules are listed in decreasing order of their importance in producing good assignments. For tables of any complexity, they will generally lead to conflicting requirements, but an assignment made to satisfy a higher-priority rule should not be changed to satisfy a lower-priority rule.

The reader should note that Rule II has lower priority than all three parts of Rule I. It might seem surprising, as column-to-column adjacency would seem to be just as useful as row-to-row adjacency. The problem is that we have no control over column-order, this being a function of the input code. Thus, when we group the entries in one row, we are just as likely to "ungroup" the entries in another row. In general, Rule II produces useful results only if the adjacency requirements of several rows can be satisfied at the same time.

Let us now apply these rules to another example to clarify their use.

Example 10.11

Design a serial parity checker for 4-bit binary characters. Starting from reset, the circuit receives 4-bit characters serially on a single input line x. At the time of the fourth bit, the output is to be 1 only if the total number of 1's in the character was even. At all other times the output is to be 0. Following receipt of the fourth bit, the circuit should return to reset to receive the next character.

Solution: At any time, the only important factors are how many bits have been received and whether the number of 1's has been even or odd. This analysis leads directly to the state diagram of Fig. 10.41, which is clearly minimal.

The corresponding state table is shown in Fig. 10.42a. Inspecting this table, we see first that Rule Ia requires that states F and G be adjacent. Rule Ib in turn requires that D and E be adjacent if F and G can be made adjacent and that B and

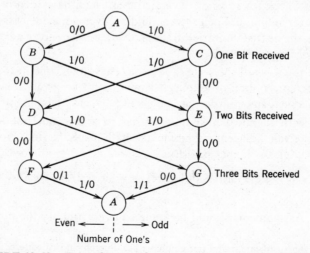

FIGURE 10.41.   *State diagram for parity checker, Example 10.11.*

C be made adjacent if D and E can be made adjacent. Thus adjacent assignments for B and C, D and E, and F and G will simultaneously satisfy all requirements of Rules Ia and b.

In setting up adjacent assignments for up to eight states, the Boolean hypercube is useful for visualization (Fig. 10.42b). The location of state A is apparently arbitrary, so we start by placing it at 000. Unless there are strong reasons to the contrary, it is best to have the reset state correspond to all flip-flops being cleared. This will usually simplify any separate reset circuit, as many flip-flops have special reset inputs available. With A located, we then place B and C, D and E, and F and G on succeeding edges of the cube, as shown. The corresponding transition table is shown in Fig. 10.42c. Note that one combination of state variables is not used since there are only seven states, so all entries for this row are don't cares.

	x^v	
q^v	0	1
A	$B, 0$	$C, 0$
B	$D, 0$	$E, 0$
C	$E, 0$	$D, 0$
D	$F, 0$	$G, 0$
E	$G, 0$	$F, 0$
F	$A, 1$	$A, 0$
G	$A, 0$	$A, 1$

$$q^{v+1}, z^v$$

(a)

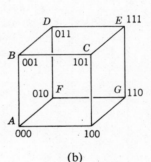

(b)

	x^v			
$(y_2y_1y_0)^v$	0	1	0	1
A 000	001	101	0	0
B 001	011	111	0	0
D 011	010	110	0	0
F 010	000	000	1	0
100	× × ×	× × ×	×	×
C 101	111	011	0	0
E 111	110	010	0	0
G 110	000	000	0	1
	$(y_2y_1y_0)^{v+1}$		z^v	

(c)

FIGURE 10.42. *State table, assignment cube, and transition table, Example 10.12.*

Top maps

y_2^v	x^v = 0	1
0	0	1
0	0	1
0	0	1
0	0	0
1	×	×
1	1	0
1	1	0
1	0	0

y_2^{v+1}

$y_2y_1y_0$	x^v = 0	1
000	0	1
001	0	1
011	0	1
010	0	0
100	×	×
101	×	×
111	×	×
110	×	×

J_{y_2}

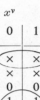

	x^v = 0	1
000	×	×
001	×	×
011	×	×
010	×	×
100	×	×
101	0	1
111	0	1
110	1	1

K_{y_2}

Middle maps

y_2^v	x^v = 0	1
0	0	0
0	1	1
1	1	1
1	0	0
0	×	×
0	1	1
1	1	1
1	0	0

y_1^{v+1}

$y_2y_1y_0$	x^v = 0	1
000	0	0
001	1	1
011	×	×
010	×	×
100	×	×
101	1	1
111	×	×
110	×	×

J_{y_1}

	x^v = 0	1
000	×	×
001	×	×
011	0	0
010	1	1
100	×	×
101	×	×
111	0	0
110	1	1

K_{y_1}

Bottom maps

y_0^v	x^v = 0	1
0	1	1
1	1	1
1	0	0
0	0	0
0	×	×
1	1	1
1	0	0
0	0	0

y_0^{v+1}

$y_2y_1y_0$	x^v = 0	1
000	1	1
001	×	×
011	×	×
010	0	0
100	×	×
101	×	×
111	×	×
110	0	0

J_{y_0}

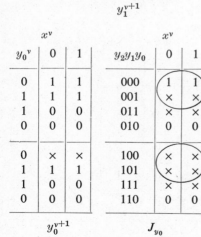
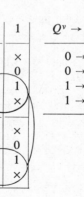

	x^v = 0	1
000	×	×
001	0	0
011	1	1
010	×	×
100	×	×
101	0	0
111	1	1
110	×	×

K_{y_0}

$Q^v \rightarrow Q^{v+1}$	J	K
$0 \rightarrow 0$	0	×
$0 \rightarrow 1$	1	×
$1 \rightarrow 0$	×	1
$1 \rightarrow 1$	×	0

FIGURE 10.43. *Excitation maps, Example 10.11.*

The translation of transition table to excitation maps for J-K flip-flops is shown in Fig. 10.43. The final equations are found to be:

$$J_{y_2} = x\bar{y}_1 + x\bar{y}_0 \quad K_{y_2} = x + \bar{y}_0 \quad J_{y_1} = y_0 \quad K_{y_1} = \bar{y}_0$$
$$J_{y_0} = \bar{y}_1 \quad K_{y_0} = y_1 \quad z = xy_2\bar{y}_0 + \bar{x}\bar{y}_2y_1\bar{y}_0 \tag{10.23}$$

We note that our rules have led us to an assignment that could hardly be simpler. We will leave it as an exercise for the reader to draw the final circuit.

It is important that the reader recognize that these rules for state assignment do not describe a complete algorithmic procedure leading directly to the best assignment. They reduce the number of alternative assignments we need consider and enable us to find a reasonably good assignment in a practical length of time. Consider again the possible assignments for Example 10.7, Fig. 10.39a. Rule Ia tells us that b and d should be adjacent, but Assignments #2 and #3 both meet this requirement. So how do we know that #3 is best? The rules give us no guidance here, but at least Rule Ic enables us to eliminate Assignment #1 from consideration.

For Example 10.11, Fig. 10.39b, Rule Ib is satisfied only by Assignment #2, and this does produce the best circuit for S-C flip-flops. But if we use J-K flip-flops, Assignment #1 turns out to be better. The reason is that on the excitation maps for J-K flip-flop, half the entries are always don't-cares, no matter what the assignment. If the present value of a state variable is 0, the K input will have no bearing on the next value. Similarly, if the present value is 1, the J input doesn't matter. The large blocks of don't cares combine with the remaining 1's and 0's in unpredictable ways which the rules cannot completely take into account. However, in this example, Assignment #1 is better then #2 for J-K's by only one gate, so the rules still get us close to the best assignment.

10.10 *Partitioning and State Assignment**

In Section 10.7, we encountered the concept of partitioning the states of a sequential circuit into equivalence classes such that two states, p and q, were in the same class ($p \triangleq q$) if and only if the following two conditions were satisfied for all inputs X.

(1) $\lambda(p, X) = \lambda(q, X)$
(2) $\delta(p, X) \triangleq \delta(q, X)$

---

* The material of this section is not essential to subsequent developments and may be omitted from a first reading.

$$x^v$$

q^v	0	1
0	7	1
1	0	2
2	1	3
3	2	4
4	3	5
5	4	6
6	5	7
7	6	0

$$q^{v+1}$$

FIGURE 10.44.   *State table, up-down counter.*

Now let us consider a partition inducing equivalence classes which satisfy only condition (2), i.e., the only requirement for states to be equivalent is for their next states to be equivalent.

Consider, for example, the state table of Fig. 10.44, which is a repeat of the state table for the up-down counter of Example 10.4. Note that the partitions $(0, 2, 4, 6)$ $(1, 3, 5, 7)$ and $(0, 4)$, $(1, 5)$, $(2, 6)$, $(3, 7)$ meet condition (2). For states $(0, 4)$, the next-states for $x = 0$ are $(3, 7)$ and for $x = 1$ are $(1, 5)$; for $(1, 5)$, the next-states are $(0, 4)$ and $(2, 6)$ for $x = 0$ and $x = 1$, respectively, etc. For $(0, 2, 4, 6)$, the next-state group is $(1, 3, 5, 7)$ for both inputs, and vice-versa. Thus, for all the states in any given block of a partition, the next-states are in a single block of the same partition for any given input.

Partitions having the property described above are called *closed*, or *preserved*, partitions. To illustrate why such partitions are useful for state assignment, let us assign one of the state variables to the two-block partition discussed above, i.e., let $y_1 = 0$ for states 0, 2, 4, 6 and let $y_1 = 1$ for states 1, 3, 5, 7. Since present states $(0, 2, 4, 6)$ always lead to next states $(1, 3, 5, 7)$, and vice-versa, if $y_1{}^v = 0$, then $y_1^{v+1} = 1$, and if $y_1{}^v = 1$, then $y_1^{v+1} = 0$. Thus, the next state equation for y_1 is simply

$$y_1^{v+1} = \overline{y_1{}^v}$$

Assigning state variables always induces partitions of the states into groups corresponding to various combinations of values of the variables. If we can assign the state variables so as to induce closed partitions, we will, as in the above example, generally obtain simpler equations than would otherwise be the case. If a partition is closed, then the block containing the next state is determined by the block containing the present state and, possibly, the input. Since there are fewer blocks than states, blocks can be identified by subsets of the state variables. If the partition is closed, the next values of the variables identifying that partition are thus determined by the present values of those variables and the inputs. The net result is to reduce the number of variables in the next-state equations, generally resulting in simpler equations.

Not only do closed partitions generally result in desirable assignments, they are also often closely related to the basic structure of the sequential circuit. In the above example, y_1 is a function only of itself, so that the circuitry developing y_1 will have no other inputs. In general, closed partitions result in the separation of the complete circuit into subcircuits, each responsible for computing some subset of the system variables and dependent only on some subset of the system variables. Figure 10.45 illustrates the manner in which an arbitrary circuit with four state variables might partition into such subcircuits. Notice that each subcircuit has as inputs some, but not all, of the state variable outputs from the other subcircuits. The flip-flop in circuit I, for example, has input equations which are functions of its own output y_0 and the overall circuit input x. Similarly, subcircuit II has as inputs only x, y_0, and its own outputs.

In general, a good partition of the states of a sequential circuit corresponds to a division of the circuit into subcircuits with as few interconnections as possible. Considerable research has been reported (4, 5, 6, 15) on the problem of finding suitable partitions, most of it directed to formal search procedures for closed partitions and other types of partitions having similar desirable properties. Unfortunately, these procedures require a much more extensive theoretical development than suitable for a book at this level. In addition, their application is tedious in the extreme and they often lead nowhere after very considerable effort has been expended.

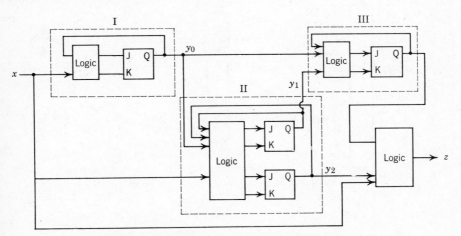

FIGURE 10.45. *Typical partitioned sequential circuit.*

Rather than concentrating on formal partitions of the states, we feel it is generally more productive to be alert for partitions of the circuit which might be suggested by the original problem formulation. Often the subcircuits will perform readily identifiable sub-functions, such as counting, shifting, keeping track of parity, etc. If the designer can recognize these subfunctions early in the process of analyzing the circuit, the state assignment problems may be considerably simplified. The 4-bit sequential parity checker of the previous section is an interesting example.

Example 10.11   (continued)

Recall that the function of the sequential parity checker is to check for even parity over sequences of four consecutive bits appearing on input line x. With the arrival of the fourth input, the output should be 1 if parity over the most recent four bits is even. Otherwise, the output will always be 0. The corresponding clock pulse must return the circuit to the reset state so that the parity check can begin anew with the next input. An external reset line will be provided to synchronize the parity checker with the first input character.

Let us forget that a state diagram has already been developed, and let us approach the problem by trying to identify subfunctions within the parity checker. Quite obviously, one function which must be accomplished is keeping track of parity as the bits arrive. From our experience with the parity checker in Chapter 9, we observe that only one flip-flop or state variable is required for that purpose. Next we notice that the circuit must behave differently when the fourth bit of a character arrives than it does for the first three. It is thus necessary to keep track of the number of bits received. A simple four-state counter which will do

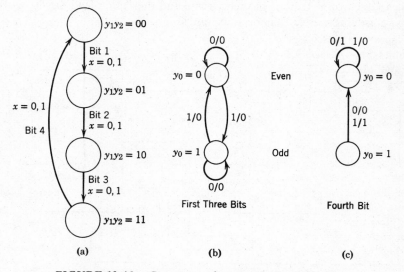

FIGURE 10.46.   *Counting and parity check subfunctions.*

xy_1y_2	$y_0 = 0$	$y_0 = 1$
0 0 0	0, 0	×
0 0 1	0, 0	1, 0
0 1 0	0, 0	1, 0
*0 1 1	0, 1	0, 0
1 0 0	1, 0	×
1 0 1	1, 0	0, 0
1 1 0	1, 0	0, 0
*1 1 1	0, 0	0, 1

$$y_0^{v+1}, z^v$$

(a)

y_1y_2	$x = 0$	$x = 1$
0 0	0 1	0 1
0 1	1 0	1 0
1 0	1 1	1 1
1 1	0 0	0 0

$$(y_1y_2)^{v+1}$$

(b)

FIGURE 10.47.   *Transition table for sequential parity checker.*

this is shown in Fig. 10.46a. In Fig. 10.46b and c we see the state diagram for the parity check flip-flop. Figure 10.46b, in which the output z is always 0, is valid for the first three inputs. The clock pulse corresponding to the fourth input resets the checker to even parity regardless of x, as shown in Fig. 10.46c. The output is specified as 1 if and only if parity over the 4-bit character has been even. Notice that the values of y_1 and y_2 are unaffected by the present values of x and y_0. However, the values of y_0^{v+1} and z^v are functions of the present values of x, y_1, y_2, and y_0 as indicated by Fig. 10.46. In order to merge the two functions into a single sequential circuit, we tabulate the values of y_0^{v+1}, as shown in Fig. 10.47a. The two rows designated by asterisks correspond to Fig. 10.46c, which is valid for bit 4 or when $y_1^v = y_2^v = 1$. The remaining rows correspond to Fig. 10.46b. There are two don't cares in the table, since y_0 will always be 0 when $y_1 = y_2 = 0$. Figure 10.47b is a transition table for the 2-bit binary counter. From these tables, the reader can verify the memory element input equations as given by Equation 10.24.

$$J_{y_2} = K_{y_1} = 1 \qquad J_{y_1} = K_{y_1} = y_2$$
$$J_{y_0} = x(\bar{y}_1 + \bar{y}_2) \qquad K_{y_0} = x + y_1y_2 \qquad (10.24)$$
$$z = y_1y_2(xy_0 + \bar{x}\bar{y}_0)$$

These equations are seen to be even simpler than those obtained earlier, with the y_2 flip-flop functioning as a T flip-flop, triggering with every pulse. With this selection of variables to perform distinct functions, the circuit breaks down into distinct subcircuits, as shown in Fig. 10.48.

Now that we have obtained a realization, let us examine the state table of the sequential parity checker. Figures 10.47a and b may be combined to form the transition table of Fig. 10.49a and the corresponding state table of Fig. 10.49b. Since y_0 is not an input to the 2-bit counter, we would expect to find a closed partition of the states into four classes, one for each combination of values of

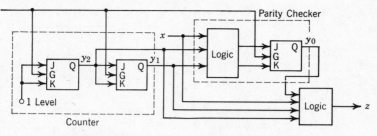

FIGURE 10.48.   *4-bit sequential parity checker.*

$y_1^{v}y_2^{v}$. This partition is (04) (15) (26) (37). That this partition is closed may be verified from Fig. 10.49c, in which the classes of next states for each class are tabulated.

The reader will note that Expressions 10.24 are somewhat simpler than those generated by Rules I, II, and III. If a student is perceptive (and lucky), partitioning can sometimes be a short route to an efficient realization.

However, obtaining a successful partition depends on the natural structure of the machine and the designer's ability to recognize this structure. As with

q	$y_0^{v}y_1^{v}y_2^{v}$	$x^{v} = 0$		$x^{v} = 1$	
q_0	0 0 0	0	01	1	01
q_1	0 0 1	0	10	1	10
q_2	0 1 0	0	11	1	11
q_3	0 1 1	0	00	0	00
q_4	1 0 0	—		—	
q_5	1 0 1	1	10	0	10
q_6	1 1 0	1	11	0	11
q_7	1 1 1	0	00	0	00

$$(y_0y_1y_2)^{v+1}$$

(a)

q^{v}	$x^{v} = 0$		$x^{v} = 1$	
q_0	q_1,	0	q_5,	0
q_1	q_2,	0	q_6,	0
q_2	q_3,	0	q_7,	0
q_3	q_0,	1	q_0,	0
q_4	—		—	
q_5	q_6,	0	q_2,	0
q_6	q_7,	0	q_3,	0
q_7	q_0,	0	q_0,	1

$$q^{v+1}, z^{v}$$

(b)

Present states	Next-states	
	$x^{v} = 0$	$x^{v} = 1$
(04)	(15)	(15)
(15)	(26)	(26)
(26)	(37)	(37)
(37)	(04)	(04)

(c)

FIGURE 10.49.   *State table for sequential parity checker.*

many aspects of engineering judgment, this ability grows with experience. We do not suggest that the reader fall back on intuitive design techniques, hoping blindly for a partitionable structure. Indeed, in many types of problems, none will exist. Instead we suggest that the reader be alert for a partitionable structure as he synthesizes each state diagram. If no such structure is evident, he will have to do the best he can with Rules I–III.

In some cases, the reader will be able to use natural structure to specify the state assignment only partially. He must then rely on Rules I–III to complete the realization. We leave it to the reader to integrate the two approaches as he gains experience.

We conclude this section with one more example.

Example 10.14

Let us consider again the modulo-8 up-down counter of Example 10.4, for which the state table is repeated in Fig. 10.5. We noted at the beginning of this section that there are two nontrivial closed partitions in this table, but let us assume that we are unaware of these and consider how our perception of the natural structure of this circuit might lead us to these partitions.

A natural partition for any counter is into blocks of states corresponding to odd counts and even counts. Such a partition will be closed if the number of states is even, since an odd count will always be followed by an even count, and vice-versa. For this circuit, this corresponds to the partition $(0, 2, 4, 6), (1, 3, 5, 7)$; and as already suggested, we will assign y_1 to this partition. We can then set up a transition table for the subcircuit developing y_1, as shown in Fig. 10.50b. The

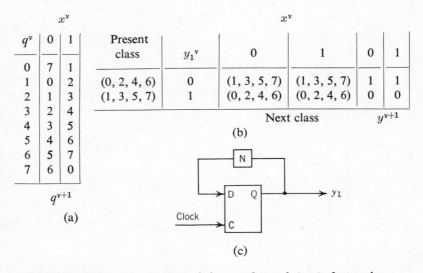

FIGURE 10.50. *State table and design of y_1 subcircuit for up-down counter.*

function of this subcircuit is to identify which class the overall circuit is in, $(0, 2, 4, 6)$ or $(1, 3, 5, 7)$. As shown, if the present class is $(0, 2, 4, 6)$, the next will be $(1, 3, 5, 7)$, and vice-versa. From this, we read off the equation

$$y^{v+1} = \overline{y_1^v} \tag{10.25}$$

as noted earlier. The corresponding circuit using a D flip-flop is shown in Fig. 10.50c.

Each state must be identified by a unique combination of state variables. If we are to identify eight states with three state variables, the second variable must divide the two four-state blocks in half, giving four two-state blocks, and the third variable must divide these in half, giving eight blocks of one state each. Following our odd-even approach, we will let $y_2 = 0$ specify the first and third states in each block of the y_1 partition and $y_1 = 1$ specify the second and fourth, i.e., y_2 induces the partitions $(0, 1, 4, 5)$ $(2, 3, 6, 7)$. The variables y_1 and y_2 together then induce the partition $(0, 4)$ $(1, 5)$ $(2, 6)$ $(3, 7)$, as shown in Fig. 10.51a.

$y_2 y_1$	
00	$(0, 1, 4, 5) \wedge (0, 2, 4, 6) = (0, 4)$
01	$(0, 1, 4, 5) \wedge (1, 3, 5, 7) = (1, 5)$
11	$(2, 3, 6, 7) \wedge (1, 3, 5, 7) = (3, 7)$
10	$(2, 3, 6, 7) \wedge (0, 2, 4, 6) = (2, 6)$

(a)

Present class	$y_2 y_1$	x^v 0	1	0	1
$(0, 4)$	00	$(3, 7)$	$(1, 5)$	1	0
$(1, 5)$	01	$(0, 4)$	$(2, 6)$	0	1
$(3, 7)$	11	$(2, 6)$	$(0, 4)$	1	0
$(2, 6)$	10	$(1, 5)$	$(3, 7)$	0	1

Next
class y_2^{v+1}

(b)

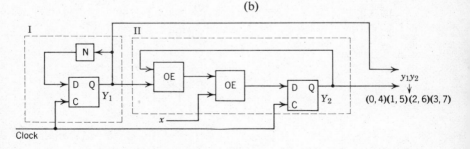

(c)

FIGURE 10.51. *Development of y_2 subcircuit of up-down counter.*

Since this is a closed partition, as noted earlier, we know that the next values of the state variables inducing it will not be functions of any other state variables. We can thus develop a transition table for the y_2 subcircuit, as shown in Fig. 10.51b. We do this by first using Fig. 10.50a to find the next state pairs corresponding to each pair of present states. Note that we do not show the next values of y_1 in this table because we have already designed a separate circuit for y_1. From this table, we can read off the equation

$$y_2^{v+1} = x^v \cdot (y_1{}^v \oplus y_2{}^v) + \bar{x}^v \cdot \overline{(y_1{}^v \oplus y_2{}^v)} = \bar{x}^v \oplus (y_1{}^v \oplus y_2{}^v) \quad (10.26)$$

The resultant interconnection of the two subcircuits inducing the partition (0, 4) (1, 5) (2, 6) (3, 7) is shown in Fig. 10.51c.

Finally, y_3 must divide those four blocks in half requiring that y_3 induce the partition (0, 1, 2, 3) (4, 5, 6, 7). Since we have already designed subcircuits to develop y_1 and y_2, it is obvious that the final subcircuit will develop only y_3. The transition table for y_3 is given in Fig. 10.52a. The complete counter circuit is shown in Fig. 10.52b with the already developed realizations of y_1 and y_2 represented by blocks I and II, respectively. It will be left to the reader to verify that the transition table of Fig. 10.52a leads to the equation

$$y_3^{v+1} = ((y_1{}^v \cdot y_2{}^v) \oplus y_3{}^v) \oplus \bar{x}^v \quad (10.27)$$

The reader may not feel that this procedure has produced a particularly economical design in view of the number of exclusive-OR gates, but this is the best assignment using D flip-flops. Any other assignment will make y_1 and y_2 functions of more variables, leading to a more complex design. We shall see in Chapter 11 that the use of T flip-flops will lead to a more economical design, but this design will be based in this same partition.

$$x^v$$

q^v	$(y_3 y_2 y_1)^v$	0	1	0	1
0	000	7	1	1	0
1	001	0	2	0	0
3	011	2	4	0	1
2	010	1	3	0	0
4	100	3	5	0	1
5	101	4	6	1	1
7	111	6	0	1	0
6	110	5	7	1	1
		q^{v+1}		y_3^{v+1}	

(a)

FIGURE 10.52. *Development of y_3 subcircuit of up-down counter.*

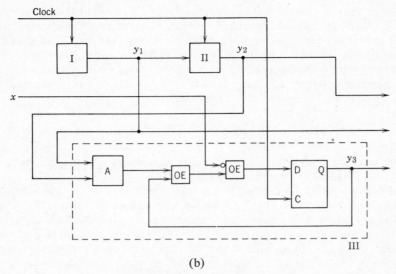

(b)

FIGURE 10.52 (*continued*).

We hope that the above example, illustrating as it does some of the more interesting and challenging aspects of partition theory, may motivate some of our readers to explore this fascinating subject in more depth (see References 6, 15, 16).

The relationship between state assignment partitions and the natural structure of the circuit function raises questions as to the desirability of state table minimization. In Example 10.5, we arrived at an eight-state table, which would require three state variables. If the reader will consider the basic structure, he will note two distinct three-state counting sequences. Perhaps a simpler and more natural structure would result with four state variables, two for each sequence. The extra state variables would require an extra flip-flop, but the simpler structure might simplify the excitation and output logic sufficiently to result in a net decrease in cost.

A special case where the use of extra flip-flops is very likely to lead to lower cost is the finite memory circuit. Recall the error detector of Example 10.3, which was originally formulated as a finite memory circuit, as shown in Fig. 10.10. If we realize the circuit in this form, the only cost, in addition to the three D flip-flops, is for logic to realize the output equation, Equation 10.5. We minimized this same table to four states in Example 10.6 and made a state assignment in Example 10.10. The equations for the selected assignment are shown in Equation 10.20. The output equations for the two cases are comparable in cost, so the saving of the flip-flop is offset by the cost of three gates to realize the excitation equations of Equation 10.20. As a rough rule of thumb, one flip-flop is equal in cost to two-to-four gates, so the basic logic cost for the two forms would be comparable. On the other hand, the four-state version will have many more interconnections between logic packages and will thus be more expensive to manufacture. In addition, use of the finite memory form initially would have saved the time spent in state reduction and assignment.

10.11 Conclusion

In this chapter, we have developed a design procedure which is considerably more dependent on the intuition and ingenuity of the designer than was the case for combinational logic circuits. This seeming lack of formality and rigor may mislead the novice designer into thinking the procedure is basically trial-and-error. But it is a carefully organized procedure, designed to lead the designer from an initial, imprecise specification to a final circuit with minimum likelihood of error along the way.

The first step, the setting up of the state diagram or state table, is the most important step. The initial specification of a circuit function is almost invariably vague and incomplete. The procedures developed in Sections 10.3 and 10.4 help clarify your understanding of the circuit function and force you to specify precisely the desired behavior in every possible circumstance. Possibly because it requires considerable time and effort to think the problem through carefully and thoroughly, there is a tendency for designers to skip this step. Instead, they try to apply the sort of intuition about structure that we discussed in connection with state assignment at the very beginning. They try to interconnect a set of hazily perceived subcircuits, such as counters, parity checkers, etc., in a trial-and-error fashion, usually producing a design that does many unexpected things. The time to apply intuition about circuit structure is after you thoroughly understand the circuit function.

If the circuit is finite memory and the transition table has been formulated accordingly, the only step left is to read the output equations directly from the transition table. As discussed above, where the finite memory form is applicable, it almost invariably produces the most economical designs and should be used.

If the circuit is not finite memory, the next step is to minimize the state table. As discussed, a minimal table does not necessarily result in the most economical design. However, it is best to start with a minimal table and consider adding more state variables only if there seems to be an obvious natural structure requiring more variables.

The final step is state assignment, which is a difficult and sometimes frustrating process. Our basic approach is to use a set of semi-empirical rules combined with a healthy dose of intuition to try to find desirable partitions.

It is quite natural to wish for rigorous procedures that will lead to guaranteed results. A great deal of research has gone into the effort to find such procedures, and papers on state assignment appear regularly in the journals reporting computer research. Perhaps some day a really good state assignment procedure will be found. For the present, we believe that the sort of "guided intuition" we have tried to illustrate in the last two sections remains the most satisfactory method.

Finally, we hope that the reader will acquire some perspective about where best to apply his efforts in this rather elaborate design process. The first step warrants the greatest attention because no amount of later effort can produce a good design from an initially incomplete or erroneous specification. The second step, state reduction, is rather tedious but has the virtue of producing guaranteed results (assuming no clerical errors) and is usually time well-spent. It is in the state assignment step that judgment must be applied. It is natural to seek simple, elegant designs and easy to spend more time on state assignment than on other parts of the design process. But engineering time is expensive, and many hours spent eliminating a gate or two can be a very bad investment unless production of the circuit is to be enormous.

The above paragraph and indeed most of this chapter is most pertinent to machines with relatively few states (≤ 32). As the total amount of information to be stored increases, so will the likelihood of uncovering a partitionable structure. For large circuits, some formal system description other than the state table would be very desirable. Such formal descriptions exist in the form of hardware register transfer languages. Such a language will be the subject of Chapter 15. As we shall see, a register transfer language description provides a starting point from which a digital system can be partitioned into subsystems. These subsystems may then be optimized using the techniques of this chapter or realized in terms of standard MSI components.

PROBLEMS

10.1. Analyze the sequential circuit given in Fig. P10.1 in the manner discussed in Section 10.1. Obtain first the flip-flop input equations and then the state diagram representation of the circuit.

10.2. A clock mode sequential circuit is to be designed featuring an external reset mechanism which will on occasion reset the circuit to state q_0. Determine a state diagram of the circuit so that it will generate an output of 1 for one clock period only coinciding with the second 0 input of a sequence consisting of exactly two 1's (no more than two) followed by two 0's. Once the output has been 1 for one clock period, the output will remain 0 until the circuit is externally reset to q_0.

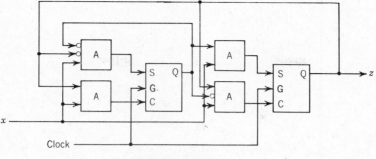

FIGURE P10.1

10.3. Derive the state diagram and state table of a circuit featuring one input line, X_1, in addition to the clock input. The only output line is to be 0 unless the input has been 1 for four consecutive clock pulses or 0 for four consecutive pulses. The output must be 1 at the time of the fourth consecutive identical input.

10.4. A sequential circuit is to be designed with two input lines x_1 and x_2 and a single output z. If a clock pulse arrives when $x_2 x_1 = 00$ the circuit is to assume a reset state which may be labeled q_0. Suppose the next 6 clock pulses following a resetting pulse coincide with the following sequence of input combinations 01—10—11—01—10—11. The output, z, is to be 1 coinciding with the sixth of such a string of six clock pulses but is to be 0 at all other times. The circuit cannot be reset to q_0 except by input 00. Define a special state q_1 to which the circuit will go once it becomes impossible for an output-producing sequence to occur. The circuit will thus wait in q_1 until it is reset. Determine a state diagram for this circuit.

10.5. A sequential circuit is to be designed in which the circuit output, z^v, is a function of only the current input x^v and the three previous inputs x^{v-1}, x^{v-2}, and x^{v-3}. A Boolean expression for z^v is given by

$$z^v = x^v \cdot x^{v-1} \cdot x^{v-2} \cdot x^{v-3} + \overline{x^v} \cdot \overline{x^{v-1}} \cdot \overline{x^{v-2}} \cdot \overline{x^{v-3}}$$

Treat the circuit as a finite input memory circuit and determine an 8-state state table representation.

10.6. A variable delay circuit is to provide a two or three clock period delay of an input signal, x_i. That is, the circuit output x_d^v is to be equal to x_i^{v-3} if the control input $r = 1$ and is to be equal to x_i^{v-2} if $r = 0$. A simplified block diagram of the circuit is shown in Fig. P10.6. Treat it as a finite memory circuit and obtain a state diagram for the variable delay sequential circuit.

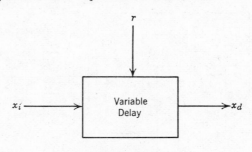

FIGURE P10.6

10.7. A sequential circuit has two inputs, x_1 and x_2. Five-bit sequences representing decimal digits coded in the 2-out-of-5 code appear from time to time on line x_1, synchronized with a clock pulse on a third line. Each five consecutive bits on the x_1 line, which occur while $x_2 = 0$ but immediately following one or more inputs of $x_2 = 1$, may be considered a code word. The single output z is to be 0 except upon the fifth bit of an invalid code word. Determine a state table of the above sequential circuit.

10.8. A clock mode sequential circuit has two input lines a and b. During any clock period, $a^v a^{v-1} a^{v-2}$ and $b^v b^{v-1} b^{v-2}$ may be regarded as 3-bit binary numbers. The output z is to be 1 if $a^v a^{v-1} a^{v-2} \geq b^v b^{v-1} b^{v-2}$. The inputs a^v and b^v are the most significant bits as implied by the format shown. Obtain a state table for this sequential circuit.

10.9. Let aRb if and only if $a = b \pm 8X$ where X is any integer. (We say that a is congruent to b modulo-8.) Show that R is an equivalence relation on the set of all integers.

10.10. Determine a minimal state table equivalent to the state table in Fig. P10.10. [*Answer:* 5 states are required.]

q^v	q^{v+1}, Z^v	
	$x = 0$	$x = 1$
1	1,0	1,0
2	1,1	6,1
3	4,0	5,0
4	1,1	7,0
5	2,0	3,0
6	4,0	5,0
7	2,0	3,0

FIGURE P10.10

10.11. Determine a minimal state table equivalent to the state table shown in Fig. P10.11.

Present State	Next State, Output			
	$x = 0$	$x = 1$	$x = 2$	$x = 3$
A	E,1	C,0	B,1	E,1
B	C,0	F,1	E,1	B,0
C	B,1	A,0	D,1	F,1
D	G,0	F,1	E,1	B,0
E	C,0	F,1	D,1	E,0
F	C,1	F,1	D,0	H,0
G	D,1	A,0	B,1	F,1
H	B,1	C,0	E.1	F,1

FIGURE P10.11

10.12. Determine a circuit whose output is equivalent to the circuit shown in Fig. P10.12 but which has a minimum number of states.

q^v	q^{v+1}, Z	
	$x = 0$	$x = 1$
q_1	$q_2,0$	$q_3,0$
q_2	$q_4,0$	$q_5,0$
q_3	$q_6,0$	$q_7,0$
q_4	$q_8,0$	$q_9,1$
q_5	$q_{10},0$	$q_{11},0$
q_6	$q_{12},1$	$q_{13},0$
q_7	$q_{14},0$	$q_{15},0$
q_8	$q_1,0$	$q_3,0$
q_9	$q_1,0$	$q_3,0$
q_{10}	$q_1,0$	$q_3,0$
q_{11}	$q_1,0$	$q_2,0$
q_{12}	$q_2,0$	$q_1,0$
q_{13}	$q_2,0$	$q_1,0$
q_{14}	$q_2,0$	$q_1,0$
q_{15}	$q_2,0$	$q_1,0$

FIGURE P10.12

10.13. The minimal state circuits satisfying the functions of previous problems are listed as follows. In each case where the circuit synthesized by the reader was not minimal, obtain a minimal machine by the formal techniques of Sections 10.7 and 10.8.

10.2   6 states
10.5   6 states
10.7 14 states

10.14. Determine a minimal state equivalent of the state table found in Problem 10.8.

10.15. Determine a realization of a clocked J-K flip-flop in terms of a clocked S-C flip-flop and appropriate combinational logic.

10.16. Obtain a circuit realization of the state table found in Problem 10.5:
(a) Using S-C flip-flops.
(b) Using D flip-flops.
(c) Using J-K flip-flops.

10.17. Obtain a circuit realization of the state table given in Fig. P10.17 using J-K flip-flops.

| | q^{v+1}, Z | |
q^v	$x = 0$	$x = 1$
A	B,0	B,1
B	F,0	D,1
C	E,1	G,1
D	A,0	C,0
E	D,1	G,0
F	F,0	A,0
G	C,1	B,0

FIGURE P10.17

10.18. Repeat Problem 10.17 using S-C flip-flops.

10.19. Repeat Problem 10.17 using D flip-flops.

10.20. Obtain circuit realizations for the minimal state tables for:
(a) Problem 10.2
(b) Problem 10.6
(c) Problem 10.14
using whichever type of flip-flop appears most suitable in each case.

10.21. A circuit with an external reset mechanism, a single input x, and a single output z is to behave in the following way. At the first clock time following reset, the output is to be $z = 0$. At the second clock time, $z^v = x^v \cdot x^{v-1}$. At the third clock time, $z^v = x^v + x^{v-1}$, and at the fourth clock time, $z^v = x^v \oplus x^{v-1}$. This sequence of output functions is then repeated every four clock periods until the circuit is reset. The circuit can be readily partitioned into two subcircuits, a 2-bit counter, and a 1-bit shift register. Take advantage of this fact to determine a state diagram and a realization of the circuit in terms of J-K flip-flops.

10.22. Obtain the state table of a serial translating circuit which performs the translations tabulated in Fig. P10.22. As indicated, there is only

X^v	X^{v-1}	X^{v-2}	Z^{v+2}	Z^{v+1}	Z^v
0	0	0	0	0	0
0	0	1	0	1	1
0	1	0	1	0	0
0	1	1	1	0	1
1	0	0	1	1	0
1	0	1	0	0	1
1	1	0	0	1	0
1	1	1	1	1	1

FIGURE P10.22

one input line, X, and one output line, Z, in the circuit. The inputs occur in characters of three consecutive bits. One character is followed immediately by another. A reset signal, which need not be included in the state table, will cause the translator to begin operation synchronized with the input. A translated output character must be delayed by only two clock periods with respect to an input character. [*Hint:* A 12-state minimal state table can be obtained by partitioning the circuit into a 3-state counter and a 2-bit shift register.]

10.23. Obtain a realization of the state table derived in Problem 10.22 using J-K flip-flops.

10.24. Repeat Problem 10.7 using an approach which partitions the circuit into a modulo-5 counter and a 2-bit counter for counting 1's appearing on line x_1. Determine a 5 flip-flop (J-K) realization of this circuit without minimizing the state table. Compare the overall complexity of this realization with a realization of a minimal equivalent of the state table determined in Problem 10.7.

BIBLIOGRAPHY

1. Huffman, D. A. "The Synthesis of Sequential Switching Circuits," *J. Franklin Inst.*, **257**, No. 3, 161–190; No. 4, 275–303 (March–April, 1954).

2. Paull, M. C. and S. H. Unger. "Minimizing the Number of States in Incompletely Specified Sequential Switching Functions," *IRE Trans. on Electronic Computers*, **EC-8**: 3, 356–357 (Sept. 1959).

3. McCluskey, E. J. and S. H. Unger. "A Note on the Number of Internal Variable Assignments for Sequential Switching Circuits," *IRE Trans. on Electronic Computers*, **EC-8**, No. 4, 439–40 (Dec. 1959).

4. Hartmanis, J. "On the State Assignment Problem for Sequential Machines, I," *IRE Trans. on Electronic Computers*, **EC-10**: 2, 157–165 (June 1961).

5. Stearns, R. E. and J. Hartmanis. "On the State Assignment Problem for Sequential Machines, II," *IRE Trans. on Electronic Computers*, **EC-10**: 4, 593–603 (December 1961).

6. Hartmanis, J. and R. E. Stearns. *Algebraic Structure Theory of Sequential Machines*, Prentice-Hall, Englewood Cliffs, N.J., 1966.

7. Karp, R. M. "Some Techniques of State Assignment for Synchronous Sequential Machines," *IEEE Trans. of Electronic Computers*, **EC-13**: 5, 507–518 (Oct. 1964).

8. Dolotta, T. A. and E. J. McCluskey. "The Coding of Internal States of Sequential Circuits," *IEEE Trans. on Electronic Computers*, **EC-13**: 5, 549–562 (Oct. 1964).

9. Torng, H. C. *Introduction to the Logical Design of Switching Systems*, Addison-Wesley, Reading, Mass., 1964.

10. McCluskey, E. J. *Introduction to the Theory of Switching Circuits*, McGraw-Hill, New York, 1965.

11. McCluskey, E. J. "Fundamental Mode and Pulse Mode Sequential Circuits," *Proc. IFIP Congress 1962*, North-Holland Publ. Co., Amsterdam, 1963.

12. Mano, M. *Computer Logic Design*, Prentice-Hall, Englewood Cliffs, N.J., 1972.

13. Harrison, M. A. *Introduction to Switching and Automata Theory*, McGraw-Hill, New York, 1965.

14. Givone, D. *Introduction to Switching Circuit Theory*, McGraw-Hill, New York, 1970.

15. Booth, T. *Sequential Machines and Automata Theory*, Wiley, New York, 1967.

16. Kohavi, Z. *Switching and Finite Automata Theory*, McGraw-Hill, New York, 1970.

Chapter 11 *Pulse-Mode Circuits*

11.1 Introduction

As a practical matter, clock pulses as such may not be available to all parts of a complex digital system. Often the clock is gated with some logic level x somewhere in the system other than at the inputs of the flip-flops. The output of such a gate would then consist of pulses meeting clock specifications but not necessarily occurring periodically. The frequency and distribution of these pulses would instead be controlled by x.

In some units of a digital system, nonperiodic pulses might be generated in response to each level change of a given signal. In a typical digital computer, we may find many such lines carrying such nonperiodic pulses. Thus, we may often be required to design sequential circuits whose inputs are nonperiodic pulses, rather than logic levels to be used in conjunction with clock pulses.

What differences in the analysis does this imply? Indeed, is it even possible to design sequential machines which are not clocked?

As it turns out, there are some interesting and important distinctions to be made between the state tables of clocked and pulse circuits, but no fundamental barrier stands in the way of design without a periodic clock. As discussed in Chapter 9, there are important restrictions on the duration of, and interval between, input pulses to flip-flops, but there is no requirement that they be periodic. Therefore, we shall find that there is no basic difficulty in designing sequential circuits subject to pulses on two or more lines,

307

provided the following criteria are satisfied:

1. All input pulses must be sufficiently wide to trigger a flip-flop. They must also satisfy the criteria for maximum pulse duration or all flip-flops must be master-slave as discussed in Chapter 9.
2. No two input pulses (irrespective of input line) will occur separated in time by less than the period corresponding to the maximum pulse repetition rate for flip-flops.

The second criterion implies that pulses may occur on only one input line at a time. Were it possible for two input pulses to occur exactly simultaneously, the inputs could be considered as an AND-coding of these pulses. This as a practical matter is impossible. Unpredictable delays would always separate the pulses slightly, and unreliable operation would result. Where pulses arrive from completely separate and independent sources, special effort may be required to synchronize them so as to satisfy the second criterion. This problem will be considered in chapter 13.

Each occurrence of a pulse on any input will constitute an input to the system and will trigger a transition from one state to another. Since simultaneous pulses are forbidden, the number of distinct inputs, and thus the number of columns in the next-state table, is equal to the number of input lines. Thus, a state-table for a typical pulse-mode circuit would appear as shown in Fig. 11.1. Between input pulses, the circuit remains in one of the states. When an input pulse arrives, a transition occurs from that present state to the next-state indicated. For example, in the circuit of Fig. 11.1, if the circuit is in state q_0 when an x_2 pulse arrives, it will move to state q_2 where it will remain until another pulse arrives. Should an x_3 pulse now arrive, the circuit will go to state q_1, where it will remain until a new pulse arrives, etc.

Some circuits may have both level inputs and pulse inputs. If the number of pulse inputs is n and the number of level inputs is m, the number of distinct input combinations is given by

$$N = n \times 2^m \tag{11.1}$$

This. then, is also the number of columns in the next-state table.

Present State	Next State		
	x_1	x_2	x_3
q_0	q_0	q_2	q_3
q_1	q_0	q_3	q_2
q_2	q_0	q_3	q_1
q_3	q_0	q_1	q_0

$x_1 \longrightarrow$
$x_2 \longrightarrow$ Pulse-Mode
$x_3 \longrightarrow$ Circuit

FIGURE 11.1.   *Typical state table for pulse mode circuit.*

11.2 Mealy Circuits and Moore Circuits

Nothing was said about the outputs in considering Fig. 11.1. In clocked circuits, the outputs, generally functions of both the inputs and state variables, are levels, since the inputs and state variables are levels. On the assumption that these outputs would in turn be inputs to other circuits subject to the same clock, it was appropriate to consider the value of these levels as meaningful only at clock time. The situation is a bit more complex for pulse-mode circuits.

If the outputs are to be functions of the inputs as well as the state variables, they must be pulses, obtained by gating the state variables with the input pulses. In that case, a distinct output may be defined for every possible combination of states and inputs.

If the outputs are functions of the state variables only, they will be levels whose value will be defined in the intervals *between* input pulses rather than at the time of the input pulses. Furthermore, the number of distinct outputs can be no greater than the number of states.

State tables and state diagrams for both types of pulse-mode circuits are shown in Fig. 11.2. The reader will notice a marked similarity between these two circuits. In fact, they represent solutions to essentially the same problem. The function of the circuits is to identify each sequence $x_1 x_2 x_1$ following an x_3 input. The x_3 pulse behaves essentially as a start, or reset, signal. In the pulse-output circuit, q_3, is reached by an input sequence $x_1 x_2$. If the next pulse to arrive is x_1, the circuits puts out a pulse and returns to state q_2 to await another start signal. If the next input is other than x_1, there is no output pulse.

The circuit in Fig. 11.2b operates in a similar fashion, except that the x_1 pulse arriving when the circuit is in state q_3 takes the circuit to a new state, q_4, which produces a 1 output *until* another pulse input occurs. Thus, one circuit produces a pulse at the time of the last pulse of the specified sequence, while the other produces a level after the sequence has been completed.

The reader will note several differences between the state diagrams of Fig. 11.2 and those of Chapter 10. Since only one input pulse can be present at a time, there is no need to list the values for all inputs with each transition. Instead, we identify the pulses which can cause each transition. In both circuits of Fig. 11.2, for example, if the circuit is initially in state q_0, an x_1 pulse will take it to q_1 and an x_2 pulse will take it to q_2. If two or more inputs can cause the same transition, they are separated by commas, e.g., x_1 or x_2 from q_2 will keep the circuit in q_2. Note carefully that this does not imply that both can occur at the same time, only that both cause the same transition.

For the pulse-input, pulse-output circuit, since the outputs are functions

	x_1	x_2	x_3
q_0	$q_1, 0$	$q_2, 0$	$q_0, 0$
q_1	$q_2, 0$	$q_3, 0$	$q_c, 0$
q_2	$q_2, 0$	$q_2, 0$	$q_0, 0$
q_3	$q_2, 1$	$q_2, 0$	$q_0, 0$

$$q^{v+1}, z$$

	x_1	x_2	x_3	z
q_0	q_1	q_2	q_0	0
q_1	q_2	q_3	q_0	0
q_2	q_2	q_2	q_0	0
q_3	q_4	q_2	q_0	0
q_4	q_2	q_0	q_0	1

$$q^{v+1}$$

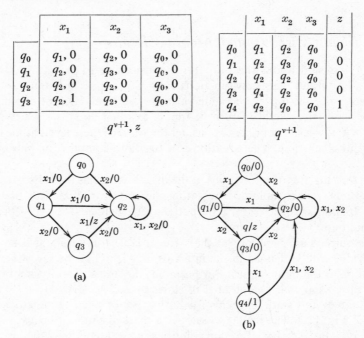

FIGURE 11.2. (a) *Pulse input-pulse output (Mealy).* (b) *Pulse input-level output (Moore).*

of the inputs, they are listed with the outputs, following a slash (/) in a manner similar to that used for clock mode state diagrams. A 0 following the slash indicates that there are no output pulses, i.e., all outputs remain at 0 during that transition. If there are any output pulses, the lines on which they will appear are listed after the slash. For example, on the transition from q_3 to q_2 in Fig. 11.2a, there will be a z pulse out if the transition is caused by an x_1 input but no pulse if the transition is caused by an x_2 input.

For pulse-input, level-output circuits, the outputs are associated with the states, so they are listed with the state designations within the circles corresponding to the states. Since it is the values of the outputs, rather than their identity, that is listed, they must be identified by a key, such as q/z for Fig. 11.2b.

Note that, in both state diagrams, the arrows corresponding to the x_3 transitions have been omitted in the interests of clarity. In every case, they would simply return the circuit to state q_0.

The circuit corresponding to Fig. 11.2a is an example of a *Mealy* circuit,[1] while that corresponding to Fig. 11.2b is an example of a *Moore* circuit.[2] In general, *the output of a Mealy circuit depends on both the present state and the inputs to the circuit. The output of a Moore circuit depends only on its present state.*

It will be illustrated later in the chapter that any input-output function expressible as a Mealy circuit may be expressed as a Moore circuit, and vice-versa. As was the case in the above example, a Moore model of a function will generally require more states than the corresponding Mealy model. This in itself is not a disadvantage of the Moore model. The design problem may call for an output which is available over a period of time rather than only momentarily as a pulse. If a Mealy circuit were used in that case, a flip-flop elsewhere in the system would be required to store the information carried by an output pulse.

11.3 Design Procedures

The design procedures for pulse-input, pulse-output circuits (Mealy) are essentially the same as for clocked circuits. In fact, if we regard the clock as input and use it to gate the outputs, we see that the clocked circuit may be regarded as a special case of the Mealy circuit.

Example 11.1

A major cycle of a certain digital system contains three subcycles which must be completed in a certain order. To check this, a sequence checker will receive a completion pulse from each subcycle and a check pulse when the major cycle is complete. When the check pulse K arrives, the sequence checker should reset and put out an error pulse if the three completion pulses A, B, C, were not received in that order. Each completion signal will occur exactly once in each major cycle. Design the sequence checker, using S-C flip-flops and AND-OR logic. Assume these pulses are synchronized to meet the criteria of Section 11.1.

Solution: There are six possible sequences, ABC, ACB, BAC, BCA, CBA, CAB, of which only the first is correct. We see that, once an incorrect sequence has been identified, the circuit can go to a state awaiting the arrival of the check pulse. For example, if the first pulse is B or C, the order in which the other two pulses arrive does not matter. This leads to the partial state diagram of Fig. 11.3a, where state q_2 is the state of the machine after the first symbol of a correct sequence and state q_5 is the state of the machine as soon as the sequence has been identified as incorrect. From q_2, the only possible inputs are B and C. Since B represents the continuation of a correct sequence, it leads to a new state q_3 as shown in Fig. 11.3b.

From state q_3, the only possible input is C leading to a new state q_4. Since a pulse on line K will always follow the three-pulse sequence to be checked, this input will occur only when the circuit is in state q_4 or q_5. Adding the arrows corresponding to K with an output from state q_5 completes the state diagram of Fig. 11.3b. ■

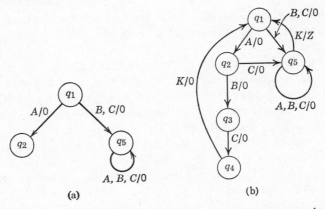

FIGURE 11.3.   *Developing the state diagram for the sequence checker.*

The reader should notice that Fig. 11.3b differs from the state diagram of Fig. 11.2a in one important respect—not all possible input arrows are included for every present state. That is, there are some combinations of inputs and present states which will never occur. For this reason, it doesn't matter what next states or outputs are specified for these combinations. That is, they are don't-cares in much the same sense as don't-cares in combinational logic. In translating the state diagram to a state table, we enter blanks in place of the outputs and next states for these don't-care combinations so that Fig. 11.3b translates to the state table of Fig. 11.4a.

A state table containing don't-care entries is defined as *incompletely specified*. In general, state table minimization is more difficult for incompletely specified state tables. This topic will be treated in detail in Chapter 12. For now, let us minimize Fig. 11.4a by intuitively extending the techniques of Chapter 10.

	Inputs			
q^v	A	B	C	K
1	2, 0	5, 0	5, 0	–, –
2	–, –	3, 0	5, 0	–, –
3	–, –	–, –	4, 0	–, –
4	–, –	–, –	–, –	1, 0
5	5, 0	5, 0	5, 0	1, 1
	q^{v+1}, z^v			

(a)

	Inputs			
q^v	A	B	C	K
1	2, 0	5, 0	5, 0	–, –
2	–, –	3, 0	5, –	–, –
3	–, –	–, –	3, 0	1, 0
5	5, 0	5, 0	5, 0	1, 1
	q^{v+1}, z^v			

(b)

FIGURE 11.4.   *State table for sequence checker.*

Consider states 3 and 4. Since $\delta(3, K)$, $\lambda(3, K)$ is a don't care, we may enter what ever values we choose. Therefore, let

and similarly let
$$\delta(3, K), \lambda(3, K) = 1, 0$$

$$\delta(4, C), \lambda(4, C) = 4, 0$$

Now rows 3 and 4 of the state table are identical. Thus, we combine the two rows replacing all next state values of 4 in the table with 3. This leads to the table of Fig. 11.4b. We conclude by inspection that this table cannot be further simplified regardless of how the remaining don't-care entries might be specified.

For state assignment of pulse-mode circuits, we apply the same procedures as for clock mode circuits. In this case, Rule Ic suggests that q_1 should be adjacent to q_3 and q_5, and the assignment shown in Fig. 11.5 satisfies these adjacency requirements. The corresponding excitation tables are shown in Fig. 11.6.

Note that the tables of Fig. 11.6 are not conventional K-maps. Since there are four input variables, a Karnaugh map would have sixteen columns. The four columns in the tables of Fig. 11.6 represent the columns $ABCK = 1000$, $ABCK = 0100$, $ABCK = 0010$, and $ABCK = 0001$. Since only one pulse can occur at a time, all the other columns except $ABCK = 0000$ of the K-map would have only don't-care entries. Thus, we can let the A columns of the excitation tables represent the $A = 1$ columns of the corresponding K-map, the B columns represent the $B = 1$ columns, etc. On this basis, we can write the excitation equation directly from these tables and save the labor of drawing four 4×16 K-maps. The equations for S and C for the two flip-flops are:

$$S_1 = B + C \tag{11.3}$$

$$C_1 = K \tag{11.4}$$

$$S_2 = A\bar{y}_1 \tag{11.5}$$

$$C_2 = K + C\bar{y}_1 \tag{11.6}$$

		Inputs			
q^ν	$(y_1 y_2)^\nu$	A	B	C	K
1	0 0	01, 0	10, 0	10, 0	–, –
2	0 1	–, –	11, 0	10, 0	–, –
3	1 1	–, –	–, –	11, 0	00, 0
5	1 0	10, 0	10, 0	10, 0	00, 1
		$(y_1 y_2)^{\nu+1}, z^\nu$			

FIGURE 11.5. *Transition table for sequence checker.*

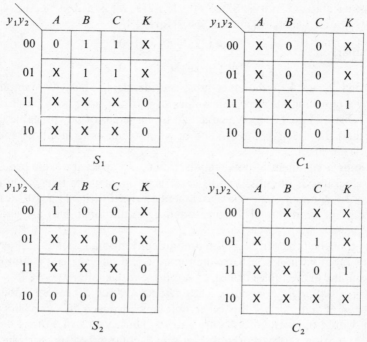

y_1y_2	A	B	C	K
00	0	1	1	X
01	X	1	1	X
11	X	X	X	0
10	X	X	X	0

S_1

y_1y_2	A	B	C	K
00	X	0	0	X
01	X	0	0	X
11	X	X	0	1
10	0	0	0	1

C_1

y_1y_2	A	B	C	K
00	1	0	0	X
01	X	X	0	X
11	X	X	X	0
10	0	0	0	0

S_2

y_1y_2	A	B	C	K
00	0	X	X	X
01	X	0	1	X
11	X	X	0	1
10	X	X	X	X

C_2

FIGURE 11.6. *Excitation tables, Example 11.1.*

It is important to realize that the above equations are for unclocked flip-flops, as discussed first in Chapter 9. Pulses corresponding to the above equations are applied directly to the Set and Clear lines rather than to a separate clock line. From the transition table, we can read off the output equation,

$$Z = K\bar{y}_2 \tag{11.7}$$

The realization of these equations is shown in Fig. 11.7.

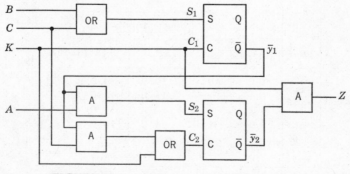

FIGURE 11.7. *Final circuit for Example 11.1.*

y_1y_2	A	B	C	D
00	1	1	X	X
01	X	X	X	0
11	X	0	X	X
10	X	X	0	0

FIGURE 11.8.   *Example excitation table, S_1.*

In the above example, we saw that there are no basically new problems in designing a Mealy model circuit. The restrictions on simultaneous inputs reduce the number of columns required in the tables, but otherwise there is little real difference in procedure from clocked design. One precaution we might note is that every product term in the excitation and output equations must contain a pulse variable in order to generate the desired pulses. Suppose that a certain design resulted in the excitation table shown in Fig. 11.8. It would appear that the best equation for S_1 would be

$$S_1 = \bar{y}_1\bar{y}_2 \tag{11.8}$$

However, this would not work since S_1 would then be a level, while a pulse is required to trigger the flip-flop. The equation that should be used is

$$S_1 = A + B\bar{y}_1 \tag{11.9}$$

Looking at the problem another way, we recall that Fig. 11.8 represents only four columns of a 16-column K-map. The column $ABCD = 0000$ is a column of 0's, since no transitions should take place in the absence of an input pulse. Thus, all product terms in the excitation equations (assuming S-of-P forms) must include at least one input variable. Furthermore, no input variable should ever appear complemented in the excitation equations since, for pulse inputs, a complemented variable corresponds to the absence of the input.

Example 11.2

Let us again consider the same basic specification given in Example 11.1, but let us formulate it this time as a Moore-model problem. We now specify that, after the third pulse is received, the sequence-checker shall put out a *level* signal with a value of 0 if the pulses were received in the correct order, a 1 otherwise. This level will remain until the K pulse resets the circuit. Since the output will not be sampled until after the third pulse, the output will be a don't-care until then. A state diagram for this version of the sequence checker is given in Fig. 11.9. Once again the circuit goes to state 5 as soon as an incorrect sequence is detected. In

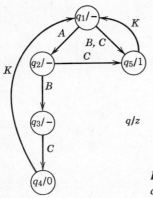

FIGURE 11.9.   *State diagram for level output sequence checker.*

this case, the output level must go to 1 as soon as state 5 is reached. The circuit reaches state 4 following a correct sequence, so the output corresponding to state 4 is 0. The output is a don't-care for the remaining states.

Figure 11.9 is translated to Fig. 11.10a and minimized in much the same way as Example 11.1. Assigning states in the same order leads to the transition table of Fig. 11.10b. Note that the outputs are listed in a separate column since they are functions of the present state only.

We leave it as an exercise for the reader to set up the excitation tables to obtain the following excitation equations:

$$S_1 = B + C \tag{11.10}$$
$$C_1 = K \tag{11.11}$$
$$S_2 = A\bar{y}_1 \tag{11.12}$$
$$C_2 = C\bar{y}_1 + K \tag{11.13}$$

and the output equation

$$z = \bar{y}_2 \tag{11.14}$$

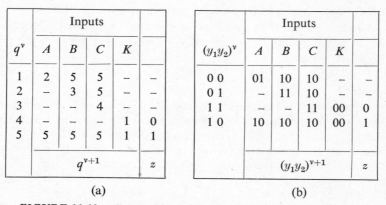

FIGURE 11.10.   *State table and transition table, Example 11.2.*

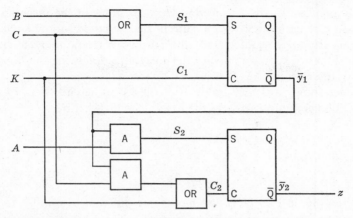

FIGURE 11.11. *Level output (Moore) circuit for Example 11.2.*

The realization of these equations is shown in Fig. 11.11. When we add an AND gate to sample Z upon the occurrence of each K pulse, we have the same circuit as found in Example 11.1. ■

11.4 Mealy-Moore Translations*

In the last example, it was possible to accomplish the function of a Mealy circuit by using a Moore circuit with the same number of states. Thus, the design of the two circuits proved to be almost identical. Let us repeat that this will not always be the case.

Let us translate the Mealy circuit whose state diagram is depicted in Fig. 11.12 to a Moore circuit with the same input-output function. Every output,

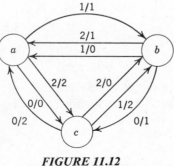

FIGURE 11.12

* This section and the following section are not essential to the continuity of subsequent chapters.

which is associated with a transition between states in the machine of Fig. 11.12, must be associated with a state in the proposed Moore circuit. Since the inputs cannot be anticipated, the outputs corresponding to inputs in Fig. 11.12 must be associated with the next states following the same inputs in the Moore circuit.

Consider some particular state, p, of the Mealy circuit. Let $\{\lambda_i\}_p$ be the set of all distinct outputs, λ_i, of the Mealy circuit given by

$$\lambda_i = \lambda(q, I_i) \tag{11.15}$$

for any input I_i and any state q satisfying

$$\delta(q, I_i) = p \tag{11.16}$$

For example, in Fig. 11.12, we have three different transitions to state c. That is,

$$
\begin{aligned}
\delta(a, 0) &= c &\text{where} &\quad \lambda(a, 0) = 0 \\
\delta(a, 2) &= c & &\quad \lambda(a, 2) = 2 \\
\delta(b, 0) &= c & &\quad \lambda(b, 0) = 1
\end{aligned}
\tag{11.17}
$$

Therefore, $\{\lambda_i\}_c = \{0, 1, 2\}_c$.

We may now define a set of states $\{p_{\lambda_i}\}$ in the Moore circuit corresponding to the single state, p, of the Mealy circuit. That is, for every $\lambda_i \in \{\lambda_i\}_p$, let p_{λ_i} be a distinct state in the Moore circuit such that

$$\lambda(p_{\lambda_i}) = \lambda_i \tag{11.18}$$

By definition, $\lambda(p_{\lambda_i})$ is the output associated with p_{λ_i} in the Moore circuit.

In Fig. 11.12, for example, the set of states $\{C_0, C_1, C_2\}$ corresponds to the output set $\{0, 1, 2\}_c$. Similarly, for states a and b, the output sets are $\{0, 1, 2\}_a$ and $\{0, 1, 2\}_b$; and the sets of states in the Moore circuit are $\{A_0, A_1, A_2\}$ and $\{B_0, B_1, B_2\}$. These states are listed in the state table of Fig. 11.13.

The next-state function for any input, I, in the Moore circuit may be defined as follows. Suppose

$$\delta(q, I) = p \tag{11.19}$$

for two states p and q in the Mealy circuit. Let $\{p_{\lambda_j}\}$ and $\{p_{\lambda_i}\}$ be the sets of states in the Moore circuit corresponding to p and q, respectively. Let p_{λ_j} be the member of $\{p_{\lambda_j}\}$ such that

$$\lambda_j = \lambda(q, I) \tag{11.20}$$

Then we may define the next-state function in the Moore circuit for input, I, and any state $q_{\lambda_i} \in \{q_{\lambda_i}\}$ as

$$\delta(q_{\lambda_i}, I) = p_{\lambda_j} \tag{11.21}$$

	$X = 0$	$X = 1$	$X = 2$	λ
A_0	C_0	B_1	C_2	0
A_1	C_0	B_1	C_2	1
A_2	C_0	B_1	C_2	2
B_0	C_1	A_0	A_1	0
B_1	C_1	A_0	A_1	1
B_2	C_1	A_0	A_1	2
C_0	A_2	B_2	B_0	0
C_1	A_2	B_2	B_0	1
C_2	A_2	B_2	B_0	2

FIGURE 11.13. *Moore circuit corresponding to the Mealy circuit of Fig. 11.12.*

Suppose the input to the Moore circuit of Fig. 11.13 is $X = 0$ when the circuit is in one of the set of states $\{A_0, A_1, A_2\}$. Recall from Fig. 11.12 that

$$\delta(A, 0) = C \quad \text{and} \quad \lambda(A, 0) = 0 \quad (11.22)$$

Therefore, we conclude that for present state, a, $\lambda_j = 0$ as defined by Equation 11.20, and that from Equation 11.21

$$\delta(A_0, 0) = \delta(A_1, 0) = \delta(A_2, 0) = C \quad (11.23)$$

The remaining next state entries in Fig. 11.13 may be determined in a similar manner.

So far we have only indicated that the circuits of Figs. 11.12 and 11.13 should have the same output sequence for any input sequence. The further refinement in notation required to complete a proof of this fact would take us too far afield. The reader who remains unconvinced can easily check the outputs of the circuit for a typical input sequence. In Fig. 11.14a, we see the sequence of outputs and sequence of states assumed by the Mealy circuit when subjected to the input sequence

$$1021002 \quad (11.24)$$

when initially in state a. Similarly, we see in Fig. 11.14b the output and next-state sequences of the Moore circuit when subjected to the same input sequences. In the latter case, the initial state may be any member of A_0, A_1, A_2.

Notice in this case that a 9-state Moore circuit was required to imitate a 3-state Mealy circuit. This Mealy circuit was deliberately chosen to have distinct outputs for each transition to each of its three states. As we have seen, the situation is not this bad for many examples of practical interest.

Initial State

Input	$\downarrow$ 1 1 2 1 0 0 2	A_0 ⎱	1 1 2 1 0 0 2	
New State	@ b a c b c a c	A_1 ⎬	B_1 A_0 C_2 B_2 C_1 A_2 C_2	
Output	1 0 2 2 1 2 2	A_2 ⎰	1 0 2 2 1 2 2	
	(a)		(b)	

FIGURE 11.14. *Input-output sequences.*

The Moore circuit of Fig. 11.13 may be translated to an equivalent Mealy circuit by merely assigning an output to each input-present-state combination equal to the output associated with the next state for that combination in the Moore circuit. The result is Fig. 11.15.

Using the minimization procedure of Chapter 10, we find that the states of the circuit in Fig. 11.15 fall into the following equivalence classes:

$$(A_0A_1A_2)(B_0B_1B_2)(C_0C_1C_2) \tag{11.25}$$

The circuit formed from these equivalence classes is the original circuit of Fig. 11.12.

In the previous examples we have considered Moore circuits as alternate versions of circuits originally specified in terms of the Mealy model. Let us now consider an example in which level outputs are initially specified.

A_0	C_0 0	B_1 1	C_2 2
A_1	C_0 0	B_1 1	C_2 2
A_2	C_0 0	B_1 1	C_2 2
B_0	C_1 1	A_0 0	A_1 1
B_1	C_1 1	A_0 0	A_1 1
B_2	C_1 1	A_0 0	A_1 1
C_0	A_2 2	B_2 2	B_0 0
C_1	A_2 2	B_2 2	B_0 0
C_2	A_2 2	B_2 2	B_0 0

FIGURE 11.15

Example 11.3

A monorail system provides shuttle service between an airport and a downtown terminal. There are two cars on the system, but only one track, except for a siding at the halfway point, as shown in Fig. 11.16. The two cars will always start from the two stations at the same time, one inbound and one outbound. The outbound car will pull into siding and wait there until the inbound car has passed by on the main track. The signal and switches controlling the siding will be controlled by three track sensors, x_1, x_2, x_3, which emit pulses when a car passes over them. When the cars are at the stations, both signals will be red, switch S_1 will be switched

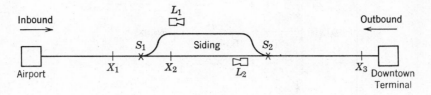

FIGURE 11.16.   *Monorail shuttle system.*

to the main line, and switch S_2 will be switched to the siding. The outbound car will go onto the siding and stop at signal light L_1. This signal will remain red until the inbound car passes over both x_1 and x_2, at which time it will turn green, and switch S_1 will switch to the siding. The inbound car will stop at signal L_2 until the outbound car moves off the siding and over x_1, at which time L_2 will turn green and S_2 will switch to the main line. When the inbound car passes over x_3 when arriving at the downtown terminal, the system will return to the original condition. Design a level output circuit to develop signals z_1 and z_2, which correspond to conditions of the signal lights and switches as follows:

$$z_1 = 0 \leftrightarrow L_1 = \text{red}, \quad S_1 \text{ to main line}$$
$$z_1 = 1 \leftrightarrow L_1 = \text{green}, \quad S_1 \text{ to siding}$$
$$z_2 = 0 \leftrightarrow L_2 = \text{red}, \quad S_2 \text{ to siding}$$
$$z_2 = 1 \leftrightarrow L_2 = \text{green}, \quad S_2 \text{ to main line}$$

Solution: The placement of the sensors and the sequence of operations are such that x pulses cannot coincide, so the Moore model is applicable. The state diagram for the system is shown in Fig. 11.17a. Starting in state q_0 (both signal lights red), the outbound car first passes over the x_3 sensor, but this causes no change of state. The inbound car then passes over x_1 and x_2 in sequence, moving the circuit to states q_1 and q_2, at which time z_1 goes to 1, causing L_1 to change to green and S_1 to switch to the siding. The outbound car then passes over x_1,

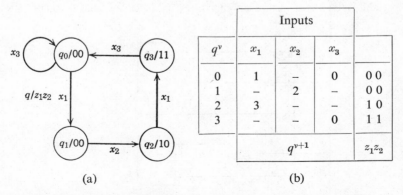

(a)   (b)

FIGURE 11.17.   *State diagram and table, monorail system controller.*

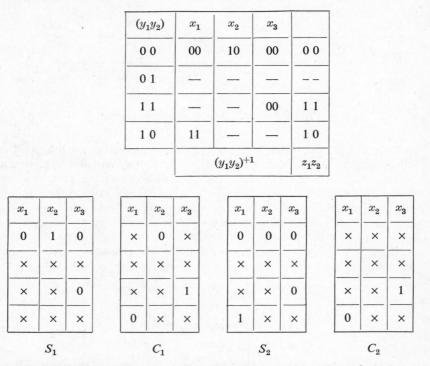

(y_1y_2)	x_1	x_2	x_3	
0 0	00	10	00	0 0
0 1	—	—	—	– –
1 1	—	—	00	1 1
1 0	11	—	—	1 0
		$(y_1y_2)^{+1}$		z_1z_2

x_1	x_2	x_3
0	1	0
×	×	×
×	×	0
×	×	×

S_1

x_1	x_2	x_3
×	0	×
×	×	×
×	×	1
0	×	×

C_1

x_1	x_2	x_3
0	0	0
×	×	×
×	×	0
1	×	×

S_2

x_1	x_2	x_3
×	×	×
×	×	×
×	×	1
0	×	×

C_2

FIGURE 11.18. *Transition table and excitation tables, Example 11.3.*

taking the circuit to q_3 and changing z_2 to 1. The cycle is completed when the inbound car passes over x_3, returning the circuit to q_0. The corresponding state table is shown in Fig. 11.17b.

It is obvious that the only possible simplification is the combination of q_0 and q_1 into a single state. Since two state variables are required and there are two output variables, the obvious state assignment is to let the state variables correspond to

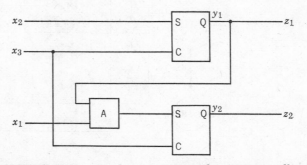

FIGURE 11.19. *Final circuit, monorail system controller.*

the outputs, i.e., $y_1 = z_1$ and $y_2 = z_2$. The resultant transition table and excitation tables are shown in Fig. 11.18. From the excitation table, we read the excitation equations:

$$S_1 = x_2 \tag{11.26}$$

$$C_1 = C_2 = x_3 \tag{11.27}$$

$$S_2 = x_1 y_1 \tag{11.28}$$

The corresponding circuit is shown in Fig. 11.19.   ■

11.5   Counters Revisited

In Chapter 10, we designed an up-down counter as a Mealy model machine in which the outputs coincided with the next state. More commonly, counters are designed such that the outputs are coded to coincide with the present state, so that the flip-flop outputs become the circuit outputs. Such circuits are natural Moore circuits. The state table of an elementary *decade* or modulo-10 counter is given Fig. 11.20a as an example. Notice the single next-state column corresponding to pulse line P. In effect, the counter counts (modulo-10) the number of pulses appearing on this line. This pulse line might be the system clock. If so, Fig. 11.20a could equally well describe a clock mode sequential circuit. This interpretation is somewhat less natural, however, since the clock is not counted as an input in clock mode circuits. Thus, the circuit would have no inputs. It would, however, still have $2^0 = 1$ next-state columns, as indicated in the figure.

q^v	P		$(y_3 y_2 y_1 y_0)^v$	P	
0	1	0	0 0 0 0	0 0 0 1	0 0 0 0
1	2	1	0 0 0 1	0 0 1 0	0 0 0 1
2	3	2	0 0 1 0	0 0 1 1	0 0 1 0
3	4	3	0 0 1 1	0 1 0 0	0 0 1 1
4	5	4	0 1 0 0	0 1 0 1	0 1 0 0
5	6	5	0 1 0 1	0 1 1 0	0 1 0 1
6	7	6	0 1 1 0	0 1 1 1	0 1 1 0
7	8	7	0 1 1 1	1 0 0 0	0 1 1 1
8	9	8	1 0 0 0	1 0 0 1	1 0 0 0
9	0	9	1 0 0 1	0 0 0 0	1 0 0 1
	q^{v+1}	z^v		$(y_3 y_2 y_1 y_0)^{v+1}$	$(z_3 z_2 z_1 z_0)^v$

(a)   (b)

FIGURE 11.20.   *Elementary decade counter.*

Assigning states in binary-coded decimal order yields the transition table of Fig. 11.20b. This state assignment effectively permits a partition very similar to the one discussed for the counter of Fig. 10.43. A T flip-flop realization of Fig. 11.20b may be obtained very simply by noting when the four flip-flops change state. Notice that y_0 changes state with every input pulse. Therefore,

$$T_{y_0} = P \tag{11.29}$$

Notice that y_1^{v+1} (the value of y_1 after a pulse on line P) differs from y_1^v if and only if $y_0^v = 1$. Therefore,

$$T_{y_1} = y_0 \cdot P \tag{11.30}$$

Similarly, y_2 will change state ($0 \to 1$ or $1 \to 0$) if and only if $y_1^v = y_2^v = 1$. Therefore,

$$T_{y_2} = y_0 \cdot y_1 \cdot P \tag{11.31}$$

For a modulo-16 counter the expression for T_{y_3} would be analogous. While y_3 changes from 0 to 1 in the case of the decade counter when $y_2 = y_1 = y_0 = 1$, we have a change from 1 to 0 when $y_3 = y_0 = 1$. Therefore,

$$T_{y_3} = (y_2 \cdot y_1 \cdot y_0 \cdot P) + (y_3 \cdot y_0 \cdot P) \tag{11.32}$$

We leave it as an exercise for the reader to obtain these same expressions using K-maps and transition lists.

As mentioned in Chapter 9, T flip-flops are not commonly available as standard integrated circuit components, but a J-K flip-flop becomes a T if the J and K inputs are permanently connected to the 1-level. A T flip-flop realization of the decade counter is given in Fig. 11.21.

Sometimes counters which count to different values, depending on one or more level inputs, are available in integrated circuit form. We conclude this chapter with an example of this type of counter.

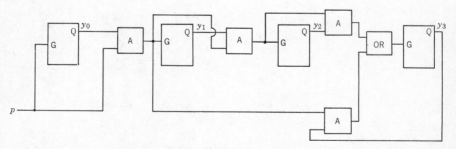

FIGURE 11.21.   *Decade counter realization.*

Example 11.4

Design a counter which will count modulo-5 if a level input $L = 1$ or will count modulo-8 if $L = 0$. The count is to be incremented each time a pulse appears on line P_1 and decremented each time a pulse appears on line P_2.

Solution: A state table for the counter is given in Fig. 11.22. Notice that there are $n \times 2^m = 4$ input columns, where the number of pulse inputs is $n = 2$ and the number of level inputs is $m = 1$. In the modulo-5 columns, we enter don't-cares

	Inputs				
	$L = 0$		$L = 1$		
q^v	P_1	P_2	P_1	P_2	
0	1	7	1	*4	0
1	2	0	2	0	1
2	3	1	3	1	2
3	4	2	4	2	3
4	5	3	*0	3	4
5	6	4	✕	✕	5
6	7	5	✕	✕	6
7	0	6	✕	✕	7
	q^{v+1}				z

FIGURE 11.22.   *Level-controlled counter.*

on the last row, assuming that the value of L will be changed only occasionally and never when the output count is being monitored. An external reset line is also assumed.

Because of the modulo-5 columns, the closed partitions leading to the binary state assignment in the counters considered thus far are not valid for this state table. However, the closure property for partitions (0246) (1357) and (04) (15) (26) (37) is violated only for the two entries marked by asterisks in Fig. 11.22. Therefore, the straight binary assignment should still yield relatively simply flip-flop input equations as well as no output logic. Assigning states in this way and using K-maps along with the T flip-flop transition list will yield the following input equations.

$$T_{y_0} = P_1 \cdot \overline{(y_2 L)} + P_2 \cdot \overline{(\bar{y}_2 \bar{y}_1 \bar{y}_0 L)} \tag{11.33}$$

$$T_{y_1} = P_1 \cdot y_0 + P_2 \cdot \bar{y}_0 \cdot \overline{(L \bar{y}_2 \bar{y}_1)} \tag{11.34}$$

$$T_{y_2} = P_1 \cdot (y_0 y_1 + y_2) + P_2 \cdot (\bar{y}_0 \bar{y}_1) \tag{11.35}$$

We leave the verification of these equations as a homework problem for the reader. ∎

PROBLEMS

11.1. Construct the state diagram and state table of a sequential circuit with two pulse inputs, A and B, and a single level output, Z. Following a pulse on line B, the output Z is to be 1 provided that there have been an even number of pulses on line A, since the previous pulse on line B. Otherwise, a B pulse will reset the output to 0. The output will not change except on the arrival of a B pulse.

11.2. Determine the minimal state table of a sequential circuit, which has three pulse inputs x_1, x_2, and x_3, and a single level output Z. The output Z is to be 1 following a pulse x_1 until a total of at least four subsequent pulses have been observed on either of lines x_2 or x_3. Otherwise, the output of the circuit is to be 0. Another x_1 will reset the machine to its initial state.

11.3. A sequential circuit is to have 3 pulse inputs A, B, and C and a single level output z. A pulse may appear on any one of the three lines at any time consistent with the pulse mode assumption. The output is to be 1 if and only if the last 3 pulses in sequence have been AAA or ABA. Determine a state table for this sequential circuit.

11.4. A pulse mode sequential circuit has four pulse inputs C, S_1, S_2, and N and two level outputs z_1 and z_2. The pulse mode assumption is satisfied so that any two pulses are sufficiently separated in time to permit the circuit to react to each pulse independently. A pulse on line C will cause both z_1 and z_2 to go to 0. A pulse on line S_1 will cause z_1 to go to 1 without affecting z_2. A pulse on S_2 will cause z_2 to go to 1 without affecting z_1. A pulse on line N will complement the values of both z_1 and z_2. Determine a state diagram and a state table for this sequential circuit.

11.5. The state table of Fig. P12.3 has been developed for a pulse-input, pulse-output circuit. The indicated assignment of flip-flop states has already been determined. Complete the design by determining minimal expressions for Z, S_{y_2}, C_{y_2}, S_{y_1}, and C_{y_1}.

$y_2\ y_1$	Present State	Next State, z		
		A	B	C
0 0	a	$c,1$	$d,1$	$d,1$
0 1	b	$d,0$	$d,1$	$c,0$
1 1	c	$d,0$	$b,0$	$a,1$
1 0	d	$c,1$	$a,0$	$b,0$

FIGURE P11.5

11.6. Using S-C flip-flops, design a sequential circuit with three pulse inputs A, B, and C and a pulse output Z. The circuit is to yield an output pulse coinciding as nearly as possible with an input pulse if and only if the two previous input pulses have been on each of the other two lines. (For example, a pulse on line A will cause an output if the previous two pulses have been on lines B and C in either order.)

11.7. Determine circuit realizations of the state tables in Fig. 11.2 using S-C flip-flops.

11.8. A 20-μsec pulse delay line and a supply of NAND gates and S-C flip-flops are available for construction of the following circuit. Using these components, design a circuit which has two pulse inputs, A and B, and a single pulse output, Z. The input pulses have widths of 0.1 μsec. For 20 μsec after each pulse on line A, each pulse on line B must generate an output pulse Z. Otherwise, the B pulses are ignored. Should one A pulse occur less than 20 μsec after another A pulse, then no further output pulses are to be generated until a third A pulse arrives. (*Hint:* Let the output of the delay line be a third input and construct a state table.)

11.9. A perfect 256-μsec pulse delay line is to be used as a memory for storing 256 bits of information in thirty-two 8-bit characters. Assume that all pulses are of 0.1 μsec duration. A circuit which will control the reading and writing of 8-bit characters to and from the memory is to be designed, using S-C flip-flops and NAND gates (negligible delay). The four inputs to the circuit are control lines, W and R, a data line, D, and a clock, C. The clock, which puts out 10^6 pulses per second, is to be used to synchronize the entry of data pulses on the delay line. Line D is a level input. Pulses will occur on lines W and R only intermediate between clock pulses. If a pulse appears on line W, the next eight bits appearing at the output of the delay line are to be replaced by pulses corresponding to the level of D at the time of the next eight clock pulses. In the absence of a write cycle, the circuit should cause data to be recirculated into the delay line. Following an R pulse, the next eight pulses appearing from memory are to constitute a circuit output. A W or an R pulse will never follow another W or R pulse by less than 8 μsec.

11.10. Determine a Moore circuit equivalent to the Mealy circuit given in Fig. P11.5.

11.11. Use K-maps to derive equations 11.33, 11.34, and 11.35.

11.12. Design in detail a counter which will count clock pulses modulo-16 if a level $L = 1$ or as a decade counter if $L = 0$.

11.13. A stamp vending machine dispenses 10¢, 15¢, and 25¢ packets of stamps and accepts nickels, dimes, and quarters. The user may insert exact change or an excess amount and receive change. It is assumed the user will not insert an excess amount if he has correct change, e.g., he might insert a quarter to pay for a 10¢ packet, but not a dime and a nickel. After inserting the coins, the user will press one of three buttons to select the packet desired. If he has inserted enough money, he will receive the packet and any change due. If he has not inserted enough, his money will be refunded. The buttons are mechanically interlocked so that only one may be pressed at a time. The coins must be inserted one at a time into a single slot. A coin detector issues one of three pulses, N, D, or Q, for nickels, dimes, or quarters, respectively. The three buttons also emit pulses, X_1 for 10¢ packet, X_2 for 15¢ packet, X_3 for 25¢ packet. You are to design a pulse output circuit to control the dispensing of the packet and the change. Output pulses should be emitted at the time of any X pulse according to the following code:

$$Z_1 = 10¢ \text{ packet} \quad Z_2 = 15¢ \text{ packet} \quad Z_3 = 25¢ \text{ packet}$$
$$Z_4 = 5¢ \text{ change} \quad Z_5 = 10¢ \text{ change} \quad Z_6 = 15¢ \text{ change}$$
$$Z_7 = \text{refund}$$

The circuit should reset following the X pulse.

BIBLIOGRAPHY

1. Mealy, G. H. "A Method for Synthesizing Sequential Circuits," *Bell System Tech. J.*, **34**: 5, 1045–1080 (Sept. 1955).

2. Moore E. F. "Gedanken Experiments on Sequential Machines," in C. E. Shannon and J. McCarthy (Eds.), *Automata Studies*, Princeton Univ. Press, Princeton, N.J., 1956.

3. Kohavi, Z. *Switching and Finite Automata Theory*, McGraw-Hill, New York, 1970.

4. McCluskey, E. J. *Introduction to the Theory of Switching Circuits*, McGraw-Hill, New York, 1965.

5. Booth, T. L. *Digital Networks and Computer Systems*, Wiley, New York, 1971.

Chapter 12 *Incompletely Specified Sequential Circuits*

12.1 Introduction

Throughout Chapter 10, we assumed that the output and next state were specified for every possible combination of input and present state. Thus, no don't-care conditions arose due to the circuit specifications, except somewhat indirectly when the number of internal states required was not a power of two. While such don't-cares are very useful in simplifying excitation functions, they are more a result of the form of realization chosen than of the actual circuit specifications. Because of this special character, such don't-cares, due to extra internal states, are sometimes known as *incidental* don't-cares.[3]

In Chapter 11, we encountered another type of don't-care, resulting from the fact that certain *sequences* of inputs may never occur due to external constraints. In the sequence checker of Example 11.1, no pulse could occur more than once in a given three-pulse sequence so that a sequence ABA, for example, could not occur. In combinational circuits, if certain input conditions could never occur, the result was *don't-care* entries in the function maps. Similarly, in sequential circuits, input sequences which cannot occur will introduce *don't-care* entries in the state tables.

Just as don't-cares offered the possibility of simple realizations for combinational circuits, it is reasonable to expect that don't-cares in state tables may make possible further simplification than might otherwise be the case. For combinational circuits, the designer in effect specifies the don't-cares as 1's or 0's in whatever manner provides the simplest realization, and the appropriate choice is generally quite obvious. Sequential circuits being generally more complex, however, the problems of assigning don't-cares are correspondingly much more difficult.

First, there are several types of don't-cares in sequential circuits. The most common is that described above, resulting from input sequences which cannot occur, in which case both the next state and the output are unspecified. The justification is the same as for combinational don't-cares; if the event will never occur, then we "don't-care" what the circuit would do in response to it. Second, we may have circuits which receive a continuous series of inputs, to which they respond with appropriate sequences of states, but the outputs are sampled only at specified times rather than continuously. When the outputs are not being sampled, they may be unspecified even though the next states are specified. Third we have situations where a given input is the *last* event of a sequence in the sense that it will always be followed by a general reset signal, which can reset the circuit from any state. For a Mealy circuit, the output will then be specified, but the next state will not.

In the case of combinational don't-cares, there are only two ways to assign the don't-cares (0 or 1); but for don't cares in a state table, there are generally many choices. Where there are two rows of a state table which are identical in every column where *both* are specified, the obvious step is to assign the don't-cares in such a manner as to make the two rows identical, in which case they are equivalent by inspection. This is what was done in Example 11.1 to combine states q_3 and q_4 (see Fig. 11.4). Note that when a given column is unspecified in *both* rows, it may be left unspecified, resulting in don't-cares in the resultant transition table.

In Example 11.1, there were only two rows that could be combined in this manner, and the resultant 4-state circuit was obviously minimal. In most cases, however, there are a number of ways that the don't-cares might be assigned, and it may not be obvious which choice will lead to the minimal state table. If the number of choices is not too large, we might assign all don't-cares in all possible ways and minimize the resultant completely specified state tables (by the methods of Chapter 10) to see which is minimal.

Example 12.1

A clock mode sequential circuit has an input x and an output z. Starting in a reset state, it stays in reset as long as $x = 0$ and the output is unspecified. An input of $x = 1$ will move the circuit from reset state, R. This will be followed by a string of 0 inputs of arbitrary length, producing $z = 1$ outputs at the time of the second, fourth, sixth, etc., 0-inputs. When the circuit is not in state R, a

1-input may occur at any time and will return the circuit to state R with an output of $z = 0$. Develop a minimal state table.

Solution: A state diagram corresponding to the above operation and the corresponding state table are shown in Fig. 12.1. There is only one unspecified entry, the output for $x = 0$ and state R.

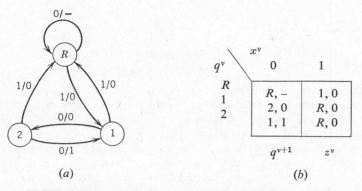

q^v \ x^v	0	1
R	$R, -$	$1, 0$
1	$2, 0$	$R, 0$
2	$1, 1$	$R, 0$
	q^{v+1}	z^v

(a)   (b)

FIGURE 12.1. *State diagram and state table, Example 12.1.*

At first glance, it might seem that the safe way to proceed would be to form two state tables by first specifying the single "don't care" as a 0 and then as 1. These tables could then be minimized, finally choosing the one with the fewest states. This results in the two completely specified state tables of Fig. 12.2. For Fig. 12.2a, we see that only states R and 1 have identical outputs, but their equivalence would imply equivalence of R and 2. Thus, the table of Fig. 12.2a is minimal; and by similar argument the table of Fig. 12.2b is also minimal.

It appears that Example 12.1 cannot be realized with less than three states. But now consider the two-state circuit specified by the state diagram and table of Fig. 12.3, in which states A and B are equivalent to states 1 and 2 of Fig. 12.1, and the function of state R is "shared" between them. First consider the circuits in either states 1 or 2, or states A or B, with a string of 0 inputs applied. Both circuits will generate as outputs an alternating string of 0's and 1's. For a

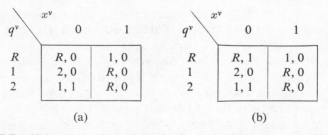

q^v \ x^v	0	1
R	$R, 0$	$1, 0$
1	$2, 0$	$R, 0$
2	$1, 1$	$R, 0$

q^v \ x^v	0	1
R	$R, 1$	$1, 0$
1	$2, 0$	$R, 0$
2	$1, 1$	$R, 0$

(a)   (b)

FIGURE 12.2. *Equivalent completely specified state tables, Example 12.1.*

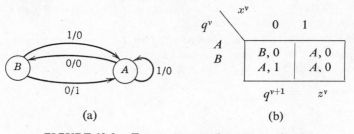

FIGURE 12.3.   *Two-state equivalent, Example 12.1.*

string of 1's applied, both circuits will produce a string of 0's. Next, consider the circuit in state 2 or state B, with a 1 applied, followed by a string of 0's. The 2-state circuit will produce a 0 followed by an alternating string of 0's and 1's. This behavior satisfies the specifications for the 3-state circuit since response to a string of 0's beginning in state R is a string of don't-care outputs. A similar analysis applies to a 1 followed by a string of 0's from states 1 or A. Thus, the circuits described by these two state diagrams are equivalent in the sense of generating the same required response to any sequence of inputs.   ■

The above example illustrates that arbitrary specification of don't-care entries in rows of a state table which are not identical may prevent minimization of the state table. As in Example 11.1, such simple and obvious specification may result in a table that is obviously minimal or so close to minimal that more complex and time-consuming efforts at minimization may not be justified. However, if absolute minimization is deemed important more powerful techniques must be developed. For this purpose, we will need some new definitions and theorems.

12.2   Compatibility

We have seen that in some cases incompletely specified states can be combined to reduce the total number of states in a circuit. However, we cannot say that such states are equivalent because the formal definition of equivalence requires that both outputs and next states be defined for all inputs. Instead, we shall say that incompletely specified states that can be combined are *compatible* with one another. When two states are identical wherever both are specified, as in the partial state table of Fig. 12.4, they can be made equivalent by assigning the unspecified entries the same values in both states.

States which are identical wherever both are specified are thus clearly compatible, but in the last section we saw a case where states that were not

q^v	A	B	C	K
3	2, –	3, –	4, 0	1, –
4	2, 1	–, –	–, 0	1, 0

$$q^{v+1},\, z^v$$

FIGURE 12.4

identical could be combined. In such cases we cannot simply specify the states so as to make them equivalent. To define compatibility we must go back to the basic concept of states having the same function in a circuit if they produce the same outputs in response to the same inputs. Since the circuit behavior is not specified for all possible input sequences, we cannot require identical outputs for all possible sequences, as was the case for equivalent states.

Definition 12.1.   Let a sequence of inputs be applied to an incompletely specified circuit S^* in initial state q. The sequence is said to be *applicable* to q if no unspecified next-state entries are encountered, except possibly as a result of the last input of the sequence.

Definition 12.2.   Two states p and q, of a circuit S are said to be *compatible* if and only if

$$\lambda_S[\delta(p, X_1 X_2 \cdots X_K), X_{K+1}] = \lambda_S[\delta(q, X_1 X_2 \cdots X_K), X_{K+1}] \quad (12.1)$$

whenever both of these ouputs are specified, for every input sequence $X_1 X_2 \ldots X_{K+1}$ applicable to both states. The expression $\delta(q, X_1 X_2 \cdots X_K)$ denotes the state of the circuit after the sequence of inputs $X_1 X_2 \cdots X_K$ is applied to the circuit, originally in state q.

Suppose the output sequences resulting from some applicable input sequence are computed for a circuit initially in compatible states p and q, leaving blank those outputs which are not specified. Each of these output sequences can be computed only as far as the next states are specified. Definition 12.2 says that at no point will specified outputs in these two sequences differ.

Theorem 10.1 was used as the basis for eliminating pairs of states which cannot be placed in the same equivalence classes. Theorem 12.1 will be used

---

* Strictly speaking, once constructed, any deterministic (nonrandom) circuit is completely specified. In Definition 12.1, we refer to the analytic determination of outputs and next states from the state table.

similarly to eliminate pairs of states which are not compatible. The proofs of most theorems in this chapter will be omitted. They are similar to but in general more tedious than the proofs presented in Chapter 10.

THEOREM 12.1.   *If two states q_i and q_j of circuit S are compatible, then the following conditions must be satisfied for every input X.*

1. $\lambda(q_i, X) = \lambda(q_j, X)$ *whenever both are specified.*

2. $\delta(q_i, X)$ *and* $\delta(q_j, X)$ *are compatible whenever both are specified.*

The processing of the implication table proceeds in exactly the same fashion as for the completely specified case. We first cross out any squares corresponding to pairs of states for which the *specified* outputs differ, i.e., for which condition 1 of Theorem 12.1 is not satisfied. In the remaining squares, we enter the numbers of pairs of states whose compatibility is implied. For example, if $\delta(q_i, X) = q_2$ and $\delta(q_j, X) = q_3$ in a certain state table, we will enter 23 in square i, j of the implication table. We enter check marks in squares corresponding to pairs of states which imply only their own compatibility. Once the initial implication table has been obtained, the elimination of further incompatible pairs of states by successive "passes" through the table proceeds in exactly the same manner as the elimination of nonequivalent states in completely specified circuits.

Example 12.2

A pulse input-pulse output sequence checker has four inputs, A, B, C, K, and two outputs Z_1, Z_2. At anytime the circuit may be in either a *check* or a *noncheck* mode. In the noncheck *mode* these four pulses may arrive in any order but only a K pulse will move the circuit into the check mode. Following this K pulse, the pulses A, B, C will arrive exactly once, in any order, followed by another K pulse. At the time of the K pulse following the sequence of A, B, C pulses, an output pulse will appear on line Z_2 if the pulses arrived in the order A, B, C and a pulse will appear on Z_1 if the order was C, B, A. If the A, B, and C pulses arrived in any other order, there will be no output pulse. The same K pulse which can generate an output will return the circuit to the noncheck mode, requiring another K pulse to return it to the check mode. The circuit outputs will be observed only at the time of a K pulse. Therefore the output must be 00 when a K pulse drives the circuit to the check mode.

Solution: A state diagram and state table are shown in Fig. 12.5. State q_1 constitutes the noncheck mode, from which a K pulse will take the circuit to q_3 the initial state of the check mode. Since only two of the A, B, or C pulses are needed to identify a sequence, the second pulse leads to one of three possible states q_7, q_8, q_9, corresponding to the three output possibilities. The output-producing K

pulse sends the circuit back to q_1. It is seen that there are many don't-cares, due to the restriction to exactly one A pulse, one B pulse, and one C pulse per sequence and the fact that the output is specified only for the input K.

The initial and final implication tables are shown in Fig. 12.6. The squares which contain check marks correspond to pairs whose compatibility is not dependent on the compatibility of any other states. The remaining squares, not crossed out in the final implication table, correspond to sets which are compatible because the implied states satisfy condition 2 of Theorem 12.1.

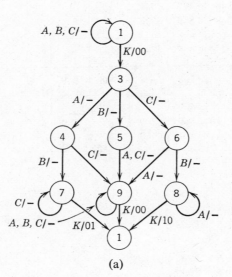

(a)

q^ν	A	B	C	K
		Input		
1	1, –	1, –	1, –	3, 00
3	4, –	5, –	6, –	–, –
4	–, –	7, –	9, –	–, –
5	9, –	9, –	9, –	–, –
6	9, –	8, –	–, –	–, –
7	–, –	–, –	7, –	1, 01
8	8, –	–, –	–, –	1, 10
9	9, –	9, –	9, –	1, 00

$$q^{\nu+1}, Z_1 Z_2$$

(b)

FIGURE 12.5. *State diagram and state table, Example 12.2.*

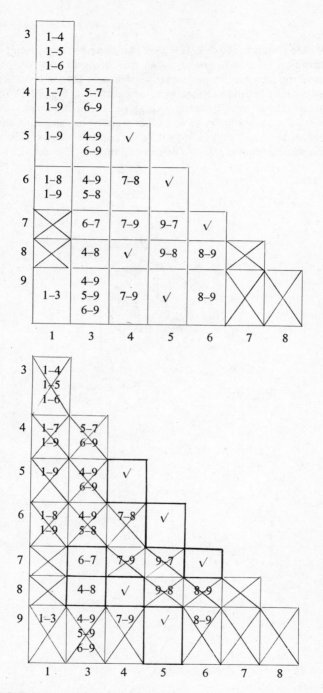

FIGURE 12.6.   *Initial and final implication tables, Example 12.2.*

The reader will recall that, in Chapter 10, equivalent pairs determined by the implication tables were combined to form equivalence classes. A similar process can be applied to compatible pairs. Although not formally proven, it should be clear that *pairs of states corresponding to squares not crossed out in the final implication table are compatible pairs.*

Definition 12.3.   A set of states, S_i, of a circuit S will be said to form a *compatibility class* if every pair of states in S_i is compatible.

Definition 12.4.   A *maximal compatible* is a compatibility class, which will not remain a compatibility class if any state not already in the class is added to the class. A single state which is compatible with no other state is a maximal compatible.

Example 12.2 (continued)

Determine the set of all maximal compatibles for the sequence checker.
Solution: The process is exactly that utilized to determine equivalence classes, as shown in Fig. 12.7. There are no compatibles indicated in columns 7 and 8. In column 6, we find that q_6 and q_7 form a compatible pair. Next we find q_5

```
8   –
7   –
6   (6, 7)
5   (6, 7) (5, 6) (5, 9)
4   (6, 7) (5, 6) (5, 9) (4, 5) (4, 8)
3   (6, 7) (5, 6) (5, 9) (4, 5) (4, 8) (3, 7) (3, 8)
1   (6, 7) (5, 6) (5, 9) (4, 5) (4, 8) (3, 7) (3, 8) (1)
```

FIGURE 12.7.   *Determination of maximal compatibles, Example 12.2.*

compatible with q_6 and q_9, but here we note a situation which we have not seen before. For completely specified circuits, if a state is equivalent to two other distinct states, then these states are equivalent to each other. But here we find that q_6 and q_9 are not compatible, so instead of having a class (5, 6, 9), we have two compatible pairs, (5, 6) and (5, 9). The completion of the process follows in the same manner. ■

In this case, the maximal compatibles all happen to be pairs, but that is a peculiarity of this problem and not the general case. What is significant is the fact that several states appear in more than one maximal compatible, as was the case with (5, 6) and (5, 9) as discussed above. It is easy to see why these pairs are compatible while states 6 and 9 are not from an inspection of the circled next-state entries in column B of Fig. 12.5b. Note that the next state for state 5 in this column is unspecified so that (at least with respect to column *B*) (56) and (59) are compatible regardless of the next states specified for 6 and 9. However, the next states for 6 and 9 are 8 and 9, respectively, and 8 and 9 are in turn incompatible since their outputs differ in column *K*. Therefore, 6 and 9 are not compatible. Similarly, state 3 is compatible with states 7 and 8 while these two states are not compatible with each other.

The above are two examples of a more formal observation that compatibility is *not* a transitive relation. Thus, compatible states cannot be grouped neatly into disjoint classes, and the formal basis of the procedures of Chapter 10 slips away.

This fact accounts for such puzzling situations as we saw in Example 12.1, and also makes the Huffman partition method inapplicable to incompletely specified circuits.

Nevertheless, our basic simplification problem is still the same: to find a circuit with fewer states which will produce the same output sequence. Since the simplified circuit need have the same output as the original circuit only for those outputs which are specified, slightly different criteria than specified by Definition 10.1 must be satisfied by the states of the two circuits.

Definition 12.5. A state p_j of a state table T is said to cover a state q_i of a state table S, written $p_j \geq q_i$, if, for any input sequence applicable to q_i and applied to both T and S initially in states p_j and q_i, respectively, the output sequences are identical whenever the output of S is *specified*.

The distinction between covering and compatibility is illustrated in Fig. 12.8. Notice that, at least for the single input sequence 0112031, q_1 and q_2 are compatible. In addition, the state p covers *both* q_1 and q_2 for the input sequence shown, and (q_1, p) and (q_2, p) are compatible pairs. Neither q_1 nor q_2, however, cover p. Thus, we observe that requiring a state r to cover a state s is more restrictive than requiring the two states to be compatible.

We further note that q_2 and q_3 are compatible, while q_1 and q_3 are not compatible. Thus, as we have observed from Example 11.1, compatibility is not a transitive property.

Input Sequence	0	1	1	2	0	3	1
Initial States	\multicolumn{7}{c}{Output Sequences}						
q_1	X	1	X	X	0	1	1
q_2	0	1	1	X	0	X	X
q_3	X	1	X	1	0	0	0
p	0	1	1	0	0	1	1

FIGURE 12.8

Definition 12.6. A state table T is said to cover a state table S if for every state q_i in S there exists some state p_j in T which covers q_i.

A circuit T which covers S necessarily performs the function of S. Given S, then, our goal is to find such a circuit, T, having a minimum number of states. Toward this goal, we present the following theorems.

THEOREM 12.2. *If an internal state p of T covers both internal states q_j and q_i of S, then states q_i and q_j must be compatible.*

Proof: If p covers both q_i and q_j, then for any applicable input sequence $X_1 X_2 \cdots X T_{K+1}$ for which

$$\lambda_S[\delta(q_j, X_1 \cdots X_K), X_{K+1}] \quad \text{and} \quad \lambda_S[\delta(q_i, X_1 \cdots X_2), X_{K+1}]$$

are both specified, we must have

$$\lambda_T[\delta(p, X_1 \cdots X_K), X_{K+1}] = \lambda_S[\delta(q_j, X_1 \cdots X_K), X_{K+1}] \quad (12.2)$$

and

$$\lambda_T[\delta(p, X_1 \cdots X_K), X_{K+1}] = \lambda_S[\delta(q_i, X_1 \cdots X_K), X_{K+1}] \quad (12.3)$$

Therefore,

$$\lambda_S[\delta(q_j, X_1 \cdots X_K), X_{K+1}] = \lambda_S[\delta(q_i, X_1 \cdots X_K), X_{K+1}] \quad (12.4)$$

and q_i and q_j are compatible.

The following is an immediate corollary of Theorem 12.2 and the definition of a compatibility class, Definition 12.3.

COROLLARY 12.3. *If a state p_i of T covers a set of states, S_i, in S, these states must form a compatibility class.*

For completely specified machines, it was only necessary to include each state of S in one equivalence class. A machine T with a state corresponding to each such equivalence class would then be equivalent to S. In the incompletely specified case, however, it may be necessary to include a state in more than one compatibility class.

Definition 12.7. A collection of compatibility classes is said to be closed if, for any class $\{q_1, q_2, \ldots, q_m\}$ in the collection and for every input X_i, all specified next states, $\delta(q, X_i)$, $\delta(q_2, X_i)$, $\ldots$, $\delta(q_m, X_i)$, fall into a single class in the collection.

We are now in a position to state the critical theorem of this section.

THEOREM 12.4. *Let the n states of an incompletely specified sequential circuit S form a collection of m compatibility classes such that each of the n* *states is a member of at least one of the m compatibility classes. Then the circuit S may be covered by a circuit T, having exactly m states, $p_1, p_2, \ldots, p_m$, such that each compatibility class of S is covered by one of the states of T, if and only if the collection of m compatibility classes of S is closed.*

From this theorem, it can be seen that our basic problem is to find a minimal closed collection of compatibility classes which includes all the states of the original circuit. It can be shown that the set of all maximal compatibles is a closed collection. Therefore, this set of maximal compatibles could be used to form the basis for an equivalent circuit. However, the resulting circuit will seldom be minimal, since there will usually be considerable overlap. Indeed, the number of maximal compatibles will often be greater than the number of states in the original circuit.

It is obvious that we will wish to delete redundant classes wherever possible. However, we must be careful not to violate the closure conditions by removing a class which is necessary to satisfy the implications of a class which is retained. Since any subset of a compatibility class is also a compatibility class, we may delete sets from classes in order to reduce closure requirements.

To facilitate discussion, we shall refer to a closed collection which contains each state of a machine S in at least one class as a *cover collection*. Once we find a cover collection, the construction of a machine covering S is straightforward.

Example 12.2 (continued)

All the compatibility classes determined in Fig. 12.7 are listed in Fig. 12.9 together with any implied classes. We first note that states q_9 and q_{10} each appear in only one compatibility class, so these classes (5, 9) (3, 10), along with the states which did not combine at all, q_1 and q_2, will be included in the cover collection. The selection of these classes leaves states q_3, q_4, q_6, q_7 and q_8 to be covered. Four of these states can be covered by including classes (4, 8) and (6, 7) in the cover collection which can be done without implying the inclusion of any other classes. It remains to cover state 3. This can be accomplished by including any of the three classes (3, 7), (3, 8), or (3) in the cover collection. We note that (3, 7) implies (6, 7) and (3, 8) implies (4, 8). Both of these implied classes have already

Comp. Classes	(6, 7)	(5, 6)	(5, 9)	(4, 5)	(4, 8)	(3, 7)	(3, 8)	(1)
Impl. Classes	√	√	√	√	√	(6, 7)	(4, 8)	√

FIGURE 12.9. *Compatibility classes and implied classes, Example 12.2.*

been included, so any of the following can be used to form the final cover collection.

(1)	(5, 9)	(4, 8)	(6, 7)	(3, 7)
(1)	(5, 9)	(4, 8)	(6, 7)	(3, 8)
(1)	(5, 9)	(4, 8)	(6, 7)	(3)

The third of these closed collections is probably preferable since it will result in more don't-cares in the final state table.

Once a closed collection of compatibility classes has been obtained, the output entries in the table are specified as required to satisfy Definition 12.5. That is, for each state p_i of T and each input, the output must agree with any outputs which are specified for the states covered by p_i. Theorem 12.4 assures us that there will be no conflicts.

Similarly, each next-state entry for present state p_i must be the state covering the class which includes all of the next-state entries for the states covered by p_i. A minimal state table covering our machine is given in Fig. 12.10. As shown, states a, b, c, d, and e are assigned to cover classes (1), (4, 8), (5, 9), and (6, 7) respectively. The outputs and next state entries are obtained in much the same way as for completely specified circuits. Consider the circled next state entry for present state d and input pulse B. In Fig. 12.5, $\delta(5, B)$ is unspecified while $\delta(9, B) = 9$. Since state 9 is covered by class (5, 9), we enter the corresponding state d as $\delta(d, B)$. The remaining entries in Fig. 12.10 may be obtained in a similar manner. ■

Comp. Class	q^ν	Inputs			
		A	B	C	K
(1)	a	$a, -$	$a, -$	$a, -$	$b, 00$
(3)	b	$c, -$	$d, -$	$e, -$	$-, -$
(4, 8)	c	$c, -$	$e, -$	$d, -$	$a, 10$
(5, 9)	d	$d, -$	$(d), -$	$d, -$	$a, 00$
(6, 7)	e	$d, -$	$c, -$	$e, -$	$d, 01$

$$q^{\nu+1}, z^\nu$$

FIGURE 12.10.   *Final minimal state table, Example 12.2.*

Let us next apply our modified minimization techniques to the 3-state table of Example 12.1 to see that they lead to the 2-state cover circuit which we determined by trial-and-error methods.

Example 12.1 (continued)

Minimize the 3-state table of Fig. 12.1b (repeated in Fig. 12.11a) by obtaining a closed collection of compatibility classes.

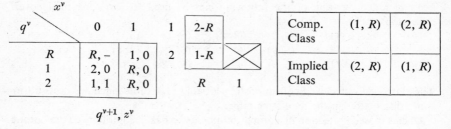

q^{v+1}, z^v

(a)   (b)   (c)

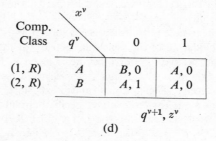

(d)

FIGURE 12.11.   *Minimization of state table of Example 12.1.*

Solution: The simple implication table of Fig. 12.11b leads to the list of maximal compatibles and implied classes of Fig. 12.11c.

As might be expected since they form a collection of maximal compatibles, the two classes provide a cover collection. The two classes imply each other so that both must be used to satisfy closure. Thus, it is the closure requirement that prevents us from arbitrarily combining q_R with one or the other of q_1 or q_2, as was originally suggested. It may seem strange that compatible classes which overlap can be used at the same time, but a comparison of the final state table of Fig. 12.11d with the original (Fig. 12.1b) shows that q_A satisfies all specifications of q_1 and q_R, and q_B does the same for q_2 and q_R. ■

In both of the above examples, the closure problems were trivial, and the choice of closed covers was fairly obvious. In some cases where there are a large number of interacting closure requirements, it may be desirable to reduce the size of some of the compatibility classes. It is apparent that any subset of a maximal compatible is also a compatibility class, and deleting one or more states from a maximal compatible may reduce the closure requirements. One such situation illustrated by the following example.

Example 12.3

Find a minimal state table equivalent to the state table of Fig. 12.12a.

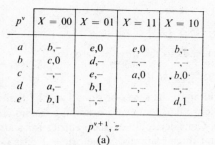

(a)

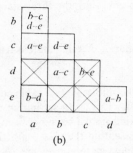

(b)

d	(de)
c	(de)
b	(bc), (bd), (de)
a	(ae), (abc), (bd), (de)

(c)

Comp. Classes	ae	abc	bd	de
Impl. Pairs	bd	bc, ae, de	ac	ab

(d)

		q^{v+1}				z^v			
		X^v				X^v			
	q^v	00	01	11	10	00	01	11	10
(ab)	1	2	3	3	1/2	0	0	0	–
(bc)	2	2	3	1	1/2	0	–	0	0
(de)	3	1	1/2	–	3	1	1	–	1

(e)

FIGURE 12.12. *Minimizing state table, Example 12.3.*

Solution: The final version of the implication table is shown in Fig. 12.12b, and the determination of maximal compatibles is shown in Fig. 12.12c.

The maximal compatibles are listed in Fig. 12.12d, with the implied pairs. The reader may easily verify that the smallest closed cover obtainable from this listing is the set

$$[(abc), (ae), (bd), (de)] \tag{12.6}$$

If we delete state c from the class (abc), we remove the implication of class (ae), thus obtaining the three-state closed cover

$$[(ab), (bc), (de)] \tag{12.7}$$

Similarly, deleting state b from (abc) leads to the three-state cover

$$[(ac), (ae), (bd)] \tag{12.8}$$

The minimal state table corresponding to the cover of Equation 12.7 is shown in Fig. 12.12e. In setting up this table we find that the overlapping classes create a situation we have not seen before. Consider the next-state entries for state q_1, representing the class (ab). In the first three columns of the original state table, the next-state entries are (bc), (de) and (e), corresponding to states q_2, q_3, and q_3, respectively. In the fourth column, the next-state entry is (b), which appears in both q_1 and q_2. Thus the next-state entry can be either q_1 or q_2, as indicated by the entry 1/2. Such entries may be considered partial don't-cares and usually provide extra flexibility in design. The other optional entries in the table arise in the same manner. ■

The above example illustrates some of the problems that can occur due to closure requirements, but the state table was given arbitrarily rather than derived from a description of some desired circuit function. The reader may wonder if such situations are likely to occur in "real-life" problems. Let us now consider a rather complex example which will lead to all three types of don't-cares and a rather difficult covering problem.

Example 12.4

An electronic ignition lock is to be designed for the purpose of preventing drunk drivers from starting their cars. The basic idea is to flash an arbitrary sequence of letters on a screen very briefly and then require the driver to punch a set of buttons in the same order. Tests have shown that better discrimination between drivers who are drunk and those who simply have poor retention of visual images is obtained by requiring the task to be performed on several short sequences rather than one long sequence. On this basis, the following scheme is proposed. There will be four buttons, labeled A, B, C, D, and three "legal" sequences, as listed in Fig. 12.13a. When the system is turned on, Sequence #1 will be displayed, after which the driver must punch the buttons in that order. If he is successful, Sequence #2 will then be displayed, and if it is successfully repeated, Sequence #3 will be displayed. Only if all three sequences are correctly followed will the ignition be turned on. If the driver makes a mistake at any point, the keyboard will be mechanically locked to prevent further operation and may be released only by removing the ignition key and starting over.

The basic block diagram of the system is shown in Fig. 12.13b.

The counter contains the number of the sequence being used, always starting with 1 when turned on. The sequence detector checks the sequence. If an error is made in any sequence, the error signal E is issued to disable the system. If a sequence is correctly completed, a pulse is issued on one of the lines P_1, P_2, or P_3, corresponding to sequence 1, 2, 3, respectively. This information is

Seq. No.	Sequence
1	A D B C
2	A C B D
3	B D A C

(a)

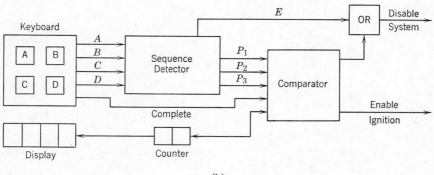

(b)

FIGURE 12.13.   *Electronic ignition lock system.*

compared by the comparator with the number of the sequence that was displayed. If they agree for #1 or #2, the counter is stepped to the next number to start the next sequence.

After agreement on Sequence 3, the ignition is turned on. If the pulse received by the comparator does not match the contents of the counter, a disabling error signal is issued by the comparator. The keyboard is so constructed that each button may be pushed only once for a given sequence. A 1-level will be supplied to the comparator by the keyboard on a line labelled *complete* after the fourth button has been depressed. The comparator will ignore any pulse inputs occuring before *complete* = 1. Develop a minimal flow table for the pulse input pulse output sequence detector.

Solution: A state diagram for the sequence detector is developed in Fig. 12.14a. The three expected input sequences lead in 4 steps from the reset state q_0 back to state q_0. Those inputs which do not fall in any of the three sequences are shown as generating output pulse E and leading to a figurative "don't-care" state. The P_1, P_2, and P_3 outputs for the three expected input sequences are don't-cares until the fourth pulse appears. We refine Fig. 12.14a slightly to form the state table of Fig. 12.14b by separating the error output E from the other three pulse outputs. Starting in state q_0, either A or B can be the start of a correct sequence, leading to states q_1 or q_6, respectively. The error output, E is 0 since no error has occurred, and the other outputs are don't cares, as previously noted. C or D are

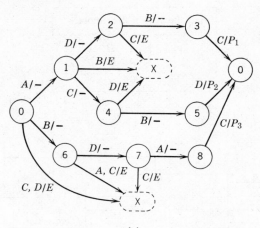

(a)

q	A	B	C	D
0	1, 0, –	6, 0, –	–, 1, –	–, 1, –
1	–, –, –	–, 1, –	4, 0, –	2, 0, –
2	–, –, –	3, 0, –	–, 1, –	–, –, –
3	–, –, –	–, –, –	0, 0, P_1	–, –, –,
4	–, –, –	5, 0, –	–, –, –	–, 1, –
5	–, –, –	–, –, –	–, –, –	0, 0, P_2
6	–, 1, –	–, –, –	–, 1, –	7, 0, –
7	8, 0, –	–, –, –	–, 1, –	–, –, –
8	–, –, –	–, –, –	0, 0, P_3	–, –, –

q^{v+1}, E, P

(b)

FIGURE 12.14. *State diagram and table, sequence detector, Example 12.4.*

"illegal" as a first input, resulting in an E pulse. Since this output will lock the keyboard, preventing further inputs, the next state is a don't care. This is an example of the third type of don't-care described at the beginning of the chapter.

In state q_1, correct inputs of C or D lead to q_2 or q_4, and an incorrect input of B produces an error output. An A input cannot occur due to the restriction that no button can be pushed more than once, so all entries are don't-cares. Operation proceeds in this general manner until the fourth input, at which time a pulse output indicates the number of the sequence just completed and the circuit returns to state q_0 to await the next sequence.

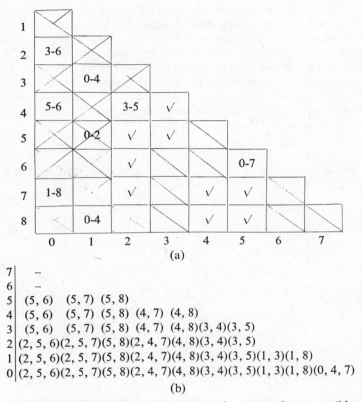

FIGURE 12.15. *Implication table and maximal compatibles, Example 12.4.*

The final implication table and the formation of the maximal compatibles are shown in Fig. 12.15. All states appear in at least one compatible pair, so no single states need be included in the list. The maximal compatibles and their implied classes are shown in Fig. 12.16. We first note that there are only two 3-state classes which do not overlap, so a minimum of four classes will be required for a cover. If we can find a 4-state cover, we will know it is minimal.

Maximal Compatibles	(2, 5, 6)	(2, 5, 7)	(2, 4, 7)	(0, 4, 7)	(5, 8)	(4, 8)	(3, 4)	(3, 5)	(1, 3)	(1, 8)
Implied Classes	(0, 7)	√	(3, 5)	(5, 6) (1, 8)	√	√	√	√	(0, 4)	(0, 4)

FIGURE 12.16. *Maximal compatibles and implied classes, Example 12.4.*

As a first step, we look for any states which appear in only one maximal compatible. A maximal compatible or some subset of a maximal compatible which contains such states must be included in the cover collection. In this case, q_0 appears only in the class $(0, 4, 7)$, so we tentatively place $(0, 4, 7)$ in the cover collection. This class implies $(5, 6)$ and $(1, 8)$. The $(5, 6)$ implication is satisfied by $(2, 5, 6)$, which in turn implies $(0, 7)$, which is satisfied by $(0, 4, 7)$. Similarly, $(1, 8)$ implies $(0, 4)$, which is satisfied by $(0, 4, 7)$. All three classes can be used, since they imply each other, thus satisfying closure. Further, they cover all states except q_3, the addition of which yields

$$(0, 4, 7), (2, 5, 6), (1, 8), (3)$$

as a 4-state minimal cover. The corresponding minimal state table is shown in Fig. 12.17. ■

Maximal Compatibles	q^ν	x^ν			
		A	B	C	D
$(0, 4, 7)$	a	$c, 0, -$	$b, 0, -$	$-, 1, -$	$-, 1, -$
$(2, 5, 6)$	b	$-, 1, -$	$d, 0, -$	$-, 1, -$	$a, 0, 2$
$(1, 8)$	c	$-, -, -$	$-, 1, -$	$a, 0,$	$b, 0, -$
(3)	d	$-, -, -$	$-, -, -$	$a, 0,$	$-, -, -$

$$q^{\nu+1}, E,$$

FIGURE 12.17.   *Minimal state table, Example 12.4.*

In the above example, we note that there are a number of states that combine without implication. In Example 12.1, we saw that a minimal circuit could not always be obtained after an arbitrary assignment of don't-cares, but in that case there were implications involved and the difficulty arose from the need to satisfy the closure requirement. But if there are no implications and thus no closure requirements, can we safely combine states by inspection? Let's try it in Example 12.4 and see.

In that example, we see that $(2, 5, 7)$ is a compatible without implication, so let us combine them as state 2 by inspection before setting up the implication table. The reduced state table and corresponding implication table after this simplification are shown in Fig. 12.18. Now the only compatible classes are $(3, 4)$ and $(4, 8)$, which cover only three of the seven states, so the other four must be included individually, resulting in a 6-state cover. This may seem surprising, since a 4-state cover is known to exist, and combining $(2, 5, 7)$ did not set up any implications.

A partial explanation is seen in the fact that the combination of states 2, 5, 7 forced 2 and 5 to assume the specifications of 7, which is incompatible

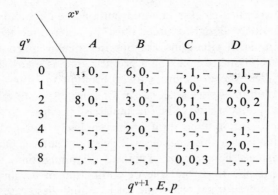

q^v	x^v A	B	C	D
0	1, 0, –	6, 0, –	–, 1, –	–, 1, –
1	–, –, –	–, 1, –	4, 0, –	2, 0, –
2	8, 0, –	3, 0, –	0, 1, –	0, 0, 2
3	–, –, –	–, –, –	0, 0, 1	–, –, –
4	–, –, –	2, 0, –	–, –, –	–, 1, –
6	–, 1, –	–, –, –	–, 1, –	2, 0, –
8	–, –, –	–, –, –	0, 0, 3	–, –, –

q^{v+1}, E, p

FIGURE 12.18. *Reduced state table and implication table, Example 12.4.*

with 6. This in turn prevents the combination of (0, 4, 7), which in turn eliminates (1, 8), etc. Thus, the problem is not just closure but the basic intransitive nature of compatibility. The conclusion is that combination by inspection of any states other than those identical in every entry, including don't-cares, may prevent state table minimization.

The reader may wish to question our overall approach to the design of the drunk detector. He might visualize a more efficient realization through considering the entire design as a unit rather than designing the sequence checker as a separate device. The state table resulting from this approach would likely have fewer don't-cares and, even though larger, might be easier to minimize. The restriction that each button could be pushed only once might seem unnecessary and even undesirable since it makes it easier to punch the correct sequence accidentally. This restriction led to many of the don't-cares in Fig. 12.14b.

Incompletely specified clock-mode and pulse-mode circuits typically arise only when separately designed units must be interconnected to form a complete system. Such situations are not infrequent, given the standard organizational technique of breaking a problem into as many separate parts as possible to allow more resources to be brought to bear and hopefully a more rapid solution. As we shall see in Chapter 13, state tables for level mode circuits will almost always be incompletely specified.

When the need for minimization of incompletely specified tables does occur, the techniques in this chapter will generally suffice. Occasionally in very complex systems there may be so many overlapping compatibility classes and interacting implications that the determination of a minimal cover is very difficult. A more formal technique for handling these cases is presented in Appendix A, but the typical designer will rarely find need for more than has been presented here.

12.3 Completion of Design*

The incompletely specified circuit presents certain problems in the determination of the final state table which do not occur in the completely specified case. First, there are often several covers available; and second, some of the classes in these covers may overlap.

Example 12.3 (continued)

The state table for Example 12.3 is shown in Fig. 12.19a, and the two minimal covers are listed in Fig. 12.19b.

	q^{v+1}				z^v			
q^v	$X^v = 00$	$X^v = 01$	$X^v = 11$	$X^v = 10$	$X^v = 00$	$X^v = 01$	$X^v = 11$	$X^v = 10$
a	b	e	e	b	–	0	0	–
b	c	d	–	–	0	–	–	–
c	–	e	a	b	–	–	0	0
d	a	b	–	–	–	1	–	–
e	b	–	–	d	1	–	–	1

(a)

$$[(ab), (bc), (de)] \qquad [(ac), (ae), (bd)]$$

(b)

FIGURE 12.19. *State table and minimal covers, Example 12.3.*

* This section may be omitted without affecting the continuity of the remaining chapters.

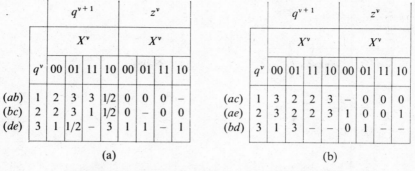

FIGURE 12.20. Reduced state tables, Example 12.3.

The reduced state tables corresponding to the two minimal covers are shown in Fig. 12.20.

Notice that the state table of Fig. 12.20a is a repeat of that of Fig. 12.12e. As discussed earlier, this table has several optional entries, which provide some extra flexibility in design. Therefore, we shall realize this state table, rather than that of Fig. 12.20b.

Since there are only three states, we simply try the three possible assignments. We also note that an assignment that places the extra (don't-care) row next to the two assigned to state 3 will provide the simplest possible output function. The transition table, next-state maps, output map, and excitation maps for one such assignment are shown in Fig. 12.21.

Notice the dual entries on the transition table. With the state assignment chosen, these result in $Y_2 = 0$ and Y_1 equal to a don't-care for these entries, as shown in the next-state maps. From the output and excitation maps, the corresponding functions are easily found.

$$z = y_2$$
$$J_{y_2} = \bar{x}_2 x_1 + y_1 x_1 \qquad K_{y_2} = \bar{x}_2 \tag{12.12}$$
$$J_{y_1} = x_1 \qquad K_{y_1} = \bar{y}_2 \bar{x}_1$$

The corresponding circuit is shown in Fig. 12.22. ■

We found a very simple set of equations for Example 12.3, but there still remains a question as to whether some other assignment or cover might produce an even simpler circuit. There is another assignment for the cover of Fig. 12.20a which maintains the same output function, and it should be checked. It turns out to be not as good as the one used above.

The other minimal cover (Fig. 12.20b) does not appear as promising for two reasons. First, the output function will be more complex, regardless of the placement of the don't-care row. Second, it does not have any optional next-state conditions. There is overlap, since state q_a appears in two classes.

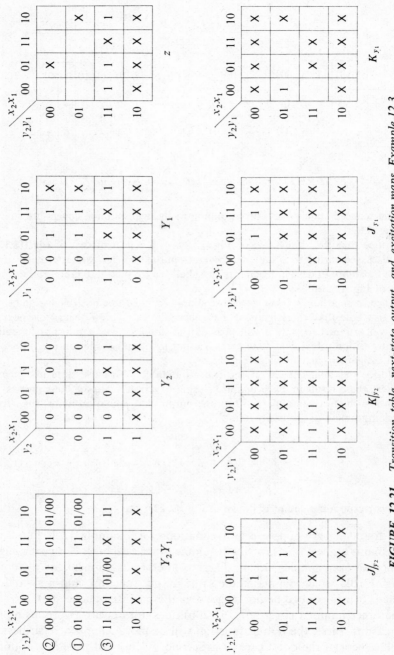

FIGURE 12.21. *Transition table, next-state output, and excitation maps, Example 12.3.*

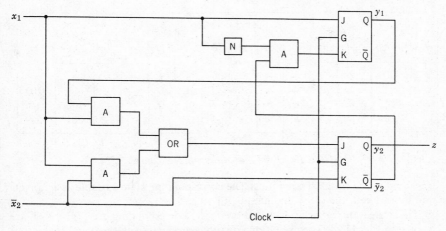

FIGURE 12.22. *Final circuit for Example 12.3.*

However, state q_a is not the single next state for any class, so no optional assignments result. As we have seen, optional assignments give us more flexibility, often resulting in additional don't-cares and thus often contributing to simpler realizations. In general, if there is more than one minimal cover, the reduced state tables should be formed for all these covers. From these, the table offering the most optional assignments and the simplest output functions should be tried first. If there is no clear-cut choice, you can either try them all or just make a random selection and hope for the best.

The reader may recall that we also found a 4-state cover for the circuit of Example 12.3. Since this would not require any more state variables, should it be tried? Generally, a minimal cover should be used. If the number of states in the minimal cover is a power of 2, using any more states will require more state variables and thus more memory. As pointed out in Chapter 10, this is not always undesirable but should be considered only in rather special cases. If more states are used, even though they do not increase the number of state variables, they will reduce the number of don't-cares, generally complicating the realization.

Example 12.4 (continued)

Complete the design of sequence detector for the electronic ignition lock.

Solution: The minimal state table is shown in Fig. 12.23, with the next state and output sections separated.

There are so many don't-cares in the next-state and P functions that the choice of assignment is unlikely to make much difference in the final circuits for the state variables and P outputs. However, the E function has only a few don't-cares, so all three possible assignments were tried for E, with Assignment #1

q^v	$(y_1y_2)^v$	A	B	C	D	A	B	C	D	A	B	C	D
a	00	c	b	–	–	0	0	1	1	–	–	–	–
b	01	–	d	–	a	1	0	1	0	–	–	–	010
c	11	–	–	a	b	–	1	0	0	–	–	100	–
d	10	–	–	a	–	–	–	–	0	–	–	001	–

| | q^{v+1} | | | E^v | | | $(P_3, P_2, P_1)^v$ | |

FIGURE 12.23.   *Final state table, Example 12.4.*

resulting in the simplest equations. The corresponding state variable assignments are indicated in the second column of Fig. 12.23.

The transition table and excitation maps corresponding to Fig. 12.23 are shown in Fig. 12.24. The excitation equations are read from these and the output equations are read directly from Fig. 12.23, as given below. The final circuit is shown in Fig. 12.25.

$(y_2y_1)^v$	A	B	C	D
00	11	01	× ×	× ×
01	× ×	10	× ×	00
11	× ×	× ×	00	01
10	× ×	× ×	00	× ×

$(y_2y_1)^{v+1}$

$(y_2y_1)^v$	A	B	C	D
00	1	0	×	×
01	×	1	×	0
11	×	×	0	0
10	×	×	0	×

S_{y_2}

$(y_2y_1)^v$	A	B	C	D
00	0	×	×	×
01	×	0	×	×
11	×	×	1	1
10	×	×	1	×

C_{y_2}

$(y_2y_1)^v$	A	B	C	D
00	1	1	×	×
01	×	0	×	0
11	×	×	0	×
10	×	×	0	×

S_{y_1}

$(y_2y_1)^v$	A	B	C	D
00	0	0	×	×
01	×	1	×	1
11	×	×	1	0
10	×	×	×	×

C_{y_1}

FIGURE 12.24.   *Transition table and excitation maps, Example 12.4.*

$$S_{y_2} = A + By_1 \Big\}$$
$$C_{y_2} = C + D \Big\} \qquad (12.13)$$

$$S_{y_1} = A + B\bar{y}_1 \Big\}$$
$$C_{y_1} = By_1 + C + D\bar{y}_2 \Big\} \qquad (12.14)$$

$$E = Ay_1 + By_2 + C\bar{y}_2 + D\bar{y}_1 \qquad (12.15)$$

$$P_1 = C\bar{y}_1 \Big\}$$
$$P_2 = D \Big\} \qquad (12.16)$$
$$P_3 = C\bar{y}_1 \Big\}$$

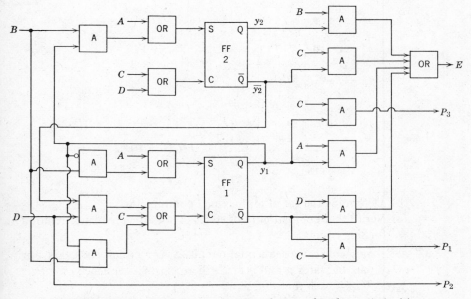

FIGURE 12.25. *Final circuit, sequence detector for electronic ignition lock.* ■

PROBLEMS

12.1. A single input line X, on which no more than two consecutive 1-bits or no more than two consecutive 0-bits may occur, is to serve as the input to two distinct sequential circuits. Both circuits are synchronized by the same clock.

(a) Determine the incompletely specified state table of a circuit with a single output, Z_1, such that $Z_1^{v+2} = X^v X^{v+1} + X^{v+2}$ (logical OR).

(b) Determine the incompletely specified state table of a circuit with a single output, Z_2, such that $Z_2^{v+2} = (X^v + X^{v+1}) \cdot X^{v+2}$. Do not use overlapping notation for the states of (a) and (b).

(c) Using the implication table, show that 3-state covers can be , obtained for each of the circuits of parts (a) and (b).

(d) Obtain the state table of a 3-state circuit with four outputs, 00, 01, 10, and 11, which will simultaneously accomplish the function of both circuits.

12.2. (a) Determine the set of all compatible pairs of states of the incompletely specified sequential circuit given in Fig. P12.2.

q^v	$X = 0$	$X = 1$	$X = 2$
		q^{v+1}, Z	
q_1	$q_3,0$	$q_5,1$	
q_2	$q_3,0$	$q_5,$	
q_3	$q_2,0$	$q_3,0$	$q_1,$
q_4	$q_2,$	$q_3,$	$q_5,$
q_5		$q_5,\hat{0}$	$q_1,$

FIGURE P12.2

(b) Determine the set of all maximal compatibles.

(c) Verify that compatibility of states is not an equivalance relation for this circuit.

12.3. A *strongly connected* sequential circuit, S, is a circuit such that, for every pair of states p and q of S, there exists a sequence of inputs $X_1 X_2 \cdots X_K$ such that $\delta(q, X_1 X_2 \cdots X_K) = p$.

Although the result may not serve a useful function, form a circuit simply by writing the two state tables of Problem 12.1 (a) and (b) as a single table. The outputs and initial states of each row of the new table remain the same as in the original tables.

(a) Show that the circuit formed as above is not strongly connected.

(b) Using the Paull-Unger implication table, obtain a 5-state equivalent of the above circuit, which is strongly connected.

12.4. Determine a minimal state equivalent of the sequential circuit described by the flow table in Fig. P12.4.

q^v	q^{v+1}, Z			
	$X = 0$	$X = 1$	$X = 2$	$X = 3$
A	B,	C,0		D,
B		E,0		
C	D,	F,	C,	
D	E,	C,0		A,
E		F,0		
F		,1	D,	B,

FIGURE P12.4

12.5. Determine a minimal state table for a circuit with a single input and a single output which is only sampled once every five clock times. When sampled, the output must indicate whether three or more of the last five input bits have been 1 (including the current bit). Also provided are a clock and a synchronizing input which resets the circuit to an initial state following each sampling instant. The latter need not be shown in the state table.

12.6. Determine the state table of a circuit which will accomplish the same function as the circuit of Problem 10.7. The only change is that input X_2 will be 1 only for the one-clock pulse immediately preceding the input of each coded character. The state table is now incompletely specified. Show that 14 states are still required in the minimal circuit.

12.7. One information input line X_1, a clock, and the outputs of a modulo-4 counter serve as the inputs to a sequential circuit. Binary-coded decimal numbers appear sequentially in 4-bit characters (lowest. order bit first) on line X_1, synchronized with the output of the counter- That is, $c_2 c_1 = 00$ when the lowest-order bit arrives, and $c_2 c_1 = 01$, 10, and 11, respectively, as the next three bits arrive. Derive the incompletely specified state table for a circuit which will translate the BCD code to excess-3 code with a delay of three clock pulses.

12.8. Obtain a minimal state table of a synchronous sequential circuit which will perform the translation shown in the figure with only a two-period delay. The process may be assumed to be synchronized by a reset signal not described in the state table. As shown, only the first two bits of the output character are sampled, so the third is redundant. Only the four valid characters of Fig. P12.8 will occur as inputs.

12.9. Determine the maximal compatibles of the circuit described by the state table in Fig. P12.9. Determine a 4-state cover of this circuit.

X^{v+2}	X^{v+1}	X^v	Z^{v+4}	Z^{v+3}	Z^{v+2}
0	0	1	–	1	0
0	1	0	–	1	1
1	0	0	–	0	1
1	1	1	–	0	0

FIGURE P12.8

q^v	q^{v+1}, Z						
	$X=1$	$X=2$	$X=3$	$X=4$	$X=5$	$X=6$	$X=7$
q_1	$q_1,0$	$q_3,1$	$q_5,$		$q_5,$	$q_5,1$	,
q_2	$q_1,0$	$q_3,1$	$q_5,1$				$q_8,0$
q_3		$q_4,$		$q_1,$	$q_1,0$		$q_5,$
q_4	$q_1,$	$q_4,$	$q_5,$		$q_1,$	$q_4,$	$q_5,1$
q_5	$q_5,0$		$q_3,0$	$q_4,1$	$q_1,0$	$q_5,$	
q_6	$q_5,1$	$q_4,0$	$q_3,1$	$q_1,0$	$q_1,$	$q_4,$	$q_5,1$
q_7	$q_1,0$	$q_2,$	,1	$q_6,1$	$q_7,1$	$q_8,0$	
q_8		$q_2,1$		$q_4,1$		$q_8,0$	$q_7,0$

FIGURE P12.9

12.10. Determine a minimal covering collection for the sequential circuit partially described by the implication table of Fig. P12.10.

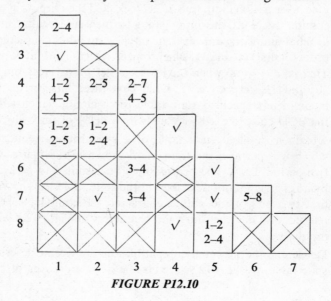

FIGURE P12.10

12.11. Repeat 12.10 for the implication table of Fig. P12.11.

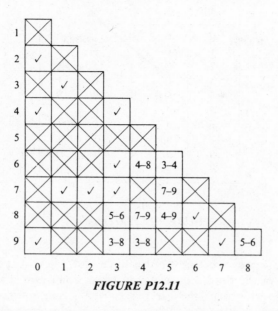

FIGURE P12.11

12.12. A clocked detector circuit is connected to a controller, as shown in Fig. P12.12. Only when $x_1 = 1$ are the outputs z_1, z_2 gated to the controller. x_1 and x_2 are level signals changing only between clock pulses. When x_1 goes to 1 for two successive clock periods, then x_2 for the next four clock periods will represent a decimal digit in the Gray code of Fig. P12.12b, bit b_0 first. At the time of the fourth bit (b_3), $z_1 = 1$ if the digit received as even, and $z_2 = 1$ if it was odd. At all other times that $x_1 = 1$, $z_1z_2 = 00$. If x_1 remains 1 for the two clock periods following receipt of a digit, another digit will follow, and the process will repeat until x_1 goes to 0, which may occur only between digits. Design the ODD/EVEN detector using J-K flip-flops.

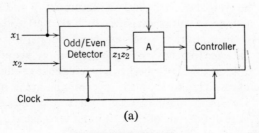

(a)

FIGURE P12.12

Dec.	Gray Code			
Digit	b_0	b_1	b_2	b_3
0	0	0	1	0
1	0	1	1	0
2	0	1	1	1
3	0	1	0	1
4	0	1	0	0
5	1	1	0	0
6	1	1	0	1
7	1	1	1	1
8	1	1	1	0
9	1	0	1	0

(b)

FIGURE P12.12 (*continued*)

12.13. In the sequence detector of Example 12.4, the next state on the final input of the third sequence could be a don't-care since the state of the detector once the car has started is unimportant. Determine whether this additional don't-care would provide any further simplification of the design.

12.14. Redesign the sequence detector of Example 12.4, removing the restriction that each button can be pushed only once.

12.15. At the Monongahela Works of the International Croquet, Badminton and Quoits Co. Ltd., packages of wickets come down a conveyor belt and pass through a weighing machine. As the packages enter the weighing machine, they pass a photocell which emits a pulse x_1 to activate the weighing machine, which then emits a pulse x_2 if the package is underweight, a pulse x_3 if it is overweight, and no signal if the weight is correct. Farther down the line, each package passes another photocell, which emits a pulse x_4 to cause overweight or underweight packages to be diverted into one of two special areas to have the weight corrected. The placement of the special stations and the spacing of the packages is such that an additional package may pass through the weighing machine before the previously weighed package reaches the x_4 photocell. A sequential circuit is required to keep track of whether a given package should be diverted to the underweight area or the overweight area, or pass straight through. Design a circuit to accept the four x pulses as inputs and emit a z_1 pulse at the time of the x_4 input if the package is underweight, a z_2 pulse if it is overweight. Although the spacing of packages on the belt is not uniform, you may assume that spacing is sufficiently controlled so that input pulses cannot overlap. Assume the S-C flip-flops used have separate direct reset inputs, which should be connected to a special reset line to reset the circuit when the process is started.

BIBLIOGRAPHY

1. Grasselli, A. and F. Luccio. "A Method of Minimizing the Number of Internal States in Incompletely Specified Sequential Networks," *IEEE Trans. on Electronic Computers*, **EC-14:** 3, 330–359 (June 1965).

2. Paull, M. C. and S. H. Unger. "Minimizing the Number of States in Incompletely Specified Sequential Switching Functions," *IRE Trans. on Electronic Computers*, **EC-8:** 3, 356–357 (Sept. 1959).

3. Hartmanis, J. and R. E. Stearns. *Algebraic Structure Theory of Sequential Machines*, Prentice-Hall, Englewood Cliffs, N.J., 1966.

4. Harrison, M. A. *Introduction to Switching and Automata Theory*, McGraw-Hill, New York, 1955.

5. Miller, R. E. *Switching Theory, Sequential Machines*, Vol. 2, Wiley, New York, 1965.

6. Moore, E. F. *Sequential Machines, Selected Papers*, Addison-Wesley, Reading, Mass., 1964.

7. Ginsburg, S. *Introduction to Mathematical Machine Theory*, Addison-Wesley, Reading, Mass., 1962.

8. Givone, D. D. *Introduction to Switching Circuit Theory*, McGraw-Hill, New York, 1970.

9. Kohavi, Z. *Switching and Finite Automata Theory*, McGraw-Hill, New York, 1970.

10. Sheng, C. L. *Introduction to Switching Logic*, Intext, Scranton, Pa., 1972.

Chapter 13 *Level Mode*
Sequential Circuits

13.1 Introduction

In Chapter 9, we saw that, if we remove the requirement for input pulses to trigger state transitions, we have the most versatile category of sequential circuits, *level mode* circuits. The general form for this type of circuit is a combinational logic circuit with feedback. Feedback around otherwise combinational networks will usually result in sequential operation.* Figure 13.1 shows a circuit which we shall now analyze in terms of the timing chart of Fig. 13.2 to see if it is indeed sequential.

As a starting condition, we assume both inputs and the output are at 0. Because of the feedback, the output is a function of itself and these assumptions are not necessarily consistent. In this case, if $Y = 0$, the output of NOR-gate 1 is $\bar{C}y = 0$, so OR-gate 2 implements $Y = S + \bar{C}y = 0 + 0 = 0$, which checks.

At time t_1, S goes to 1. This causes the output, Y, of OR-gate 2 to go to 1 after a short propagation delay. In turn, $\bar{y}$ goes to 0 and $\bar{C}y$ to 1, again after short propagation delays. OR-gate 2 is now implementing $1 + 1 = 1$, so when S returns to 0 at time t_2, there is no further change. Indeed, the $\bar{C}y = 1$ term has "locked in" the output at 1, so that S has no further effect.

---

* Not all combinational circuits with feedback are sequential. See Ref. 1, p. 141.

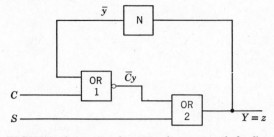

FIGURE 13.1. *Combinational circuit with feedback.*

Next, with $S = 0$, C goes to 1 at t_3. This drives $\bar{C}y$ to 0 and Y to 0 in turn. Then $\bar{y}$ goes to 1, locking in $\bar{C}y$ at 0. Thus, when C returns to zero at t_4, there is no further change and we are back at the starting condition.

Consider the condition of the circuit immediately before t_1 and immediately after t_2. At both times, the inputs are the same, $S = C = 0$, but the output is different. Thus, the output is not dependent solely on the inputs, and the circuit is sequential. Thus, we do not have to use flip-flops or other memory devices to produce sequential circuits. Indeed, if the reader will compare Fig. 13.1 with Fig. 9.3, he will recognize that this circuit is the basic S-C flip-flop, redrawn to emphasize the form of the basic model. It might also be noted that the form of Fig. 9.3 seems to have two feedback variables, while Fig. 13.1 has only one. This illustrates one of the special characteristics of level mode circuits—that it is somewhat arbitrary just which variables you identify as feedback variables, i.e., state variables. In clock mode and pulse mode circuits, the state variables are the outputs of flip-flops, so there is no question as to their identity. In level mode circuits, the feedback loops form continuous paths and which part of the loop you wish to identify with the state variables is open to choice. Fortunately, as we shall see, this apparent ambiguity causes no problems in analysis or design.

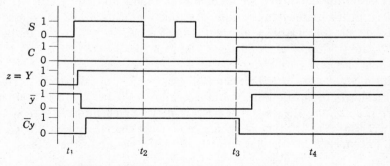

FIGURE 13.2. *Timing chart for circuit of Figure 13.1.*

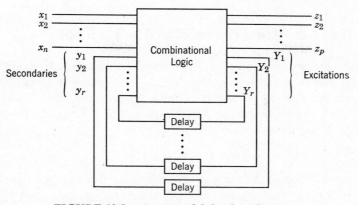

FIGURE 13.3.   *Basic model, level mode circuit.*

The general model for the level mode circuit is repeated in Fig. 13.3. Recall that the delay elements are not as a rule specific elements inserted in the feedback path for that purpose, but rather represent a "lumping" of the distributed delays in the combinational logic elements into single delay elements, one for each feedback variable. These delay elements may be considered as providing short-term memory. When an input changes, the delay enables the circuit to "remember" the present values of the variables, $y_1, y_2, \ldots, y_r$, long enough to develop new values of $Y_1, Y_2, \ldots, Y_r$, which in turn, after the delay, become the next-state values of the y's. Note the distinction between the y's and the Y's. In the steady-state condition, they are the same, but during transition they are not. The y's shall be referred to as the *secondaries* and the Y's as the *excitations*. Both correspond to the state variables in the clocked and pulse circuits, and either may be referred to by that name when the distinction between them is not important.

In the timing chart of Fig. 13.2, we note that changes in the secondaries and excitations, i.e., changes in state, may occur in response to any change in the value of any input. Furthermore, in that analysis, we allowed only one variable to change at a time and allowed no further changes to occur until all variables had stabilized. When such constraints on the circuit inputs can be enforced, we will say that the circuit is operating in the *fundamental mode*. This assumption about input changes considerably simplifies our design problems. Although the fundamental mode is a method of operation, not a special type of circuit, we shall refer to circuits designed with this assumption as fundamental mode circuits. In the next sections, we shall consider only fundamental mode circuits. In Section 13.9, we shall consider the special problems introduced by elimination of the fundamental mode restrictions.

13.2   Analysis of a Fundamental Mode Circuit

We shall now see that a fundamental mode circuit can be described by transition and flow tables in essentially the same manner as clocked and pulse circuits. The transition table for a fundamental mode circuit is a tabulation of the excitations and the outputs. For the circuit of Fig. 13.1, the equation for the excitation is

$$Y = S + \bar{C}y \tag{13.1}$$

Since the excitation is the same as the output in this case, this is also the equation for z. The map corresponding to this equation is shown in Fig. 13.4. Just as with the clocked and pulse circuits, we list the inputs across the top and the secondaries (state variables) down the side, one row for each of the 2^r combinations of secondaries. Since the excitations are the next values of the state variables in fundamental mode circuits, no distinction will be made between excitation maps and transition tables. The excitation maps will differ from the transition tables only in that they will be maps of single state variables, while the transition tables will show all state variables. For only one state variable, they are identical.

Since we must assume that both inputs will not change simultaneously from their unexcited or 0 states, don't-care conditions are in the $SC = 11$ column. This is reflected in the specification that both inputs of an S-C flip-flop must not be pulsed simultaneously. The circled entries represent *stable* states, that is, states for which $y = Y$. For example, for entry $SCy = 000$, $Y = 0$. Since Y represents the next value of y, this is a stable condition. This corresponds to the starting condition in the timing table of Fig. 13.2. Similarly, $SCy = 010$ is a stable entry since a 0 output will be unaffected if C goes to 1. The square $SCy = 100$ corresponds to the case where S goes to 1 while the flip-flop output is 0. As specified by Equation 13.1, the resultant value of Y is 1. Since this is not equal to the present value of y, it represents

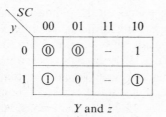

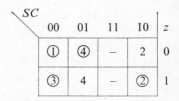

FIGURE   13.4.   *Transition table for circuit of Fig. 13.1.*

FIGURE 13.5.   *Flow table for circuit of Fig. 13.1.*

an unstable value and is not circled. After a delay, y changes to 1 and the circuit moves to the circled stable entry at $SCy = 101$. As S returns to 0, the circuit assumes the stable state indicated in square $SCy = 001$ without further change in output. Finally, C going to 1 moves the circuit through unstable entry $SCy = 011$ to stable entry $SCy = 010$ and then to $SCy = 000$ when C returns to 0.

We have seen that more than one stable state may be found on the same row of a fundamental mode transition table. It is convenient, in fact, to let each possible combination of inputs and secondary states to be numbered as a separate state. Doing this to the transition table of Fig. 13.4 results in the *flow* (state) table of Fig. 13.5. Note the distinction between this flow table and those of clocked and pulse mode circuits, for which the states are only a function of the state variables. Also note that the circled entries indicate the final destination when the state variables must change, i.e., the stable state into which the circuit will settle when all state variables have stabilized.

13.3   Synthesis of Flow Tables

We have just seen how an existing circuit can be analyzed in terms of transition and flow tables. In the design process, we must first go from a verbal statement of the desired circuit performance to a flow table, and thence to transition table and actual circuit. The likelihood of error in the process of translating the problem statements into a flow table is minimized by initially going to a *primitive* flow table.

Definition 13.1.   A *primitive flow table* is a flow table in which only one stable state appears in each row.

The primitive flow table corresponding to the flow table of Fig. 13.5 is shown in Fig. 13.6. This table describes the same response as the reduced flow table of Fig. 13.6. Starting in stable state ① with $SC = 00$, S going to 1

SC				
00	01	11	10	z
①	4	—	2	0
3	—	—	②	1
③	4	—	2	1
1	④	—	—	0

FIGURE 13.6.   *Primitive flow table for circuit of Fig. 13.1.*

takes us through 2 to ②. When *S* returns to 0, the transition is through 3 to ③. Again, changes in *S* will move the circuit back and forth between ② and ③. When *C* goes to 1, the transition is through 4 to ④, and *C* going to 0 takes the circuit through 1 to ①.

It is often convenient in fundamental mode problems first to represent the problem as a timing chart rather than a state diagram. This procedure is followed in the following example.

Example 13.1

Design a fundamental mode circuit with two inputs, x_1 and x_2, and a single output, z. The circuit output is to be 0 whenever $x_1 = 0$. The first change in the input x_2, occurring while $x_1 = 1$ will cause the circuit output to go to 1. The output is then to remain 1 until x_1 returns to 0. Changes in inputs will always be separated sufficiently in time to permit the circuit to stabilize before a second input transition takes place. Some typical input-output sequences are illustrated in Fig. 13.7.

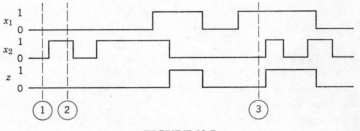

FIGURE 13.7

Solution: The first step of the design process is to determine a primitive flow table, for which we must assume some convenient starting stable state. In this case it is most convenient to choose a state for which $x_2x_1 = 00$. For this state, the output will be 0, regardless of which input was the last to change. This initial state is indicated as ① in Figs. 13.7 and 13.8. We also indicate the output corresponding to the stable state. For now, we shall indicate the outputs only for stable states on the flow table. The outputs for unstable states will be considered later.

Since both inputs are not allowed to change simultaneously, we enter a dash in the $x_2x_1 = 11$ column of the first row of the flow table. This will eventually result in a don't-care condition in the circuit realization. Dash marks will similarly be entered in each row of the primitive flow table in that column differing in both inputs from the inputs for the stable state of that row.

If x_2 goes to 1 while x_1 is still 0, there will still be no circuit output, so we let the circuit go to a stable state ②, with $z = 0$. If x_1 goes to 1 while $x_2 = 0$, the circuit goes to stable state ③, also with $z = 0$. From state ②, x_2 returning to 0 will take the circuit back to ①, as will x_1 going to 0 from state ③. From state

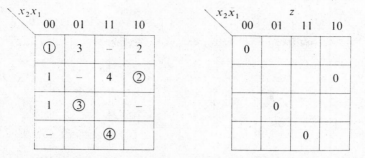

FIGURE 13.8.   *Partial primitive flow table, Example 13.1.*

②, x_1 going to 1 will take the circuit to a new state, ④, also with $z = 0$. We have so far determined the partial flow table of Fig. 13.8. The uncircled entries, or unstable states, merely represent the circuit destination following an input transition.

Both states ③ and ④ correspond to the last change having been from $x_1 = 0$ to $x_1 = 1$, so that a change in x_2 should cause the output to go to 1. Thus, from state ③, x_2 going to 1 takes us to ⑤, with $z = 1$. Similarly, from state ④, x_2 going to 0 takes us to ⑥, with $z = 1$. From state ④, x_1 going to 0 will return the circuit to the previous state, ②. These additions to the primitive flow table are shown in Fig. 13.9.

From either states ⑤ or ⑥, the output should remain 1 as long as x_1 remains 1, regardless of further changes in x_2. So we provide states ⑦ and ⑧, both with $z = 1$, such that the circuit will cycle between ⑤ and ⑦ or ⑥ and ⑧, for consecutive changes in x_2. A change of x_1 to 0 will take ⑥ or ⑦ to ①, and ⑤ or ⑧ to ②. The completed primitive flow table is shown in Fig. 13.10. ■

x_2x_1	00	01	11	10
	①	3	--	2
	1	--	4	②
	1	③	5	--
	--	6	④	2
			⑤	
			⑥	

x_2x_1	00	01	z 11	10
	0			
				0
		0		
			0	
			1	
		1		

FIGURE 13.9

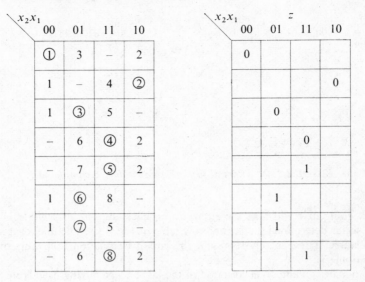

FIGURE 13.10

13.4 Minimization

It would, of course, be possible to implement Fig. 13.10 as is, utilizing eight internal states corresponding to the eight possible combinations of three secondaries. If the primitive flow table were identical to the state tables of Chapter 10 or 11, we might consider application of the various techniques developed in those chapters. However, if we compare the primitive flow table of Fig. 13.10 with a typical state table, we see a significant difference, namely the absence of a present-state column in the flow table.

This difference arises from the fact that a clocked circuit is stable only between clock pulses, and the state of the circuit at these times is independent of the present inputs. Therefore, all the stable states may be grouped in a

q^v	q^{v+1}				z					States				Output		
	00	01	11	10	00	01	11	10	00	01	11	10	00	01	11	10
1	1	2	=	→3	0	–	–	–	①	2	–	→3	0	–	–	–
2	5	2	4	–	–	0	–	–	5	②	4	↓	–	0	–	–
3←	1	–	8	3	–	–	–	0	1	–	8	③	–	–	–	0

(a) State Table	(b) Primitive Flow Table

FIGURE 13.11. *Partial state and primitive flow tables.*

Present State	Next State 00	01	11	10	Output 00	01	11	10
①	①	2	–	3	0	–	–	–
②	5	②	4	–	–	0	–	–
③	1	–	8	③	–	–	–	0

(a)

Present State	Next State 00	01	11	10	Output 00	01	11	10
⑨	⑨	2	8	⑨	0	–	–	0
②	5	②	4	–	–	0	–	–

(b)

FIGURE 13.12.   *Primitive flow table with present state column added.*

single column, corresponding to no clock input. A fundamental mode circuit will be stable any time all the inputs are stable, and the state will depend on the values of the inputs. Thus, a present-state entry may appear in any column of the flow table.

The distinctions may be further clarified by considering a partial state table and a partial primitive flow table, as shown in Figs. 13.11a and b. First, consider that the circuit of Fig. 13.11a is in state q_1, and the inputs are 10 when a clock pulse arrives. The transition is from q_1 to unstable 3 during the clock interval and thence to stable q_3 when the clock interval is complete, as shown by the dotted arrow in Fig. 13.11a. Next, assume that the circuit of Fig. 13.11b is in stable ① with inputs 00. If the inputs go to 10, the circuit moves across to unstable 3 and then down to stable ③, as shown.

Since there is only one stable state in each row of the primitive flow table, we can regard these as the present states for the individual rows and group them in a present state column, as shown in Fig. 13.12a. Now the transition from ① to ③ may be regarded as a transition from a present to a next state. Thus, we see that there is little real distinction between the state table for a clocked circuit and the primitive flow table for a fundamental mode circuit. The only distinction is that for *every row of the flow table there must be one input combination for which the next state is the same as the present state.* This condition could be imposed on clocked circuits if the designer so desired. This would only limit his flexibility, having no effect on the determination of a covering circuit.

We also see that states may be combined in the same manner on both state and flow tables. In Fig. 13.11a, states q_1 and q_3 are obviously compatible and could be combined. Similarly, states ① and ③ in Fig. 13.12a are compatible and could be combined. Should they be combined, they would be replaced by a state, say, ⑨, which would be stable in both columns 00 and 10, as indicated in Fig. 13.12b. That is, the next state would be the same as the present state in both columns.

We have established a precise analogy between the minimization problems for clocked circuits and fundamental mode circuits. We may proceed then to utilize all of the techniques of Chapters 10 and 11 on Example 13.1.

q^v	x_2x_1				x_2x_1			
	00	01	11	10	00	01	11	10
①	①	3	–	2	0			
②	1	–	4	②				0
③	1	③	5	–		0		
④	–	6	④	2			0	
⑤	–	7	⑤	2			1	
⑥	1	⑥	8	–		1		
⑦	1	⑦	5	–		1		
⑧	–	6	⑧	2			1	
	q^{v+1}				z			

FIGURE 13.13

Example 13.1 (Continued)

We now proceed to reduce the number of rows in the primitive flow table of the earlier section of Example 13.1. We repeat Fig. 13.10 as Fig. 13.13 in the present-state, next-state format justified by the above discussion.

The implication table corresponding to this flow table is shown in Fig. 13.14a. It is obtained in exactly the same way as for the clocked circuits. We enter an X for any pair with conflicting outputs, a $\sqrt{}$ for any pair compatible without implication, and the implied pairs in the remaining squares. We then eliminate pairs which imply incompatible pairs to produce the final implication table of Fig. 13.14b.

The maximal compatibles are easily determined to be

$$(12), (13), (24), (5678) \tag{13.2}$$

The smallest closed collection may be found by inspection to be

$$(13), (24), (5678) \tag{13.3}$$

The 3-state table corresponding to this closed collection is shown in Fig. 13.15. Note that the combining of states may result in more than one stable state in a single row. In the primitive flow table, we see that, for present-state ① and input $x_2x_1 = 00$, the next state is also ①. Similarly, in state ③, for input $x_2x_1 = 01$, the next state is also ③. Thus, when states ① and ③ are combined into state (a), the next state must be the same as the present state, for both $x_2x_1 = 00$ and $x_2x_1 = 01$. The reasoning is similar in arriving at the pairs of stable states in the other rows of the minimal table.

In row ① of the primitive table, the entry for $x_2x_1 = 11$ is a don't-care, as this would correspond to a forbidden double transition from the stable ① entry. In row ③, however, the entry in this column is an unstable 5, corresponding to the transition from ③ to ⑤. When states 1 and 3 combine into state (a), the entry in this column must be an unstable c, corresponding to a transition to the

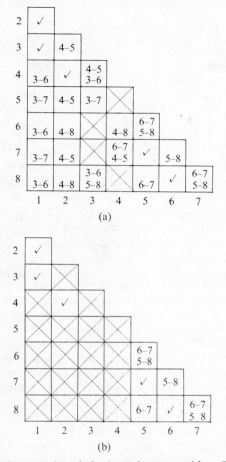

FIGURE 13.14. *Initial and final implication tables, Example 13.1.*

class containing state ⑤. This does not mean that a double transition, from $x_2x_1 = 00$ to $x_2x_1 = 11$ is any more "legal" than it was before. If the circuit is operated properly in the fundamental mode, the only transition through this unstable c will be from the stable ⓐ corresponding to $x_2x_1 = 01$. The other cases where the circuit is to go to a new stable state are handled in a similar manner, as shown in Fig. 13.15.

Now let us reconsider our earlier assumption that the outputs are of no concern during the transitions. If an output variable is to change value as a result of a state change, then this variable is a don't-care during the transition. Assume an output variable is to change from 0 to 1 in a certain transition. If a 0 is entered as the value during the transition, then the change of variable will not take place until the end of the transition. If a 1 is entered, the change will take place at the

	q^v	x_2x_1 00	01	11	10	x_2x_1 00	01	11	10
(13)	ⓐ	ⓐ	ⓐ	c	b	0	0	–	0
(24)	ⓑ	a	c	ⓑ	ⓑ	0	–	0	0
(5678)	ⓒ	a	ⓒ	ⓒ	b	–	1	1	–
			q^{v+1}					z	

FIGURE 13.15. *Minimal flow table, Example 13.1.*

start of the transition. Since it makes no difference exactly when in the transition the output change occurs, the entry is a don't-care. Referring to the final flow table of Fig. 13.15, we see that all transitions to or from state ⓒ involve a change in output. The outputs during these transitions thus remain unspecified.

If an output variable is *not* supposed to change as the result of a transition, the situation becomes a bit more complicated. It may be that the outputs are sampled by whatever other circuits may be involved only between transitions, in which case the transition outputs are don't-cares. But if the outputs during transitions may affect other circuits, we usually must specify them. Refer again to Fig. 13.15 and consider the transition from ⓐ to b to ⓑ, as the input goes from 00 to 10. If the output were a 1 for unstable b, as might occur if we leave it unspecified, a 1 pulse might appear on the output line during the transition. If the circuit driven by z were, for example, a pulse circuit, this pulse might cause some undesired transition. Thus, the output corresponding to unstable b must be specified as 0. Similarly, the output for unstable a in row ⓑ is specified as 0.

Several more problems must be considered before an adequate circuit can be constructed for Example 13.1. Before considering these difficulties, let us develop another fundamental mode circuit of some practical interest. ■

Example 13.2

Certain forms of digital computer control units[11] make liberal use of a fundamental mode sequential circuit called a control delay. This circuit has a control pulse input, x, a clock input, C, and a control pulse output, z, as illustrated in Fig. 13.16a. Pulses on line x will always be separated by several clock periods. Whenever a pulse occurs on line x, it will overlap a clock pulse and be of approximately the same width as a clock pulse. That is, line x will only go to 1 after the clock has gone to 1 and will return to 0 only after the clock has returned to 0. For each input pulse, there is to be an output pulse on line z coinciding with the next clock pulse following the x pulse. Thus, each x pulse results in a z pulse delayed by approximately one clock period, as shown in Fig. 13.16b.

The input line x originates at the output of a similar control delay with perhaps additional gate and line delay in between. The control delay output will be a

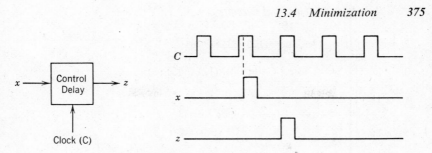

FIGURE 13.16.   *Control delay.*

very slightly delayed version of the clock due to internal gate delay. Thus, we assume that no two inputs will change simultaneously so that the *fundamental mode* constraint is satisfied. Develop a minimal state table for this circuit.

Solution: State ①, representing a condition between clock pulses and at least two clock periods after the last pulse on line x, is a natural starting state. From here, the next input change can only be from 0 to 1 on line C. We let this change take the circuit to state ② as shown in Fig. 13.17a. From here, the next change may be the leading edge of a pulse on line x, in which case the circuit goes to state ③. If no x pulse occurs, the circuit returns to state 1 when C goes to 0. We have assumed that the clock will always return to 0 ahead of a coincident x pulse. Thus, the circuit must go from state ③ to state ④, as shown in Fig. 13.17. Next, x will return to 0, taking the circuit to state ⑤. We have assumed that

xC	00	01	11	10
	①	2	–	–
	1	②	3	
			③	

(a)

xC	00	01	11	10
	①	2	–	–
	1	②	3	–
	–	–	③	4
	5	–	–	④
	⑤			

(b)

xC	00	01	11	10	00	01	11	10
	①	2	–	–	0			
	1	②	3	–		0		
	–	–	③	4			0	
	5	–	–	④				0
	⑤	6	–	–	0			
	1	⑥	–	–		1		

z

(c)

FIGURE 13.17.   *Developing flow table for control delay.*

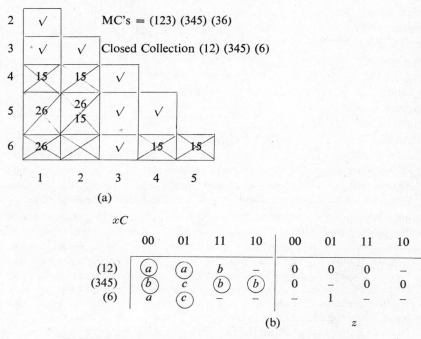

FIGURE 13.18.   *Minimal flow table for control delay.*

another control pulse will not occur until at least one clock period after an output has been generated. Therefore, the circuit will go from state ⑤ to state ⑥ with the arrival of the next clock pulse and remain in that state for the duration of that clock pulse. The output $z = 1$ while the circuit is in state ⑥. A complete tabulation of outputs for all stable states is given in Fig. 13.17c. Following the output pulse, the circuit returns to state ①.

Using the implication table of Fig. 13.18a, the primitive flow table is readily minimized to form the flow table of Fig. 13.18b. The outputs corresponding to unstable states are specified in this table to prevent unwanted output transitions. ∎

13.5   Transition Tables, Excitation Maps, and Output Maps

The next step in the design is to assign specific combinations of the secondaries to the rows of the reduced flow table. Here we have all the state assignment problems of the clock and pulse mode circuits, plus some special problems peculiar to the fundamental mode. Let us proceed for now as though there were no new problems. As an example, let us construct a transition table corresponding to the flow table just developed for the control delay. We first assign the combinations of secondary values $y_2y_1 = 00$ to $\textcircled{a}$, $y_2y_1 = 01$ to $\textcircled{b}$, and $y_2y_1 = 11$ to $\textcircled{c}$. Substituting these values for the stable states in Fig. 13.18b results in the partial transition table of Fig. 13.19a. This figure is a tabulation of the excitations, Y_2Y_1, as a function of the inputs and the present values of the secondaries, y_2y_1. The excitations for unstable states must be the same as for the corresponding stable states. For example, the inputs 11 cause a transition from state $\textcircled{a}$ to state $\textcircled{b}$ through unstable state b. To cause this change to take place in the physical circuit, we enter $Y_2Y_1 = 01$ in place of unstable state b. Thus, when xC takes the value 11, the excitations are logically specified as 01. After a circuit delay, the secondaries take on the values $y_2y_1 = 01$. Now the secondaries and excitations are the same (01), and the circuit is stable.

Excitation values corresponding to all unstable states are entered in completed transition table of Fig. 13.19b. In succeeding problems, we will obtain the transition table in one step by making no distinction between stable

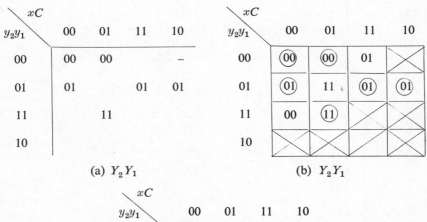

(a) Y_2Y_1 (b) Y_2Y_1

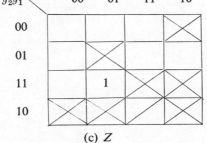

(c) Z

FIGURE 13.19.   *Transition table and output map for control delay.*

and unstable states in the assignment of excitation values. The output table of Fig. 13.19c is a direct translation of the minimal state table.

For convenience in reading off the excitation equations, we may separate the transition table into separate *excitation maps* for the excitation variable, as shown in Fig. 13.20a. From these maps and the output map, we read the equations:

$$Y_2 = \bar{x}Cy_1 \tag{13.4}$$

$$Y_2 = x + \bar{y}_2y_1 + Cy_1 \tag{13.5}$$

$$z = y_2 \tag{13.6}$$

The corresponding circuit is shown in Fig. 13.20b.

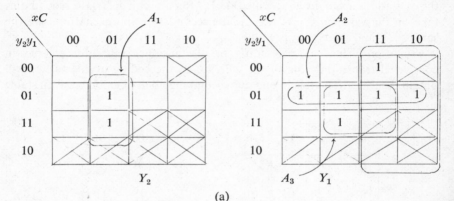

(a)

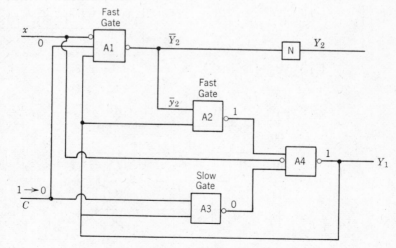

(b)

FIGURE 13.20. *Excitation maps and circuit for control delay.*

If this circuit were clock or pulse mode, we would also check the other two possible assignments to see which is minimal. However, as mentioned above, fundamental mode circuits have some special problems, so let us look at the functioning of this circuit in some detail before worrying about an absolute optimum.

Let us assume that the circuit is in the stable state $xCy_2y_1 = 0111$, corresponding to the stable $\textcircled{c}$ state in the second column, third row of the state table. The corresponding values at various points in the circuit are shown in Fig. 13.20b. Note that the 1 values for Y_2 and Y_1 are produced by gates $A1$ and $A3$, respectively. Now assume that C goes from 1 to 0, as shown. This should take the circuit through the unstable a entry in the third row to the stable $\textcircled{a}$ in the first column, first row, with both state variables changing to 0.

Recalling that a NAND gate will have a 1 out if any input is 0, we see that the $C = 0$ signal will drive the outputs of $A1$ and $A3$ to 1. Providing $A2$ does not change too fast, the 1 at the output of $A3$ will then drive Y_1 to 0 at the output of $A4$. This value of Y_1, fed back to the input of $A2$, will lock its output at 1, in turn stabilizing Y_1 at 0, as required.

But now assume that the gates have unequal delays, as shown in Fig. 13.20b. Then the change in C will first drive the output of $A1$ to 1, which will in turn drive $A2$ to 0 *before* the output of $A3$ goes to 1. As a result, Y_1 will not change and the circuit will make an erroneous transition to $Y_2 Y_1 = 01$, the stable $\textcircled{b}$ state in the first column.

The difficulty just described occurred because both secondaries were required to change simultaneously following a single input change but did not change at the same time due to unequal circuit delays. Such a situation is termed a *race* because the nature of the transition may depend on which variable changes fastest. Sometimes races may be "fixed" by introducing extra delays, but this approach is expensive and doesn't always work. In the next section, we shall investigate the nature of these races more carefully and show that they can be eliminated by a better state assignment.

13.6   Cycles and Races

It is possible for a circuit to assume more than one unstable state prior to reaching a new stable state. If for a given initial state and input transition such a sequence of unstable states is unique, then it is termed a *cycle*. For example, in the table and corresponding Y map of Fig. 13.21, the circuit will cycle through three unstable states during a transition from $\textcircled{1}$ to $\textcircled{4}$. Notice in the Y map that the next state for $y_2y_1x_2x_1 = 0001$ is not $y_2y_1 = 10$. Instead, the machine proceeds from 00 to 01 to 11 to 10. Each of the transitions involves a change in only one secondary. Thus, the circuit proceeds

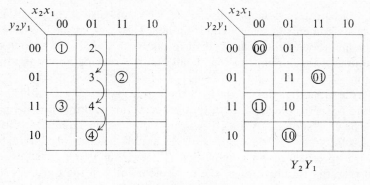

FIGURE 13.21

reliably through each of the unstable states to the final stable state. The cycle may be entered from stable states ② and ③ as well.

Where a change of more than one secondary is specified by the Y map, the resulting situation is termed a *race*. The race illustrated in Fig. 13.22 is a *noncritical race*. Suppose the inputs change to 01 while the machine is in state $y_2y_1 = 00$. If y_2 changes first, the machine goes to unstable state 10. If y_1 changes first, the machine goes to 01. For either of these unstable states, the Y map shows the next state to be stable ②. Thus, no matter what the outcome of the race, the stable state ②, as designated by the Y map, is reached.

The situation encountered in the control delay, as discussed in the last section, is an example of a *critical* race, a situation in which there are two or more stable states in a given column, and the final state reached by the circuit depends on the order in which the variables change. In other words, the desired result might occur or might not occur depending on the actual circuit delays.

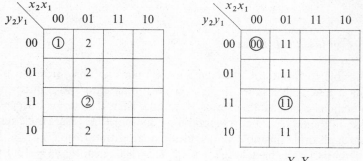

FIGURE 13.22

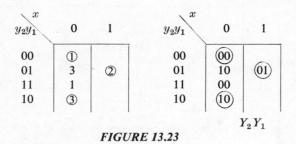

FIGURE 13.23

Another critical race is illustrated in Fig. 13.23. There the circuit is supposed to move from stable ② ($y_2y_1 = 01$) in the $x = 1$ column to stable ③ ($y_2y_1 = 10$) in the $x = 0$ column. If Y_1 changes first, the excitations will be $Y_2Y_1 = 00$, leading to an erroneous transition to stable ①. If Y_2 changes first, the excitations will be $Y_2Y_1 = 11$, leading to unstable 1 in the third row and thence to stable ①. Thus, no matter what the delays, the wrong transition will occur.

It might seem that the first type of critical race would be less troublesome since it can, in theory, be eliminated by proper adjustment of delays. In practice, however, constraining delays is difficult and expensive so that both types of critical races are to be avoided. In some problems, this may be accomplished by a judicious assignment of state variables. In others, it becomes necessary to use more than a minimal number of states.

Example 13.2 (Continued)

We are now in a position to construct a more satisfactory version of the control delay. It is only necessary to eliminate the critical race in column 00 of Fig. 13.19b and replace it with a cycle through the spare state. A new transition table is given in Fig. 13.24. The only changes are in the lower two squares of column

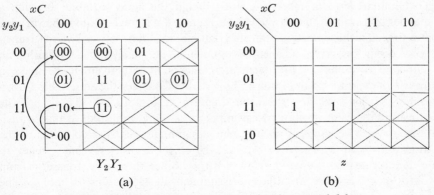

FIGURE 13.24. *Revised transition for the control delay.*

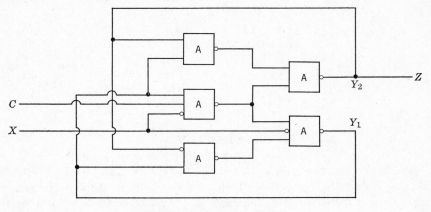

FIGURE 13.25.   *Control delay.*

$xC = 00$. Thus, from stable state (11), the circuit will cycle as indicated by the arrows when C goes to 0. In terms of the excitation maps, the only change is an additional 1 in the third row. The new equation for Y_2 is:

$$Y_2 = y_2 y_1 + \bar{x} C y_1 \tag{13.7}$$

There is no change in the equation for Y_1. To avoid a possible output pulse during the cycle, an additional 1 in the output map must be specified, as shown in Fig. 13.24b, but this does not change the equation for z. The revised circuit for the control delay shown in Fig. 13.25 will function properly as a control delay provided the input pulses satisfy the fundamental mode constraints as previously assumed. ■

13.7   Race-Free Assignments

To eliminate critical races, we must assign the state variables in such a manner that transitions between states require the change of only one variable at a time. To this end, it is desirable that stable states between which transitions occur be given adjacent assignments, i.e., assignments differing in only one variable. To find such assignments, it is convenient to make a *transition diagram* on the Boolean hypercube. Consider the state table shown in Fig. 13.26a and the transition diagram in Fig. 13.26b. To set up the transition diagram we start by assigning (a) to 00. Noting that (a) has transitions to (c), we give (c) an adjacent assignment, at 01, and indicate the transition by an arrow from (a) to (c). Next, the transition from (c) back to (a) is indicated by an arrow from (c) to (a). Since there is a transition from (c) to (b), (b) is given an adjacent assignment at 11, leaving 10 for (d). The diagram is then completed by filling in arrows for all the other transitions.

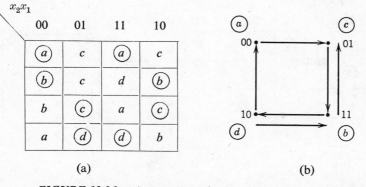

FIGURE 13.26. *Assignments of states to prevent critical races.*

In this case, all transitions are between adjacent states, so this assignment is free of critical races.

A slightly different situation is illustrated in Fig. 13.27. Now d has a transition to c, resulting in a diagonal transition, i.e., a change of two variables. Note that this diagonal transition cannot be eliminated since d has transitions to three other states and cannot be adjacent to all of these with only two state variables. However, this diagonal transition is seen to represent a noncritical race and will cause no problems. Whenever there is only one stable state in a column, transitions into that column need not be considered in choosing a state assignment since critical races cannot occur unless there are at least two stable entries in a column. Thus, the same assignment will be valid for this state table.

In Fig. 13.28, we see a more difficult situation. Here we have several critical races, as indicated by the diagonal transitions, and it is clear that no

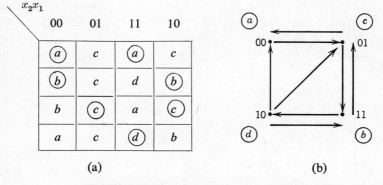

FIGURE 13.27. *Assignment with noncritical race.*

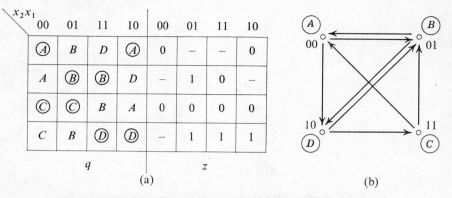

q $\qquad$ z

(a) $\qquad$ (b)

FIGURE 13.28. *State assignment with critical races.*

permutation of the assignment can eliminate all diagonal transitions. For example, the first row requires both $\widehat{B}$ and $\widehat{D}$ to be adjacent to $\widehat{A}$, while the second row requires $\widehat{B}$ and $\widehat{D}$ to be adjacent. Clearly, not all these requirements can be satisfied.

In situations such as the above, the critical races can be eliminated only by the use of spare states. If the number of states in the original state table is not a power of two, spare states will automatically be available. This was the case in the control delay, where there were only three states and the spare state was used for a cycle to avoid the critical race. If the original number of states is a power of two or if a satisfactory assignment is not found using available spare states, spare states must be created by adding extra state variables.

There are two basic techniques for using the spare states, *shared row* assignments and *multiple row* assignment. In the multiple row assignment, as the name implies, each state is assigned two or more combinations of state variables, i.e., two or more rows in the transition table. Figure 13.29 shows a universal multiple row assignment for four-state circuits. The extra rows are created by using three state variables. The four states are arbitrarily numbered 1 to 4, and their assignments are indicated by their location on

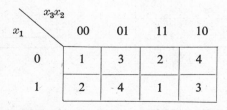

FIGURE 13.29. *Universal assignment for 4-state tables.*

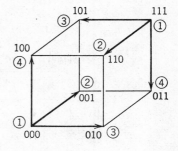

FIGURE 13.30.   *Universal 4-state assignment on hypercube.*

the K-map of the state variables. There are two state variable combinations assigned to each state, with the two assignments for each state being logical complements.

The structure of this assignment is shown more clearly in Fig. 13.30. The two combinations assigned to each state are on diagonally opposite corners of the cube. Both assigned combinations for a state are adjacent to one of the combinations for each of the other states. This is illustrated by the arrows in Fig. 13.30, indicating adjacent transitions from either ① state to any of the other states.

The development of the transition table for the universal assignment for the state table of Fig. 13.28 is shown in Fig. 13.31. We start by setting up an augmented state table (Fig. 13.31a) with the eight state variable combinations listed and the corresponding state assignment. Since state numbers are arbitrary, we let $1 \rightarrow A$, $2 \rightarrow B$, etc. Although there is no logical difference, we denote the two representations of each state as A_1, A_2, B_1, B_2, etc., for ease of reference. For each row assigned to a given state, the stable entries are the same as in the original state table. For example, state Ⓐ is stable in the first and fourth columns, so rows Ⓐ₁ and Ⓐ₂ are stable in the same columns. Similarly the outputs are the same for each row assigned a given state. For state Ⓐ, the output row is 0, –, –, 0, so the output rows for Ⓐ₁ and Ⓐ₂ are both 0, –, –, 0.

The transitional entries are filled in to provide adjacent transitions. State Ⓐ makes transitions to Ⓑ and Ⓓ. Ⓐ₁ is adjacent to Ⓑ₁ and Ⓓ₂, and Ⓐ₂ is adjacent to Ⓑ₂ and Ⓓ₁, so the transitional entries are filled in accordingly. We can now see why the external behavior of the circuit is not changed by the use of multiple states. Consider the circuit to be in state Ⓐ, with a stable output of 0 and either $x_2 x_1 = 00$ or $x_2 x_1 = 10$. It makes no difference whether the circuit is actually in Ⓐ₁ or in Ⓐ₂. From state Ⓐ, it should go to Ⓑ or Ⓓ. From Ⓐ₁, it will go to Ⓑ₁ or Ⓓ₂; from Ⓐ₂, it will go to Ⓑ₂ or Ⓓ₁. Either way, the new stable state will correspond to Ⓑ or Ⓓ of the original state table. The same argument holds for any other state transitions.

$y_3y_2y_1$	x_2x_1 00	01	11	10	00	01	11	10
$\textcircled{A_1}$ 000	$\textcircled{A_1}$	B_1	D_2	$\textcircled{A_1}$	0	–	–	0
$\textcircled{B_1}$ 001	A_1	$\textcircled{B_1}$	$\textcircled{B_1}$	D_1	–	1	0	–
$\textcircled{D_1}$ 011	C_1	B_1	$\textcircled{D_1}$	$\textcircled{D_1}$	–	1	1	1
$\textcircled{C_1}$ 010	$\textcircled{C_1}$	$\textcircled{C_1}$	B_2	A_1	0	0	0	0
$\textcircled{D_2}$ 100	C_2	B_2	$\textcircled{D_2}$	$\textcircled{D_2}$	–	1	1	1
$\textcircled{C_2}$ 101	$\textcircled{C_2}$	$\textcircled{C_2}$	B_1	A_2	0	0	0	0
$\textcircled{A_2}$ 111	$\textcircled{A_2}$	B_2	D_1	$\textcircled{A_2}$	0	–	–	0
$\textcircled{B_2}$ 110	A_2	$\textcircled{B_2}$	$\textcircled{B_2}$	D_2	–	1	0	–

States z

(a)

$y_3y_2y_1$	x_2x_1 00	01	11	10
000	$\textcircled{000}$	001	100	$\textcircled{000}$
001	000	$\textcircled{001}$	$\textcircled{001}$	011
011	010	001	$\textcircled{011}$	$\textcircled{011}$
010	$\textcircled{010}$	$\textcircled{010}$	110	000
100	101	110	$\textcircled{100}$	$\textcircled{100}$
101	$\textcircled{101}$	$\textcircled{101}$	001	111
111	$\textcircled{111}$	110	011	$\textcircled{111}$
110	111	$\textcircled{110}$	$\textcircled{110}$	100

$Y_3Y_2Y_1$

(b)

FIGURE 13.31. *Development of transition table for state table of Fig. 13.28.*

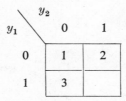

FIGURE 13.32.   *Universal shared-row assignment for 3-state tables.*

The final transition table, Fig. 13.31b, is obtained by replacing the state entries with the assigned combinations of state variables.

The multiple-row assignment method always requires extra state variables since the number of rows must be at least twice the number of states. Adding state variables will increase the cost, so *whenever extra rows are available because the number of states is not a power of two, the shared-row approach is generally preferable.* In this method, the extra rows are not assigned to any specific state but instead are *shared* as required to convert any critical races into cycles of adjacent transitions. For 3-state tables, any assignment such as the one shown in Fig. 13.32 will lead to a critical-race-free shared-row realization. Any transitions between 1 and 2 or 1 and 3 are adjacent; transitions between 2 and 3 cycle through the spare row. We have already used this assignment for the control-delay example.

For 4-row tables, there will be no spare rows to share unless we add a state variable. But we have already seen a universal 4-row assignment with three state variables, so either this approach or the shared-row approach can be used for such tables. Almost all 5-state machines can also be handled using three state variables. A universal assignment for *n* states is an assignment which will work for every *n* state flow and any numbering of the states. Universal assignments for 5 to 8 states and 9 to 12 states are shown in Fig. 13.33.[12] Since there is only one row per state and the numbering is arbitrary, we do not indicate any state numbers in Fig. 13.33 but simply indicate by ×'s the rows to which states should be assigned. For 5 to 8 states, simply assign one ×'ed row from Fig. 13.33a to each state, and similarly for 9 to 12 states for Fig. 13.33b.

Although universal assignments require an extra variable to allow for all possible cases, shared-row assignments which do not require extra variables can often be found for specific tables, particularly if there are several spare states available without adding extra variables. A useful concept in finding such assignments is the *destination set*. In any column of a state table, a destination set consists of any stable state and any rows which make transitions into that state. For example, in the state table of Fig. 13.34, in the first column, the stable states are ①, ③, ⑤; and row 4 leads to ①,

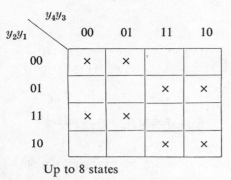

Up to 8 states

(a)

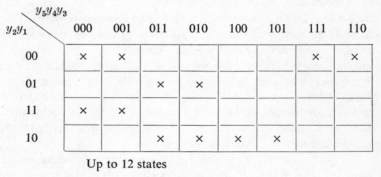

y_2y_1 \ $y_5y_4y_3$	000	001	011	010	100	101	111	110
00	×	×					×	×
01			×	×				
11	×	×						
10			×	×	×	×		

Up to 12 states

(b)

FIGURE 13.33. *Universal shared-row assignments.*

	\ x_2x_1 00	01	11	10	00	01	11	10
1	①	3	4	①	0	–	0	0
2	5	4	②	②	–	1	1	0
3	③	③	2	1	1	1	1	–
4	1	④	④	④	–	1	0	1
5	⑤	⑤	⑤	4	0	1	1	1

States

FIGURE 13.34. *Example 5-state table.*

row 2 to ⑤. The destination sets for this column are thus (1, 4) (2, 5) (3). Similarly, the destination states for the second, third, and fourth columns are (1, 3) (2, 4) (5), (1, 4) (2, 3) (5), and (1, 3) (2) (4, 5), respectively. The destination sets are important because, to avoid critical races, the members of each set must either be adjacent or so located in relation to the spare states that cyclic transitions for all sets in a given column may be made without interference.

A standard assignment for 5-state tables is shown in Fig. 13.35a. We term it "standard" rather than "universal" because it will work for almost every table but will fail for a few rare cases, as we shall see. For this case we arbitrarily assign the five states as shown in Fig. 13.35b on the K-map, and, Fig. 13.35c on the Boolean hypercube. In the first column of the state table, the destination sets are (1, 4) (2, 5) (3), requiring transitions from 4 to 1 and from 2 to 5. We see in Fig. 13.35c that this is impossible with this assignment as the two cycles must use the same spare rows enroute to different destinations. Rows may be *shared* only in the sense that they may be used for different cycles in different columns. In any single column, a transitional row may be part of a cycle to only one destination.

This problem occurs only in columns having (2, 2, 1) destination sets, i.e., columns having 2, 2, and 1 members in three destination sets. It can be eliminated by permuting the assignment so that the two 2-member sets do

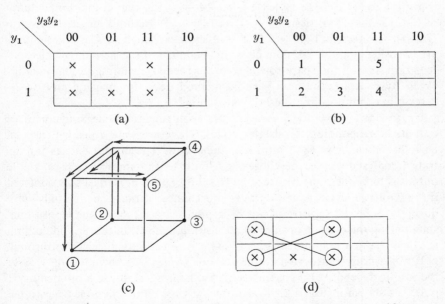

FIGURE 13.35.   *Standard shared-row assignment for 5-state tables.*

not occupy diagonally opposite vertices of the hypercube. In terms of the assignment map, the 2-member sets must not occupy the connected rows as shown in Fig. 13.35d. There are 15 possible $(2, 2, 1)$ column configurations and, as long as all 15 do not occur in the same table, it is possible to avoid assigning any $(2, 2, 1)$ sets in the "forbidden" pattern. This assignment will thus fail only in the unlikely event of a state table with 15 or more columns, including all 15 possible $(2, 2, 1)$ columns.[12]

The easiest way to use this assignment is to construct any arbitrary $(2, 2, 1)$ group of destination sets that does *not* occur in the destination sets of the state table and assign them to the "forbidden" pattern. This guarantees that none of the $(2, 2, 1)$ columns which are in the state table can be assigned to that pattern.

For the table of Fig. 13.34, we see that the $(2, 2, 1)$ pattern $(1, 2)$ $(3, 5)$ (4), for example, does not appear among the destination sets, so we assign $(1, 2)$ and $(3, 5)$ to the diagonal locations, as shown in Fig. 13.36a. For ease of reference, the spare rows are labeled α, β, γ. To set up the augmented state table (Fig. 13.34b), we start by filling in the stable states and the corresponding outputs, just as for a multiple-row assignment.

The transitional entries are filled in with reference to the map of the assignment (Fig. 13.36a) to determine if the transition is direct or whether a cycle through spare states must be provided. State ① makes transitions to ③ in column 01 and to 4 in column 11. The transition to ③ is direct, but the transition to ④ uses the cycle ① → α → 4 → ④, as seen in the third column of Fig. 13.36b. Transitions ② → ④ and ② → ⑤ are both direct. Transition ③ → ① is direct, but ③ → γ → 2 → ② cycles through γ. The other entries are filled in similarly. We note here an advantage of the shared row method. It is rare that all the spare row entries are needed, resulting in many don't-cares. By contrast, the multiple row method does not provide any don't-cares other than those in the original state table.

Up to now, very little has been said with regard to the assignment of outputs corresponding to unstable states. There are some rules which must be followed, and we shall introduce them now. The output entries for the transitional states must be chosen to eliminate any spurious pulses. If the output is to be the same before and after the transition, it must stay constant throughout the cycle. If the output is to change, it must change only once during a cycle. Therefore, it can be a don't-care for only one transitional state during the cycle. Consider the transition from ④ in the second column to ① in the first column, in Fig. 13.36b, a transition which cycles through x. If we let both transitional outputs be don't-cares, as shown in Fig. 13.37a, the result, depending on groupings of don't-cares, might be a momentary 1 pulse, as shown in Fig. 13.37b. To prevent this, either the first transitional state must be specified as 1, or the second as 0, as shown in Figs. 13.37c and

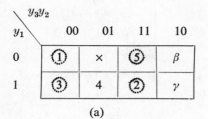

(a)

$y_3y_2y_1$	x_2x_1 00	01	11	10	00	01	11	10
① 000	①	3		①	0	–	0	0
③ 001	③	③	γ	1	1	1	1	–
④ 011	α	④	④	④	1	1	0	1
α 010	1	–	4	4	–	–	0	–
β 100	–	–	–	–	–	–	–	–
γ 101	–	–	2	–	–	–	1	–
② 111	5	4	②	②	0	1	1	0
⑤ 110	⑤	⑤	⑤	α	0	1	1	–

(b)

$y_3y_2y_1$	x_2x_1 00	01	11	10
000	⟨000⟩	001	010	⟨000⟩
001	⟨001⟩	⟨001⟩	101	000
011	010	⟨011⟩	⟨011⟩	⟨011⟩
010	000	× × ×	011	011
100	× × ×	× × ×	× × ×	× × ×
101	× × ×	× × ×	111	× × ×
111	110	011	⟨111⟩	⟨111⟩
110	⟨110⟩	⟨110⟩	⟨110⟩	010

$Y_3Y_2Y_1$

(c)

FIGURE 13.36. *Valid 5-state assignment for Fig. 13.34.*

13.37d, respectively. Then choice between the latter two should be made on the basis of which seems likely to result in the simplest output equations.

The transition table follows directly from the augmented state table by replacing each state by the corresponding state variable combination, as shown in Fig. 13.36c.

States: ④ → α → 1 → ①   ④ → α → 1 → ①
Output: 1   –   –   0   1   0   1   0
            (a)   (b)

States: ④ → α → 1 → ①   ④ → α → 1 → ①
Output: 1   1   –   0   1   –   0   0
            (c)   (d)

FIGURE 13.37. *Transitional output assignment in cyclic transition.*

There are no generally applicable 3-variable assignments for 6 and 7 state tables, but suitable assignments can often be found through careful analysis of the destination sets. This may be illustrated by the following example.

Example 13.4

Obtain a critical-race-free state assignment for the table of Fig. 13.38.

x_1x_2

	00	01	11	10
	3	6	①	①
	②	3	1	②
	③	③	③	5
	3	④	④	1
	2	⑤	6	5
	⑥	⑥	⑥	2

FIGURE 13.38. *Example 6-state table.*

Solution: The destination sets are $(1, 3, 4)$ $(2, 5)$ (6), $(1, 6)$ $(2, 3)$ (4) (5), $(1, 2)$ (3) (4) $(5, 6)$, and $(1, 4)$ $(2, 6)$ $(3, 5)$. First we note a destination set with more than two members, $(1, 3, 4)$, in the first row. In such cases, it is often possible to cycle within the destination set, e.g., ① → 4 → ③ or ④ → 1 → ③. In view of the flexibility offered by such cycles, it is usually best to consider the 2-member destination sets first. In this case, we note that state 2 is a member of four such sets $(2, 5)$, $(2, 3)$, $(1, 2)$, and $(2, 6)$. In a 3-variable assignment, a single state can be adjacent to only 3 other states, so these four destination states require that ② be adjacent to one of the spare states, which we shall designate α. Of the four states paired with ② in destination sets, ⑤ and ⑥ are not part of the $(1, 3, 4)$

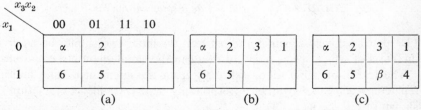

FIGURE 13.39. *Development of race-free assignment for Fig. 13.38.*

set, so we place them adjacent to the (2, α) pair, giving the initial assignment shown in Fig. 13.39a. The remaining state paired with ②, ① and ③ must be adjacent to ② or α, so we locate them as shown in Fig. 13.39b. Finally, noting that state ④ can cycle to ③ through state ①, we place 1 and the remaining spare state, β, as shown in Fig. 13.39c. We then check the destination sets for all columns to see that all transitions can be made without critical races with the assignment of Fig. 13.39c. The development of the augmented state table and transition table is shown in Fig. 13.40. ∎

$y_3 y_2$

y_1	00	01	11	10
0	α	2	3	1
1	6	5	β	4

(a)

$x_2 x_1$

$y_3 y_2 y_1$		00	01	11	10
α	000	–	6	1	2
6	001	⑥	⑥	⑥	α
5	011	2	⑤	6	⑤
2	010	②	3	α	②
1	100	3	α	①	①
4	101	1	④	④	1
β	111	–	–	–	5
3	110	③	③	③	β

(b)

$x_2 x_1$

$y_3 y_2 y_1$	00	01	11	10
000	× × ×	001	100	010
001	⟨001⟩	⟨001⟩	⟨001⟩	000
011	010	⟨011⟩	001	⟨011⟩
010	⟨010⟩	110	000	⟨010⟩
100	110	000	⟨100⟩	⟨100⟩
101	100	⟨101⟩	⟨101⟩	100
111	× × ×	× × ×	× × ×	011
110	⟨110⟩	⟨110⟩	⟨110⟩	111

$y_3 y_2 y_1$

(c)

FIGURE 13.40. *Assignment, augmented state table and transition table for Fig. 13.38.*

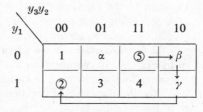

FIGURE 13.41. *5-state assignment with triple transition.*

Shared row assignments utilizing cycles tend to be more economical than multiple-row assignments, but we often pay for this economy in slower operation. An assignment such as the universal 4-state assignment of Fig. 13.29 is known as a *single transition time* (STT) assignment, since each transition involves the change of only a single variable. By contrast, consider the standard 5-state assignment as shown in Fig. 13.41, and assume that a transition is required from $\text{(5)} \to \beta \to \gamma \to \text{(2)}$. This requires that y_2 first change from 1 to 0, then y_1 from 0 to 1, and finally y_3 from 1 to 0, thus requiring three transition times to complete the change. Not every state table will call for this particular transition (Fig. 13.34 does not); but since it can occur, this assignment is characterized as a *three transition time* (3TT) assignment. The standard 8-state and 12-state assignments of Figs. 13.33a and 13.33b are 3TT and 5TT assignments, respectively.

Since speed of operation is often very important, the designer may wish to find assignments that are faster than the standard shared row assignments. Unfortunately, for tables with more than four states, faster assignments are either very difficult to find or require an excessive number of state variables, or both. In most cases, the designer will probably do best to use one of the assignment techniques discussed in this section. For those cases when speed is critical to the extent that a considerable expenditure in design time and extra hardware is warranted, the reader is referred to Reference 13 for a number of techniques for finding fast assignments.

13.8 Hazards in Sequential Circuits

In Fig. 13.42, we show the flow table and transition table for a sequential circuit. The circuit is free from critical races and so, presumably, will function properly. The minimal equations for the state variables are:

$$Y_2 = y_2 x_2 + y_2 \bar{y}_1 x_1 + y_1 x_2 \bar{x}_1 \tag{13.8}$$

and

$$Y_1 = \bar{y}_2 \bar{x}_2 y_1 + y_2 y_1 x_1 + y_2 x_2 x_1 + y_1 x_2 \bar{x}_1 \tag{13.9}$$

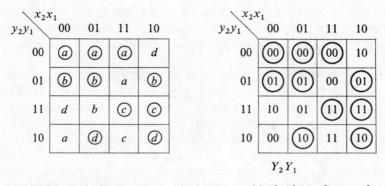

FIGURE 13.42.   *Flow table and transition table for hazard example.*

The logic circuit for Y_1 is shown in Fig. 13.43. The logic for Y_2 is only indicated symbolically since it is not pertinent to the present discussion.

Let us assume that the circuit is in the stable state (c) at $y_2y_1x_2x_1 = 1111$. Under this condition, both gates A2 and A3 will develop 1 at their outputs, producing $Y_1 = 1$ at the output of the OR gate. Now let x_1 change to 0. We see that the circuit should move to the stable state (c) at $y_2y_1x_2x_1 = 1110$, that is, neither Y_1 or Y_2 should change. But let us consider the circuit action in detail. When x_1 goes to 0, the outputs of gates A2 and A3 will go to 0, and the output of gate A4 will go to 1 due to $\bar{x}_1$ going to 1. However, because of the delay in the inverter, $\bar{x}_1$ will not go to 1 as fast as x_1 goes to 0, so that there will probably be a short interval when the outputs of all AND gates

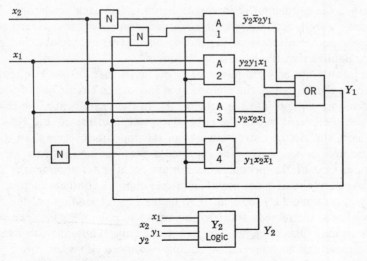

FIGURE 13.43.   *Circuit illustrating static hazard.*

will be 0. The output of the OR gate will thus momentarily go to 0. Since y_2 does not change, the circuit will momentarily see the state variable input $y_2 y_1 = 10$ and may make an erroneous transition to stable state $\widehat{d}$ at $y_2 y_1 x_2 x_1 = 1010$.

Next, let us assume that same starting state, $y_2 y_1 x_2 x_1 = 1111$, and now let x_2 go to 0. Now the circuit should go through unstable b to stable $\widehat{b}$ at $y_2 y_1 x_2 x_1 = 0101$. Again, Y_1 should not change. As before, gates A2 and A3 are 1 initially. When x_2 goes to 0, gate A3 goes to 0 and, at the same time, the transition in Y_2 is started. After delay through the Y_2 logic, y_2 goes to 0, taking A2 to 0, and A1 goes to 1 due to $\bar{y}_2$ going to 1. However, as before, the delay through the inverter will cause all gates to be 0 momentarily, in turn causing Y_1 to go to 0 momentarily. The circuit may thus make an erroneous transition to stable a at $y_2 y_1 x_2 x_1 = 0001$.

The above are examples of malfunctions caused by *static hazards*.

Definition 13.2. Let there be a state transition in a fundamental mode circuit such that one or more of the secondary variables is required to remain constant. If it is possible, due to unequal delays in the circuit, for one or more of the constant state variables to change momentarily, then a *static hazard* is said to exist.

Note that a static hazard is a characteristic of the combinational logic. In the example given earlier, if y_1 were a normal input rather than a feedback variable, Y_1 would still go to 0, but only momentarily. In normal combinational design or in clocked or pulse circuits, static hazards are not of concern since momentarily erroneous signals are not generally troublesome. However, if a momentarily incorrect signal is fed back to the input, as in the above example, a permanent error can occur. A static hazard will not always cause a malfunction, depending on the various delays, but reliable design requires that the hazard be eliminated.

The hazard can be seen quite clearly on the K-map of the Y_1 circuit, Fig. 13.44a. In the first case, the change in x_1 moves the circuit out of the cubes $y_2 y_1 x_1$ and $y_2 x_2 x_1$ and into the cube $y_1 x_2 \bar{x}_1$. In the second case, the change in x_2 moves the circuit out of the cube $y_2 y_1 x_1$ and into the cube $\bar{y}_2 \bar{x}_2 y_1$. As we have seen, the circuit moves out of the starting cubes faster than it moves into the final cubes. Each cube represents one product input to the output OR gate. One of the product-terms must be at 1 to provide a 1 output. Whenever the circuit must move from one cube to another, there is a possibility of a momentary interval when neither corresponding product is at 1. On this basis, the remedy is fairly obvious. Any pair of 1's between which a transition may take place must be in a single cube. Thus, the two hazards in this example can be removed by adding two more cubes, as shown in Fig. 13.44b.

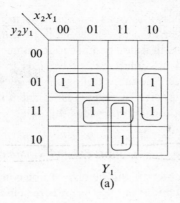

Y_1
(a)

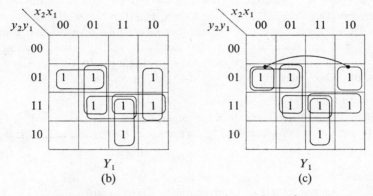

Y_1
(b)

Y_1
(c)

FIGURE 13.44.   *Removal of static hazards.*

We note that there is another static hazard between the 1's in $y_2y_1x_2x_1 = 0100$ and $y_2y_1x_2x_1 = 0110$, so that these two 1's should also be in a single cube. On the other hand, the two 1's in $y_2y_1x_2x_1 = 0110$ and $y_2y_1x_2x_1 = 1110$ need not be in a single cube. From the flow table, we see that both of these 1's correspond to stable states and vertical transitions between stable states are not possible in fundamental-mode circuits. On this basis, the minimal hazard-free realization is that shown in Fig. 13.44c. The resultant circuit will have two more AND gates than the original, but it will be hazard-free.

The situation illustrated here, where the output is supposed to remain 1, is known as a *static hazard in the 1's*. It should be obvious that the same thing could happen when the output is supposed to remain 0, in which case we have a *static hazard* in the 0's. It is also possible to have a hazard when the output is supposed to change. When certain patterns of delay occur, the circuit may go momentarily to the new value and then back to initial value before making the permanent transition, that is, $0 \to 1 \to 0 \to 1$ instead of $0 \rightarrow 1$. This situation is known as a *dynamic hazard*. Like the static hazard the dynamic hazard is seldom of real concern in combinational circuits but may cause malfunctions in fundamental-mode sequential circuits.

The detection of all possible types of hazards in a given circuit is very complicated and has been the subject of extensive investigation.[3] However, as designers, all we need is some rule which is sufficient to ensure freedom from hazards. The following theorem provides this rule.

THEOREM 13.1. *A second-order S-of-P circuit that is free of all static hazards in the 1's will be free of all static and dynamic hazards.*

The proof of this theorem is too involved to be considered here. Note that this is not a necessary condition, i.e., hazard-free circuits of higher order are possible. However, the conditions are much more complex for higher-order circuits. For purposes of designing hazard-free fundamental-mode circuits, we need only restrict ourselves to second-order sums-of-products, ensuring that any pair of adjacent 1's on the excitation maps, corresponding to a legitimate input transition, are included in a single product-cube. For P-of-S design, the same rules may be applied to design on the 0's.

Example 13.3

Develop a hazard-free realization for the excitation map shown in Fig. 13.45a. The minimal equation, representing the groupings shown in Fig. 13.45a, is given by

$$Y_1 = \bar{x}_1 y_2 + x_1 x_2 y_1 + x_1 \bar{x}_2 \bar{y}_2 \qquad (13.10)$$

We see that there are two hazards, between 5 and 13 and between 11 and 15. The revised groupings of Fig. 13.45b eliminate these hazards and, therefore, all static and dynamic hazards. The resulting Boolean expression,

$$Y_1 = \bar{x}_1 y_2 + x_2 y_1 y_2 + x_1 y_1 \bar{y}_2 + x_1 \bar{x}_2 \bar{y}_2 + x_1 x_2 y_1 \qquad (13.11)$$

is not minimal, but it is hazard-free. ■

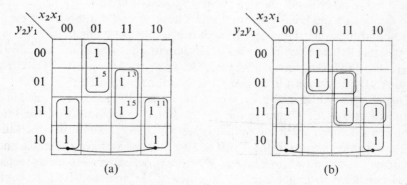

FIGURE 13.45. *Elimination of static hazards by redundant products.*

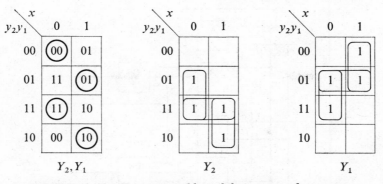

FIGURE 13.46. *Transition table and design maps for counter.*

Like races, the static and dynamic hazards are due to unequal delays in various signal paths. Still another type of hazard arises when delays result in one excitation changing before the circuitry generating another excitation is even aware of the input change. In Fig. 13.46, we show the transition table, excitation maps, and output map for a sequential circuit designed to count modulo-4 the number of changes in input x.

The excitation equations with static hazards eliminated are given by Equations 13.12 and 13.13.

$$Y_2 = y_1\bar{x} + y_2x + y_1y_2 \tag{13.12}$$
$$Y_1 = y_1\bar{x} + \bar{y}_2x + \bar{y}_2y_1 \tag{13.13}$$

Since we have carefully avoided all critical races and all static and dynamic hazards, the resulting circuit should work as intended. Or will it?

Consider the realization of these equations as shown in Fig. 13.47. Assume the circuit is in the stable state ⑪, for which the signals on the various lines are as shown. Now let x go to 1, which should change Y_1 and take the circuit to stable ⑩. The input $x = 1$ ($\bar{x} = 0$) drives AND-3 to 0 and thus Y_1 to 0. The change in Y_1 in turn takes the output of AND-1 to 0. If the circuit is functioning properly, AND-2 will change to 1 at the same time AND-3 changes to 0, so that the subsequent change of AND-1 will not affect Y_2. But suppose that the delay in AND-2 is very long so that its output is still 0 when the output of AND-1 goes to 0. Then OR-1 will see all 0 inputs and Y_2 will go to 0. Instead of seeing itself at the stable combination $xy_2y_1 = 110$, the circuit now sees itself at the unstable condition $xy_2y_1 = 100$ and thus makes an erroneous transition to stable ⑪.

At first, this might look very much like the static hazard discussed in connection with Fig. 13.42, but in that case there was only one excitation involved, so that the problem could be corrected by adding gates in the circuitry generating that excitation variable. Here the problem involves an

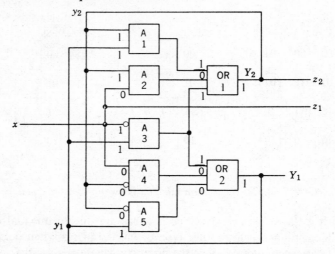

FIGURE 13.47. *Essential hazard in input-change counter.*

interaction between two excitation variables. The variable Y_1 has changed and propagated back through the circuitry for Y_2 before that circuitry has completed responding to the initial input change. This type of hazard, which is peculiar to sequential circuits, is known as an *essential hazard.* As the name suggests, an essential hazard is created by the basic logical structure, or specification, of the circuit and is not affected in any way by the particular form of realization used. The hazard can be detected on the original state table.

Definition 13.3. An *essential hazard* exists in a state table if there is a stable state from which three consecutive changes in a single input variable will take the circuit to a different state than the first change alone.[4]

This behavior will occur if any two adjacent columns of the state table exhibit either of the two patterns shown in Fig. 13.48a and 13.48b. In both

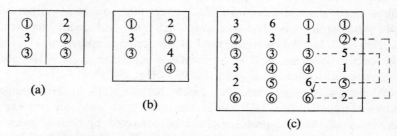

FIGURE 13.48. *Flow tables indicating essential hazards.*

cases, if the circuit starts in ①, a single change of variable takes the circuit to ② and two more changes take it to ④. The reader should note that the numbering of the states and ordering of the rows in these two tables is quite arbitrary; the patterns may appear quite different in a larger flow table. For example, the transition indicated in the table of Fig. 13.48c, ③ → ⑤ → ⑥ → ②, corresponds to the pattern of Fig. 13.48b.

The reader will note that for the hazard described in connection with Fig. 13.47 to occur the delay through AND-2 would have to be longer than the cumulative delay along the path AND-3, OR-2, AND-1. The reader may feel that such a combination of delays would be very unlikely, but it is not. In typical IC gates, the standard production tolerances on delay specification allow a range of three-or-four-to-one between minimum and maximum delays times.

Essential hazards can be found in many circuits of considerable practical importance including almost all counters. Since the effect of these essential hazards cannot be eliminated by modifying logic, they can be handled only by carefully controlling the number of gates in each feedback loop. Very often realizations can be worked out with pairs of cross-coupled gates making up loops which must respond very rapidly. Gates on a single chip will usually have less variance in delay times than typical gates on different chips. Thus, essential hazards are less likely to cause trouble in a level mode circuit implemented in a single IC package.

Occasionally it will be necessary to insert a pair of inverters in a feedback loop so that the secondaries do not change until the input change has propagated to all parts of the circuit. The deliberate insertion of delay is not at all desirable, since speed is generally of the essence, but there is often no other remedy for the essential hazard.

The reader may ask why hazards and races do not occur in clock or pulse mode circuits. Recall from the discussion of Chapter 9 that flip-flops are designed so as to ensure that no resultant changes in the flip-flop excitations would be seen until after the end of the clock pulse. Furthermore, the interval between clock pulses was required to be long enough to allow all changes to stabilize. Taken in this light, we may regard clocking and the use of flip-flops as means of introducing controlled delays to eliminate the timing problems associated with level-mode circuits.

13.9   General Level Mode Circuits

We now must consider the special problems occurring when the restrictions are removed that only one input may change at a time and that no further changes may occur until the circuit has stabilized. Essentially this means that we will now allow simultaneous changes in two or more inputs. While no two events are ever precisely simultaneous, if one change occurs and then a second follows while the circuit is still responding to the first, the changes are simultaneous in the sense that the circuit is trying to respond to both at the same time. In a sense, the speed of response of the circuit defines the finest division by which we can measure time. If the separation of two changes is less than this finest division, then the changes are simultaneous within this frame of reference.

Whenever two circuits inputs come from two unrelated and unsynchronized sources, simultaneous input changes can occur. There are many practical situations where we must deal with signals from unrelated sources. As an example, consider a computer-controlled manufacturing process. The inputs will be derived from sensors monitoring the process at a variety of points. From our knowledge of the process, we can determine the signal rates of individual variables and select our logic circuitry accordingly. But there is no way we can guarantee that two or more signals may not change at the same time. In such circumstances, the best solution may be to synchronize the inputs to an internal clock, but this will also be an unrelated source, so the same problem of simultaneous signal changes still exists.

Let us begin by considering what might go wrong in a specific multiple input transition. Note that relaxing the fundamental mode restrictions does not necessarily mean that every possible change may occur regardless of the present stable state. In many practical situations, there may be only one or two points within the flow table from which multiple input changes might occur. For each stable state in the flow table, the designer must consider whether a multiple change might occur and, if so, analyze carefully the possible functioning of the circuit, as in the following.

Note that the table of Fig. 13.49 has no critical races. Although no assignment is indicated, all transitions will be adjacent with the rows arranged as shown. Suppose that the only simultaneous input change which might be expected in Fig. 13.49a is a change in $x_2 x_1$ from 00 to 11 when in stable state $\textcircled{a}$, as indicated.

Because of random delays in logic paths, not all gates will see the changes in both inputs at the same time. Thus, at the outset, there may be gates which see the inputs in any one of the four combinations of values. This is depicted graphically by darkening the first row of Fig. 13.41. Some delay combination may, for example, cause a critical gate to see only the change of x_1 from 0 to 1. Thus, the circuit might go, momentarily at least, to state $\textcircled{b}$, and different points in the circuit might take on values corresponding to the second row of the table. Now suppose there is another gate which still sees the inputs as 00 but sees the state variables as corresponding to the second row. This gate

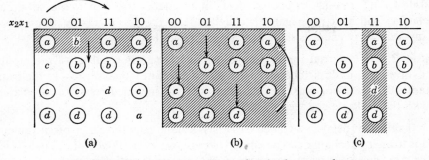

FIGURE 13.49. *Possible effects of multiple input changes.*

might thus send the circuit to state c, making it possible for some points in the circuit to take on the values in the third row of the table. Then a gate seeing the inputs correctly as 11 might take it to d, where it might stabilize, or go on to a if some other gate sees the inputs as 10, as shown in Fig. 13.49b. Thus, depending on delays, the circuit might momentarily pass through any point in the state table as suggested by the shading of the entire table in Fig. 13.49b. Any arbitrary combination of input and secondary values could conceivably be assumed by some section of some realization of the flow table at some point in time shortly after the input change. On the other hand, many of the combinations of values might be remote if not impossible in any particular realization. Eventually, the effect of the dual input change will have propagated throughout the circuit. Once this has occurred, the circuit will be restricted to states in column 11, as depicted in Fig. 13.41c. Since there are 3 stable states in this column, a malfunction would seem to be a real possibility.

The situation illustrated above is very similar to the essential hazard in that not all gates receive full information about the input change at the same time. The essential hazard is a special case in that difficulty can arise even with only a single input change.

Let us now consider tactics by which we might limit the chaos suggested by the maximally pessimistic analysis above. First, if there is only one stable state in a column, there can be no malfunction. Sooner or later all changes propagate sufficiently to bring the circuit into the proper column; and if there is only one stable state in that column (noncritical race), the circuit must terminate there. Unfortunately, this is not a condition over which we generally have any control, as the number of stable states in a column is determined by the basic function the circuit is to perform.

If the multiple change is into a column with two or more stable states, reliable operation will be insured if the following two rules are satisfied.

Rule 1. There must be enough delay in each feedback loop so that all input changes will have propagated through at least the first level of gating before any secondary changes have propagated around a loop.

Rule 2. If a partial input change, i.e., a change in a subset of the inputs involved in a multiple change, can cause the circuit to leave a row in the state table, this behavior must be consistent with the intended behavior associated with the multiple change.

If sufficient delay is included to satisfy Rule 1, then the range of movement due to an input change can be limited to two rows of the flow tables. To do this, we must specify a transition table so that no races critical or noncritical are initiated in a row where a multiple input change can occur. The situation is illustrated in Fig. 13.50. If the x_2 change is seen before the x_1 change $(00 \rightarrow 10)$, there is no change of state, so the circuit stays on the proper row until the x_1 change is seen. If the x_1 change is seen first $(00 \rightarrow 01)$, y_1 may change, so that part of the circuit may see the state as on the second row. Because of the delay in feedback, no gate will see the y_1 change until it sees the inputs as 11. The circuit is thus initially limited to the shaded area of Fig. 13.50a, and then, when all gates see the 11 input, to the shaded area in Fig. 13.50b.

The feedback delay prescribed by Rule 1 thus insures that the circuit moves into the proper column, but what happens then will depend on the satisfaction of Rule 2. Figure 13.50c illustrates one manner in which this rule

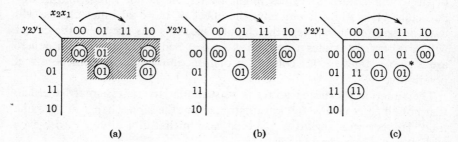

(a)   (b)   (c)

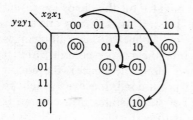

(d)

FIGURE 13.50. *Limiting the range of multiple input changes.*

might be satisfied. Notice that both unstable states in row 00 specify a transition to stable state ⑴ . Thus, no matter how the dual input change might initially be interpreted, the circuit will always stabilize at the point marked with the asterisk, *provided Rule 1 is satisfied*. We have just illustrated the most common way in which Rule 2 can be satisfied. That is, the intended destination for any multiple input change must be the same as the final destination reached by an equivalent sequence of single input changes. In this case, the intended destination of the change $00 \rightarrow 11$ is the same as the final destination of the sequential change $00 \rightarrow 01 \rightarrow 11$.

It is important to note that both rules must be satisfied. Figure 13.50d shows a version of the table for which Rule 2 is not satisfied and for which the circuit is about equally likely to wind up in either ⑴ or ⑽ even though Rule 1 may be satisfied. In terms of the design process, the designer must consider Rule 2 at the time the state table is set up. He will then insure that Rule 1 is satisfied only after the complete circuit has been designed by selecting gates to achieve the proper delay reationships.

Let us now consider a complete design example.

Example 13.5

Because of gate delays, line delays, and communication between units with separate clocks, a control pulse to a section of a computer system may arrive out of synchronism with the clock for that section. In order to avoid unpredictable operation in the clocked circuits of the computer, such a control pulse must be routed through some type of pulse synchronizer. This pulse synchronizer will have two inputs, the computer clock line C and the control pulse line P. Pulses on line P may arrive at random times relative to clock pulses. The inputs are pulses, but because they may violate the pulse mode restriction of one input pulse at a time, we must consider them as consecutive changes in levels and apply level mode techniques.

Corresponding to every pulse on line P, the synchronizer is to emit a pulse on output line Z, synchronized with a clock pulse. Pulses on line P are of approximately the same duration as clock pulses but will occur less frequently, i.e., always separated by several clock periods. This is an example of a situation in which some, but not all, multiple input changes may occur.

Solution: As with fundamental mode design, it is possible to start with a primitive flow table, but in this case we shall find it convenient to proceed directly to a table with more than one stable state per row. Let state ① be a quiescent stable state following an output pulse, i.e., a situation in which no further output should occur until another input appears on P. As illustrated in Fig. 13.51a, the circuit merely moves back and forth between the stable states labelled ① as the clock goes off and on in the absence of an input control pulse. When a pulse appears on line P, the circuit goes to stable state ②, as indicated in Fig. 13.51b. Whether the ultimate circuit senses single changes or multiple

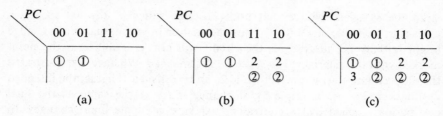

FIGURE 13.51. *Developing a flow table for the pulse synchronizer.*

changes on row ①, it will eventually stabilize in row ② when a clock pulse arrives, given satisfaction for Rule 1.

Note that the output, though not shown, must remain 0 for all stable states shown in Fig. 13.51b. It might seem that a 1 output could be specified for ② with input 11, since this corresponds to the direct change 00 → 11 and would apparently produce a full-width output pulse. But to do so would violate our rule that we must not specify different behavior for a simultaneous change than for any sequence of single changes, since we do not wish an output if either *P* or *C* changes alone. In turn, we must specify a stable ② with 0 output for input 01, else we would have a partial output when the pulses overlapped and *P* ended first. The circuit can move out of row ② only when both inputs return to 0, as shown in Fig. 13.51c.

The net effect of the above specification is to ensure that no output will occur until after the circuit has sensed both inputs at 0 following a pulse on *P*. In the worst case, as shown in Fig. 13.52, this means that an output pulse may be delayed relative to a control pulse by more than one clock period. Given the minimum interval of several clock periods between control pulses, this delay presents no problem and is necessary to achieve reliable operation in response to multiple input changes.

Once the circuit has reached stable state ③, the next clock pulse will take it to state ④, generating an output of 1 for the duration of the clock pulse. The assumed interval between control pulse inputs precludes the possibility of multiple input changes on rows ③ and ④ and permits the don't-care entries shown in Fig. 13.53. In the output table, the values for *z* corresponding to unstable states have been specified where necessary.

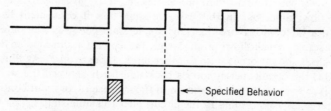

FIGURE 13.52. *Worst case synchronizer delay.*

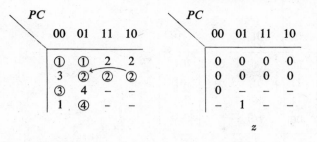

FIGURE 13.53.   *Flow table for synchronizer.*

The only possible multiple input change which has not been discussed (and found to satisfy Rules 1 and 2) is identified by an arrow in Fig. 13.53. If the circuit initially only senses the change in P (from 1 to 0), the circuit will leave row ② via unstable 3 and move on to stable state ④ when the complete input change is sensed. This will result in the shaded output pulse in Fig. 13.52. If the circuit actually interpreted the changes from $PC = 10$ to $PC = 01$ as taking place simultaneously, the circuit will wait a full clock period to put out the unshaded pulse in Fig. 13.52. Clearly the two possible forms of circuit behavior are quite different; but does it matter? In both cases, the output is a single pulse synchronized with a clock pulse and of clock pulse duration. From the problem statement, we conclude that it makes no difference which type of circuit behavior actually takes place. This is an interesting special case of the satisfaction of Rule 2. The circuit "does something different," but not in any significant sense. ∎

We now turn our attention to obtaining a circuit realization. States are assigned without difficulty yielding the transition table of Fig. 13.54a. Individual Karnaugh maps are included to illustrate the minimal multiple output realization for Y_2 and Y_1 as given by Equations 13.14 and 13.15. Notice the inclusion of the term $y_2 y_1$ to avoid a static hazard in Y_2. Equation 13.16 for output z is taken directly from Fig. 13.53. A logic block diagram

$$Y_2 = \bar{P}\bar{C}y_1 + y_1 y_2 + y_2 C \tag{13.14}$$

$$Y_1 = \bar{P}\bar{C}y_1 + \bar{y}_2 y_1 + P \tag{13.15}$$

$$z = y_2 \bar{y}_1 \tag{13.16}$$

implementation of these excitation equations is given in Fig. 13.55. If the statistical variation in delay is reasonably small, the circuit will satisfy Rule 1 without the addition of redundant gates.

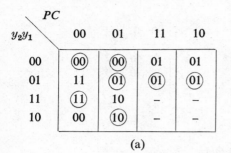

y_2y_1 \ PC	00	01	11	10
00	⓪⓪	⓪⓪	01	01
01	11	⓪①	⓪①	⓪①
11	①①	10	–	–
10	00	①⓪	–	–

(a)

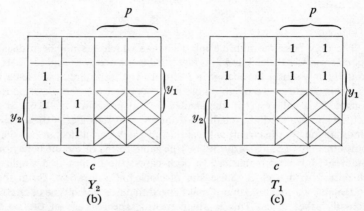

(b) Y_2 (c) T_1

FIGURE 13.54. Transition table for synchronizer.

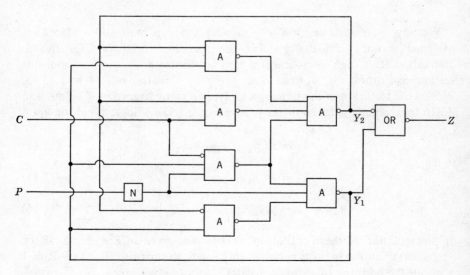

FIGURE 13.55. Pulse synchronizer.

PROBLEMS

13.1. Determine the primitive flow table of a fundamental-mode sequential circuit with two inputs, x_2 and x_1. The single output, z, is to be 1 only when $x_2x_1 = 10$ provided that this is the fourth of a sequence of input combinations 00 01 11 10. Otherwise the output is to be 0. Include a tabulation of outputs assuring that no spurious 1 outputs will occur during transitions between two states with 0 outputs. Both inputs will not change simultaneously.

13.2. A controller is to be designed for an automatic parking lot gate of the type where the driver inserts a coin in a slot and the gate then opens long enough to let him enter. When the driver inserts the coin, a coin detector will emit a pulse P, which will cause the gate control signal z to go to 1 to open the gate. As the car passes through the gate, it will pass by a photocell which will cause a signal x to go to 1. When the car is past the gate, x will go back to 0, at which time z should go to 0, closing the gate. To prevent persons from defeating the system by blocking off the photocell to keep the gate open, there will be a treadle switch in the pavement immediately in front of the gate which will emit a pulse T whenever car wheels pass over it. The car passing through will trip this switch twice. If it is triggered a third time before x has gone back to 0, z should go to 0 to close the gate. Assume that the switch is located relative to the photocell so that x will go to 1 before the front wheels hit the switch, and the rear wheels will hit the switch before x goes back to 0. Determine a primitive flow table of a circuit to develop the control signal z.

13.3. A circuit has a single input line on which pulses of various widths will occur at random. Construct the primitive flow table of this circuit with a single output, z, on which a pulse will occur coinciding with every fourth input pulse. As this circuit is part of a random process generation scheme, the power on state is of no interest.

13.4. A sequential circuit has two level inputs and two level outputs. The

z_2z_1	Count	Next Count
00	0	1
01	1	2
10	2	3
11	3	0

FIGURE P13.4

outputs are coded to count in the straight binary code, modulo-4, as shown in Fig. P13.4. The count is to increase by one for each 0-to-1 transition of either input occurring when the other input is at the 0 level. The two inputs will not change at the same time. Determine a primitive flow table of a circuit to meet this specification.

13.5. Determine a minimal state equivalent of the primitive flow table given in Fig. P13.5. Suppose that the don't-care outputs were all specified such that unstable states always have the same output as

	x_2x_1				z x_2x_1			
00	01	11	10	00	01	11	10
ⓐ	f	i	k	0	–	0	–
ⓑ	f	g	j	0	–	0	–
ⓒ	d	h	j	0	–	0	–
c	ⓓ	i	k	–	1	–	1
a	ⓔ	i	l	–	1	–	–
a	ⓕ	h	j	–	1	–	1
a	e	ⓖ	k	0	–	0	–
c	f	ⓗ	k	0	–	0	–
a	d	ⓘ	j	0	–	0	–
c	f	i	ⓙ	–	1	–	1
a	d	h	ⓚ	–	1	–	1
b	d	h	ⓛ	0	–	0	0

FIGURE P13.5

similarly lettered stable states. Thus, the state table becomes completely specified. Repeat the minimization after this modification and compare the two solutions. Would this approach in general prevent the determination of a minimal table? Why not?

13.6. Determine a minimal row equivalent of the flow table of Fig. P13.6.

	q x_2x_1				z x_2x_1			
00	01	11	10	00	01	11	10
①	2	–	5	0	0	–	0
–	②	3	–	–	0	0	–
–	4	③	9	–	–	0	0
1	④	–	–	–	1	–	–
–	–	6	⑤	–	–	0	0
–	7	⑥	8	–	0	0	0
1	⑦	–	–	0	0	–	–
1	–	–	⑧	0	–	–	0
1	–	–	⑨	0	–	–	0

FIGURE P13.6

13.7. Determine minimal state equivalents of the primitive flow tables determined in:
(a) Problem 13.1.
(b) Problem 13.2.

13.8. Assign secondaries to the minimal flow table of Fig. P13.8. Construct a transition table which avoids critical races. Obtain Boolean expressions for each excitation variable.

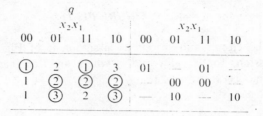

q							
$x_2 x_1$				$x_2 x_1$			
00	01	11	10	00	01	11	10
①	2	①	3	01	—	01	---
1	②	②	②	--	00	00	...
1	③	2	③	—	10	---	10

FIGURE P13.8

13.9. Assign the eight combinations of values of three secondaries to the six rows of the flow table in Fig. P13.9. Determine a transition table which avoids critical races. Do not minimize the flow table.

q				z			
$x_2 x_1$				$x_2 x_1$			
00	01	11	10	00	01	11	10
①	2	–	5	0			
②	②	4	6	1	1		
–	2	④	6			1	
1	2	–	⑤				0
1	⑥	8	⑥		0		1
–	6	⑧	5			0	0

FIGURE P13.9

13.10. Obtain a minimal flow table covering the table of Fig. P13.9. Assign secondary states to rows of the minimal state table using only two secondaries. If possible, obtain a transition table for this assignment that avoids critical races.

13.11. Make a secondary assignment for the flow table of Fig. P13.11. Construct a transition table without critical races.

	q x_2x_1				z x_2x_1		
00	01	11	10	00	01	11	10
ⓐ	ⓐ	d	b	0	1		
a	ⓑ	c	ⓑ		0		0
ⓒ	a	ⓒ	d	1		1	
c	b	ⓓ	ⓓ			1	1

FIGURE P13.11

13.12. For the circuit and secondary assignment of Fig. P13.12, obtain expressions for Y_2, Y_1, and z, which are free from all static and dynamic hazards.

Y_2Y_1	q x_2x_1				z x_2x_1			
	00	01	11	10	00	01	11	10
00	ⓐ	ⓐ	ⓐ	b	1	1	0	0
01	a	ⓑ	c	ⓑ	1	1	1	0
11	ⓒ	ⓒ	ⓒ	d	0	0	1	1
10	c	—	a	ⓓ	0	—	0	1

FIGURE P13.12

13.13. For the flow table shown, determine a state assignment free of critical races, using three state variables.

	x_1x_2			
	0 0	0 1	1 1	1 0
	Ⓐ, 0	Ⓐ, 0	E, 0	D, 0
	D, 0	E, 0	Ⓑ, 0	Ⓑ, 0
	A, 0	Ⓒ, 0	F, 0	Ⓒ, 0
	Ⓓ, 0	G, 0	Ⓓ, 0	Ⓓ, 0
	D, 0	Ⓔ, 0	Ⓔ, 0	F, 0
	Ⓕ, 0	A, 0	Ⓕ, 0	Ⓕ, 0
	Ⓖ, 1	Ⓖ, 0	B, 0	C, 0

q, z

FIGURE P13.13

13.14. For the flow tables shown in Fig. P13.14, determine assignments free of critical races. Assume the tables are minimal.

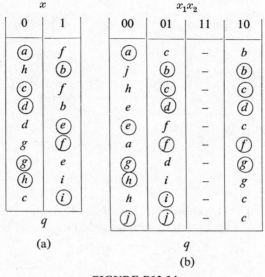

FIGURE P13.14

13.15. For the flow tables of Fig. P13.15, determine assignments free of critical races and determine excitation equations free of combinational hazards.

x_1x_2

00	01	11	10
ⓐ,0	ⓐ,0	d,0	e,0
ⓑ,0	ⓑ,0	d,0	ⓑ,0
b,0	ⓒ,0	ⓒ,0	e,0
e,–	ⓓ,0	e,0	b,0
ⓔ,0	ⓔ,0	c,0	e,0

q,z

(a)

x_1x_2

00	01	11	10
①,0	2,–	①,1	5,–
1,–	②,0	②,0	4,–
③,0	5,–	4,–	③,1
3,–	5,–	④,0	④,0
⑤,0	⑤,0	1,–	⑤,1

q,z

(b)

FIGURE P13.15

13.16. Write hazard-free excitation and output expressions for the following flow tables. Make certain that there are no transient output pulses.
(a) Figure P13.9.
(b) Figure P13.11.

13.17. Inspect the flow tables of Fig. P13.17 for essential hazards. If any exist, indicate them by specifying the single and triple transitions

$x_2 x_1$

00	01	11	10
ⓐ	b	–	ⓐ
ⓑ	ⓑ	c	a
–	–	ⓒ	d
b	–	–	ⓓ

(a)

$x_2 x_1$

00	01	11	10
ⓐ	–	b	ⓐ
–	c	ⓑ	d
d	ⓒ	b	–
ⓓ	–	f	ⓓ
–	ⓔ	f	ⓔ
–	c	ⓕ	e

(b)

FIGURE P13.17

fulfilling the condition of Definition 13.3. For example, for the flow table of Fig. 13.48b, the essential hazard would be indicated

$$①→2→②→3→③→④$$
$$\underbrace{\quad}_{1}\underbrace{\quad}_{2}\underbrace{\quad}_{3}$$

13.18. A troublesome problem associated with a switch is contact bounce. There are generally many switches and relays in the console and input-output sections of computers. It is very difficult and expensive to construct switches and relays so that the contacts do not bounce. This problem is illustrated in Fig. P13.18. At t_1, the switch moves

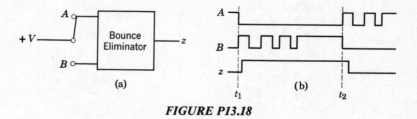

(a)   (b)

FIGURE P13.18

from A to B. It makes initial contact at B and then bounces several times, producing a series of several short pulses before settling permanently at B. We may assume it does not bounce back far enough to contact A. At t_2, the switch is moved back to A, with similar results.

Design a "bounce eliminator" to produce a z output as shown, following B but eliminating the multiple transitions.

13.19. Another common feature of computer consoles is a step control, which enables an operator or service engineer to step the computer through a program one cycle at a time. For this purpose, we require a circuit where a push-button can be used to parcel out the clock pulses, one clock pulse per activation of the push-button. The timing diagram is shown in Fig. P13.19. We assume that the push-button signal has

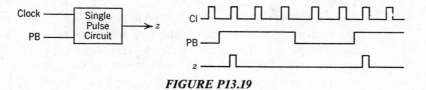

FIGURE P13.19

had the bounce eliminated by the circuit of Fig. P13.18. The timing of the PB signal is random, except that the duration and interval may be assumed to be very long compared to clock duration and interval. Design a fundamental mode circuit which will accomplish the above function.

13.20. A fundamental-mode circuit has two inputs and two outputs. The two outputs should indicate which of the inputs has changed most recently and whether the number of times it has changed is odd or even.
 (a) How many secondaries will be required?
 (b) Obtain a minimal flow table of such a circuit.
 (c) Obtain a race-free transition table.
 (d) Design a circuit to realize the transition table of c which is free of static and dynamic hazards.
 (e) If any essential hazards exist, indicate them on the flow table.

13.21. A fundamental-mode circuit is to be designed which will identify a particular sequence of inputs and provide an output to trigger a combination lock. The inputs to the circuit are three switches labeled X_3, X_2, and X_1. The output of the circuit is to be 0 unless the input switches are in the position $X_3X_2X_1 = 010$ where this position occurs at the conclusion of a sequence of input positions $X_3X_2X_1 = 101\ 111\ 011\ 010$. Note that two switches are not required to be switched simultaneously.

 Using only NAND gates, design a fundamental-mode circuit which will have an output of 1 only at the conclusion of the above described sequence of switch positions. A new correct sequence may begin every time the switches are set to $X_3X_2X_1 = 101$. Be sure that the circuit is free from hazards and critical races.

13.22. Complete the design of the circuit of Example 13.1, starting with the final flow table of Fig. 13.15. Make an assignment free of critical races, determine hazard-free equations, and draw the final circuit.

13.23. Complete the design of the controller for the parking lot gate, previously considered in Problems 13.2 and 13.7b.

13.24. Let us now consider a more elaborate version of the parking lot gate of Problem 13.2. Assume that there are two lanes entering a parking garage and that two gates, as described in Problem 13.2, are arranged side-by-side. After these gates, the two lanes merge into a single lane ramp to the parking floors. The operation of the gates will be just as described in Problem 13.2 except that we wish to control the spacing of cars as they go up the ramp. For this purpose, we locate another treadle switch a short distance up the ramp. After a car passes through either of the gates, its front wheels should pass over the ramp treadle before either gate can open again. Let the output pulse of the ramp treadle be R. If there are two cars at the gates at the time of an R pulse, they should be released in alternate order. If two cars arrive at the gates at the same instant, they may be released in either order, but take care in your design to ensure that both cannot be released at the same time. Design a circuit to meet these requirements.

13.25. A detector circuit is to be designed for a reaction tester. At random intervals, a pulse X_1 will be emitted by a pulse generator. This X_1 pulse will turn on a red light for the duration of the pulse. When the person being tested sees the light come on, he is to push a button which emits a pulse X_2, which may be assumed to be shorter in duration than the X_1 pulse. The output of the detector, Z, is to be 0 unless a complete X_2 pulse occurs during the X_1 pulse, in which case the output is to go to 1 and remain there until a pulse R appears on a reset line. Design a detector to meet these requirements. Include the reset line in your design.

13.26. On television quiz show, three contestants are asked a question, all at the same time. If a contestant thinks he knows the answer, he pushes a button. Whoever pushes his button first gets to answer the question. Design a detector circuit to turn on a light in front of the contestant who pushed his button first. Let the signals from the three push buttons be X_1, X_2, X_3 and let the outputs controlling the corresponding lights be Z_1, Z_2, Z_3. Include provision for a reset line controlled by the M.C. Since the contestants are racing each other, two or more signals will occasionally coincide. The circuit may resolve these "ties" in a random fashion, but the circuit must be designed so that no two lights can ever come on at the same time.

13.27. Repeat Problem 13.26, except now assume that we wish an indication of the order in which the contestants pushed their button. There will be three lights in front of each contestant, numbered 1, 2, 3. The appropriate light should come on in front of each contestant, indicating whether he was first, second, or third to push his button.

BIBLIOGRAPHY

1. Miller, R. E. *Switching Theory*, Vol. 1, Wiley, New York, 1965.

2. McCluskey, E. J. "Fundamental and Pulse Mode Sequential Circuits," *Proc. IFIP Congress, 1962*, North Holland Publ. Co., Amsterdam, 1963.

3. McCluskey, E. J. *Introduction to the Theory of Switching Circuits*, McGraw-Hill, New York, 1965.

4. Unger, S. H. "Hazards and Delays in Asynchronous Sequential Switching Circuits," *IRE Trans. on Circuit Theory*, CT-6, 12 (1959).

5. Huffman, D. A. "The Design and Use of Hazard-Free Switching Networks," *J. ACM*, **4**, 47 (1957).

6. Huffman, D. A. "The Synthesis of Sequential Switching Circuits," *J. Franklin Institute*, **257**, 161–190, 275–303 (March–April 1954).

7. Huffman, D. A. "A Study of the Memory Requirements of Sequential Switching Circuits," *Technical Report No. 293*, Research Laboratory of Electronics, M.I.T., April 1955.

8. Caldwell, S. H. *Switching Circuits and Logical Design*, Wiley, New York, 1958.

9. Eichelberger, E. B. "Hazard Detection in Combinational and Sequential Switching Circuits," *IBM J. Res. and Dev.*, **9**: 2 (1965).

10. Maley, G. A. and J. Earle. *The Logic Design of Transistor Digital Computers*, Prentice-Hall, Englewood Cliffs, N.J., 1963.

11. Hill, F. J. and G. R. Peterson. *Digital Systems: Hardware Organization and Design*, Wiley, New York, 1973.

12. Saucier, G. A. "Encoding of Asynchronous Sequential Networks," *IEEE Trans. on Computers*, **EC 16**: 3, (1967).

13. Unger, S. H. *Asynchronous Sequential Switching Circuits*, Wiley, New York 1969.

14 *A Second Look at Flip-Flops and Timing*

14.1 *Introduction*

Only two complete circuit examples, the control delay and the pulse synchronizer, were presented in the last chapter. We shall now consider the design of the most commonly used of all level mode circuits, the master-slave J-K flip-flop. This may surprise the reader who is accustomed to thinking of a flip-flop as a clock mode circuit. Once designed to satisfy the clock mode criteria of one state change per clock pulse, a flip-flop *is* a clock mode circuit. While it is being designed to satisfy this criteria, we must regard the flip-flop as level mode. The clock line must be treated as any other level input.

We allot a short chapter to this subject to permit a thorough treatment. In Chapter 9, we mentioned *clock skew*. In this chapter, we shall examine in detail the effect of this phenomenon and consider its implications as we examine alternatives in the design of the J-K flip-flop. When we reach the point of determining a realization of the transition table, we shall introduce a new approach for eliminating static and dynamic hazards. In the case of J-K flip-flop, this approach will lead to the efficient realization used in most integrated circuit families.

The remainder of the book will not depend on the material of this chapter, and many readers may wish to skip this chapter. Others may find it satisfying to see that formal level-mode procedures can lead directly to time-honored designs. The student who is likely to become involved in the design of high-speed clock-mode circuits will find that this material will provide needed insight into the capabilities and limitations of available flip-flops. A thorough understanding of memory element timing will allow the reader to anticipate problems which might otherwise appear after a circuit has been constructed.

14.2   Clock   Skew

Sometimes a particular physical realization of a sequential circuit will cause the clock mode assumption to be violated even though master-slave (trailing-edge triggering) or edge-triggered (leading-edge triggering) flip-flops are used. Consider the clock mode sequential circuit of Fig. 14.1. Each flip-flop shown

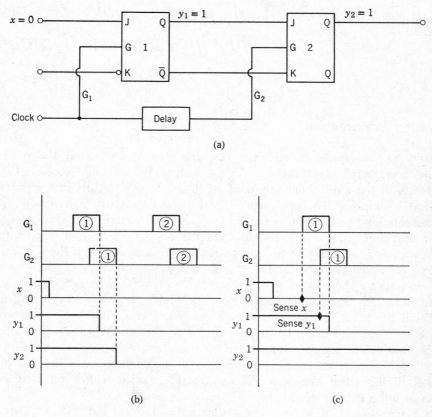

FIGURE 14.1

will be assumed to be master-slave so that any state change will take place immediately after the trailing edge of clock pulse reaches its G input terminal. In effect, the circuit is a simple 2-bit shift register. Assume that, because of some peculiarity in wiring the clock distribution lines, a delay has appeared between the clock line entering flip-flop 1 and the clock line entering flip-flop 2. This may have been caused by a long wire or by a pair of inverters inserted to help handle heavy loading on the clock line. In high-speed circuits, wire length can become quite critical.

The initial values of the flip-flops are shown in Fig. 14.1b to be $y_1 = y_2 = 1$, and the input is 0. If the circuit were working properly, the next clock pulse would place a 0 in flip-flop 1, and the second pulse would propagate the 0 on to y_2. In Fig. 14.1b, we observe the problem caused by the delay in the clock line. Notice that y_1 quite properly goes to 0 upon the trailing edge of clock pulse 1. This value is propagated directly to the input of flip-flop 2 while the delayed clock pulse, G_2, is still 1. In a typical master-slave flip-flop, this new value of J will be gated right into the master element. When G_2 returns to 0, the new value is passed onto the slave, and y_2 goes to 0 as shown. Thus, a state change has taken place at clock pulse 1 which should not have occurred until the trailing edge of pulse 2.

Where typical J-K flip-flops are used, great care must be taken to avoid the above behavior. One must assure that the clock inputs to all flip-flops are synchronized. Another approach might be to use a somewhat more complicated flip-flop. Suppose that J and K were sampled at the leading edge of the clock pulse so that subsequent changes in these values would have no effect until the arrival of the next clock pulse. In that case, the original value of y_1 would be sensed at the input of flip-flop 2 by pulse 1, and the circuit would perform properly as illustrated in Fig. 14.1c. As we shall see in the next section, adding this edge-sensing capability to the master slave flip-flop will considerably complicate its design.

14.3   A Flow Table for a J-K Master-Slave Flip-Flop

In this section, we shall obtain a flow table for a master-slave flip-flop. Departing slightly from our basic procedure of the previous chapter, we shall introduce as we proceed some constraints which will tend to channel the design toward an expected result. Whether one would expect these constraints as part of a predesign specification is arguable, but each can be defended in terms of good design practice.

First we assume that the concept of a master element and a slave element has meaning apart from the final realization. To enforce this property, we assume two outputs, the master output z_1 and the slave output z_2. Once the

CJK

	000	001	011	010	100	101	111	110	z_1z_2
	a	a	a	a			b	b	0 0
					b	b	b	b	0 1

FIGURE 14.2.   *Developing J-K master-slave flow table.*

device is constructed, only the slave output will be available outside of the integrated circuit package. We begin the flow table by defining a stable state a corresponding to $z_2 = z_1 = 0$. As long as the clock C remains 0, the circuit will stay in state a, as indicated in Fig. 14.2. When the clock goes to 1, the output z_1 of the master element should be set to 1 provided that $J = 1$. Because of the defined function of the J-K flip-flop, the value of K is immaterial. Therefore, we enter unstable state b in the last two columns of the first row of Fig. 14.2. This will cause the circuit to go to stable state b, for which the outputs are defined as $z_2z_1 = 01$. We want no further activity until the clock returns to 0, hence the four b entries on the $C = 1$ half of the second row in Fig. 14.2.

We can now choose between two approaches to the completion of the flow table in Fig. 14.2. One will lead to a standard master-slave flip-flop and the second will lead to a more complicated edge-sensing master-slave. For

CJK

	000	001	011	010	100	101	111	110	z_2z_1
	a	a	a	a	a	a	b	b	0 0
	c	c	c	c	b	b	b	b	0 1
	c	c	c	c					1 1

(a)

CJK

	000	001	011	010	100	101	111	110	z_2z_1
	a	a	a	a	a	a	b	b	0 0
	c	c	c	c	b	b	b	b	0 1
	c	c	c	c	c	d	d	c	1 1
	a	a	a	a	d	d	d	d	1 0

(b)

FIGURE 14.3.   *Flow table for standard master-slave J-K flip-flop.*

now we follow the design process for the standard master-slave flip-flop by filling out the first row with stable state ⓐ entries. Thus, a 1 will be propagated into the master element if J changes to 1 while $C = 1$. Once the master element has been set to 1, no further activity will take place until $C \to 0$, regardless of changes in J and K. When $C \to 0$, the circuit goes to stable state ⓒ, for which $z_2z_1 = 11$, as shown in Fig. 14.3a. We complete the table by providing for a return to state ⓐ after the next occurrence of $C = K = 1$. This requires a fourth stable state ⓓ corresponding to $z_2z_1 = 10$, as given in Fig. 14.3b.

We shall continue with the realization of this table later. Before we do this, let us first develop the flow table for an edge-sensing master-slave flip-flop as Example 14.1.

Example 14.1

Obtain the flow table for an edge-sensing master-slave flip-flop.

Solution: Notice that two blank spaces remain in the first row of Fig. 14.2. If the flip-flop is to be an edge-sensing master-slave, it is important to distinguish between $J \to 1$ followed by $C \to 1$ and $C \to 1$ followed by $J \to 1$. In the latter case, the circuit must not be allowed to go to state ⓑ. To assure this, it will be necessary to define another state for the circuit to assume as soon as $C \to 1$ while $J = 0$. This will require entering the corresponding unstable states in the blank spaces of the first row of Fig. 14.2.

CJK									
000	**001**	**011**	**010**	**100**	**101**	**111**	**110**		z_2z_1
ⓐ	ⓐ	ⓐ	ⓐ	e	e	b	b		0 0
a	a	a	a	ⓔ	ⓔ	ⓔ	ⓔ		0 0
				ⓑ	ⓑ	ⓑ	ⓑ		0 1

(a)

CJK									
000	**001**	**011**	**010**	**100**	**101**	**111**	**110**		z_2z_1
ⓐ	ⓐ	ⓐ	ⓐ	e	e	b	b		0 0
a	a	a	a	ⓔ	ⓔ	ⓔ	ⓔ		0 0
c	c	c	c	ⓑ	ⓑ	ⓑ	ⓑ		0 1
ⓒ	ⓒ	ⓒ	ⓒ	f	d	d	f		1 1
c	c	c	c	ⓕ	ⓕ	ⓕ	ⓕ		1 1
a	a	a	a	ⓓ	ⓓ	ⓓ	ⓓ		1 0

(b)

FIGURE 14.4.   *Flow table for edge-sensing J-K master-slave flip-flop.*

We continue from Fig. 14.2 by entering unstable state e in the blank spaces of the first row. This leads to the partial flow table of Fig. 14.4a. Remembering the discussion of Section 13.9, we see that the stable state reached following a multiple input change on the first row would be unpredictable. To avoid multiple changes, *we must impose the restriction that both J and K be stable when C goes to 1 and that the circuit will stabilize following the leading edge of the clock pulse before a subsequent change in J or K will take place.* If we assume that the maximum possible clock skew is known, it will always be possible to enforce this restriction, although sometimes a reduction in clock frequency may be required. With this restriction, the first row of Fig. 14.4a is essentially fundamental mode.

State e serves as a waiting state to assure that changes in J or K while the clock is 1 will not affect the master element. When the clock returns to 0, the circuit returns to state a. From state b, the circuit will contine to c as before when C returns to 0. The lower three rows of the completed flow table as shown in Fig. 14.4b are analogous to the upper three rows. State f has been added to hold the output at 1 if a signal on line K occurs while the clock is 1. A three-state-variable realization of Fig. 14.4 can be constructed. We shall leave this as a homework problem for the reader. ∎

14.4 Another Approach to Level Mode Realization

Before trying to obtain a realization of the standard J-K flip-flop, let us introduce a new model on which level mode realizations can be based. Where the standard realization would require the addition of several gates to eliminate static and dynamic hazards, a realization based on the model of Fig. 14.5 might be preferable. The realization of only one excitation variable, Y_i, is shown, but each excitation will be realized by a similar network. The inputs may be any of the circuit inputs or secondaries.

The excitation, Y_i, is the output of a pair of a cross-coupled NAND gates, looking very much like an S-C flip-flop. As shown in the figure, we denote the outputs of the input NAND gates as $\bar{f}_1 \cdots \bar{f}_k$ and $\bar{g}_1$ to $\bar{g}_m$, (k and m have no necessary relation to the number of variables), and then define

$$\overline{S_{y_i}} = \bar{f}_1 \cdot \bar{f}_2 \cdots \bar{f}_k \tag{14.1a}$$

or

$$S_{y_i} = f_1 + f_2 + \cdots + f_k \tag{14.1b}$$

and

$$\overline{C_{y_i}} = \bar{g}_1 \cdot \bar{g}_2 \cdots \bar{g}_m \tag{14.2a}$$

or

$$C_{y_i} = g_1 + g_2 + \cdots + g_m \tag{14.2b}$$

With these definitions, we see that the output of the lower of the cross-coupled gates is given by

$$\overline{C_{y_i} \cdot y_i}$$

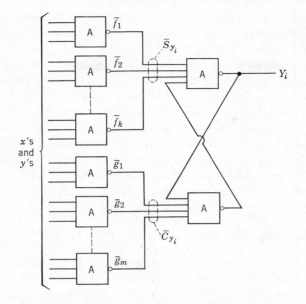

FIGURE 14.5. *Alternate level mode realization.*

so that

$$Y_i = \overline{\overline{S_{y_i}} \cdot \overline{\overline{C_{y_i}} \cdot y_i}}$$
$$= S_{y_i} + \overline{C_{y_i}} \cdot y_i \qquad (14.3)$$

Equation 14.3 we recognize as the next-state equation of the S-C flip-flop, so that the proposed modification consists of storing the state variables in unclocked flip-flops in a manner reminiscent of pulse-mode design except that there are no pulses involved.

Now consider a level-mode circuit with a given state variable equal to 1 and an input change such that this state variable is not supposed to change. If the 1 is produced by one gate before the input change and a different gate after, it is possible for the state variable to go to 0 momentarily during the change, a phenomenon known as a static combinational hazard. In the standard level-mode circuit, since this momentary change is fed back directly, it can cause an erroneous transition.

In the alternate design of Fig. 14.5, since the value of Y_i is stored in a flip-flop, there is no need for any signal controlled by the inputs to remain at 1 to hold Y_i; all that is required is that C_{y_i} remain at 0 whenever Y_i is to be 1. That is, no matter what combination hazards may exist in S_{y_i}, no transitions of Y_i from 1 to 0 can occur unless C_{y_i} goes to 1. With AND-OR design, we "create" 1's with each product function (the f's and g's in Fig. 14.5) so there is no possibility of a combinational hazard creating an erroneous 1 in the S_{y_i} or C_{y_i} functions. By a similar argument, static hazards in the 0's cannot occur since S_{y_i} will be 0 whenever Y_i is to be 0.

Not only does this design method eliminate static hazards, it may produce simpler designs. Because of the flip-flops, 1's are *required* in the S_{y_i} or C_{y_i} functions only for transitions where the state variables are to change. Where Y_i is to remain 0 or 1, C_{y_i} or S_{y_i}, respectively, are don't-cares. Thus, if the number of transitions involving changes of state variables is small, the alternate design may be simpler than the classic design of Chapter 13. Unfortunately, this alternate method does not eliminate problems due to races or essential hazards. Both of these phenomena are inherent in the flow table structure and independent of the mode of circuit realization.

14.5   Realizing the Standard J-K Flip-Flop

A glance at the flow table of Fig. 14.3b indicates that no problem will be caused by either of the multiple input changes which can occur. The natural state assignment of $z_2 = y_2$ and $z_1 = y_1$ will yield a transition table free from cycles and races. It remains to determine an economical realization free from static and dynamic hazards, in which the essential hazard associated with switching the clock on and then off is carefully controlled.

A transition table for the above state assignment is given in Fig. 14.6. Using this table as a K-map for both Y_2 and Y_1, we determine the following straightforward realization of the flip-flop.

$$Y_2 = y_1\bar{C} + y_2C + {}^*y_2y_1 \tag{14.4}$$

$$Y_1 = y_1\bar{C} + \bar{y}_2y_1 + y_1\bar{K} + \bar{y}_2CJ \tag{14.5}$$

The term y_2y_1 as marked by the asterisk is included to eliminate a static hazard. A circuit implementation of these expressions will require 8 gates,

y_2y_1 \ CJK	000	001	011	010	100	101	111	110
00	(00)	(00)	(00)	(00)	(00)	(00)	01	01
01	11	11	11	11	(01)	(01)	(01)	(01)
11	(11)	(11)	(11)	(11)	(11)	10	10	(11)
10	00	00	00	00	(10)	(10)	(10)	(10)

$$Y_2Y_1$$

FIGURE 14.6.   *Transition table for master-slave flip-flop.*

including one to cover the $y_1\bar{C}$ term in both expressions, three inverters, and a total of 23 gates inputs.

Although the above implementation can be made to function satisfactorily, let us now try the approach discussed in the preceding section. The transition table of Fig. 14.6 can be directly translated using Equation 14.3 to form the

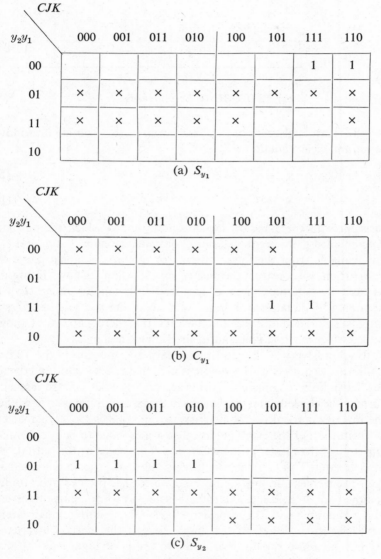

FIGURE 14.7.   *Excitation maps.*

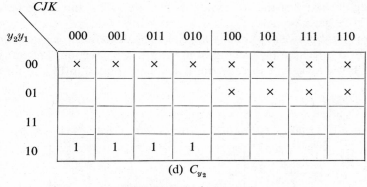

(d) C_{y_2}

FIGURE 14.7 (*continued*)

four S and C Karnaugh maps of Fig. 14.7. From these we obtain the following simple set and clear equations.

$$S_{y_1} = C \cdot J \cdot \bar{y}_2 \qquad C_{y_1} = C \cdot K \cdot \bar{y}_2 \qquad (14.6)$$

$$S_{y_2} = \bar{C} y_1 \qquad C_{y_2} = \bar{C} \cdot \bar{y}_1 \qquad (14.7)$$

Consistent with the standard flip-flop notation, in which the outputs of the cross-coupled gates are the Q and $\bar{Q}$ outputs, we note that $\bar{y}_2$ and $\bar{y}_1$ are directly available from the lower outputs of the cross-coupled pairs. With this observation, we construct the master slave flip-flop as shown in Fig. 14.8. In this case, the alternate method does lead to a simpler design than that corresponding to Equations 14.4 and 14.5, requiring two less inverters and 4 less inputs. Let us emphasize again, however, that there is no way of knowing in advance which method will lead to the simpler design.

This flip-flop is very similar to the standard J-K flip-flop in the TTL and DTL IC lines. As discussed, this device is race-free and all static and dynamic hazards have been eliminated. However, an essential hazard associated with turning the clock on and off remains in the circuit. Suppose, for example, that the inverter marked by the asterisk in Fig. 14.8 is slow. In that case, a 0-to-1 transition of the clock might cause a new value to be propagated to the output y_1 of the master element before a 0 appeared at the output of the indicated inverter. This could cause the new value to propagate immediately into the slave element, in direct contradiction to the specified operation.

As pointed out in the last chapter, there is no way to eliminate this essential hazard by modifying the logic design. Each manufacturer must carefully tailor the electronic design of the device so that the marked inverter will always respond more rapidly than the two levels of logic making up the master element. This can be accomplished conveniently since there is only

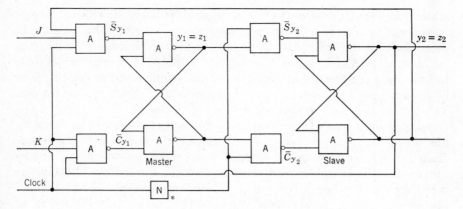

FIGURE 14.8. *Master-slave J-K flip-flop.*

one essential hazard which can be troublesome. The reader will observe that the effect of the above essential hazard can be overcome if the designated inverter is replaced by two inverters, one on the clock input line and one between the clock line and the clock input to the master element. Unfortunately this only transfers the problem to the essential hazard associated with the 1-to-0 clock transition.

Obtaining a realization of the edge-sensing master-slave device characterized by the 6-row flow table of Fig. 14.4 will be left as a homework problem for the reader. Because three state variables are required, the circuit will be more complicated than the one obtained in Fig. 14.8. Although it is a superior device, most manufacturers have not found justification for adding the edge-sensing master-slave to their product line.

14.6 *Analysis of Races and Hazards*

In Chapter 13, we attempted to design systems which included no races or hazards. We found that this was not always possible. It is necessary to design essential hazards deliberately into many important circuits in order to make them function. In addition, we observed that it is sometimes necessary to allow for simultaneous changes in the values of more than one input. Problems which could arise in this connection were termed function hazards. The proper functioning of circuits containing function hazards or essential hazards will in some way be dependent on gate delay.

After completing the design in Section 14.5, we noticed an essential hazard in the design of the J-K flip-flop. Although we might have picked it up earlier, the fact that we didn't is not atypical. Often some type of race or hazard will turn up in a design in a way not anticipated by the designer. In other cases, an essential hazard could knowingly be included, but the designer might be unsure how serious a problem it might be. It would be desirable to have some computational technique which could be used to identify cases of uncertain behavior and evaluate the likelihood that they will cause trouble. This technique should be applicable to the final design.

One approach would be an outright simulation of the circuit as designed. Simulation is an important tool and will be the topic of Section 16.5. Eichelberger[1] has developed a computational technique which will usually obtain the desired information more explicitly and with considerably less computer running time than a full circuit simulation. Eichelberger's method utilizes a 3-value logical calculus which we shall take time to introduce at this point.

In Chapter 4, we observed that a Boolean algebra must have a number of elements given by 2^n for some integer n. Thus, the three-value calculus cannot be described by a Boolean algebra. Post algebras[2] are possible for any number of elements. However, we shall sidestep any algebraic formulation of the three-value logic, since algebraic manipulation will not be required. For our purposes, we require only that the three-value computational system be closed under the operations $+$, $\cdot$, and NOT.

In addition to 0 and 1, the third element will be denoted $\times$. This element may be regarded as an indeterminate value or unknown value in Eichelberger's method. That is, when the output of a gate is $\times$, it is not known whether the value should be 1 or 0. We define the complement of $\times$ very simply as $\overline{\times} = \times$, thus violating the postulates of a Boolean algebra. We assure ourselves of a closed system by defining the mappings AND and OR as given in Fig. 14.9b and c. An intuitive feel for Fig. 14.9 can be achieved by thinking of $\times$ as an unknown value. Clearly, if A is unknown, $\bar{A}$ is unknown. $0 + \times$ and $\times + \times$ will always be determined by the unknown value $\times$. $1 + \times = 1$ regardless of the value of $\times$. The situation is analogous for the AND operation.

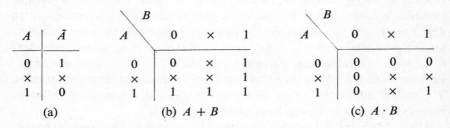

FIGURE 14.9.   *3-value logic defined.*

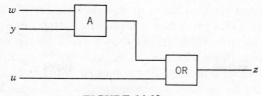

FIGURE 14.10

Example 14.2

The value of y in the simple circuit of Fig. 14.10 is unknown while w and u will always be known as 0 or as 1. For what combination of values of w and u will the output, z, depend on the unknown value of y?

Solution: If we let $y = \times$, we have $z = w \cdot \times + u$. From Fig. 14.9, we see that $z = \times$ if and only if $w = 1$ and $u = 0$. ∎

In any level mode circuit, including those designed according to the model described in Section 14.3, Boolean expressions can be obtained expressing all excitations in terms of the inputs and secondaries. Eichelberger's method utilizes these expressions to compute the circuit response to changing inputs as specified by the following two-step procedure. The technique, which utilizes the 3-value calculus, will identify the existence of any possibility of a sequential circuit reaching an erroneous state due to the existence of a critical race or hazard. The computation must be performed corresponding to each input change where problems are suspected.

Step 1. With the changing input variables assigned the values $\times$ and all other inputs and secondaries as originally specified, evaluate the excitation functions to determine if one or more have changed from 1 or 0 to $\times$. If so, change the corresponding secondary values to $\times$ and repeat the process until no additional changes in the excitation functions are determined.

Following Step 1, some of the secondaries may have stabilized at $\times$. The question arises as to whether these secondaries will assume the desired values after the input change is completed in Step 2.

Step 2. With the changing inputs assigned their new values (0 or 1) and all other variables equal to their values at the end of Step 1, evaluate all excitations. If one or more of the excitations changes from $\times$ to 0 or 1, change the corresponding secondaries accordingly; and recompute the excitations. Repeat the process until no further changes are noted.

The results after Steps 1 and 2 may be evaluated using Theorem 14.1.

THEOREM 14.1. *If an excitation is 1 (0) following steps 1 and 2 for a given input change and a given initial state, then this excitation must stabilize at 1 (0) for the same initial state and input change in the physical circuit regardless of gate delays.*

If an excitation remains equal to × following Step 2, then the circuit is delay dependent. This procedure will reveal as suspect any circuit containing any race or hazard. Conclusions drawn from this procedure are more conservative than those which might be provided by an actual 3-value simulation. If the procedure identifies a hazard which was unnoticed by the designer, it may be possible to revise the circuit design. Where the design cannot be modified, as with an essential hazard, for example, it may be necessary to resort to an actual simulation. The following is a very simple example of the procedure.

Example 14.3

From Fig. 14.11b we see that the elementary D flip-flop of Fig. 14.11a has been implemented without eliminating a static hazard. Carry out Steps 1 and 2 of the above procedure for the circuit of Fig. 14.11a, and repeat the process for a modified design with the static hazard eliminated. Consider only the input change where initially $D = C = y_1 = 1$, and C changes to 0. From Fig. 14.11b, the reader can observe that the static hazard may well affect the result of this input change.

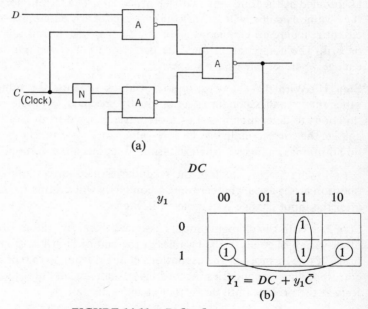

(a)

$$Y_1 = DC + y_1\bar{C}$$

(b)

FIGURE 14.11.   *D flip-flop.*

Solution: The Step 1 and 2 calculations for the 0 flip-flop described are given in Fig. 14.12a. The values listed for Y_1 are computed from the expression listed in the figure. When C assumes an × in Step 1, both DC and $y_1\bar{C}$ are ×, so Y_1 takes on this value. When C goes to zero, $Y_1 = y_1\bar{C} = \times$.

	D	C	Y_1		D	C	Y_2
Initial	1	1	1		1	1	1
Step 1	1	×	×		1	×	1
Step 2	1	0	×		1	0	1

$$Y_1 = DC + Y_1\bar{C} \qquad Y_1 + DC + Y_1\bar{C} + Y_1D$$

$$\text{(a)} \qquad\qquad\qquad\qquad \text{(b)}$$

FIGURE 14.12

The addition of the term y_1D as given in Fig. 14.21b eliminates the static hazard. Thus, Y_1 remains 1 as C goes to × and then to 0. In more complicated circuits, a critical race or hazard which would be identified by this procedure may not be so easily spotted on a Karnaugh map. ■

As a second example, let us investigate the effect of the essential hazard which was observed to exist in the master-slave J-K flip-flop of Fig. 14.8. In a complex circuit such as this, the application of Eichelberger's method is simplified by cutting the feedback loops at the outputs of the gates producing the excitation. Although it is not immediately evident in this circuit, there are only two state variables, y_1 and y_2, produced by the upper gates of the cross-coupled pairs. Cutting the loops at these points results in the combinational logic network of Fig. 14.13. The outputs are the excitations Y_2 and Y_1 and the corresponding secondary inputs are y_2 and y_1. For convenience, two reference points in the network are labeled f_A and f_B. The reader will notice that these

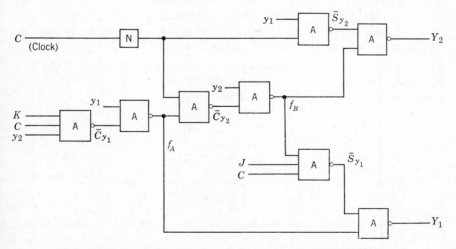

FIGURE 14.13. *J-K master-slave flip-flop with feedback loops cut.*

points correspond to the outputs of the lower gates of the two cross-coupled pairs. In order to apply Eichelberger's procedure, we must determine excitation expressions for Y_2 and Y_1. First we determine expressions for f_A and f_B by referring to the figure.

$$f_A = \overline{y_1 \cdot \overline{K \cdot C \cdot y_2}}$$
$$= \bar{y}_1 + K \cdot C \cdot y_2 \tag{14.8}$$
$$f_B = \bar{y}_2 + \bar{C} f_A$$
$$= \bar{y}_2 + \bar{C}\bar{y}_1 + \bar{C}KCy_2 = \bar{y}_2 + \bar{C}\bar{y}_1 \tag{14.9}$$

Now we can use these expressions in developing expressions for Y_2 and Y_1.

$$Y_2 = \bar{f}_B + \bar{C} y_1$$
$$= \overline{\bar{y}_2 + \bar{C} \cdot \bar{y}_1} + \bar{C} y_1$$
$$Y_2 = y_2 \cdot (C + y_1) + \bar{C} y_1$$
$$= y_2 C + y_1 \bar{C} + y_1 y_2 \tag{14.10}$$
$$Y_1 = f_A + JC f_B$$
$$= \bar{y}_1 + KC y_2 + JC\bar{y}_2$$
$$= y_1 \bar{K} + y_1 \bar{C} + y_1 \bar{y}_2 + JC\bar{y}_2 \tag{14.11}$$

The reader may note that Equations 14.10 and 14.11 are the same as the hazard-free expression for Y_2 and Y_1 (Equations 14.4 and 14.5) obtained by the conventional methods. This should not be taken as an indication that the two methods lead to identical designs; indeed, we have already noted that the circuits are different. What it does reflect is that both circuits have the same function and can, with appropriate analysis and manipulation, be represented by identical Boolean forms.

We are now ready to use these expressions in terms of the 3-value calculus to verify the essential hazard present when the clock C changes from 0 to 1. Initially we let $y_1 = y_2 = 0$, $J = 1$, and $K = C = 0$.

$C\ y_2\ y_1$	$y_1\bar{K} + y_1\bar{C} + y_1\bar{y}_2 + JC\bar{y}_2 = Y_1$		$y_2 C + y_1\bar{C} + y_1 y_2 = Y_2$	
0 0 0		0		0
× 0 0	$0 + 0 + 0 + \times$	×	$0 + 0 + 0$	0
× 0 ×	$\times + \times \cdot \bar{\times} + \times + \times$	×	$0 + \times \cdot \bar{\times} + 0$	×
× × ×	$\times + \times + \times + \times$	×	$\times + \times + \times$	×
1 × ×	$\times + 0 + \times + \times$	×	$\times + 0 + \times$	×

Step 1 corresponds to the row `× 0 0`; Step 2 corresponds to the row `1 × ×`.

FIGURE 14.14. *Input change calculation.*

The Step 1 and Step 2 computations are tabulated in Fig. 14.14 for C changing from 0 to × to 1. As an aid to following the computation the values of each product term of the Boolean expressions are given at each step. Notice in this case that three step-1 iterations are required before both y_1 and y_2 are stable at ×. Both of these values remain × after Step 2, thus verifying the essential hazard.

As we have said, the result of this test is necessarily conservative. The circuit may perform satisfactorily anyway. In this case, it would be necessary to resort to a simulation to obtain a better estimate of the reliability of the circuit. In other cases, a design change might be based on the results of this computation.

14.7 Summary

In this chapter, we have utilized the understanding of fundamental mode circuits gained in Chapter 13 to explore in detail the function of a master-slave flip-flop. It is unlikely that the reader will ever be called upon to design such a device. He is much more likely to have occasion to use them in situations where the validity of the clock mode assumption may be called into question. He should now be able to perceive when this might be the case and to take whatever action might be appropriate.

The discussion in this chapter has been restricted to the J-K flip-flop. Other types of master-slave flip-flops can be designed using the same procedure. These designs are included as homework problems.

PROBLEMS

14.1. Develop a level mode flow table for an edge-sensing D flip-flop. Assume that the D input will be stable at the time of the leading edge of the clock pulse.

14.2. Develop the level mode flow table for a master-slave T flip-flop. Recall that the T flip-flop is regarded as a pulse mode circuit. The only input is the pulse line, T.

14.3. Obtain a level mode flow table for an edge-sensing master-slave D flip-flop. That is, the value of D present at the time of the leading edge of a clock pulse will become the flip-flop output at the time of the trailing edge of the pulse.

14.4. Use the technique of Section 14.3 to obtain a hazard-free realization of the transition table given in Fig. 13.42.

14.5. Repeat problem 14.4 for the transition table of Fig. 13.54.

14.6. Obtain a realization of the flow table obtained in Problem 14.1 which is free from races and from static and dynamic hazards. Attempt to manipulate the form of the excitation equations to obtain the reailzation given in Problem 9.1.

14.7. Obtain a realization of the edge-sensing master-slave J-K flip-flop given in Fig. 14.4. Make sure the circuit is free from races and static and dynamic hazards. Are there essential hazards in this flow-table? If so, identify them in the state table and find the critical gates in the circuit realization. Discuss the probability of circuit malfunction due to each of these essential hazards.

14.8. Tabulate the functional values of the following Boolean expressions for all combinations of inputs using the three value logic of section 14.6.

$$\text{(a) } A \oplus B \qquad \text{(b) } \bar{A} + B$$

14.9. Use Eichelberger's method to identify the hazard in the circuit of Fig. 13.42. Recall that the hazard will have an effect as circuit values change from $y_2 y_1 x_2 x_1 = 1111$ to $y_2 y_1 x_2 x_1 = 1110$.

BIBLIOGRAPHY

1. Eichelberger, E. B. "Hazard Detection in Combinational and Sequential Switching Circuits," *IBM Journal of Research and Development*, 90, (March 1965).

2. Harrison, Michael. *Introduction to Switching and Automata Theory*, McGraw-Hill, New York, 1965.

Chapter

15 *Description of*
Large Sequential Circuits

15.1   Introduction

Digital computers and other large digital systems are examples of sequential circuits, but they clearly differ in size and complexity from the circuits discussed in the previous six chapters. The primary components, excluding memory, of most computers and peripheral systems are many registers of several flip-flops each. In such systems, the total number of states can thus reach into the thousands. Clearly, state diagrams and state tables are not practical mediums for description of such large sequential circuits. In this chapter, we shall develop a more suitable technique for describing sequential circuits composed primarily of large numbers of registers.

Most large sequential circuits are *clock mode*, and the development of this chapter will be restricted to clock mode. We begin by arbitrarily partitioning the clock mode model into two separate sequential circuits as shown in Fig. 15.1. The upper portion consists of data registers together with interconnecting combinational logic. The lower sequential circuit provides a sequence of control signals which cause the appropriate register transfers to take place with the arrival of each clock pulse. The numbers of control signal lines and control input lines are normally small in comparison to the numbers of data input and output lines and interconnecting data lines within block A.

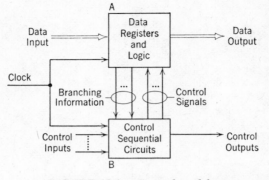

FIGURE 15.1.   *Control model.*

The purpose and advantage of the above model have yet to be made clear. This we shall accomplish in the next several sections. Consistent with our technique throughout the book, we shall explore the model in more detail through a series of examples.

15.2 Clock Input Control

In this section, we shall be concerned with clocking the flip-flops in block A of Fig. 15.1. Until now, we have inferred that a sequential circuit only operates in the clock mode if the clock or toggle inputs of all flip-flops on the circuit are tied *directly* to the system clock, as depicted in Fig. 15.2a. Under certain circumstances, particularly when D type flip-flops are used, this approach to clocking can be unduly restrictive. A simple illustration of a sequential circuit with direct clocking is given in Fig. 15.2b. If control signal

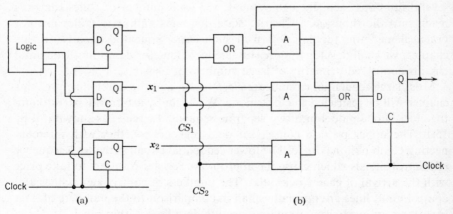

FIGURE 15.2.   *Direct clocking.*

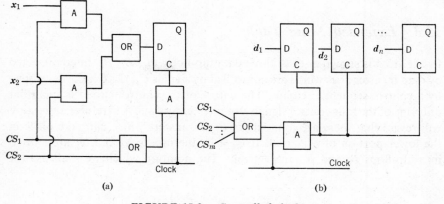

FIGURE 15.3.   *Controlled clocking.*

$CS_1 = 1$, x_1 is to be shifted into the flip-flop shown. If $CS_2 = 1$, x_2 is shifted in. Regardless of the control signals which might be present, it is always necessary that the value to be stored in the D flip-flop after a clock pulse be present at the flip-flop input before the clock pulse arrives. Therefore, if both control signals are 0, the current value of the flip-flop must be routed back to the D input. This is accomplished by the upper AND gate in Fig. 15.2b.

The network in Fig. 15.2b can be simplified by a departure from the direct clocking procedure. Suppose that a clock pulse is allowed to reach the toggle input of a flip-flop only if the contents of the flip-flop are to be modified. For the example under discussion, then, a change will occur only if $CS_1 \lor CS_2 = 1$. In Fig. 15.3a, this signal is used to gate the clock pulse into the flip-flop toggle. Thus, the circuits of Figs. 15.2b and 15.3a accomplish the same function. Although only one inverter is saved by the latter approach, the savings can be more dramatic in more complicated systems, particularly where several flip-flops are combined together to form a register. Often the same control signals are used to gate data into the different flip-flops of a register. In this case, the clock control gating can be common to all flip-flops of the register, as illustrated in Fig. 15.3b. (The gating of the data is not shown.) If the approach of Fig. 15.2 were used, this saving could not be achieved.

It should be emphasized that using control signals to gate clock pulses will not preclude designating a sequential circuit as clock mode. While certain flip-flops are controlled so that they can be triggered only during certain clock periods, no flip-flop can change state except on the arrival of a clock pulse. Thus, all state changes in the overall sequential circuit will coincide with clock pulses so that the sequential circuit may be designated clock mode. We shall find that allowing control of the clock line will simplify our discussion in succeeding sections.

15.3   Extended State Table

In Fig. 15.4 is shown a set of three data flip-flops, B_0, B_1, B_2, interconnected by data transfer lines which are controlled by signals C_1, C_2, C_3, C_4, developed by a control sequential circuit. The nature of the control circuit is such that only one of the four control signals can be 1 at any time. A careful examination will reveal that there are four sets of data transfers which can take place in the lower portion of Fig. 15.4. If $C_1 = 1$, the inputs X_0 and X_1 are shifted into flip-flops B_0 and B_1, respectively. The notation which we shall use for

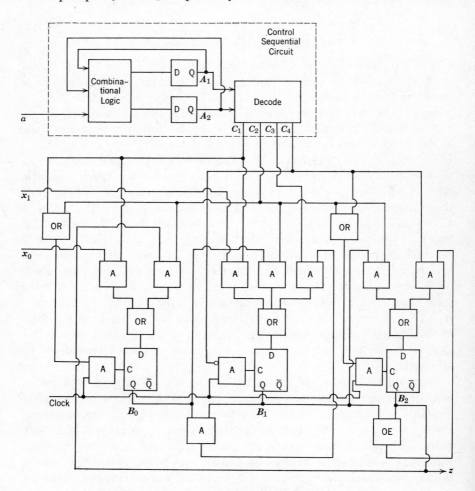

FIGURE 15.4.   *Example of control of data transfers.*

this transfer is given by expression 15.1. (The contents of B_2 are unchanged.)

$$1. \quad B_0 \leftarrow X_0; \qquad B_1 \leftarrow X_1 \tag{15.1}$$

If $C_2 = 1$, the information in the three flip-flops is shifted right with the contents of the rightmost flip-flop shifted around into B_0. For the present, we shall denote this transfer similarly as given in expression 15.2.

$$2. \quad B_1 \leftarrow B_0; \qquad B_2 \leftarrow B_1; \qquad B_0 \leftarrow B_2 \tag{15.2}$$

If $C_3 = 1$, the transfer is

$$3. \quad B_1 \leftarrow B_0 \wedge B_1 \tag{15.3}$$

and if $C_4 = 1$ the transfer is

$$4. \quad B_2 \leftarrow B_1 \oplus B_2 \tag{15.4}$$

Note the use of the symbol $\wedge$ for the AND operation. We shall find this symbol convenient as we expand our transfer notation.

The above describes the transfers that can take place, but the sequence of transfers and, therefore, the sequence of outputs, will depend on the sequence of control signals. Since there are four possible combinations of control signals, we may characterize the control sequential circuit as a Moore circuit with four states, which we may designate as C_1, C_2, C_3, C_4, each producing a 1 on the corresponding control output line. Let us further assume that the sequencing between these states is controlled by a sequence control signal, a, as shown in the state table of Fig. 15.5. This is consistent with Fig. 15.4, where a is shown as an input to this 4-state control sequential circuit. To this state table, we add a list of the transfers initiated by the control signals corresponding to each state.

Together with the output equation $z = B_2$, this extended state table provides a complete description of the desired sequential circuit behavior from which the complete circuit could be generated. There are three data flip-flops, and at least two control flip-flops will be required. Therefore, a conventional description in the manner of Chapter 10 would require a 32-state table, which would clearly be far less convenient as a design tool than the compact table of Fig. 15.5.

q^v	q^{v+1}		Transfer corresponding to q^v
	$a = 0$	$a = 1$	
C_1	C_3	C_2	$B_0 \leftarrow X_0; B_1 \leftarrow X_1$
C_2	C_1	C_1	$B_1 \leftarrow B_0; B_2 \leftarrow B_1; B_0 \leftarrow B_2$
C_3	C_4	C_4	$B_1 \leftarrow B_0 \wedge B_1$
C_4	C_2	C_1	$B_2 \leftarrow B_1 \oplus B_2$

FIGURE 15.5. *Extended state table.*

A conventional state table description of the entire circuit would be considerably bulkier. A 32-state table would be required. We shall leave it as a homework problem for the reader to obtain such a table.

15.4   A Program Description

Although Fig. 15.5 seems to be a convenient description of the corresponding circuit, a similar description for a circuit with a large number of control states and many control inputs could still be unwieldy. It is the purpose of this section to develop a still briefer sequential circuit description which will take the form of a computer program.

The list of transfers in Fig. 15.5 very much resembles the transfer or replacement statements in a computer program. Individual flip-flops correspond to scalar variables. New values of these variables are the expressions at the right of each transfer arrow. Transfers listed on the same line will be assumed to take place simultaneously. A vector notation to be introduced in the next section will serve to describe simultaneous transfers more compactly.

Missing from the list of transfer statements of Fig. 15.5 are the branch statements required to sequence through the transfer statements in a data dependent order. However, while the branching information is not found in the transfer list, it is included in Fig. 15.5 in the form of the state table. As each transfer is accomplished, the control circuit goes to its next state to enable the proper transfer at the next clock pulse. From states C_1 and C_4, a two-way branch (two possible next states) is executed as a function of input a. From state C_2, the next state is always C_1, much like a GO TO statement in FORTRAN.

It remains only to adopt an appropriate branching notation to permit the expansion of the transfer list into a complete program description of the sequential circuit. We shall use a variation of the branching notation of APL.* A branch step will be included following each transfer. The steps will be numbered to agree with the control state numbers. The transfer and subsequent branch will carry the same number. The branch will be omitted only if control is always to proceed to the next numbered step. An individual branch step will be denoted in the general form of Expression 15.5. The s_i's

$$\rightarrow ((f_1 \times s_1) + (f_2 \times s_2) + \cdots + (f_n \times s_n)) \tag{15.5}$$

are numbers of steps in the hardware program, and the f_i's are Boolean

---

* The instructor who notices the difference between the approach here and that in the authors' other book, *Digital Systems: Hardware Organization and Design*, may wish to read the relevant portion of the preface.

expressions which must satisfy Equations 15.6 and 15.7 for all combinations

$$f_i \wedge f_j = 0 \qquad\qquad \text{where } i \neq j \qquad (15.6)$$

$$f_1 \vee f_2 \vee f_3 \vee \cdots \vee f_n = 1 \qquad\qquad (15.7)$$

of variables which are not don't-cares. Thus, one and only one of the f_i will always be 1, so that the result of the arithmetical computation in expression 15.5 will be the number of some step in the hardware program. In particular, if $f_i = 1$, expression 15.5 will reduce to

$$\rightarrow (s_i)$$

Following the conventions described above, Fig. 15.5 may be translated into the following *control sequence*.

1. $B_0 \leftarrow X_0$; $B_1 \leftarrow X_1$
   $\rightarrow ((\bar{a} \times 3) + (a \times 2))$
2. $B_1 \leftarrow B_0$; $B_2 \leftarrow B_1$; $B_0 \leftarrow B_2$
   $\rightarrow (1)$
3. $B_1 \leftarrow B_0 \wedge B_1$
4. $B_2 \leftarrow B_1 \oplus B_2$
   $\rightarrow ((\bar{a} \times 2) + (a \times 1))$

Throughout the rest of this chapter, we shall use the term *control sequence* to identify hardware descriptions of this form. This is a complete description of the circuit of Fig. 15.4 containing all information found in Fig. 15.5. The efficiency of this new method of circuit description is not particulary apparent for the above example. Notice that only one branch statement (Step 3) was omitted because the next step (Step 4) is always taken in sequence. This situation will be much more common in larger systems. Similarly, where there are a large number of inputs to the control unit, the f_i's at any given step may be functions of only a few of these inputs.

15.5 Synthesis

A principal advantage of a program description of hardware must necessarily be its use as a synthesis tool. For large systems, it is the only approach other than trial and error manipulation of the logic block diagram. Language is used as a design tool for large systems such as computers and peripheral equipments.[1] In this chapter, we shall show that it can be used effectively over a range of less complex systems which are still too bulky to describe in state table form. We shall even find it useful as a means of quickly obtaining unambiguous descriptions of certain smaller sequential circuits for which we obtained state tables in Chapter 10.

Before we move to discard all of our state-table-related tools, let us remind ourselves that our new medium is a *language and only a language*. It will have no apparent algebraic structure in terms of which state minimization can be performed. Nor does the language itself provide any constraints which will guide the designer to a minimal circuit. With practice, the designer will readily arrive at an optimal list of program steps for certain types of examples. For other examples, particularly where certain flip-flops can serve a variety of functions, it would be unrealistic to expect to obtain a minimal realization without resorting to state table minimization. Even if the minimal number of flip-flops are used, the fact that don't-care conditions are not reflected in the language may cause a realization to be more complicated than necessary. With the recent trend toward very low-cost integrated circuits (see Chapter 16), minimization would seem to become progressively less important. This observation should not be over emphasized, however, or the result can be sloppy design requiring many more gates than necessary. The notion of state table minimization should always be kept waiting in the wings for occasional use at least on parts of sequential circuits.

As a first example, let us obtain a design of the 3-bit serial code translater of Problem 10.22. This problem is presented below as Example 15.1.

Example 15.1

Obtain the control sequence of a serial translating circuit which performs the translations tabulated in Fig. 15.6. As indicated, there is one data input line, x, and one output line, z, in the circuit. The inputs occur in characters of three consecutive bits. One character is followed immediately by another. A reset input, $r = 1$, will cause the translator to begin operation synchronized with the input. A translated output character will be delayed by only two clock periods with respect to an input character.

x^v	x^{v-1}	x^{v-2}	z^{v+2}	z^{v+1}	z^v
0	0	0	0	0	0
0	0	1	0	1	1
0	1	0	1	0	0
0	1	1	1	0	1
1	0	0	1	1	0
1	0	1	0	0	1
1	1	0	0	1	0
1	1	1	1	1	1

FIGURE 15.6. *Code translation.*

Solution: The output translation cannot be performed until all three input bits have been received, at time t_v. Therefore, as the first two bits arrive, we shall shift them into a 2-bit shift register B consisting of bits B_0 and B_1. At time t_v, x^{v-2} will be in B_1 and x^{v-1} will be in B_0. When x_v arrives, all three output bits will be computed simultaneously. The immediate output will be z^v, and outputs z^{v+1} and z^{v+2} will be stored in B_1 and B_0, respectively. At time t^{v+1}, z^{v+1} can be read from B_1, z^{v+2} will be shifted from B_0 to B_1, and the first input bit of the next character will be shifted into B_0. At t^{v+2}, the output z^{v+2} will be read from B_1, the previous input bit will be shifted into B_1, and the next input bit will be shifted into B_0, setting the state for the translation of the next character at the following clock time. The operation of this circuit can thus be described in terms of a repetitive three-step sequence, the first two steps being shifts, the third a translation and parallel register entry, as shown below.

1. $B_0, B_1 \leftarrow x, B_0; \qquad z = B_1$

    $\rightarrow (r \times 1 + \bar{r} \times 2)$

2. $B_0, B_1 \leftarrow x, B_0; \qquad z = B_1$

    $\rightarrow (r \times 1 + \bar{r} \times 3)$

3. $B_0, B_1 \leftarrow f_2(x, B_0, B_1), f_1(x, B_0, B_1); \qquad z = f_0(x, B_0, B_1)$

    $\rightarrow (1)$

The three output functions of Step 3 may be obtained by converting the table of Fig. 15.6 into three K-maps, one for each output, as shown in Fig. 15.7. In correlating these maps to the table, note that x^{v-2} is stored in B_1, x^{v-1} in B_0.

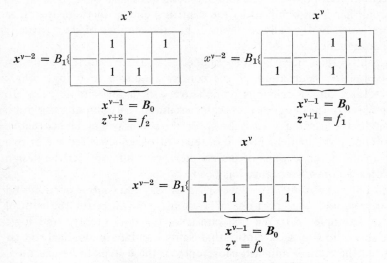

FIGURE 15.7. *Translation functions.*

The resultant functions are tabulated in Equations 15.8, 15.9, and 15.10. Note, from Step 3 of the sequence, that z^{v+1} is initially stored in B_1 and z^{v+2} in B_0.

$$f_0 = B_1 \tag{15.8}$$

$$f_1 = (x^v \wedge \bar{B}_1) \vee (x^v \wedge B_0) \vee (\bar{x}^v \wedge \bar{B}_0 \wedge B_1) \tag{15.9}$$

$$f_2 = (\bar{x}^v \wedge B_0) \vee (B_0 \wedge B_1) \vee (x^v \wedge \bar{B}_0 \wedge \bar{B}_1) \tag{15.10}$$

The reset mechanism has been incorporated into the three-step sequence. Whenever the input $r = 1$, control is returned to Step 1 to begin the input of a new character. The reader will notice that the single putput, z, is specified separately on each line of the control sequence. This is necessary because the output is not a fixed function of some data register or input line, as was the case for the circuit of Fig. 15.4. Rather, the function is different for different steps in the sequence, i.e., it is also a function of the control signals. It is common in large systems for the output to be taken directly from a data register to be a fixed function of several data registers. In such cases, the outputs need be listed only once in the control sequence. ■

The above example makes evident a very important difference between the control sequence method of design and the state diagram method presented in Chapter 10. Setting up the state diagram requires only an understanding of the required sequencing of inputs and outputs; no preliminary guesses as to the numbers of state variables or to the internal structure are required. By contrast, the control sequence approach essentially starts with a proposed basic structure. This method thus requires considerable intuitive understanding of possible circuit structures on the part of the designer, and it is very likely that the resultant design will not be minimal in the formal sense of Chapter 10. Once a circuit has been obtained by the control sequence method, it can, at least theoretically, be formally minimized, but the number of states usually makes this impractical.

Once a control sequence has been obtained, the creative portion of design effort is complete. Translation of the control sequence to a logic block diagram requires only a religious compliance with a few simple rules. In fact, this portion of the design process can be reduced to a computer program. At this point in the learning process, however, the reader will find it instructive to follow in detail the implementation of the control sequence for one or two examples. In the latter sections of the chapter, we shall consider the design complete once a control sequence has been obtained.

We shall find it convenient to develop separately the *control* portion and the *data storage* and *logic* portions of a circuit realization for the control sequence of Example 15.1. We first construct the data storage and logic portion as shown in Fig. 15.8. Notice that Steps 1 and 2 are identical so that one line from the control unit will cause both of these steps to be executed. A second control line will cause the execution of Step 3. These lines are clearly

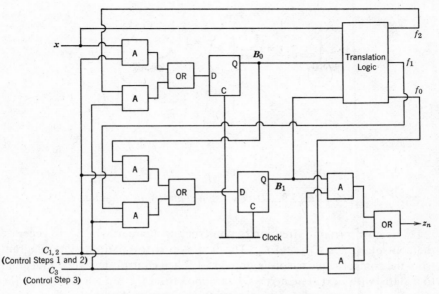

FIGURE 15.8. *Data unit for code translator.*

marked as inputs at the left of Fig. 15.8. Notice in the control sequence that the contents of both flip-flops are changed at every step. Thus, the clock line may be connected to the T input of flip-flops B_0 and B_1 without any clock control gating.

The circuit output z will usually be the output of an OR gate if it is a function of the control signals. At any control step, that is, at any *clock pulse* time, one of the inputs to this OR gate will carry a data value while the others inputs will be 0. Each control step signal and the output value corresponding to that step serve as the input to one of the AND gates of the two level AND-OR network generating z. The control signal corresponding to the step being executed is 1 so the output value corresponding to this step is routed through the network to the output z. A similar two-level AND-OR network is placed at the input of each flip-flop so that the new flip-flop values corresponding to the current step are routed to the appropriate D inputs. An AND gate will appear in the input network to a flip-flop for each control step, which actually causes that flip-flop to be updated. In this case, there are only two first-level AND gates, one corresponding to Steps 1 and 2 and the other corresponding to Step 3. Since these two signals cover all three control steps, we have a check on our earlier observation that clock control was not required. To avoid unnecessarily complicating Fig. 15.8, the realization of the Step-3 functions, f_0, f_1, and f_2, are represented by a single block.

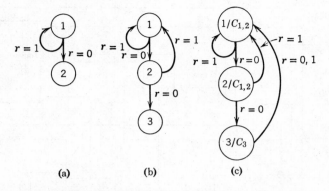

(a)   (b)   (c)

FIGURE 15.9.   *Developing a control state diagram.*

Let us now turn our attention to constructing a circuit which will generate the control signals $C_{1,2}$ and C_3. The first step is to obtain a state diagram from the control sequence of Example 15.1. Each control step will correspond to a state, with next state arrows determined from the branch portion of the step. We begin by representing control Step 1 as state 1 with the appropriate arrows to next states, as shown in Fig. 15.9a. Adding next state arrows from state 2, as specified by Step 2 of the control sequence, yields Fig. 15.9b. Adding the next state arrow for state 3 results in the completed state diagram of Fig. 15.9c. The control circuit is a Moore circuit, providing a level output during the duration of each control step, to gate the appropriate signals to the flip-flops at clock time. On the state diagram, the control signals corresponding to each step are shown as outputs for each state, as shown in Fig. 15.9c. Note that only one control signal can be 1 at a time.

Once a state diagram has been obtained, there are two possible ways to obtain circuit for the corresponding control unit. One is to minimize the corresponding state table and proceed to a circuit design by the formal techniques of Chapter 12.

Example 15.2

Determine a minimal realization of the control sequential circuit of Fig. 15.9c.

Solution: The reader can easily verify that Fig. 15.9c is a minimal state sequential circuit. The state assignment depicted in 15.11a was obtained using Rules I and II of Chapter 10. From this, we can easily obtain the D flip-flop input equations given by Equation 15.11 and the output equations given by Equation 15.12.

$$Y_1^{\nu+1} = \bar{r} \cdot \bar{y}_2 \qquad Y_2^{\nu+1} = \bar{r} \cdot \bar{y}_2 \cdot y_1 \qquad (15.11)$$

$$C_{1,2} = \bar{y}_2 \qquad\qquad C_3 = y_2 \qquad\qquad (15.12)$$

An implementation of these expressions is given in Fig. 15.10.   ∎

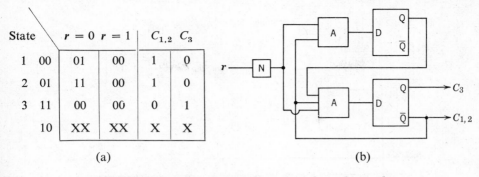

State	$r = 0$	$r = 1$	$C_{1,2}$	C_3
1 00	01	00	1	0
2 01	11	00	1	0
3 11	00	00	0	1
10	XX	XX	X	X

(a)   (b)

FIGURE 15.10.   *Minimal control sequencer for code translator.*

The data logic and storage unit of Fig. 15.8, together with the minimal control sequencer of Fig. 15.10, is probably as efficient a realization as can be found for the code translator. A minimal state table, determined by the formal techniques of Chapter 10, has 12 states, which agrees with the

$$(3 \text{ control states}) \times (4 \text{ data states})$$

used here. The fact that the number of states agrees does not mean that the complete circuits obtained by the two methods would be identical, but they would be of the same order of complexity.

Although the above technique will always produce the minimal control circuit, under some circumstances a preferable realization can be obtained through a "brute force" technique of assigning a separate flip-flop to each state and using a branching network to route control to the proper next step. A branching network corresponding to

$$\rightarrow (r \times 1 + \bar{r} \times 2) \qquad (15.13)$$

which represents the branch portion of Step 1 in the control sequence of Example 15.1, is given in Fig. 15.11a. Prior to the clock pulse which executes Step 1, the output of the flip-flop is 1. This signal serves as one of the inputs to both AND gates in the branching network. The other input of each is one of the branch functions of Expression 15.13. In general, there will be one gate for each branch function in a branch expression. Since one and only one of the branch functions can be 1, only the corresponding output of the branch network can be 1. In Fig. 15.11a, the ouput labelled Step 1 will be 1 if $r = 1$ and the flip-flop output $Q = 1$. If $r = 0$ and $Q = 1$, the output labelled Step 2 will be 1. This latter signal will be routed to the D input of the Step-2 flip-flop. If $r = 0$, the clock pulse which executes Step 1 will

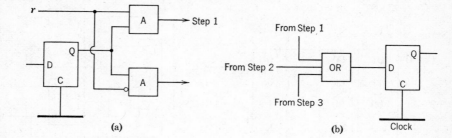

(a)

(b)

Clock

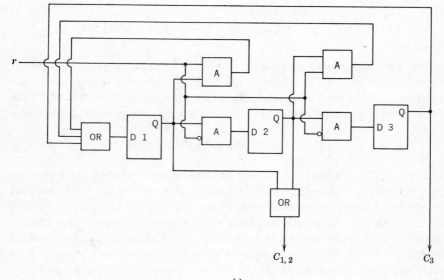

(c)

FIGURE 15.11.   *Alternate control sequencer for code translator.*

propagate the signal into the Step-2 flip-flop, thus passing control to Step 2. Where the output of more than one branch network can be routed to the input of a control flip-flop, OR gates must be provided at the input of control flip-flops, as illustrated in Fig. 15.11b. If any of the illustrated branch signals from Steps 1, 2, or 3 are 1 when a clock pulse arrives, the Step-1 flip-flop is set to 1.

The flip-flops for Steps 1, 2, and 3 are all shown in Fig. 15.11c, together with the interconnecting branch networks. Also shown are the output signals C_3 and $C_{1,2}$. The latter signal is formed easily by connecting the outputs of flip-flops 1 and 2 through an OR gate. We shall refer to the realization of the control sequential circuit for a control sequence as a *control sequencer* throughout the remainder of the chapter. The reader should note that with this second method the control sequencer may be obtained as a direct translation of the control sequence, thus avoiding the intermediate step of finding a state diagram. The state diagram is needed only if we try to obtain a minimal state and minimal flip-flop realization.

Since the minimal control sequencer obtained by the formal minimization techniques was clearly more economical than that obtained by direct translation, the reader may wonder why the latter technique might be preferable. In some cases, even with the division of the circuit into data and control sections, the number of states in the control sequence may be so large as to make formal minimization impractical. Further, even though minimization produces a circuit with the minimum number of flip-flops, this does not necessarily mean that the overall circuit is minimal. Although not so for the above example, in many cases minimizing the number of states may require so much additional combinational logic that the overall cost is increased. Finally, even though it may be more expensive initially, the one-flip-flop-per-step approach may be chosen to ease the problems of testing, fault diagnosis and maintenance. A sequencer configured as a physical implementation of the control flow chart can be most helpful in understanding the sequence.

Before considering another example, it is desirable to take a closer look at the problem of specifying outputs. In design by the classical state-table method, the entire circuit is specified together, so a uniform system of output specification is practical. In the control sequence method, we divide the circuit into two sections, the *data storage and logic* section and the *control sequencer* section. The outputs of the complete circuit may be dependent on either or both sections, and the appropriate output notation may vary accordingly.

In the first example considered, the circuit of Fig. 15.4, the output was a function only of the data storage and logic section, namely $z = B_2$. The *value* of the output will, of course, change as B_2 changes, but the *identity* of the line z is fixed and need be specified only once. In such cases, where the output is solely a function of the data storage and logic section, we shall specify the outputs by a special statement preceding the first statement of the control sequence. In this case, the statement would be:

$$\text{OUTPUT: } z = B_2$$

In other cases, there may be several outputs, or the outputs may be functions of several flip-flops in the output storage section. Thus, other possible output specification statements might be as shown below.

$$\text{OUTPUT: } z_1 = B_2, z_2 = A_0$$

$$\text{OUTPUT: } z = A_0 \wedge B_2$$

In Example 15.1, the output was a function of both the output and the control sections, being derived from different data flip-flops during different steps in the sequence. Here, and this will be the standard procedure for such cases, we list an output specification with each step of the sequence. The actual *value* of the output will depend on the contents of the flip-flops during that step; the output specification step serves only to *identify* the function from which the output is derived.

Although we have not yet seen an example, there are cases where an output is derived solely from the control sequencer section; in such cases, an actual value for the output is specified. For example, in Example 15.5, we shall see a statement:

$$3. \quad \textbf{\textit{read}} = 1$$

This signifies that the output line read is set to logical 1 during step 3; at times when it is not specified, it is assume to be at logical 0.

This last example raises an important point, the distinction between statements of the form

$$3. \quad \textbf{\textit{read}} \leftarrow 1$$

and statements of the form

$$3. \quad \textbf{\textit{read}} = 1$$

The first statement is a *transfer* statement, specifying that the value 1 is transferred into the flip-flop **read**, which will remain at that value until reset. By contrast, the second statement is a *connection* statement, specifying that a line labelled **read** is connected to logical 1 during step 3. Similarly, the statement

$$\text{OUTPUT:} \quad z = B_2$$

does not specify the transfer of any information to z, but rather specifies that a line labelled z is connected to the output of flip-flop B_2. This distinction, between the transfer of information and the connection of line, is critical and should be kept carefully in mind by the reader.

Let us now apply the control sequence method to the parity checker of Example 10.10 to provide a direct comparison of an implementation of a control sequence with a classical solution to the same problem.

Example 15.3

Determine a control sequence for a sequential circuit which will check parity serially over successive sequences of 4 bits on input line x. A reset input line r is provided which, if $r = 1$, will cause the circuit to regard the next input on line x as the first bit of a 4-bit sequence. The only output, z, is to be 0 except upon the fourth bit of a 4-bit sequence over which parity has been even.

Solution: Only one data flip-flop will be required. It will be labelled p and assigned the function of keeping track of parity. The first three steps of the control sequence are identical, storing the cumulative parity value in p at each step unless $r = 1$.

$$
\begin{aligned}
&\textbf{1.}\quad p \leftarrow (p \oplus x) \wedge \bar{r}; \quad z = 0\\
&\qquad \rightarrow (r \times 1 + \bar{r} \times 2)\\
&\textbf{2.}\quad p \leftarrow (p \oplus x) \wedge \bar{r}: \quad z = 0\\
&\qquad \rightarrow (r \times 1 + \bar{r} \times 3)\\
&\textbf{3.}\quad p \leftarrow (p \oplus x) \wedge \bar{r}; \quad z = 0\\
&\qquad \rightarrow (r \times 1 + \bar{r} \times 4)\\
&\textbf{4.}\quad p \leftarrow 0; \quad\quad z = \overline{p \oplus x}\\
&\qquad \rightarrow (1)
\end{aligned}
$$

At Step 4, p is always initialized to 0, and the output is 1 if and only if parity over the four bits has been even.

Since the output is a function of both the data storage and the control sequence, it is specified at each step. A situation such as encountered here, where an output is normally 0 and takes on another value only occasionally, is fairly common. To simplify notation, we will adopt the convention that such an output need be specified only for steps where it may take on a value other than zero; for steps where not specified, it will be assumed to be 0. If he wishes, however, the designer may specify the 0-outputs in the interests of clarity.

The reader can easily supply the data logic and control sequencer realizations for the above sequence. A minimal control sequencer will be a modulo-4 counter with a reset line added. In fact, the overall realization will be the same as can be obtained by adding a reset line to Fig. 10.48. This latter realization was determined in Section 10.10 by anticipating a natural partition of memory elements. ■

The above example suggests that there may be a relation between a partition of a sequential circuit and its control sequence. It can be shown that any control sequence for a sequential circuit which does not contain data branches defines a closed partition of the states of the circuit with an equivalence class corresponding to each step in the sequence. It will be left as a problem for the reader to match the equivalence classes of Fig. 10.49 to the steps listed in Example 15.3. As was emphasized in Chapter 10, a closed partition of the states will always permit the grouping of a subset of the circuit memory elements so that the input networks to these elements are not functions of values stored outside the set of flip-flops. In the present context, the flip-flops forming this subset constitute the control sequencer.

There are many possible ways to lay out the data and logic hardware to satisfy a design requirement prior to developing a control sequence. Not all such layouts will contain a minimal number of flip-flops. An optimal hardware realization of the complete sequential circuit could correspond to more than one control sequence. Presumably one should exist corresponding to every distinct closed partition of the states. In particular, there must be sequences corresponding to the two trivial partitions in which (1) all states are in the same equivalence class or (2) every state is in a class by itself. In the latter case, the control sequencer is the complete sequential circuit. In the former, there is only one state (and therefore no memory elements) in the control sequencer.

A one-step control sequence would be a most unwieldy way to describe a large system. For the relatively simple 4-bit parity checker, the one-step sequence is worth examining. Expression 15.14 is a one-step control sequence defining a sequential circuit which satisfies the problem specification of the 4-bit parity checker equally as well as the four-step sequence obtained in Example 15.3.

$$p \leftarrow (p \oplus x) \wedge \bar{r} \wedge \overline{N_2 \wedge N_1} \qquad N_2 \leftarrow (N_1 \oplus N_2) \wedge \bar{r} \qquad N_1 \leftarrow N_1 \wedge \bar{r}$$

$$z = \overline{(p \oplus x)} \wedge N_1 \wedge N_2 \tag{15.14}$$

We know that a minimal realization includes 8 states. Therefore, two more flip-flops must be added to the data storage and logic block, since there is no memory in the control block. These two flip-flops N_1 and N_2 count continuously as a modulo-4 counter unless a reset signal appears. Notice that the expressions for z and the next state of p are now functions of N_1 and N_2. Thus, z is 0 except when $N_1 = N_2 = 1$, and parity is accumulated in p except when $N_1 = N_2 = 1$. Since the minimal control unit for the 4-bit sequence is effectively a modulo-4 counter, the realization of Expression 15.14 will be very much the same as the two realizations which we have already discussed.

As we consider control sequences for more complicated systems in the next two sections, we shall not find it practical to compare approaches in the detail we have just done for the 4-bit parity checker. The designer will be left on his own to choose an approach which will lead to an efficient hardware realization. Hopefully the discussion of this section will provide the reader with greater confidence in the building blocks with which he must work.

15.6 Vector Operations

We have already presented sufficient notation to permit writing a control sequence for any sequential circuit. This notation is convenient where it is necessary to specify the next state expression for every memory element separately. In larger systems, it is more convenient to treat sets of flip-flops as registers. As we shall see, it is usually not necessary to write separate next state expressions for individual flip-flops within a register. To take advantage of this fact and greatly simplify control sequences for large systems, we will introduce some notation for logical operations on registers and transfers into registers. To remain consistent with *Digital Systems*,[1] we shall use portions of APL (A Programming Language), a language which was first developed by Iverson.[5]

In the previous section, we denoted scalar variables or flip-flops in two ways, by boldface italic lower-case letters and by boldface italic capital letters with a subscript. We now define a register as a vector variable, i.e., a one-dimensional array of flip-flops, to be designated by a boldface italic capital letter (or short string of boldface italic capital letters.) Just as is the case in a software program, it is necessary somehow to specify the number of elements or flip-flops in a register. As in FORTRAN, this will be done by a DIMENSION statement to be found at the beginning of the control sequence of any sequential circuit containing a register. For example, the statement

$$\text{DIMENSION } A(3), \, B(8)$$

at the beginning of a control sequence indicates a sequential circuit consisting of at least 11 flip-flops, 3 in A and 8 in B, in addition to any designated by lower-case scalar variables. In most of the examples in this chapter, the number of flip-flops in each register will be set forth in the discussion so that the dimension statements are omitted for brevity.

We have used 1's and 0's to represent logical constants throughout the book. Constant vectors whose elements are 1's and 0's will also have a place in the language we are developing. Occasionally it will be convenient to transfer a constant into a register as illustrated for the 4-bit register A in expression 15.15. Constant vectors will also be used artificially in

$$A \leftarrow (1, 0, 0, 1) \tag{15.15}$$

another connection to be discussed shortly. A particular constant vector which we shall find it convenient to use often is $\epsilon(n)$, which is a vector of n 1's.

Sometimes we shall want to pick out individual flip-flops within a register and treat them independently. This can be accomplished easily by numbering the flip-flops in the register from *left to right*, starting with zero. We can identify a particular flip-flop in a register by affixing its number as a subscript of the symbol indicating the register. Thus, the leftmost flip-flop would be A_0, the rightmost bit A_3, as shown in Fig. 15.12.

FIGURE 15.12.   *Numbering of flip-flops in a register.*

We may combine registers into a single vector which may be treated for some operations as a single large register by writing the individual register symbols consecutively and separating them by commas. This process will be called *catenation*. If, for example, A is a 4-bit register and B is a 3-bit register, then C as given by Expression 15.16 is an 8-bit register. A few of the individual element equivalencies are $C_4 = p$, $C_5 = B_0$, and $C_7 = B_2$.

$$C = A, p, B \tag{15.16}$$

As suggested above, it is always possible to form a vector from a subset of the flip-flops of a register by subscripting and catenation. For example, if we wished to transfer the contents of the odd numbered flip-flops of the 6-bit register D into the 3-bit register B, we could designate this operation as given in Expression 15.17.

$$B \leftarrow D_1, D_3, D_5 \tag{15.17}$$

In some cases, it will be convenient to abbreviate this operation somewhat so we define the notation of Expression 15.18 to mean *compression*. The vector R on the left of the slash must be a

$$B \leftarrow R/Y \tag{15.18}$$

constant vector of 1's and 0's. The vector Y on the right is a register with the same number of elements as R. By definition, R/Y forms a register consisting of those flip-flops of Y for which the corresponding bits of R are 1's. As an example, we may rewrite Expression 15.17 as given by 15.19. Two particular forms of compression which will be used often involve selecting

$$B \leftarrow (0, 1, 0, 1, 0, 1)/D \tag{15.19}$$

some specified number of the first (leftmost) or last (rightmost) flip-flops from a register. To do this, we use the prefix and suffix vectors α and ω, respectively. We define α^j/Y to be a vector composed of the first j flip-flops of Y, and we define ω^k/Y to be a vector composed of the last k flip-flops in Y. Thus,

$$\alpha^3/A = A_0, A_1, A_2$$

and

$$\omega^2/A = A_4, A_5$$

where A is a 6-bit register.

15.7   Logical Functions of Vectors

If the vector notation just introduced is really to be of value in simplifying our program description of hardware, we must be able to express compactly

Boolean operations involving registers. One convenient notation is the *reduction operation* over AND and OR, denoted $\wedge/A$ and $\vee/A$, respectively. The operation $\wedge/A$ is defined to mean the ANDing together of all the elements of A. That is,

$$\wedge/A = A_0 \wedge A_1 \wedge A_2 \wedge \cdots \wedge A_{pA-1} \tag{15.21}$$

Similarly,

$$\vee/A = A_0 \vee A_1 \vee A_2 \vee \cdots \vee A_{pA-1} \tag{15.22}$$

where pA is the number of elements in A.

The three logical operators may also be applied to vectors on an element by element basis. For example,

$$\bar{A} = \bar{A}_0, \bar{A}_1, \ldots, \bar{A}_{pA} \tag{15.23}$$

and

$$A \wedge B = A_0 \wedge B_0, A_1 \wedge B_1, \ldots, A_{pA-1} \wedge B_{pB-1} \tag{15.24}$$

Clearly the latter operation is defined only if $pA = pB$. However, we shall allow an $\wedge$ or an $\vee$ operation between a scalar and a vector and define it to mean an operation between the scalar and each element of the vector. For example,

$$a \wedge B = a \wedge B_0, a \wedge B_1, \ldots, a \wedge B_{pB-1} \tag{15.25}$$

Occasionally it will be necessary to rearrange the elements within a vector. The operation $n \uparrow A$ signifies rotating the elements of A n positions to the left, and $n \downarrow A$ signifies rotating A n elements to the right. In the case of $n \uparrow A$, the rightmost n bits of A are replaced by the leftmost n bits in order, and similarly with $n \downarrow A$. These two operations will be referred to as the *rotate* operations. If $n = 1$, n may be omitted, with the 1-bit rotate operations abbreviated to read $\uparrow A$ and $\downarrow A$.

The *shift left*, $n \updownarrow A$, and *shift right*, $n \updownarrow A$, operations are closely related to the rotate operations. In the case of $n \updownarrow A$, however, the n elements vacated at the right are replaced by 0's. In $n \updownarrow A$, the leftmost n bits are replaced by 0's.

Example 15.4

Compute the vectors to be placed in the register on the left of each of the following, transfer statements.

$$A \leftarrow 2 \downarrow R$$

$$B \leftarrow \updownarrow R$$

$$C \leftarrow (\wedge/R) \vee S$$

$$D \leftarrow \overline{R \wedge S}$$

Let $\mathbf{R} = (1, 0, 0, 1, 0, 1, 1)$ and $S = (1, 1, 0, 0, 0, 0, 1)$

Solution: (a) $2 \downarrow R = 2 \downarrow (1, 0, 0, 1, 0, 1, 1)$

$\qquad\qquad\qquad = (1, 1, 1, 0, 0, 1, 0)$

(b) $\overset{\uparrow}{_0} R = 1 \overset{\uparrow}{_0} (1, 0, 0, 1, 0, 1, 1) = (0, 0, 1, 0, 1, 1, 0)$

(c) $(\wedge/R) \vee S = (\wedge/1, 0, 0, 1, 0, 1, 1) \vee S$

$\qquad\qquad\qquad = (1 \wedge 0 \wedge 0 \wedge 1 \wedge 0 \wedge 1 \wedge 1) \vee S$

$\qquad\qquad\qquad = 0 \vee S = S$

(d) $\overline{R \wedge S} = \overline{(1, 0, 0, 0, 0, 0, 1)}$

$\qquad\qquad\quad = (0, 1, 1, 1, 1, 1, 0)$ ∎

With the addition of one more notational device, we shall have a language sufficiently powerful to permit the efficient expression of control sequences for most sequential circuits. Most of the more complex combinational logic expressions which are needed in the description of a digital system are both familiar and frequently used. For this reason, it is convenient to provide specific abbreviations for these functions so that they need not be written out in detail every time they appear in a control sequence.

As an example, consider the decoding function which was discussed in Chapter 8. Suppose one step of a control sequence calls for using the contents of a 3-bit register B to select a particular bit from an 8-bit register A. That is, if $B = 0, 0, 0$, bit A_0 is desired, and so on. Assuming the selected bit is to be placed in a flip-flop, a detailed expression for the transfer statement is given by 15.26.

$$r \leftarrow ((\bar{B}_0 \wedge \bar{B}_1 \wedge \bar{B}_2) \wedge A_0) \vee ((\bar{B}_0 \wedge \bar{B}_1 \wedge B_2) \wedge A_1) \vee ((\bar{B}_0 \wedge B_1 \wedge \bar{B}_2) \wedge A_2)$$

$$\vee ((\bar{B}_0 \wedge B_1 \wedge B_2) \wedge A_3) \vee ((B_0 \wedge \bar{B}_1 \wedge \bar{B}_2) \wedge A_4) \vee ((B_0 \wedge \bar{B}_1 \wedge B_2) \wedge A_5)$$

$$\vee ((B_0 \wedge B_1 \wedge \bar{B}_2) \wedge A_6) \vee ((B_0 \wedge B_1 \wedge B_2) \wedge A_7) \qquad (15.26)$$

Suppose instead that we assign a name to the 8-bit output vector of a three-input decoding network and use this name in place of the logical expression for the vector in the control sequence. That is, we shall use the notation DECODE (B) to refer to the output of the combinational logic network shown in Fig. 15.13. In the future, we shall use DECODE (R) to represent the 2^n output leads of any n-to-2^n decoding network with R as the n-bit input vector. We may now use the compact expression 15.27 in place of 15.26.

$$r \leftarrow \vee/(\text{DECODE } (B) \wedge A) \qquad (15.27)$$

Used in this way, DECODE (B) looks very much like a function subroutine in FORTRAN. For this reason, we shall refer to it as an example of a *combinational logic subroutine*. In reference 1, a means of generating a logic network defined by a combinational logic subroutine from a sequence of program steps is introduced. For our purposes, we shall conform to the

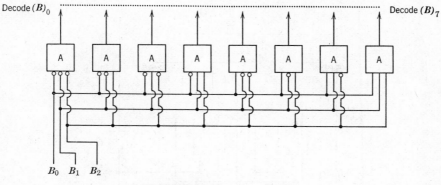

FIGURE 15.13.   Decode (B).

notion of a subroutine only insofar as its nomenclature appears in the control sequence. We shall allow the network itself to be defined by any convenient means including the logic block diagram. Often we shall use the subroutine notation for a network, leaving it to the reader to obtain a network realization.

Another network for which the combinational logic subroutine notation is useful is the adder. We may use the notation of Expression 15.28 to represent the addition of an n-bit data register D and an n-bit accumulater AC to form an $(n + 1)$-bit vector which is placed in AC and an overflow flip-flop ℓ. We may agree that ADD refers to the simple ripple carry adder of

$$\ell, AC \leftarrow \text{ADD} (AC, D) \tag{15.28}$$

Figure 8.7, or to some more sophisticated adder if desired.

We shall have occasion to use a third combinational logic subroutine to increment the contents of a register, that is, adding 1 to the binary number stored in the register (Bit 0 is assumed to be the most significant bit.) The logic network required to increment the contents of an arbitrary 4-bit register R is given in Fig. 15.14a. The output of this network is the 4-bit vector INC(R). The transfer statement given by 15.29 will cause R to be incremented when it is executed in a control sequence. As was pointed out in Chapter 11, a counter is most

$$R \leftarrow \text{INC}(R) \tag{15.29}$$

efficiently implemented using T flip-flops. If a register is used only as a counter, i.e., it appears on the left of no other transfer statements, this more economical approach may be employed. In this case, $CT \leftarrow \text{INC} (CT)$ may be used to represent a control signal which will gate a clock pulse into the counter CT, as illustrated in Fig. 15.14b. With this interpretation of the control signal, a standard counter fabricated in a single MSI package may be used.

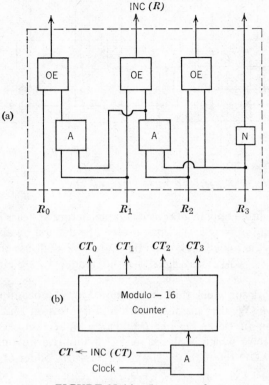

(a)

(b)

FIGURE 15.14.   *Increment logic.*

15.8   Applications

In order to illustrate further the value of our register transfer language and to improve the reader's facility with its use, we shall consider two more design examples. Both of these systems will be required to handle so much information that it would not be practical to apply the classical sequential circuit design methods. First we shall design a very simple controller for a tester for combinational logic circuits fabricated in integrated circuit form.

Example 15.5

Design a controller to be used in conjunction with a digital magnetic tape cassette recorder to test combinational logic circuits with up to eight inputs and up to eight outputs. Usually the circuits to be tested will be fabricated in integrated circuit form with only the input and output leads available to the tester. The system is only required to perform fault detection. That is, once it has been determined that some output does not take on its correct value for a given input vector, the circuit under test is rejected and the test sequence is started over again for the the the next copy of the circuit.

Pairs of 8-bit characters, consisting of an input vector followed by the corresponding correct output vector are stored on tape. The controller must be capable of sequencing through at least 256 such pairs. The system will not be required to operate at a particularly fast clock rate, so that the tape cassette unit and the controller may be assumed to be driven synchronously by the same clock. During each clock period in which the tape device receives a signal on the control line *read*, it will read an 8-bit character from the tape and place it in its output register *TAPE*. Following a clock period in which the tape device receives a signal on line *rewind*, the tape is rewound. At the completion of a rewind operation or when the system is first turned on, the tape reader responds with a one clock period signal on line *ready*.

The controller must provide the appropriate control signals to the tape unit. It must place each input vector in a register connected to the inputs of the circuit under test and compare the eight outputs with the correct output vector. Whenever a discrepancy between the actual output and the correct output is detected, an output signal z is to be turned on and remain on until the system is reset. A manually generated signal on a line labelled *go* will appear for one clock period after a new device to be tested has been placed in the test jig.

Solution: The memory elements, both vector and scalar, required in the data and logic unit for the *test controller* are depicted in Fig. 15.15. There are three 8-bit registers. The register *TV* will be connected to the eight input leads of the circuit under test. The register *YD* is used to store the correct values expected on the eight output lines. The third vector *COUNT* will be used to count modulo-256 the number of test vectors applied to each circuit under test. When the count reaches zero, a signal on line *rewind* will cause the tape cassette to be rewound and will alert the operator that the circuit under test is good and that a new copy can be connected. For convenience, the lines connected to the outputs of the circuit under test will be labelled as vector *Y*. The flip-flop z (connected to the similarly labeled output line) is set to 1 whenever a fault is detected. The remaining signal lines from the controller to the cassette unit are not shown since they originate in the control sequencer. Only the data and logic unit is shown in Fig. 15.15.

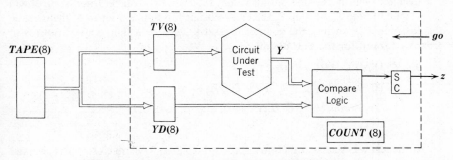

FIGURE 15.15. *Data and logic unit for test controller.*

We start the control sequence description with a list of the inputs and those outputs which are not dependent on the control sequencer.

INPUT: *TAPE*(8), *go*, *ready*

OUTPUT: z

We have not found it necessary before to define the inputs and outputs other than in the general description of the circuit and/or the block diagram. However, with complex circuits, with many internal and external signals, it is desirable to specify clearly those signals which communicate with the "outside world," as opposed to those which are found only within the circuit. The actual control sequence begins at a point prior to the start of testing of a new device. The only transfer associated with the first two steps resets the error output z to 0. These steps are branches which hold control at these two steps until a *ready* signal appears to indicate that the tape cassette is rewound and a *go* signal indicates than a new circuit is ready for test.

1. $\rightarrow (ready \times 2 + \overline{ready} \times 1)$

2. $z \leftarrow 0$

$\rightarrow (go \times 3 + \overline{go} \times 2)$

Each individual test begins at Step 3. First, a *read* $= 1$ causes a test input vector to be placed in *TAPE*. This vector is then transferred to *TV*, exciting the test circuit. Next, the corresponding expected output vector is transferred to

YD through *TAPE*.

3.   *read* $= 1$ (2 clock periods)

4. $TV \leftarrow TAPE$

5.   *read* $= 1$ (2 clock periods)

6. $YD \leftarrow TAPE$

At Step 7, *COUNT* is incremented, and at the same time the actual outputs are compared to the correct values. The exclusive-OR symbol is used here as an abbreviation of the logical equivalent in terms of AND, OR, and NOT. If there is disagreement in any one output value, $(\vee/Y \oplus YD)$ will be 1 and control will continue to Step 8. At Step 8, the error output flip-flop z is set to 1, the cassette is instructed to rewind, the counter is reset to zero, and then control returns to Step 1 to wait for the next circuit.

7. $COUNT \leftarrow INC\ (COUNT)$

$\rightarrow (\vee/Y \oplus YD) \times 8 + \overline{\vee/(Y \oplus YD)} \times 9)$

8. $z \leftarrow 1,\ \textbf{\textit{rewind}} = 1,\ COUNT \leftarrow 0, 0, 0, 0, 0, 0, 0, 0$

$\rightarrow (1)$

If the test was successful, control proceeds to Step 9 where *COUNT* is checked to determine if all 256 test vectors have been applied. If there are more tests to be applied, control then returns to Step 3 to summon the next test input. At the completion of a test sequence, the cassette is instructed to rewind at Step 10 and control returns to step 1.

9. $\rightarrow ((\overline{\vee/COUNT}) \times 10 + \vee/COUNT \times 3)$

10. *rewind* $= 1$

$\rightarrow (1)$

The control sequencer circuit is shown in Fig. 15.16. Just as is often the case with circuits designed by the conventional methods, starting the operation requires a reset signal which is not included in the initial specifications. In this case, the reset signal would SET flip-flop #1 and CLEAR all the others. Setting flip-flop #1 (FF-1) enables the gates receiving the *ready* signal. As long as *ready* remains 0, succeeding clock pulses keep FF-1 set. When *ready* goes to 1, the next clock pulse clears FF-1 and sets FF-2. This generates control signal C_2, which goes to the C or K input of flip-flop z, so that it will be cleared by the next clock pulse. C_2 also enables the gates receiving *go*, causing FF-2 to clear and FF-3 to set when *go* $= 1$. The reader will note that output line from each flip-flop is labeled with the transfer it is to enable, as well as with the control signal number. In fact, the control signal number is often omitted, since the number of the flip-flop makes it self-evident.

Steps 3 and 5 are examples of cases where outputs are generated solely by the control sequencer, as discussed earlier, with the *read* line to be set to 1 during these steps. As shown in Fig. 15.16, this is accomplished simply by OR'ing C_3 and C_5 together and connecting the output of the OR gate to the *read* line. Thus, *read* $= 1$ during Steps 3 and 5, and is 0 during all other steps. The *rewind* signal is generated by a similar connection at Steps 8 and 10. The reader is strongly urged to go through this diagram carefully in conjunction with the control sequence to be sure that the correspondence between the sequence and each element of the circuit is thoroughly understood. ■

We shall conclude this section with another example. This example will require more storage for information than encountered until now. We shall meet this new requirement by introducing a matrix notation to represent an array of registers. Suppose we have a large number of registers, each having the same number of bits. Rather than assigning a separate symbol to each such register, it would seem convenient to use the same symbol for all of the registers in the set and to distinguish between individual registers numerically. Organized in this way, the set of registers can be referred to as a memory. Since we are already using subscripts to designate individual bits within a register, we must resort to superscripts to identify registers within a memory. For example, an m-word memory $\mathbf{M}$ will consist of registers $\mathbf{M}^0, \mathbf{M}^1, \ldots,$ $\mathbf{M}^{m-1}$. The mechanism for retrieving information from a memory will vary with the physical implementation.

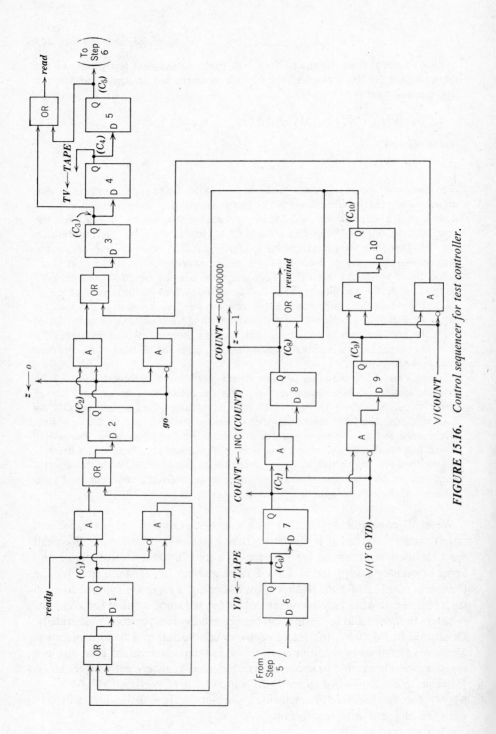

FIGURE 15.16. *Control sequencer for test controller.*

In this section, we shall assume a solid-state memory in which the level outputs of all registers are continuously available. By ANDing the level outputs of each register with an output line from an address decoder and ORing the resultant vectors, we make it possible to route the level outputs of an addressed register to a set of memory output terminals. This combinational logic operation is expressed by 15.30, where *AR* is an *n*-bit address register.

$$(\mathbf{M}^0 \wedge \text{DECODE } (AR)_0) \vee (\mathbf{M}^1 \vee \text{DECODE } (AR)_1) \vee \cdots$$

$$\wedge (\mathbf{M}^{2^n-1} \wedge \text{DECODE } (AR)_{2^n-1}) \tag{15.30}$$

Expression 15.30 may be used as the right side of a register transfer. In some contexts, a *read from memory* operation may be regarded as complete when this vector is clocked into a data register. For convenience, we shall abbreviate Expression 15.30 by the combinational logic subroutine SELECT as given in Expression 15.31.

$$\text{SELECT } (\mathbf{M}, \text{DECODE } (AR)) \tag{15.31}$$

In this example, we shall also introduce a new "conditional transfer" notation, designed to handle cases where there are several possible transfers at a given point in a sequence, the choice dependent on other data. This will be the last new element of notation introduced.

Example 15.6

A simple machine tool controller is to be designed using a read-only-memory to store sequences of tool positions. Eighteen-bit numbers are used to specify the tool position in three dimensions in a way which need not concern us here. There are four possible sequences of positions, any of which may be requested by an operator at any time. Each sequence is 256 words long. The system hardware must include an 18-bit register, *PR*, for storing the current tool position. The tool electronics continually monitor this register through three digital-to-analog converters.

Communications with the operator are provided by a start-stop flip-flop, *ss*, and a 2-bit sequence selector register, *SSR*, in which the bit combinations 00, 01, 10, and 11 correspond to sequences *A*, *B*, *C*, or *D*, respectively. Setting the flip-flop *ss* to 1 will cause the controller to begin reading out the sequence specified by *SSR*. If the operator resets *ss*, the controller will terminate the sequence and clear *PR* to all 0's. If the controller completes a sequence, i.e., reads out all 256 words of the sequence in order, it will reset *ss* and *PR*. The operator may change the settings of *ss* and *SSR* at any time, except that a synchronizing mechanism is provided to prevent changes during a clock pulse.

It is to be noted that we are handling input in a different manner than in previous examples. Here input changes may occur at any time, and those changes are immediately stored, but no action is taken on them until an appropriate point in the sequence. These input lines will not appear directly anywhere in the control

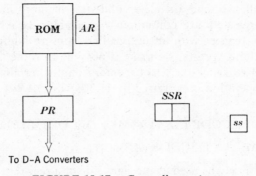

FIGURE 15.17.   *Controller registers.*

sequence and will not be listed with the sequence. Their existance will be understood in much the same way the existence of a reset line may be understood.

Solution: The only required storage registers in addition to those described above are a 10-bit address register for the read-only-memory and the read-only-memory itself. These devices may be presented by the vector **AR** and the matrix **ROM**, respectively. The complete register configuration is given in Fig. 15.17. The read-only-memory (**ROM**) contains 1024 eighteen-bit words in all. Sequence *A* is stored in locations 0–255 (decimal). In decimal, the first addresses of sequences *B*, *C*, and *D* are 256, 512, and 768, respectively.

The control sequence for the machine tool controller follows. We start with an output specification, recalling that the outputs of **PR** drive the *D*-to-*A* converters, which in turn drive the tool. Note that there is no input specification, since there are no direct inputs, as noted above. Step 1 waits for *ss* to go to 1, to move control to Step 2, to place the first address of the appropriate sequence in **AR**.

OUTPUT: *PR*

1. $(ss \times 2 + \overline{ss} \times 1)$

2. $AR \leftarrow \overline{(\epsilon(10) *(\overline{SSR_0} \wedge \overline{SSR_1}))} \vee ((0, 1, \epsilon(8)) *((\overline{SSR_0} \wedge SSR_1))$
    $\vee ((1, 0, \overline{\epsilon(8)} *(SSR_0 \wedge \overline{SSR_1})) \vee ((1, 1, \overline{\epsilon(8)} *(SSR_0 \wedge SSR_1))$

3. $PR \leftarrow$ SELECT (ROM, DECODE(*AR*))

4. $AR \leftarrow$ INC (*AR*)
    $\rightarrow (ss \times 5 + \overline{ss} \times 7)$

5. $\rightarrow ((\vee/\omega^8/AR) \times 3 + \overline{(\vee/\omega^8/AR)} \times 6)$

6. $ss \leftarrow 0$

7. $PR \leftarrow \overline{\epsilon(18)}$
    $\rightarrow (1)$

Statement 2 is an example of the conditional transfer notation we mentioned above. At this point, one of four numbers, 0, 256, or 768, is to be transferred into **AR**, depending on **SSR**. The general form of the notation is shown in Equation 15.32.

$$A \leftarrow (B * f_1) \vee (C * f_2) \tag{15.32}$$

where f_1 and f_2 are logical functions and the asterisk may be translated as "if." Thus, the statement means, "Transfer the contents of **B** to **A** if $f_1 = 1$, or transfer the contents of **C** to **A** if $f_2 = 1$." Note that in such statements only one of the logical functions may take on the value 1 at a time. In Step 2 above, this condition is satisfied since the functions are the four possible combinations of the bits of **SSR**.

Step 3 is an example of our just developed read-from-memory notation. The contents of the location in **ROM** specified by **AR** are placed in **PR**. The address in **AR** is incremented in Step 4, readying the system to read the next tool control vector in sequence. Step 4 also includes a check of the flip-flop *ss*. If the operator has cleared *ss* to 0, control proceeds to Step 7 to place vector of 0's in **PR** and then back to Step 1 to await a request for another tool sequence.

Step 5 returns control to Step 3 unless the reading of a 256-word sequence has just been completed. This requires a rather simple logical expression, since the last eight bits of the first address of each sequence are all 0's. At the completion of a tool sequence, Steps 6 and 7 clear *ss* and place a vector of 0's in **PR** and return control to Step 1.

That portion of the control sequencer circuit implementing Step 2 is shown in Fig. 15.18. The reader can appreciate here why we characterized this sort of transfer earlier as a "conditional." The gating logic has the same form as that used to implement conditional branches in previous circuits, but there is no branch of control to alternate next steps. We will leave it to the reader to complete the design of the control sequencer for the machine tool controller. ■

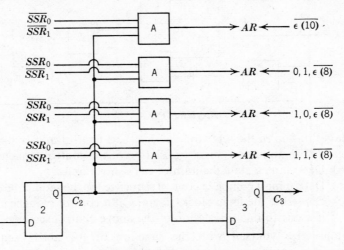

FIGURE 15.18

15.9 Summary

It is the opinion of the authors that the increasing availability of economical integrated circuit realizations of complex functions will mean that the typical designer will be asked to design ever more sophisticated systems. Usually these systems will involve more elements than can be conveniently handled by state table methods. In this chapter, we have presented an approach which can be used in systems of any size. There is nothing inherent in our language approach which will assure a minimal realization. Still, it offers many advantages over intuitive methods which might be devised by the designer for individual jobs. Most important perhaps is its value in communications about a system between individuals and within an organization.

In this chapter, we have considered relatively small systems just beyond the reach of state techniques. In reference 1, we extend our approach to larger systems, in particular the complete digital computer.

PROBLEMS

15.1. Consider the flip-flop labeled c in Fig. P15.1 with its associated input logic as a control sequencer, and consider flip-flop a to be the only

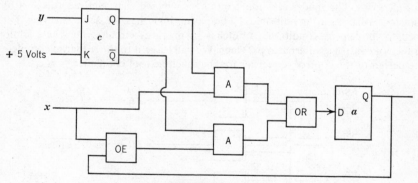

FIGURE P15.1

data flip-flop in the sequential circuit. The control unit is a minimal realization and not a direct translation of the control sequence.
(a) Determine a state diagram of the control unit.
(b) Determine a complete control sequence for the sequential circuit.
(c) Construct the logic block diagram of a control sequencer which is a one to one translation of the above control sequence.

15.2. Familiarize yourself with the function of the above sequential circuit by studying the control sequence determined in Problem 15.1.

Now synthesize the complete state diagram of a sequential circuit which will accomplish the same task. Determine a minimal realization of this state diagram using the techniques of Chapter 10, and compare its complexity with Fig. P15.1.

15.3. Construct a simpler but equivalent version of the branching network given in Fig. P15.3.

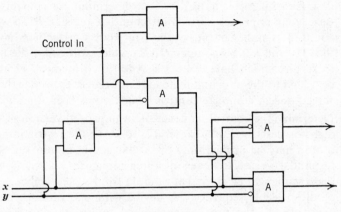

FIGURE P15.3

15.4. A straightforward realization of a design language step consisting of only a branch is given in Fig. P15.4. This circuit is effectively a

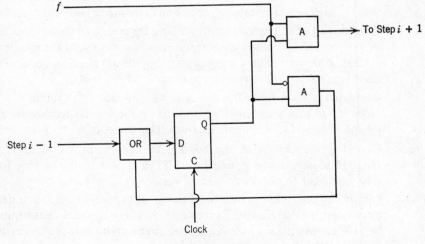

FIGURE P15.4

2-state sequential circuit. Use the techniques of Chapter 10 to realize this circuit in terms of

(a) A clocked S-C flip-flop

(b) A clocked J-K flip-flop

The inputs are the branch function f and a line from the step $i - 1$ flip-flop. The output leads to step $i + 1$.

15.5. Obtain an alternate realization of the 2-out-of-5 code checker described in Problem 10.7 by first determining a control sequence. Base your approach on a 3-bit counting register **BITS** which will count the number of bits of the 5-bit code word which have arrived thus far and a 2-bit counter **ONES** which will count the number of these bits which have been 1's. Additional memory elements may be added to the data and logic unit, if needed. Consider the problem solved once a complete control sequence has been obtained.

15.6. Determine the control sequence of an improved version of the drunk detector described in Chapter 12. Remove the restriction that each button may be pushed only once. As before, require each driver to respond correctly to each of three sequences. Now, however, let these sequences be chosen at random from a prespecified set of eight of the possible sequences. A device with one random output x is available. At the time of each clock pulse, the probability is .5 that $x = 0$ and .5 that $x = 1$. The values of successive outputs are totally uncorrelated. A given sequence may be allowed to appear more than once in a set of three sequences. The switch panel may be assumed to include a synchronizing mechanism so that a signal will be active for only one clock period each time a button is pushed. Specify storage elements as required and determine a control sequence.

15.7. Determine a control sequence describing the code translator specified in Problem 12.8. Compare the resulting realization with the minimal version. Can you make a general statement about the effect of don't-cares on the control sequence approach?

15.8. Construct a state table for the control sequencer of Example 15.5 (Fig. 15.16) and a minimal version, using the formal techniques of Chapters 10 and 12. Compare the overall hardware of the two versions.

15.9. Complete the design of the control sequencer for Example 15.6 (Fig. 15.18) on a one-to-one basis. Then repeat Problem 15.8 for this sequencer to obtain a minimal version.

15.10. Obtain the control sequence for a modified version of the integrated circuit tester described in Example 15.5. The improved tester must be able to test clock mode sequential circuits with test sequences of up to 16 vectors in length.

15.11. A random sequence generator is driven by an external clock source. This generator provides a level output, z, which is constant between clock pulses, and may or may not change, on a random basis, when triggered by a clock pulse. A special-purpose computer is to be designed employing a 1-megahertz clock. This computer is to provide an output clock to drive the random process at a frequency of 1 kilohertz. The computer must also compute the number of level changes in the random process each second. The computer must also compute and display the average number of level changes per second over the first 2^8 seconds following the depression of its start button. Specify the necessary hardware, determine the control sequencer, and construct a block diagram of the control sequencer for this special-purpose computer. Accomplish division by shifting.

15.12. A control sequence is to be obtained for digital system B whose storage elements are laid out in Fig. P15.12. The function of digital

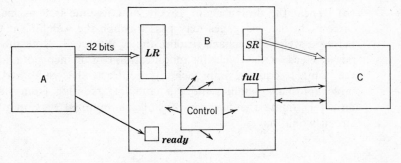

FIGURE P15.12

system B (to be designed) is to facilitate the transfer of information from system A to system C. System A might be a computer, and system C might be some sort of peripheral memory, such as a magnetic tape unit. In that case, system B will perform part of the function of a magnetic tape controller.

The output from system A will be a 32-bit vector, which we shall refer to as a word. System C will accept only 8-bit vectors or bytes as input. System B then must sequentially disassemble the 32-bit word into four 8-bit bytes. The principal hardware elements in system B are a 32-bit register *LR* and an 8-bit register *SR*. Also included are a single flip-flop labeled *ready*, associated with *LR*, and a single flip-flop labeled *full*, associated with *SR*.

We assume that the clock source is provided by system C so that systems B and C are synchronized. During any clock period for which the flip-flop *full* is 1, system C will read a byte from register *SR*. In the clock period immediately after it places a byte in *SR*, system B will set *full* to 1. During the next clock period, system C will utilize the byte in *SR* and simultaneously clear *full* to 0. System B may place another byte in *SR* during the same clock period. Since systems B and C are synchronized, it is possible to reliably read from *SR* and place a new byte in *SR* during the same clock period. The flip-flops in *SR* must be master-slave for this scheme to work. In effect, we have assumed that system C can react in one clock period to simplify the design of system B.

Whenever a word is placed in *LR*, the flip-flop, *ready*, will be set to 1 by system A. After the fourth byte has been read from *LR*, system B will clear *ready* to zero.

15.13. (a) Repeat the design of the circuit of Problem 12.15 by the control sequence method. Compare the circuit complexity for the two designs.

(b) Repeat the design again, except now assume that as many as three additional packages may pass through the weighing machine before a weighed package reaches the x_4 photocell. Note the comparative ease with which this modification can be incorporated with the control sequence approach. By contrast, the traditional state table approach would require a complete redesign from start to finish. Estimate the number of states a minimal version of this circuit would have.

BIBLIOGRAPHY

1. Hill, F. J. and G. R. Peterson. *Digital Systems: Hardware Organization and Design*, Wiley, New York, 1973.

2. Friedman, T. D. and S. C. Yang. "Methods Used in Automatic Logic Design Generator (ALERT)," *IEEE Transactions on Computers*, 593 (Sept. 1969).

3. Schorr, H. "Computer Aided Digital System Design and Analysis Using a Register Transfer Language," *IEEE Transactions on Electronic Computers*, 730–737 (Dec. 1964).

4. Gentry, M. "A Compiler for Computer Hardware Expressed in Modified APL," Ph.D. dissertation, University of Arizona, June 1971.

5. Iverson, K. E. *A Programming Language*, Wiley, New York, 1962.

Chapter 16 *LSI-MSI*

16.1 Introduction

In some quarters, it has been contended that the use of *medium and large scale integrated circuits* (MSI and LSI) will invalidate traditional approaches to logical design and cause them to be replaced with new techniques. The authors do not fully agree with this point of view. It is true that the cost, uses, construction, and much of the design process for digital systems have been revolutionized by LSI and MSI. Many of the latter stages of the design process such as mask-making, P.C. board layout, back panel wiring, and testing have been automated. The initial, most creative, portions of the design process have been least affected.

Boolean algebra remains the basic method of description of combinational logic circuits. The design of combinational logic for implementation via custom LSI will not differ basically from the design for discrete component implementation. Where standard MSI parts are used, the designer's flexibility will be limited somewhat. Some sort of "macro-algebra" whose basic functions are realized by LSI parts is an intriguing idea, but none has been invented. Most special designs will require both individual gates and MSI parts. Boolean manipulation will be required for efficient interconnection of these parts. Examples will be presented in Section 16.3.

The more complex digital systems grow (partly due to MSI and LSI), the greater the importance of a thorough understanding of sequential circuits becomes. Timing assumptions associated with the three types of sequential circuits are critical. Timing problems must be anticipated in advance. Design modification are expensive. As we shall see in Section 16.5, simulation becomes an important tool in verifying system timing in advance.

The logical designer will probably be less interested in sequential circuit synthesis when designing in terms of MSI. Very likely he will approach the problem on a larger scale, utilizing a system design language such as presented in Chapter 15. The designer of MSI parts will make substantial use of available level mode as well as clock mode synthesis techniques. Since very large numbers of these parts are manufactured, achieving an optimal design is important.

The final section of this chapter will consist of a brief introduction to test sequence generation. If a fault exists in an MSI or LSI part, a test must be applied which will cause an effect of this fault to appear on an output lead. Particularly for LSI, the number of output leads will be small compared with the number internal connections between gates. For sequential circuits with a large number of states, the determination of a test sequence can be a difficult process. No completely automatic test generation scheme exists for sequential circuits. The cost of test generation and testing can be a significant portion of the overall cost of a part.

16.2 Definitions

Most readers of this book are familiar with the notion of an integrated circuit (IC). This subject was touched briefly in Chapter 5. Concisely, we define an IC as a circuit consisting of active and passive elements fabricated on a single semiconductor chip and mounted in an individual package (TO-5 can, dual-in-line package, etc.). How complex, then, is an IC? Where do ordinary IC's leave off, and where does MSI (medium-scale integrated circuit) begin? These questions are less easily answered.

The pictorial diagram of Fig. 16.1 will be helpful in unraveling some of the confusion surrounding the terms introduced above. Strictly speaking the term integrated circuit encompasses both MSI and LSI. We shall use the term IC to refer to a small number of gates or memory elements mounted in a single package, with all gate input and output leads connected to package leads. Thus, we draw a clear distinction between discrete component logic circuits, IC's, and MSI as shown in Fig. 16.1. No such clear distinction can be made between MSI and LSI. Usually one thinks of an MSI part as a small portion of a digital system mounted in a single package. Such small subsystems might include counters, decoders, parity checkers, etc. Fabricating large portions of a computer on a single semiconductor chip is LSI.

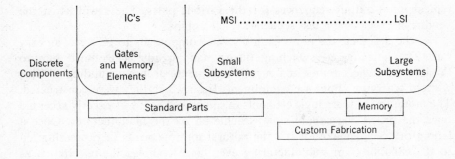

FIGURE 16.1. *Forms of integrated circuit technology.*

Identifying large subsystems with sufficiently few input-output leads can be difficult. Adders, decoders, calculators, sections of memory, or complete arithmetic units are possibilities.

More important than drawing a fine line between MSI and LSI is the separation of the concepts of standard integrated circuit parts and custom fabricated integrated circuits. As depicted in Fig. 16.1, standard parts tend to represent fewer gates than custom circuits; but this is not universal. Any digital circuit which a semiconductor manufacturer may consider to have profit potential will be manufactured and distributed as a standard integrated circuit part. Designs which subsequently turn out to be unprofitable are discarded. As certain designs become accepted, they are copied by other manufacturers.* Thus, the trend is toward ever greater standardization. Standard MSI parts include binary decoders, BCD decoders, decade counters, binary counters, shift registers, comparators, parity generators, etc. Sections of memory and read only memory are the principle examples of standard LSI parts. The uniformity within these circuits make for efficient fabrication and testing processes.

When a large number of copies of a given digital system are to be manufactured, then custom fabrication of parts of the system as MSI or LSI may be considered. The designer should always consider the use of standard parts before resorting to custom fabrication. The tradeoffs governing the choice of approach will be the topic Section 16.4. One the designer has selected custom fabrication, he submits some description of his design to a semiconductor manufacturer. Most vendors will accept logic block diagrams. The process could be simplified for both customer and vendor if design information could be transferred in some form of system design language such as discussed in Chapter 15. Many vendors will ask for (but not demand)

---

* This is not as sinister as it sounds. A customer is reluctant to adopt a part for a large volume design unless the part can also be obtained from a *second source*.

typical input-output sequences for the various parts. These will assist the vendor in generating a test sequence for the chips.

For the designer, the jump from discrete components to IC's was an easy one. The language of switching theory had already accustomed him to thinking of logic circuits and memory elements as basic building blocks, without worrying about the components from which they were constructed. The mounting of these logic elements in single packages was eagerly accepted once the processing capability was achieved by the manufacturer. Once a laboratory could make an IC, the natural improvements in processing led to the capability for manufacturing ever more complex digital circuits as integrated circuit parts. While welcoming this capability, the designer has not always been certain how best to utilize it. We continue to attack this problem in the next several sections.

6.3 Design with Standard MSI Parts

The use of standard parts more powerful than individual gates tends to simplify the design process rather than complicate it. A portion of the design activity is reduced to searching the vendor's catalog for parts which can be adapted to the task at hand. The experienced designer can make this determination in short order. Where substantial portions of the overall function are accomplished by standard parts, the necessary Boolean manipulation is correspondingly lessened. Even where the MSI parts are sequential, their interconnection is usually a combinational logic problem. The approach to combinational logic in this context does not differ from the approach used in design with discrete components.

To convince the reader of the power and simplicity of design with MSI parts, we present some typical examples. The first example combines two MSI parts to provide for processing a larger number of bits than can be handled by the individual devices. The second example will be that of a simple counter decoder for the generation of control signals.

Example 16.1

A comparator is to be designed which will compare the magnitudes of two 8-bit numbers A and B. The two outputs will take of the values $z_2, z_1 = 1, 0$ if $A > B$. If $A = B$, then $z_2, z_1 = 1, 1$; and if $A < B$, then $z_2, z_1 = 0, 1$. Comparators which compare the magnitudes of 4-bit numbers are available as MSI parts.

Solution: Let A_2 and B_2 be binary numbers made up of the four most significant bits of A and B, respectively. Similarly, let A_1 and B_1 be binary numbers given by the least significant four bits of A and B. The eight bits of A_2 and B_2 may be used as inputs to a 4-bit MSI comparator. The outputs of this comparator are

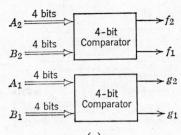

(a)

f_2	f_1		g_2	g_1	
0	0		0	0	
0	1	$A_2 < B_2$	0	1	$A_1 < B_1$
1	1	$A_2 = B_2$	1	1	$A_1 = B_1$
1	0	$A_2 > B_2$	1	0	$A_1 > B_1$

(b)

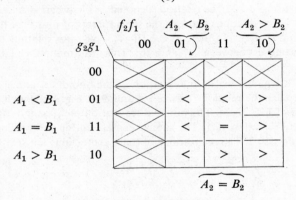

(c) $A \mathcal{R} B$

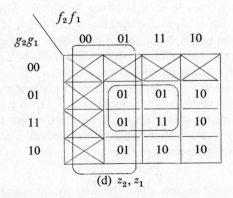

(d) z_2, z_1

FIGURE 16.2. *Design of 8-bit comparator.*

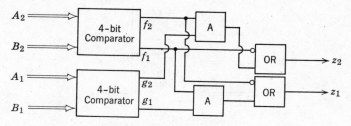

FIGURE 16.3. *8-bit comparator.*

labeled f_2 and f_1. A similar comparator for the least significant bits has outputs g_2 and g_1. These comparators are symbolized in Fig. 16.2a. The interpretation of the outputs is given in Fig. 16.2b.

To design the comparator completely, it is only necessary to determine the outputs z_2 and z_1 as functions of f_2, f_1, g_2, and g_1. In general, it may be necessary to consider the circuit input bits in the design completion step, but not in this case. As given in Fig. 16.2c, the relation (denoted $\mathscr{R}$) between A and B is completely defined by the relations between the separate sets of four bits. If $A_2 > B_2$, then $A > B$; and if $A_2 < B_2$, then $A < B$. If $A_2 = B_2$, the relation between A and B is the same as between A_1 and B_1. This information is tabulated in Fig. 16.2c.

To obtain a Karnaugh map representation of the outputs z_2 and z_1, it is only necessary to replace the relations between A and B of Fig. 16.2c with the coded representations of Fig. 16.2d. The minimal realization of z_1, as indicated on the map, is given by Equation 16.1. The similar expression for z_2 (not illustrated) is given by Equation 16.2. The realizations of these expressions are

$$z_1 = f_2 + f_1 g_1 \tag{16.1}$$
$$z_2 = f_1 + f_2 g_2 \tag{16.2}$$

added to Fig. 16.2a to yield the complete design given in Fig. 16.3. ∎

The reader has no doubt observed that no special MSI techniques were required to design the comparator in the above example. The analysis was of the sort which might be employed in any combinational logic design problem. The resulting circuit is probably cheaper and no doubt simpler to wire than a comparable design using individual gates only.

The next example is a sequential circuit. As is commonly the case, the design will be accomplished without making use of the state table of the overall sequential circuit. As of this writing, the notion of mapping the state table of a standard part into the state table of a sequential circuit to be designed remains an interesting research problem.

Example 16.2

A very simple control sequencer is to be designed for a manufacturing process. Level signals are to appear sequentially on 25 lines z_0 to z_{24}. One and only one of the lines will be logical 1 during each clock period. The control sequence is repetitive so that z_0 goes to 1 again following the clock period in which z_{25} was 1. A 4-bit counter and a 4-to-2^4 line decoder are available as MSI parts.

Solution: We base the design on the 4-bit counter, which has only two inputs, one for the clock and the other a reset, R. The counter has four outputs, C_8 C_4 C_2 C_1, which represent a 4-bit binary number. Every clock pulse causes the count to be incremented unless $R = 1$. If $R = 1$, the clock pulse will reset the counter to zero.

A sequence of 16 levels is easily generated by connecting the counter outputs to the 4-bit decoder. The decoder has 16 outputs, one corresponding to each 4-variable minterm. As is customary within the industry, a distinct output line of the decoder will be 0 corresponding to each count. The other 15 outputs are 1. Because of this convention, we label the decoder outputs as $\bar{d}_0$ to $\bar{d}_{15}$. If 16 signals were required instead of 25, the configuration of Fig. 16.4 would be the complete design.

It is necessary to add another flip-flop to permit counting to 24. The output, y, of this flip-flop becomes the most significant bit. A state table of this two state machine is given in Fig. 16.4b. Each time the count reaches 15 ($\bar{d}_{15} = 0$), the flip-flop, y, is set to 1; and the counter is reset to zero. The 24th clock pulse establishes $d_8 = 1$. The 25th pulse clears y as given in Fig. 16.4b (the square corresponding to $\bar{d}_8 = y = 1$) and resets the 4-bit counter to zero.

From the state table, the set and clear equations for flip-flop, y, are easily shown to be given by Equation 16.3. The counter

$$S_y = d_{15} \qquad C_y = d_8 \tag{16.3}$$

is normally reset when the count reaches 15, so it is only necessary to connect R to $d_8 y$ as given by Equation 16.4.

$$r = d_8 y \tag{16.4}$$

The first 16 control levels are given by Equation 16.5:

$$z_0 = d_0 \cdot \bar{y} = \overline{\bar{d}_0 + y}$$
$$z_1 = d_1 \cdot \bar{y} = \overline{\bar{d}_1 + y}$$
$$\cdot \qquad \cdot \tag{16.5}$$
$$\cdot \qquad \cdot$$
$$\cdot \qquad \cdot$$
$$z_{15} = d_{15} \cdot \bar{y} = \overline{\bar{d}_{15} + y}$$

The last 9 are given by Equation 16.6:

$$z_{16} = d_0 y = \overline{\bar{d}_0 + \bar{y}} \tag{16.6}$$
$$\cdot$$
$$\cdot$$
$$\cdot$$
$$z_{24} = d_8 y = \overline{\bar{d}_8 + \bar{y}}$$

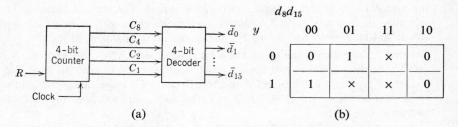

(a)   (b)

FIGURE 16.4. *Design of 25-line control sequencer.*

A complete logic design of the control unit is given in Fig. 16.5. If the complete circuit were constructed from individual gates, the most economical approach would not use the 25 AND gates at the output of the decoder. They would likely be combined into the decoding function. In this case, the decoder is packaged as a single part and cannot be modified. In spite of these 25 AND gates, the cost of the two MSI parts would almost certainly be such that the hybrid configuration of Fig. 16.5 would be the cheapest approach. ■

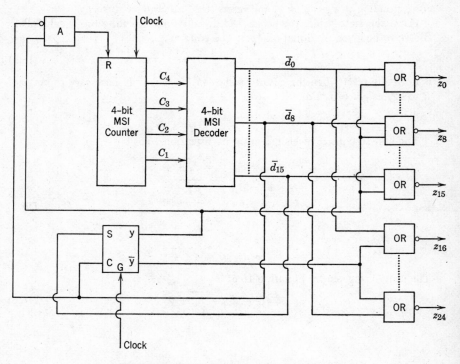

FIGURE 16.5. *25-level control sequencer.*

The more complex a system, the more likely that it will finally be fabricated using MSI parts. In the previous chapter, we conjectured that most complex systems could best be approached in terms of a design language. One would hope to be able to realize a design specified as a control sequence in terms of a collection of MSI parts. There is one problem. The principal elements of data and logic units, *registers*, are not readily implemented as single standard MSI parts. The number of leads which can be connected to any part is limited, and each flip-flop in a register will require at least two and possibly four input-output leads. Thus, a register or part of a register mounted in a 14-pin dual-in-line package would contain between two and five flip-flops only. Similarly, combinational logic subroutine functions of large vectors cannot be implemented as single parts.

The observations of the above paragraph do not weaken the argument that large systems should initially be formulated as control sequences, nor do they in any way lessen the value of using MSI parts to the maximum extent possible. The realization of the control sequence using MSI is merely slightly less elegant than might be hoped. Very often it will consist of a few MSI parts interconnected by a larger number of individual gates just as was the case in the previous two examples. In the following example, we shall formulate a control sequence while planning the ultimate realization in terms of MSI parts.

Example 16.3

An interface is to be developed between a computer and a electrically driven typewriter which is to serve as an output device only. Information will be supplied to the typewriter on six lines to be labeled as input vector X. Of the 64 possible input characters each will represent a symbol to be typed (or a space) except the character $\overline{\epsilon(6)}$, which will cause a carriage return and line feed. As illustrated in Fig. 16.6, the interface unit must generate a one clock period signal on a line labeled *ready* each time it is ready to receive a character. When a character is on lines X ready to be read, the computer will place a pulse on line *read*. A synchronizer must be included which will generate a signal *sread* synchronized with the clock and of one clock period duration each time a pulse appears on line *read*.

The typewriter input will consist of 64 lines represented by the vector Z. The binary character representation of the number k on the input lines must cause a

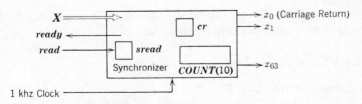

FIGURE 16.6.   *Layout of typewriter interface.*

one clock period signal to be placed on line Z_k. This will cause the symbol represented by character k to be typed. In particular, a signal on line Z_0 will cause a carriage return. Following the initiation of a typing stroke, the interface must allow 16 ms for the stroke to be completed before issuing another *ready* signal. In the event of a carriage return, a 1 sec interval must be allowed. An interface between the computer and typewriter must be designed using standard MSI parts to the best possible advantage.

Solution: Let us assume that a 4-bit decoder is available as a standard MSI part but no 6-bit decoder could be found in any of the vendors' catalogs. A reliable method of timing the wait prior to the generation of a signal on line *ready* involves the use of a counter. If we use a relatively slow 1 khz clock, a 4-bit counter will complete its count in 16 ms. Similarly, a 10-bit counter will complete its count in just slightly more than the 1 second which must be allowed for a carriage return. We shall assume that a 10-bit counter which can be used for both functions is available as a standard MSI part. This device is shown in Fig. 16.6 together with a flip-flop *cr*, which will be set to 1 when a carriage return character is detected. This flip-flop is necessary in the event that the vector X should change for some reason before the carriage return has been completed.

Before attempting to construct a logic block diagram for the interface unit, let us formulate it as a control sequence. The output vector Z must be $\overline{\epsilon(64)}$ (all zeros) at all times except for one clock period for each character received. During that clock period, one and only one of the lines will be 1. The control sequencer will issue a one clock period signal on a line to be designated *typ* when this output is to appear. Thus, the output vector Z is defined as a preliminary statement to the control sequence.

1. OUTPUT $Z = (typ \wedge \overline{X}_0 \wedge \overline{X}_1 \wedge \text{DECODE}(\omega^4/X))$, $(typ \wedge \overline{X}_0 \wedge X_1 \wedge$

   $\text{DECODE}(\omega^4/X))$, $(typ \wedge X_0 \wedge \overline{X}_1 \wedge \text{DECODE}(\omega^4/X))$, $(typ \wedge X_0 \wedge X_1 \wedge$

   $\text{DECODE}(\omega^4/X))$

This combinational logic expression which is the catenation of four 16-bit vectors represents the output at all times. Once again, notice the all bits of the vector will be 0 unless $typ = 1$, and then only one bit will be one. The combinational logic subroutine DECODE represents the 4-bit MSI decoder. We find that 64 AND gates and 4 NOT gates are required to fashion a 6-bit decoder from a 4-bit decoder.

The remaining steps of the control sequence are as follows. Step 2 specifies waiting for the synchronized read signal while

2. $COUNT \leftarrow \overline{\epsilon(10)}$; $cr \leftarrow 0$

   $\rightarrow (sread \times 3 + \overline{sread} \times 2)$

3. $cr \leftarrow \overline{X}_0 \wedge \overline{X}_1 \wedge \text{DECODE}(\omega^4/X)_0$; $COUNT \leftarrow \text{INC}(COUNT)$; $typ = 1$

4. **COUNT** ← INC(**COUNT**)

> → (5 × (∨/((cr ∧ α⁶/**COUNT**), ω⁴/**COUNT**)) + 4 × (∨/((cr ∧ α⁶/**COUNT**),
> ω⁴/COUNT)))

5. **ready** = 1, → (2)

keeping the counter at zero. We assume that the counter is designed so as to be
incremented by each clock pulse unless a 1 appears on a reset, which we shall
connect to the control line **COUNT** ← $\overline{\epsilon}$(10). Once a character has been provided,
Step 3 sets **cr** in the event of a carriage return and issues a control signal on line
typ to enable an output. Although no control signal corresponding to **COUNT** ←
INC(**COUNT**) is needed, we are reminded that the counter is incremented by

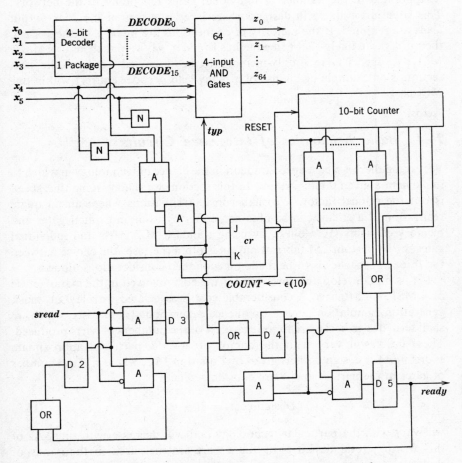

FIGURE 16.7.   *Logic block diagram for typewriter interface.*

both Steps 3 and 4. Step 4 is repeated until the last four bits of *COUNT* are 0's or, if *cr* = 1, until all 10 bits are zero. After sufficient time has elapsed, a *ready* signal is supplied at Step 5, and control returns to Step 1 to wait for another character.

An abbreviated logic block diagram of both the control sequencer and the data and logic unit is given in Fig. 16.7. The number of gates represented by each block is given to provide an indication of the extent we were able to take advantage of MSI parts. Notice that the control sequencer realizes Steps 2–5 in 1–1 fashion. In this case, this is probably the simplest approach in terms of minimizing the number of connections. ■

Although the use of the MSI counter and 4-bit decoder permits a significant saving in the number of wired connections required, the reader may still be disappointed in the number of individual gates remaining in the network. This situation is likely in designs where large numbers of input or output leads are required. If the majority of connections are internal to the system, then it is more likely that someone has found a way to put them in a single MSI package. The read-only-memory to be discussed again in the next section is an example of the use of a many-lead decoder entirely within one MSI part.

16.4 Basic Economics of Integrated Circuits

Very often there is a cost-speed tradeoff in the choice of technology in which to implement a given digital system. In this section, we shall assume that speed is not the critical factor. We shall consider situations where an adequate realization of a circuit can be obtained either by using individual gates and memory elements (IC's only), by using standard MSI parts and individual gates, or by custom LSI fabrication. As is often the case, the choice between the three approaches is to be made on economic considerations alone.

Let us first develop an expression for the cost involved in the manufacture of *n* MSI or LSI parts. A considerable cost is associated with layout, mask generation, simulation, and test sequence generation. These costs, which we shall term design costs, are fixed regardless of the numbers of parts produced. These costs will vary with the complexity of the part. An approximate expression for design costs is given by Equation 16.7, where *g* is the number of gates in the part. *A* and *B* are constants which

$$\text{Design costs} = A + Bg \tag{16.7}$$

will vary with the particular technology and will decrease with time. As of this writing, total design costs for a part are measured in thousands of dollars

A per unit cost is associated with each part produced. This cost covers material, overhead, production labor, and testing costs. Thus, the total cost of *n* parts is given by Equation 16.8. The term $C + Dg$ is measured in pennies even for a fairly complex part.

$$\text{Cost of } n \text{ parts} = A + Bg + (C + Dg)n \qquad (16.8)$$

For a custom LSI part where n may be a few hundred or less, the design cost may be the dominant term. For a standard MSI part, where production may reach the tens of thousands, the design cost can often be neglected. Thus, the cost of a standard MSI part may take the form of Equation 16.9.

$$\text{Cost of standard part} = C + Dg \qquad (16.9)$$

Example 16.4

A logic designer has developed a complex combinational logic network composed of 85 gates. Ultimately a demand for *m* of these networks is projected. Two possible approaches to realizing the design are under consideration. One involves the use of a standard 90-gate MSI part which is very similar to the desired network and costs *K* dollars per part. A special 30-gate network connected around the usable portion of the MSI part must be used to obtain the desired realization. The possible custom fabrication of the network in a single package also under consideration. Determine the most economical approach to obtaining *m* copies of the network. Assume that the cost of individual gates is $.05 per gate.

Solution: The component cost for *m* copies of the hybrid network is given by Equation 16.10.

$$\text{Hybrid component cost} = (K + 30 \cdot (.05)) \cdot m = (K + 1.5) \cdot m \quad (16.10)$$

An additional cost for wiring must be assessed for the hybrid approach. *This is a significant factor, in that wiring costs will often exceed the cost of parts.* In this case, let us assume a wiring cost of *W* dollars per network. Thus, the total cost for the *m* hybrid networks is given by Equation 16.11.

$$\text{Hybrid cost} = (K + W + 1.5)m \qquad (16.11)$$

For a fixed number of gates (85), the expression (16.8) for the cost of *m* custom parts reduces to the form of Equation 16.12.

$$\text{Cost of } m \text{ customs parts} = p + Qm \qquad (16.12)$$

This expression is plotted in Fig. 16.8. To provide a comparison, Equations 16.10 and 16.11 are plotted in the same figure. Clearly the custom approach is to be preferred for very large *m* and the hybrid approach for small *m*. The crossover point will occur for a smaller *m* due to wiring costs which are associated with the hybrid approach but which are not needed if custom packages are employed.

Suppose, for example, that a particular semiconductor manufacturer quotes a price of $2000 plus $2.00 per part for the custom parts. Suppose this same

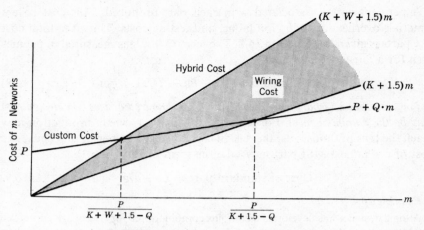

FIGURE 16.8. *Custom vs. hybrid MSI.*

manufacturer will supply the standard parts for $2.00 per part. That is, $K = \$2.00$. If we assume a cost of $1.50 more to wire the hybrid network than to wire the custom part to a circuit board, the custom approach is to be preferred for all values of m satisfying the inequality.

$$2000 + 2m < 5m$$
$$m > 667$$
(16.13)

If the similar standard part were not available, the custom approach would probably be chosen for a much smaller m. ∎

The above example represented a rather simple situation, but it will serve as a point of departure for more realistic cases. Typically, a digital system will be partitioned into several subnetworks. Standard MSI parts may be available approximating some but not all of the subnetworks. The partitioning itself may depend on whether or not custom circuits are economically attractive.

Because of the significant general interest in read-only memories, this topic was introduced in Chapter 8. Of the schemes devised by which the customer can specify some of the interconnections between devices on an IC chip at a cost less than the complete tooling cost for a special part, the ROM is by far the most widely used. For this reason, we return briefly to the subject.

As was discussed in Chapter 8, read-only memories may be used to store control sequences in microprogrammable control units or to replace two levels of combinational logic. The delay through the two logic levels in a transistor ROM can be as much as ten times the delay through two levels of some of the faster lines of logic available. Thus, one would use the ROM to realize combinational logic only in applications where speed was not an overriding consideration. ROMs are usually packaged as separate units. Thus, the use of an ROM implies a hybrid approach, with combinational logic and storage elements on separate chips. If the ROM is not ruled out by either of these considerations, its use becomes a question of economics.

There are two basic approaches to the use of semiconductor ROM's prevailing as of this writing. One type is electrically or mechanically programmable after the completed package has been supplied to the customer or in some cases the distributor. The other type must be programmed before it is packaged, and the data is permanent once the device leaves the factory. The customer-programmable type of ROM may be regarded much like a standard LSI part, with the cost of entering data treated as a wiring cost. The per-unit cost is much greater for this type of ROM, but there is no fixed set-up charge. As suggested by Fig. 16.8, the programmable ROM will be the likely choice where only a few copies of the part are required.

Let us now turn our attention to a more detailed consideration of the costs associated with the fixed ROM. There are two steps in the preparation for manufacture of this type of read-only memory. The first step consists of preparing for the production of a standard chip complete with decoder and a transistor at every intersection of a bit line and a word line. The second step consists of utilizing a user-supplied list of the desired ROM contents to develop a mask for the final metallization process. Using this mask will cause the transistors at the word line-bit line intersections to be connected or not connected as appropriate. The cost of a ROM may thus be broken down into 3 parts, the Step 1 and Step 2 preparation costs, and an incremental cost associated with the manufacture of each part.

A part of the Step 1 cost will be fixed as denoted by the constant A. A part of that cost will be associated with the decoder and will be approximately proportional to the number (2^n) of decoder output lines. The remaining Step-1 cost will be proportional to the total number of bits in the ROM. Thus, Equation 16.14 is an expression for Step 1 preparation cost. The number of bits per word is given by b. Typically, $B \gg C$.

$$\text{Step 1 cost} = A + B \cdot 2^n + C(2^n \cdot b) \qquad (16.14)$$

The cost of producing the metallization mask has a fixed component and a component proportional to the total number of bits as given by Equation 16.15.

$$\text{Step 2 cost} = D + E \cdot (2^n \cdot b) \qquad (16.15)$$

The incremental cost associated with each ROM produced is largely a function of package cost and the labor cost of bonding leads from the chip to the pins on the package. This cost is approximately proportional to the total number of pins as given by Equation 16.16.

$$\text{Incremental cost} = F \cdot (n + b) \tag{16.16}$$

We shall make no attempt to suggest values for the six constants in the above expressions. These costs are still in a state of flux as technology improves. The form of these expressions can be of considerable value to the designer as he attempts to decide whether an ROM is a suitable approach for a given application. Where required, the particular constants may be estimated from current price lists.

Certain standard ROM configurations will be available from the various vendors. Some examples might be 512×8, 256×16, or 16×16. The list will continue to grow. At some point in time, a designer will be able to obtain as a standard part almost any configuration where the number of words and the number of bits per word are powers of two. If a standard ROM can be adapted to a given application, most of the Step-1 cost can be avoided. This cost will be spread over a large number of users of the same configuration. Typically it will be reflected as a small fixed part of the listed price.

The Step-2 cost and the incremental cost must be paid by the individual user. If p identical copies of a particular ROM are needed, the per unit cost is given by Equation 16.17.

$$\text{per unit ROM cost} = \frac{D + E(2^n \cdot b)}{p} + F(n + b) \tag{16.17}$$

Notice that the cost of the decoder is largely a Step-1 preparation cost and is not reflected in Equation 16.17.

For smaller values of p, the designer must compare the cost of the fixed ROM with the cost of wiring a network of IC packages and with the cost of a programmable ROM. The cost (often higher than IC costs) of sockets, PC boards, and labor associated with the conventional network must all be considered. Much external wiring can be avoided by using an ROM. The final metallization step accomplishes this wiring in a brute force but very systematic manner. Unfortunately, the crossover point will occur at a larger value of p for single output functions ($b = 1$).

The most common use of ROM's with large b is in microprogrammable control units. This application requires sufficient functions of each address (bits per word) to specify the origin and destination of a register transfer and the address of the next instruction. A particularly large number of bits, b, is required where substantial flexibility is allowed in the conditional branch operation (See Chapter 8 of Reference 1). In this case, the ROM may be inviting even where $p = 1$. For such applications, one might consider alternative methods of establishing the contents of the ROM, which do not require the making of a special mask. One such method enters information by passing large currents through the bit lines to melt connections for addressed words. This sort of approach would be cheaper for one of a kind applications but more expensive for large p.

16.5   MOS/LSI

We return to MOS technology in this section after a brief discussion in Chapter 5 for two reasons. The types of MOS networks which we shall discuss in this chapter are implemented almost exclusively as large scale integrated circuits. Unlike the CMOS of Chapter 5, the single channel MOS of this chapter offers few advantages over bipolar circuits when packaged as individual gates or standard MSI networks. Single channel MOS has an inherent speed disadvantage. The circuits function at different voltage and impedance levels than bipolar circuits. Therefore, inputs to an MOS network must pass through special interface circuits. A set of even more complex interface circuits must be placed on each MOS chip to permit the chip outputs to drive bipolar circuits. This interface cost is acceptable for an LSI network but not for smaller MOS configurations. Another reason for waiting until now to consider single channel MOS is that clocking plays a special part in their implementation. By now, the reader should have a sound understanding of clocking in bipolar circuits. This should be helpful in understanding MOS clocking and avoiding confusion between the two philosophies.

The advantages of MOS/LSI are cost, density, and lower power dissipation. A smaller number of transistors and no passive elements are required in an MOS design as compared to the design of a bipolar element to accomplish the same function. More processing steps are required in the fabrication of bipolar circuits than MOS, and MOS circuits can be made to consume less power. Consequently, an MOS array can be produced more economically, with greater yield, and with many more gates per unit area than a similar bipolar array. At this writing, nearly all of the notably complicated LSI circuits, including single-chip calculators, have been realized as MOS devices. The hand-held calculator is particularly compatible with MOS technology since speed is relatively unimportant.

Unless the reader is employed by a manufacturer of MOS devices, he is most likely to encounter this technology in one or both of two ways. He may have occasion to use standard configurations such as ROM's or random access memories. If he decides to implement a complete system in terms of LSI, he will most likely describe it in hardware language (Chapter 15) and provide this description to the vendor. The MOS manufacturer will use this description to develop a set of masks from which as many copies of the circuit as desired can be produced.

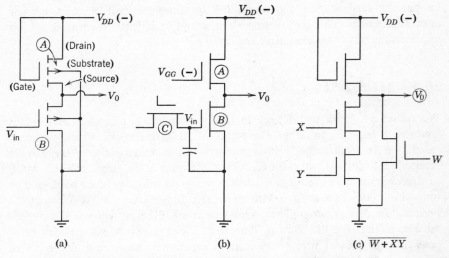

FIGURE 16.9.   *P-channel enhancement mode logic elements.*

The principal approaches to MOS/LSI are *two phase* and *four phase dynamic* and *three phase static*. Within each of these families, there are several subfamilies, each with its own advantages and disadvantages. Each manufacturer has made his own choice of approaches in terms of which to implement his designs. Space will not permit a consideration of more than one approach from each of the three major categories.

We see in Fig. 16.9a a simple inverter constructed from two *p*-channel enhancement mode MOSFETS. For convenience, the upper device (A) has the gate, source, drain, and substrate leads all explicitly labelled. If the input voltage of this type of device is negative, current will flow from source to drain. This is always the case in the upper device which serves as a load resistor whose value will typically be greater than 10 Kohm. Device (B) will be turned off if $V_{in} = 0$ and will be turned on if V_{in} is negative. Device (B) has a much higher transconductance and lower source to drain resistance so that $V_0 = V_{DD}$ (—), if $V_{in} = 0$; and $V_0 = 0$ if V_{in} is negative. Thus, the device of Fig. 16.9 functions in much the same way as a basic bipolar inverter. The dc input impedance of this device is in the order of 10^{14} ohms. Thus, the effect of dc loading will place no restriction on fan-out if this device is used to drive other devices of the same type.

The inverter of Fig. 16.9b differs in only one material respect from the device discussed above, but three additional changes have been made in the illustration. First, the subtrate connections are not shown. We shall assume them to be connected to ground throughout the remainder of the section. A third MOSFET, labeled (C), which may be regarded as part of the predecessor gate, has been added to isolate the point labeled V_{in}. As illustrated, there is always a capacitance (less than 20 pf) between the gate of device B and ground. If device Ⓒ is turned off, there exists no path of less than 10^{14} ohms between point V_{in} and ground. Thus, once a voltage level has been established at point V_{in}, it will remain there without reinforcement for as much as 1 ms. This suggests that *power need be applied to the circuit only at periodic intervals*. It is this feature which accounts for many of the economic and size advantages of *dynamic* MOS. Control on application of power can easily be accomplished by separating the gate of device Ⓐ from the drain as shown. Thus, by letting V_{GG} go to 0, the load resistance is made infinite, and no current is drawn from the power supply by this gate. Figure 16.9c illustrates the MOS implementation of a Boolean function.

The simplest way to apply power only at periodic intervals is to connect the gate of the load MOSFET to a source of clock pulses, as illustrated in Fig. 16.10. It is important that the voltage on the gate lead of a MOSFET not be changing while this voltage is controlling the charging of the input

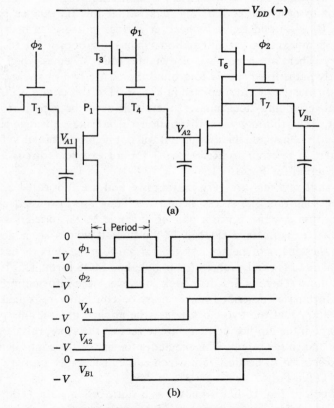

FIGURE 16.10. *2-phase shift register.*

capacitor of a succeeding circuit. This can be insured by using at least two separate clock phases. The timing of the two clock phases, ϕ_1 and ϕ_2, which synchronize the flow of data in Fig. 16.10a is shown in Fig. 16.10b.

This timing diagram will be analyzed in terms of negative logic with 0 volts representing logical 0 and $-V$ representing logical 1. Initially V_{A1} and V_{A2} are at $-V$ and $V_{B1} = 0$. When the first of the illustrated ϕ_1 pulses occurs, transistors T_3 and T_4 are turned on, opening a path between points P_1 and V_{A2}. Since $V_{A1} = -V$, T_2 is turned on, the voltage at point P_1 is 0, and the capacitor at V_{A2} is discharged to 0 volts. When the leftmost ϕ_2 pulse appears, T_6 and T_7 are turned on. Since V_{A2} is now zero, T_5 is turned off and the capacitor at point V_{B1} is charged to $-V$ via the path from $V_{DD}(-)$ through T_6 and T_7. Thus, after one complete clock period, a data bit has been advanced from point V_{A1} to V_{B1}. By adding more inverters in cascade, a dynamic shift register is formed. Two inverters are required for each bit stored.

The disadvantage of the elementary dynamic shift register of Fig. 16.10 as compared to other MOS shift registers is that low impedance paths exist between $V_{DD}(-)$ and ground whenever a clock phase is turned on. Thus, there is significant power dissipation which is proportional to the clock frequency. There are a number of dynamic MOS schemes which eliminate completely paths from supply to ground. Some are two-phase and some are four-phase like the scheme depicted in Fig. 16.11. If this circuit is utilized with four nonoverlapping clock phases, there will never be a path from supply to ground. Thus, the only power dissipation will be leakage through paths of the order of 10^{14} ohms, and this dissipation will not be proportional to frequency. The reader is referred to Reference 3 for an excellent discussion of the timing diagram for Fig. 16.11.

MOS shift registers are very inexpensive and are fabricated as standard parts for semi-random access memories. They are also used extensively within custom designs. Since a pair of dynamic MOS inverters stores data and in fact constitutes a clocked D flip-flop, it would seem desirable to organize these pairs of elements as general purpose registers so that arbitrary sequential circuits can be fabricated in dynamic MOS form. This is possible, and it is done. There is one drawback, however—the information in any D flip-flop formed as described above must be constantly refreshed. That is, it is necessary that the two clock phases be applied at least once each millisecond; and if the flip-flop contents are to remain the same, this data must be gated around to the inputs. This precludes the use of the controlled clocking scheme discussed in Section 15.2 which can considerably simplify the input logic for D flip-flops.

The circuit of Fig. 16.12a is known as a static cell or static D flip-flop. The information stored in this device will be preserved indefinitely even though

1 bit

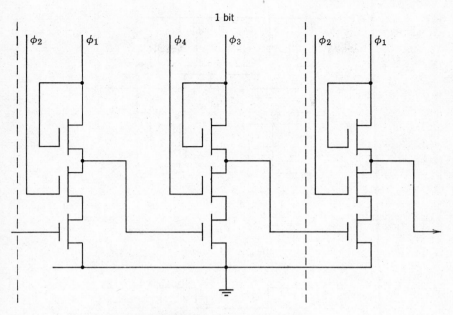

FIGURE 16.11. *4-phase shift register.*

no clocking is applied. The key to the function of this circuit is that clock inputs ϕ_2 and ϕ_3 are normally negative or logical 1. In this way, they create cross-coupling paths between the inputs and outputs of the two inverters. Thus, the output V_0 will not change as long as the third clock phase ϕ_1 remains 0.

One mode of operation is illustrated in Fig. 16.12b. Here it is assumed that clock phases ϕ_2 and ϕ_3 are periodic while ϕ_1 is a controlled clock line on which a pulse will appear only when new information is to be gated into the flip-flop. Notice that initially the input line $V_{in} = 0$ while the contents of the flip-flop represented by V_0 is a logical 1. The first pair of pulses on lines ϕ_2 and ϕ_3 have no effect since the two gate capacitors can easily store their values without reinforcement for the period of time when $\phi_2 = \phi_3 = 0$. Coincident with the second pair of these pulses, a control pulse occurs on line ϕ_1. With T_1 on and T_3 off, point V_A takes on the value of V_{in}, in this case 0. Next, ϕ_2 returns to $-V$ slightly before ϕ_3. This is very important since it allows the output of inverter A to change V_B before inverter B affects V_A. Thus, V_B goes to $-V$ in this case and V_0 goes to 0. When ϕ_3 finally goes on again, V_0 is coupled back to V_A to hold that point at 0.

The circuit of Fig. 16.12a dissipates power whenever ϕ_2 and ϕ_3 are $-V$. More complicated static MOS cells can be devised with less power dissipation.

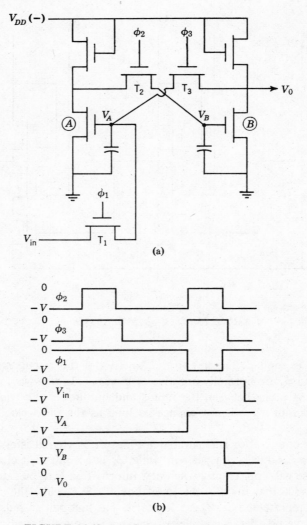

(a)

(b)

FIGURE 16.12.   *3-phase static MOS D flip-flop.*

The cross-coupling feedback lines can be added to most types of dynamic cells to permit them to hold data in the static mode.

In this section, we have only attempted to provide an introduction to MOS/LSI. The designer who is contemplating ordering the custom fabrication of a system in MOS form may wish to understand more thoroughly the family of circuits which will be employed. This information is most easily obtained in consultation with individual vendors.

16.6   Simulation

Whenever an implementation of a digital system includes a custom MSI or LSI part, the system must *"work"* when it is first connected. It is not feasible to make a series of design modifications on the circuit while a prototype is under test on the bench. Each such design change would require the replacement of at least one costly diffusion or metallization mask. In addition, a time delay of days or weeks would be associated with the fabrication of each new part. Consequently, a trial-and-error design on the bench could take the designer weeks and bankrupt his company in the process.

Since design errors are inevitable, some means must be provided for checking out a system before it is actually fabricated. The answer, of course, is *simulation*. How one might approach the simulation of a sequential circuit will be our concern in this section. Very little need be said about the simulation of purely combinational logic. In that case, simulation would usually be only the computation of functional values for all possible input combinations.

There are many approaches to the simulation of sequential circuits. The approach chosen will be a function of the level of detail of circuit performance which must be monitored. As a general rule, the greater the level of detail provided or the greater the precision of the simulation, the greater the computer time consumed. This computer time can be significant when one considers the simulation of a circuit containing 100 or even 1000 gates. Some of the alternative approaches are tabulated in Fig. 16.13a. Figure 16.13b contains a hypothetical plot of the useful information which might be obtained

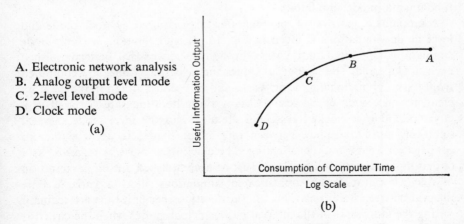

A. Electronic network analysis
B. Analog output level mode
C. 2-level level mode
D. Clock mode

(a)

(b)

FIGURE 16.13.   *Alternative simulations.*

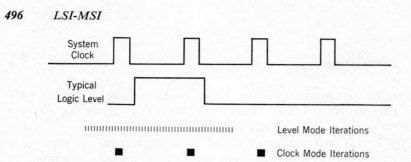

FIGURE 16.14.   *Timing in clock and level model simulations.*

from each type of simulation versus the amount of computation time consumed. No scale is presented as the plot merely presents an intuitive estimate based on some experience with these simulations.

The most accurate simulation would clearly be based on some circuit analysis program such as *SCEPTRE* (point *A* of Fig. 16.13b). By accurately simulating the internal function of individual circuit elements, the precision of the output of all gates as functions of time could not be questioned. Unfortunately, the cost of running such programs is prohibitive for large logic networks.

At the opposite end (point *D*) of the curve in Fig. 16.13b is the clock mode simulation. This type of simulation is the least costly and should be used wherever it will provide the insight required. Unfortunately, not all MSI or LSI parts are designed to operate in the clock mode. Also, there may be concern as to whether some clocked systems can actually be operated at the desired clock frequency. This can only be determined through the use of a fundamental mode simulation.

A graphic comparison of the computer time required by clock mode and level mode simulations is given in Fig. 16.14. By definition, a clock mode simulation implies that one iteration through the network will be accomplished each clock period. That is, output values and next state values are computed from the combinational logic networks and new values stored in the memory elements once each clock period. Many more iterations are required in a clock period if a clocked system is simulated in the level mode. Several iterations must take place between each input change to allow signals to propagate through multilevel gating. The clock itself must be regarded as a circuit input, so several iterations must be accomplished during the duration of a clock pulse. Some sophisticated simulations allow a variable time interval between iterations. Thus, fewer iterations per unit time are required during the latter part of the interval between clock pulses when the circuit is relatively stable. Various other time-saving techniques are possible.

More network elements must be considered in a level mode simulation. A flip-flop which can be treated as a unit in a clock mode simulation must be treated as eight or ten separate gates in a level mode simulation. Thus, each iteration might be more time-consuming. Overall, one should expect at least a 10 to 1 (conceivably 100 to 1) increase in computer time when resorting to a level mode as opposed to a clock mode simulation.

Various approaches to a level mode simulation are possible. An accurate gate model would permit representation of a continuum of voltage levels between the 0 and 1 logic values. Such an approach is represented by point *B* in Fig. 16.13b. Where races and hazards are possibilities, this type of simulation has merit. Alternatively, gate outputs may be restricted to the logic levels 0 and 1 (Point *C*). We shall treat this case in more detail in the next several paragraphs.

In simulating a level mode circuit, it is convenient to consider the output of each gate (not just the secondaries) as state variables. Physically, the block labeled *connection matrix* in Fig. 16.15a is nothing more than a network of wires. In a computer program, this function is simulated by a shuffling of values from the array storing gate outputs to the gate input storage array. This shuffle can be organized in a variety of ways. We leave this problem to the individual programmer.

The model for a single gate is given in Fig. 16.15b. Implementing this model requires 3 steps during each iteration through the network. The first step involves computing the logical function specified for a particular gate for the current gate inputs. The second step models the gate delay, and the third step determines an updated value of the gate output. The delay and rise time

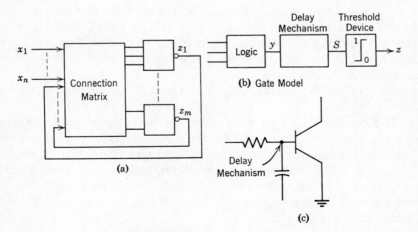

FIGURE 16.15. *Level mode model.*

Y	S^v	S^{v+1}
0	$S_{\min}$	$S_{\min}$
0	$\alpha > S_{\min}$	$\alpha - 1$
1	$\beta < S_{\max}$	$\beta + 1$
1	$S_{\max}$	$S_{\max}$

S^{v+1}	z^{v+1}
$S^{v+1} < 0$	0
$S^{v+1} \geq 0$	1

(a) Delay counter   (b) Output threshold

FIGURE 16.16.

of any practical logic circuits are influenced by a number of time constants. In the simplest possible gate model which we shall consider here, rise time is neglected altogether. That is, the gate output may jump from 0 to 1 or 1 to 0 in a single iteration. We imagine delay to be controlled by a single time constant, depicted at the base of a transistor in Fig. 16.11c. A discrete approximation of this nonlinear integration is a saturating up-down counting mechanism as specified in Fig. 16.16a.

$S_{\min}$ is a small negative integer, and $S_{\max}$ is a small positive integer. The combined effect of the delay counter and output threshold may be explained as follows. If the logical function, y, of the gate inputs is 0 for iteration, v, the count, S, is reduced. If $y = 1$, S is increased. As specified in Fig. 16.16a, S cannot exceed $S_{\max}$ or assume a value less than $S_{\min}$. In Fig. 16.16b, we see that the new output will be 1 if $S^{v+1} \geq 0$. If the $S^{v+1} < 0$, the output will be 0. Suppose, for example, that S is initially saturated at $S_{\min}$ and $z = 0$. The output will change to 1 only after $y = 1$ for $S_{\min}$ successive iterations. The situation is similar where the delay counter is initially saturated at $S_{\max}$. Thus, $S_{\min}$ and $S_{\max}$ represent nominal gate delay values. A gate modeled by Fig. 16.16 will exhibit most of the important properties of an actual circuit. Very narrow pulses, for example, can disappear in a network modeled in this way. The principal advantage of the counting scheme is efficiency of programming. It can be implemented on small fixed-point machines, no multiplication or division is required, and the counts can be stored as 8-bit bytes.

Each iteration through the complete network is characterized by the flow chart of Fig. 16.17. The important point to notice in Fig. 16.17 is that each step is accomplished for all gates before the next step is begun. Thus, all gate inputs remain unchanged until the output of every gate is updated. Gates may, therefore, be considered in any order irrespective of network connections. After some number K iterations, the program enters a supervisory routine, which permits changes in the circuit inputs $x_1 \cdots x_n$ and performs various housekeeping operations. We shall not concern ourselves with the details of this routine.

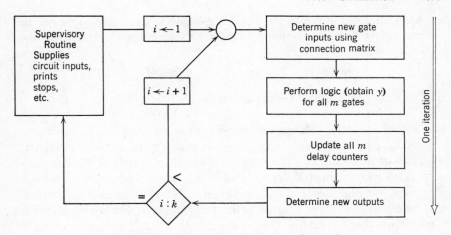

FIGURE 16.17. *Flow chart for level mode simulation.*

As pointed out earlier, many approaches to level mode simulation are possible. We have only attempted to show what might be done using one very economical gate model.

Example 16.5

Illustrate the sum, S, and the output, z, of the AND gate in Fig. 16.18 for several iterations with inputs as shown. Assume the delay model of Fig. 16.17. Let $S_{max} = 2$ and $S_{min} = -3$.

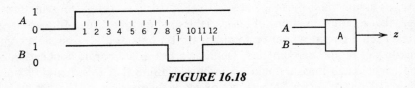

FIGURE 16.18

Solution: We assume that S is initially equal to S_{min}. The value of S begins to change when the input A goes to one at iteration 1. The output changes to 1 after iteration 3 as shown in Fig. 16.19, while S continues to increase until saturating at iteration 5. We note that the new output will not affect any other gate until iteration 4.

Iteration	1	2	3	4	5	6	7	8	9	10	11	12	13	14	15
S	-3	-2	-1	0	1	2	2	2	1	0	-1	0	1	2	2	2

z

FIGURE 16.19

At iteration 8, the narrow pulse on line B begins to affect S. We note that the output pulse is narrower than the input pulse. This can happen for pulses narrower than two gate delays. ■

16.7   Test Sequence Generation

The testing phase for discrete components or small-scale integrated circuits (individual gates) was usually regarded as a peripheral activity, seemingly added as an afterthought. The attitude toward the testing of MSI and LSI parts is necessarily much different. In Fig. 16.20, we see a representation of what might be regarded as a typical MSI part. The 100 gates in the network may be assumed to be connected in most any manner. The network may be combinational or sequential. One of the inputs x_i may represent a clock, or the circuit may be designed in the level mode. As is commonly the case, we have shown more inputs than outputs.

The black box diagram of Fig. 16.20 may seem unrevealing, but it bears a meaningful resemblance to the dual-in-line package as it is presented for test. In particular, all measurements must be taken on the four output leads, z_1, z_2, z_3, and z_4. Thus, if there is a failure in the network, its effect must be forced to appear on one of these lines. That is, an output of the faulty network must be caused to differ from what is known to be the value of the same output of the good network. All signals introduced into the circuit to facilitate these must be placed on the 10 input lines shown.

Let us assume first that the network of Fig. 16.20 is strictly combinational logic, that is, four functions of ten variables. The most straightforward way to test such a circuit would be to check the outputs as all 2^{10} possible combinations of inputs were applied. Defining a test as a separate application of input signals and measurement of outputs, we find that this method would require 1024 tests to verify the functioning of the network of Fig. 16.20.

Now let us conjecture that we have somehow determined that there are no more than 500 failures which can occur in the network of Fig. 16.20. Since the network is combinational, it may be possible to apply a set of inputs which will cause the effect of any given failure to appear at the output. In this case, no more than 500 tests, one for each failure, would be required. Very often, one test will cause an erroneous output in the event of any one of several different failures. Thus, the number of tests required might be

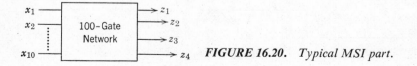

FIGURE 16.20.   *Typical MSI part.*

considerably less than 500. Various approaches have been developed for finding an approximately minimal set of tests for complete testing of a combinational logic network.[9,10] Perhaps the best known is the D-algorithm developed by Roth. Space will not permit a detailed presentation of these methods. The reader is referred to reference 6. We shall merely present a short example to suggest what may be involved.

Example 16.6

Suppose input 2 of the leftmost AND gate in Fig. 16.21 is stuck at zero (SA0) as shown. Thus, the output of this AND gate will always be 0. The output

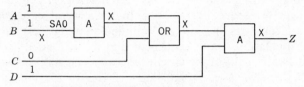

FIGURE 16.21.   Test determination.

of this gate in the good network will be 1 only if $A = B = 1$. Thus A and B must be 1 in any test for this fault. Assuming $A = B = 1$, we place an X at the output of the leftmost AND gate to indicate that this value in the good network will differ from the value in faulty network shown. It is now necessary to choose values of C and D which will cause this value conflict to be propagated to the network output. Using the 3-value logic of Chapter 14, we see that the NOR gate output will be X if and only if $C = 0$. From here, we see that the circuit output will be X if and only if $D = 1$.

A test has now been determined for the fault under consideration. We note, however, that the same test will detect the following additional faults: leftmost AND gate output SA0 Input 1 to NOR gate SA0, NOR gate output SA1, and input 1 of rightmost AND gate SA1. ■

In the above example, only circuit failures which could be modeled as stuck inputs and outputs of individual gates were considered. This is the common practice. It is also common practice to assume in developing a set of tests that no more than one failure will occur at one time. This is referred to as the *single fault assumption*. It may be justified in part by the following observations.

1. An arbitrary multiple fault is much less likely to occur than a particular single fault.
2. Massive multiple faults resulting from processing problems will be easily detected by most any set of tests.

Perhaps the most compelling justification of the single fault assumption is that it is simply the only way to proceed. In a circuit which could have 500 single faults, 2^{500} combinations of multiple faults would be possible. Clearly the enumeration of multiple faults is impractical.

So far, we have discussed only combinational logic networks. The necessity of relying on the single fault assumption is even stronger for sequential networks. The option of trying all possible combinations of input values does not exist in the latter case. The outputs are a function of the state of the network as well as the input values. A long sequence of inputs may be required to drive a sequential circuit to a particular state.

As yet no algorithmic procedure has been devised for determining optimal test sequences for sequential circuits. In the general case, this is a very difficult problem. It is the authors' opinion that the best approach is an interactive simulation. Input sequences may be suggested by the designer based on his knowledge of the circuit function. The good network as well as all single fault networks are then simulated (to some extent in parallel) for these sequences. Very likely the first sequence suggested will not cause the effect of every fault to be driven to the output during the simulation. The designer then uses the results of this simulation together with his knowledge of the network to suggest a continuation of the input sequence. The simulation is continued, and the process is repeated until a test sequence which will detect all faults is obtained.

The simulation may be either clock or fundamental mode. A clock mode simulation can make use of the 3-value logic to handle indeterminate states which can arise in the simulation of faulty memory elements. The initial state of a network is indeterminate at the beginning of the testing process. This situation can be simulated by assigning initial X values to the memory element outputs.

The reader is referred to the literature[5,8] for further details of these simulations. Our purpose here has been merely to stress the importance of test sequence generation and stimulate the reader's interest in the topic.

PROBLEMS

16.1. A standard MSI part is available which adds two 4-bit binary numbers. Thus, it has eight inputs and five outputs. Use this part in the design of a hybrid circuit which will add two binary coded decimal digits. This circuit should have five output bits, four representing the BCD sum digit and one representing a possible carry.

16.2. An available MSI part has four inputs and one output. The output will be 1 if and only if an even number of the input bits are 1.

A second MSI part, an exclusive-OR element, is also available. Use these two parts in the design of a hybrid configuration which will check for even parity over seven bits.

16.3. A 4-bit (modulo-16) clock mode counter is available as an MSI part. Its two inputs include the clock and a level reset line. The count is represented in binary on four output lines. The count increases with each clock pulse unless the reset line is 1. In that case, the count is reset to zero. Use this part in the design of a hybrid decade counter (counts 0–9).

16.4. Use the techniques of Chapter 10 to determine a minimal state and minimal flip-flop sequencer for Example 16.3. Compare your realization with the one given in Fig. 16.7 for overall economy.

16.5. Assume that the overall costs of manufacturing a finished digital product are made up of fixed costs (essentially independent of the approach used), wiring costs, and IC component costs. Ignore the fixed costs and assume that the wiring costs are proportional to the number of IC's used. Rework Example 16.4 assuming a wiring cost of $.50 per IC package. Determine a crossover value of K in terms of total cost between the individual gate approach and the hybrid MSI approach.

16.6. Assume wiring costs as given in Problem 5. Suppose that a custom part can be supplied to satisfy the design discussed in Example 16.4 at a cost of $2000 plus $2.00 per copy. Assume $K = $2.00. Derive an inequality, similar to 16.13, relating the total cost of the hybrid approach to the cost of custom parts.

16.7. A standard ROM consisting of 256 eight-bit words is under consideration for a particular application. The manufacturer quotes a standard Step-2 metallization charge of $200 plus $.03 per bit. He quotes an incremental production charge of $.02 per part. If only 100 copies of the ROM are required, compute the per-unit cost to the user. Compute this per-unit cost if 1000 copies were required.

16.8. Construct a detailed circuit diagram of a two-phase MOS implementation of the sequential circuit described by the output and next state equation

$$Z = y^{v+1} = X \oplus y^v$$

The circuit has one output, Z, one input, X, and one state variable y. Let clock phase ϕ_1 control all input logic and let ϕ_2 control the load and output of an inverting D flip-flop.

16.9. Repeat Problem 16.8 for four-phase MOS.

16.10. Use the technique of Section 16.6 to manually step through a simulation of the circuit in Fig. P16.10. Let $S_{max} = 2$ and $S_{min} = -3$ for all gates.

    (a) Assume that the circuit is initially stable with $y_1 = 1$ and $D = C = 0$. Begin the simulation as C goes to 1. After 6 iterations, C returns to 0. Continue the simulation for a total of 15 iterations.

    (b) Assume that the circuit is initially stable, with $D = C = y_1 = 1$. Let C go to 0 at the beginning of the simulation. Carry out the simulation for 15 iterations.

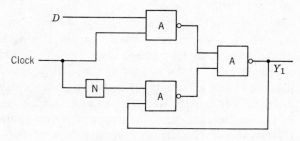

FIGURE P16.10

16.11. A single inverter is to be simulated using the technique of Section 15.5. Let $S_{max} = 3$ and $S_{min} = -4$. Suppose the gate is initially stable with an input of logical 0. Manually calculate 15 iterations of the simulation beginning at the onset on a narrow input pulse.

    (a) pulse duration = 3 iterations

    (b) pulse duration = 5 iterations

    (c) pulse duration = 9 iterations

16.12. Devise a minimal set of tests which will test for all single faults which can occur in the circuit of Fig. 16.21. Consider all stuck inputs and stuck outputs.

16.13. Use trial-and-error techniques to devise any input sequence which will test for all faults in the simple fundamental mode sequential circuit of Fig. P16.10.

BIBLIOGRAPHY

1. Hill, F. J. and G. R. Peterson. *Digital Systems: Hardware Organization and Design*, Wiley, New York, 1973.

2. Harrison, M. A. *Introduction to Switching and Automata Theory*, McGraw-Hill, New York, 1965.

3. Carr, W. N. and J. P. Mize. *MOS/LSI Design and Application*, McGraw-Hill, New York, 1972.

4. Kohonen, T. *Digital Circuits and Devices*, Prentice-Hall, Englewood Cliffs, N.J., 1972.

5. Blair, K. "Penetrate the Mystique of MOS Specs," *Electronic Design*, **20**, N. 8 (April 1972).

6. Boysel, L. L. and J. P. Murphy. "Four Phase LSI Offers A New Approach to Computer Design," *Computer Design*, **9**, No. 4 (April 1970).

7. Correia et al., "Minimizing the Problem of Logic Testing by the Interaction of a Design Group with User Oriented Facilities," *Proceedings of the 1970 Design Automation Symposium*, 100.

8. Hill, F. J. and C. Meyer. "Interaction with a Simulation for the Determination of Fault Detection Sequences for LSI Circuits," *IEEE Region 6 Conference Proceedings* (April 1969).

9. Chang, H. Y., Eric G. Manning and G. Metze. *Fault Diagnosis of Digital Systems*, Wiley-Interscience, New York, 1970.

10. Roth, J. P., W. G. Bouricius, and P. R. Schnieder. "Programmed Algorithms to Compute Tests to Detect and Distinguish between Failures in Logic Circuits," *IEEE Transactions on Electronic Computers*, EC-16, 567–579 (Oct. 1967).

11. Seshu, S. "An Improved Diagnosis Program," *IEEE Transactions on Electronic Computers*, **EC-14**, 76–79 (Jan. 1965).

12. Chang, H. Y. "An Algorithm for Selecting an Optimum Set of Diagnostic Tests," *IEEE Transactions on Computers*, **EC-14**, 706–711 (Oct. 1965).

13. Galey, J. M., R. E. Norby and J. P. Roth. "Techniques for Diagnosis of Switching Circuit Failures," *IEEE Transactions on Communications and Electronics*, **33**: 74, 509–514 (1964).

17.1 Introduction

In Chapters 6, 7, and 8, we dealt with general techniques applicable to the design of combinational circuits of all types, with emphasis on two-level, or second-order, design. In this chapter, we return to combinational design to consider some special classes of logic functions which do not lend themselves readily to second-order design and some special techniques for dealing with such functions.

One particular type of function which is not economically realized in second-order form is that in which the function value is determined only by the *number* of literals which take the value 1. Such functions are called symmetric functions and will be discussed in detail in the next section. Section 17.5 deals with the general problem of breaking up Boolean functions into higher-order representations. These representations are called decompositions. We shall see that symmetric functions are readily decomposable functions. Later we shall see that symmetric functions are also conveniently realizable in terms of threshold logic elements. Threshold logic is the topic of Chapter 18.

In Section 17.7, we turn to another special type of combinational circuit, the *iterative* circuit, made up of a cascade of identical subcircuits. We shall see that the design of such circuits has much in common with sequential circuit design. The concluding sections of this chapter develop the concepts of lattices and unate functions. These ideas are fundamental to threshold logic and, indeed, to many current topics of switching theory research.

17.2   Symmetric Functions

In Chapter 8, we encountered the expression

$$f(c, x, y) = c\bar{x}\bar{y} + \bar{c}x\bar{y} + \bar{c}\bar{x}y + cxy \tag{17.1}$$

in the generation of a sum bit in binary addition (see Fig. 8.2a). This function is an example of a *symmetric function*.

As we learned, symmetric functions may often be most economically realized with more than two levels of logic. Notice that the value of Expression 17.1 is 1 whenever an odd number of the variables are 1. The n-variable odd parity check function is the extension of this pattern to n variables. An efficient multilevel NAND gate realization of this latter symmetric function was also given in Chapter 8. Still other realizations of symmetric functions in terms of threshold logic elements are discussed in Chapter 18.

In general, a symmetric function may be defined as follows.

Definition 17.1. A switching function $f(x_1^{j_1}, x_2^{j_2}, \ldots, x_n^{j_n})$ is *symmetric* with respect to the literals $x_1^{j_1}, x_2^{j_2}, \ldots, x_n^{j_n}$, if and only if it remains unchanged after any permutation of these literals, where the j_i may take on only the values 0 and 1, and

$$x_i^{j_i} = x_i \quad \text{if } j_i = 0 \qquad \text{and} \qquad x_i^{j_i} = \bar{x}_i \quad \text{if } j_i = 1 \tag{17.2}$$

Consider, for example, expression 17.1. Since it contains only three variables, we may check exhaustively all six permutations of the variables to determine if the expression remains unchanged. In Fig. 17.1, we see Expression 17.1 as it appears after each permutation of $\{c, x, y\}$. It should

Permutation	Expression
1 Identity: $c \to c, x \to x, y \to y$	$c\bar{x}\bar{y} + \bar{c}x\bar{y} + \bar{c}\bar{x}y + cxy$
2 $c \to x, x \to y, y \to c$	$x\bar{y}\bar{c} + \bar{x}y\bar{c} + \bar{x}\bar{y}c + xyc$
3 $c \to y, y \to x, x \to c$	$y\bar{c}\bar{x} + \bar{y}c\bar{x} + \bar{y}\bar{c}x + ycx$
4 $c \to c, x \to y, y \to x$	$c\bar{y}\bar{x} + \bar{c}y\bar{x} + \bar{c}\bar{y}x + cyx$
5 $c \to x, x \to c, y \to y$	$x\bar{c}\bar{y} + \bar{x}c\bar{y} + \bar{x}\bar{c}y + xcy$
6 $c \to y, x \to x, y \to c$	$y\bar{x}\bar{c} + \bar{y}x\bar{c} + \bar{y}\bar{x}c + yxc$

FIGURE 17.1. *Permutations of the variables of Expression 14.1.*

be clear to the reader that the function is unchanged in each case. The function

$$x\bar{y}z + \bar{x}yz + \overline{xy}z \tag{17.3}$$

is also symmetric, but in this case the literals of symmetry are x, y, and $\bar{z}$ rather than x, y, and z.

This becomes clear if we let

$$q = \bar{z}$$

Thus, Expression 17.3 becomes

$$x\bar{y}\bar{q} + xy\bar{q} + \overline{xy}\bar{q}$$

which is clearly unchanged by a permutation of the variables.

The reader may notice that functions symmetric with respect to some complemented literals and some uncomplemented literals may not be easily identified. Symmetric functions have some very interesting properties which are expressed in the following set of theorems.

As was mentioned previously, the value of a symmetric function depends only on the number of literals of symmetry which are 1. This is set down formally in the following key theorem, which was originally stated by Shannon.[8]

THEOREM 17.1. *A necessary and sufficient condition that a switching function of n variables be symmetric is that it may be specified by a set of integers $\{a_k\}$ where $0 \le a_k \le n$, such that if exactly a_m $(m = 1, 2, \ldots, k)$ of the literals of symmetric have the value 1, then the function has the value 1, and not otherwise.*

Proof:

A. (Necessity) Suppose that a function is 1 when the first a_m literals are 1 but is 0 when some other a_m literals are 1. Suppose each of the a_m literals of the first set is permuted to one of the literals of the second set. Clearly, a different function will result. Thus, the function would not be symmetric.

B. (Sufficiency) Suppose that a function is 1 if and only if exactly a_m of the literals are 1. Following any permutation of the literals, the function will still be 1 when any a_m of the literals are 1. Similarly for any a_m in $\{a_k\}$, so the function is symmetric.   Q.E.D.

We may now express any symmetric function in the form

$$S^n_{\{a_k\}}(x_1^{j_1}, x_2^{j_2}, \ldots, x_n^{j_n}) \tag{17.4}$$

Thus, expression 17.1 may be written as

$$S^3_{1,3}(c, x, y) \tag{17.5}$$

and expression 17.3 as

$$S_1{}^3(x, y, \bar{z}) \tag{17.6}$$

Similarly, the n-variable even parity check function may be written

$$S_{0,2,4,\dots,n}^n(x_1, x_2, \dots, x_n) \tag{17.7}$$

if n is even. If n is odd, the last subscript is $n - 1$.

17.3 Boolean Combinations of Symmetric Functions

Theorem 17.1 not only permits a convenient notation for symmetric functions but it also provides for a considerable simplification of the algebraic operations on symmetric functions.

THEOREM 17.2.

$$S_{a_1,\dots,a_i,b_1,\dots,b_m}^n(x_1{}^{j_1}, x_2{}^{j_2}, \dots, x_n{}^{j_n}) + S_{a_1,\dots,a_i,c_1,\dots,c_k}^n(x_1{}^{j_1}, x_2{}^{j_2}, \dots, x_n{}^{j_n})$$
$$= S_{a_1,\dots,a_i,b_1,\dots,b_m,c_1,\dots,c_k}^n(x_1{}^{j_1}, x_2{}^{j_2}, \dots, x_n{}^{j_n}) \tag{17.8}$$

Proof: The proof follows immediately from Theorem 17.1. That is, a function which is 1 when $a_1, \dots, a_j, b_1, \dots, b_{k-1}$ or b_m of the literals are 1 or when $a_1, \dots, a_j, c_1, \dots, c_{k-1}$ or c_k of the same literals are 1 is 1 when $a_1, \dots, a_j, b_1, \dots, b_m, c_1, \dots, c_{k-1}$ or c_k of the literals are 1.

THEOREM 17.3.

$$[S_{a_1,\dots,a_i,b_1,\dots,b_m}^n(x_1{}^{j_1}, x_2{}^{j_2}, \dots, x_n{}^{j_n}) \cdot [S_{a_1,\dots,a_i,c_1,\dots,c_k}^n(x_1{}^{j_1}, x_2{}^{j_2}, \dots, x_n{}^{j_n})]$$
$$= S_{a_1,\dots,a_i}^n(x_1{}^{j_1}, x_2{}^{j_2}, \dots, x_n{}^{j_n}) \tag{17.9}$$

Proof: Similar to that of Theorem 17.2.

Example 17.1

$$S_{1,2,4}^5(V, W, X, Y, Z) + S_{2,3,4}^5(V, W, X, Y, Z) = S_{1,2,3,4}^5(V, W, X, Y, Z)$$

while

$$[S_{1,2,4}^5(V, W, X, Y, Z)] \cdot [S_{2,3,4}^5(V, W, X, Y, Z)] = S_{2,4}^5(V, W, X, Y, Z)$$

It is possible to state another theorem which follows directly from Theorem 17.1. Since the general statement of the theorem is rather awkward, we shall instead illustrate it by example. ∎

Example 17.2

$$\overline{S_{1,2,3,6}^6(x_1, x_2, \dots, x_6)} = S_{0,4,5}^6(x_1, x_2, \dots, x_6) \tag{17.10}$$

Notice that $S_{0,4,5}^6$ is 1 when 0, 4, and 5 of the literals $x_1, x_2, \dots, x_6$ are 1. The functions $S_{1,2,3,6}^6$ is never 1 when $S_{0,4,5}^6$ is 1, but the former function is 1 for every number of 1-literals for which $S_{0,4,5} = 0$. The complement of any symmetric function of any number of variables may be obtained in this way. ∎

The theorem illustrated by Example 17.2 concerns the complement of a symmetric function. The next theorem concerns the function obtained by complementing the variables of a symmetric function.

THEOREM 17.4.

$$S_{a_1,a_2,\ldots,a_k}^n(x_1^{j_1}, x_2^{j_2}, \ldots, x_n^{j_n})$$
$$= S_{n-a_1,\ldots,n-a_k}^n(x_1^{1-j_1}, x_2^{1-j_2}, \ldots, x_n^{1-j_n}) \quad (17.11)$$

Proof: If a_i of the literals $x_1^{j_1}, x_2^{j_2}, \ldots, x_n^{j_n}$ are 1, then $n - a_i$ of these literals are 0, and $n - a_i$ of the literals $x_1^{1-j_1}, x_2^{1-j_2}, \ldots, x_n^{1-j_n}$ are 1. Thus, the two sides of Equation 17.11 are equal for any assignment of variable values.

Example 17.3

$$S_{2,3}^4(A,B,C,D) = A,B,\bar{C},\bar{D} + A\bar{B}C\bar{D} + A\bar{B}\bar{C}D + \bar{A}BC\bar{D} + \bar{A}B\bar{C}D + \bar{A}\bar{B}CD$$
$$+ ABC\bar{D} + AB\bar{C}D + A\bar{B}CD + \bar{A}BCD$$
$$= S_2^4(\bar{A},\bar{B},\bar{C},\bar{D}) + S_1^4(\bar{A},\bar{B},\bar{C},\bar{D}) = S_{4-2,4-3}(\bar{A},\bar{B},\bar{C},\bar{D}) \quad (17.12)$$

THEOREM 17.5. *Any symmetric function* $S_{a_1,a_2,\ldots,a_k}^n(x_1^{j_1}, x_2^{j_2}, \ldots, x_n^{j_n})$ *may be represented by the conjunction of a subset of the following* elementary symmetric functions:

$$\sigma_0{}^n = x_1^{1-j_1} \cdot x_2^{1-j_2} \cdots x_n^{1-j_n}$$
$$\sigma_1{}^n = x_1^{j_1} \cdot x_2^{1-j_2} \cdots x_n^{1-j_n} + x_1^{1-j_1} \cdot x_2^{j_2} \cdot x_3^{1-j_3} \cdots x_n^{1-j_n}$$
$$+ \cdots + x_1^{1-j_1} \cdot x_2^{1-j_2} \cdots x_{n-1}^{1-j_{n-1}} \cdot x_n^{j_n}$$

$$\vdots$$

$$\sigma_n{}^n = x_1^{j_1} x_2^{j_2} \cdots x_n^{j_n}. \quad (17.13)$$

Proof: Each elementary symmetric function, σ_i, is 1 if and only if exactly i of the literals are 1. Thus, by Theorems 17.1 and 17.2, any symmetric function may be formed as the conjunction of a subset of these functions.

Example 17.4

$$S_{0,2,4}^4(A, B, C, \bar{D}) = S_0^4(A, B, C, \bar{D}) + S_2^4(A, B, C, \bar{D}) + S_4^4(A, B, C, \bar{D})$$
$$= \bar{A}\bar{B}\bar{C}D + (AB\bar{C}D + A\bar{B}CD + A\bar{B}\bar{C}\bar{D} + \bar{A}BCD + \bar{A}B\bar{C}\bar{D}$$
$$+ \bar{A}\bar{B}C\bar{D}) + ABC\bar{D}$$
$$= \sigma_0^4 + \sigma_2^4 + \sigma_4^4$$

Example 17.5

A certain communication system utilizes 16 parallel lines to send four 4-bit characters at a time. Each of these 4-bit characters has a parity bit to establish even parity over the four bits of that character. Parity is to be checked at a certain relay point in the communications system, and retransmission of all four characters is to be demanded by the relay point if there is odd parity over two or more of the characters.

Write, in symmetric function notation, the expression for a combinational circuit whose output is 1 whenever retransmission is to be demanded. Let w_1, w_2, w_3, w_4; x_1, x_2, x_3, x_4; y_1, y_2, y_3, y_4; and z_1, z_2, z_3, z_4 represent the bits of the four characters.

Solution: It is easily seen that the overall expression is not a symmetric function Consider, for example, the following sets of bits, which might be received at the relay point.

$w_1\ w_2\ w_3\ w_4$	$x_1\ x_2\ x_3\ x_4$	$y_1\ y_2\ y_3\ y_4$	$z_1\ z_2\ z_3\ z_4$
1 1 1 0	1 1 1 0	1 1 0 0	0 0 0 0

$w_1\ w_2\ z_3\ w_4$	$x_1\ x_2\ z_4\ x_4$	$y_1\ y_2\ y_3\ y_4$	$z_1\ z_2\ w_3\ x_4$
1 1 0 0	1 1 0 0	1 1 0 0	0 0 1 1

For the first set of bits, the functions should be 1 since the first two characters have odd parity. The second set of bits has the same number of 1's, but the output should be 0 since all characters have correct parity.

It is, however, straightforward to write the parity check function over each set of four bits as a symmetric function. For example, odd parity over w_1, w_2, w_3, and w_4 is given by

$$P_w = S_{1.3}^4(w_1, w_2, w_3, w_4)$$

Similarly, the function which determines from the four parity checks whether two or more of these is 1 is also symmetric. Thus, overall function may be expressed as:

$$f = S_{2.3.4}^4[S_{1.3}^4(w_1, w_2, w_3, w_4),\ S_{1.3}^4(x_1, x_2, x_3, x_4)$$
$$S_{1.3}^4(y_1, y_2, y_3, y_4),\ S_{1.3}^4(z_1, z_2, z_3, z_4)] \quad \blacksquare$$

We have said nothing so far about the physical implementation of symmetric functions. In terms of NAND or NOR gates, symmetric functions enjoy no advantage over other functions with respect to economic realization. In applications which will tolerate relays or the linear single threshold devices to be discussed in the next chapter, symmetric functions may be realized quite economically.

17.4 Higher-Order Forms

In Fig. 6.21, we saw an example of a function in which a higher-order realization was more economical than the best second-order realization. Nevertheless, we have so far concentrated on second-order circuits, primarily on the assumption that their ease of design and higher speed, would outweigh any possible cost disadvantages. This assumption is quite generally justified, but there are certainly situations where the cost factor will require the use of

CD \ AB	00	01	11	10
00	1		1	
01		1		1
11	1		1	
10		1		1

FIGURE 17.2. $S_{0,2,4}^4(A, B, C, D)$.

higher-order circuits. Perhaps the simplest and most commonly applied method of finding higher-order forms of switching functions is factoring. Factoring generally involves starting with a second-order form, looking for common terms, and then applying the distributive laws to factor them out.

One particular class of functions for which factoring frequently pays off is the class of symmetric functions, discussed in the last section. Let us consider the even parity function of four variables, $S_{0,2,4}^4(A, B, C, D)$. The K-map of this function is shown in Fig. 17.2.

We see that this function, as is the case with all parity functions, results in a "checkerboard" pattern on the K-map. From the standpoint of a second-order realization, this pattern is highly undesirable. There is not a single possible combination of 0-cubes into larger cubes, so an S-of-P realization will require eight 4-input AND gates and one 8-input OR gate, for a total of nine gates with 40 inputs. We might note that this function is the worst possible case for second-order realization. Deleting any minterms would obviously simplify the function, and moving or adding any minterms will result in larger cubes, also simplifying the function. From this observation, we can show that the most expensive possible nonredundant second-order realization of a function of n variables will have

$$N_g = 2^{n-1} + 1 \tag{17.14}$$

gates with

$$N_i = (n + 1)2^{n-1} \tag{17.15}$$

inputs.

We have already demonstrated in Chapter 8 a less expensive higher-order realization of the 8-variable parity check function. Let us now approach this problem from the point of view of factoring. To start, we shall first write the canonic expression for $S_{0,2,4}^4(A, B, C, D)$, which can be read directly from the map:

$$
\begin{aligned}
f(A, B, C, D) = {} & \bar{A}\bar{B}\bar{C}\bar{D} + AB\bar{C}\bar{D} + \bar{A}B\bar{C}D + A\bar{B}\bar{C}D \\
& + \bar{A}\bar{B}CD + ABCD + \bar{A}BC\bar{D} + A\bar{B}C\bar{D}
\end{aligned}
\tag{17.16}
$$

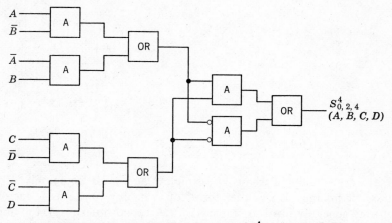

FIGURE 17.3.   *Factored realization of $S_{0,2,4}^4 (A, B, C, D)$.*

The map is of further assistance as we search for common factors in the various terms. We note, for example, that the product $\bar{C}\bar{D}$ is common to the two minterms in the first row, which can thus be factored to the form

$$\bar{A}\bar{B}\bar{C}\bar{D} + AB\bar{C}\bar{B} = \bar{C}\bar{D}(\bar{A}\bar{B} + AB)$$

Similarly, $\bar{C}D$ is common to the minterms in the second row, CD to the minterms in the third row, and $C\bar{D}$ to the fourth-row minterms. The function can then be factored to the form

$$f(A, B, C, D) = \bar{C}\bar{D}(\bar{A}\bar{B} + AB) + \bar{C}D(\bar{A}B + A\bar{B})$$
$$+ CD(\bar{A}\bar{B} + AB) + C\bar{D}(\bar{A}B + A\bar{B}) \qquad (17.17)$$

The four terms in Equation 17.17 may be combined in pairs to form

$$f(A, B, C, D) = (\bar{C}\bar{D} + CD)(\bar{A}\bar{B} + AB) + (\bar{C}D + C\bar{D})(\bar{A}B + A\bar{B})$$
$$(17.18)$$

Finally, noting the relationship of the factors in Equation 17.18 to the exclusive-OR, we have the form

$$f(A, B, C, D) = \overline{(C \oplus D)}\overline{(A \oplus B)} + (C \oplus D)(A \oplus B) \qquad (17.19)$$

This equation leads to the realization of Fig. 17.3, requiring nine gates with 18 inputs, plus two inverters. We have thus eliminated 22 inputs at the cost of two inverters, generally a quite worthwhile saving.

If the circuit is to be realized with NAND logic, we can eliminate the inverters at the cost of two more inputs, by a technique known as "bundling."[3] This technique is applicable whenever a signal derived from a NAND (or NOR) circuit is to be inverted at the input to some other gate, as shown in

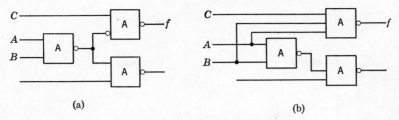

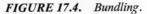

(a)

(b)

FIGURE 17.4. *Bundling.*

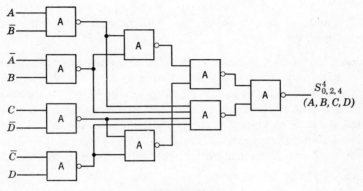

$$S^4_{0,2,4}$$
$$(A, B, C, D)$$

FIGURE 17.5

Fig. 17.4a. This places two inversions in series, which cancel, so that the inputs to the first gate can be "bundled" together and passed on directly to the input of the second gate, as shown in Fig. 17.4b. Applying this technique to the circuit of the parity checker results in the circuit of Fig. 17.5 having nine gates with 20 inputs.

17.5 Simple Disjoint Decomposition

Simplification of circuits by factoring is a satisfactory technique in many cases, but it is largely a trial-and-error process, dependent on a good deal of experience and often just plain luck in spotting common factors in the function. It is evident that a more systematic method of finding simpler designs of higher order would be very useful. In essence, the process of simplification involves breaking a complex single function into a number of related simpler functions. The general name for this process is *functional decomposition*. The methods to be presented here, which we believe to be the most generally useful, are primarily the work of Ashenhurst[4] and Curtis.[5]

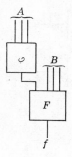

f   **FIGURE 17.6.**   *Simple disjoint decomposition.*

The first type of decomposition we shall consider is the simple disjoint decomposition.

Definition 17.2.   Let $x_1, x_2, \ldots, x_n$ be a set of n switching variables and let A and B be disjoint subsets of X such that $A \cup B = X$. If $F[\varphi(A), B]$ takes on the same functional values as $f(X)$ wherever the latter are specified, then F and φ are said to form a *simple disjoint decomposition* of f. The set of variables, A, shall be known as the *bound* variables, the set, B, as the *free* variables. The partitioning of the variables shall be indicated by $B \mid A$. The basic form of circuit realization of the simple disjoint decomposition is shown in Fig. 17.6.

Example 17.5

Obtain a simple disjoint decomposition of

$$f_1(x_1, x_2, x_3, x_4) = x_1\bar{x}_2x_3x_4 + x_1x_2x_3\bar{x}_4 + \bar{x}_1\bar{x}_2x_3x_4 + \bar{x}_1x_2\bar{x}_3x_4 \quad (17.20)$$

Solution:

$$f_1(x_1, x_2, x_3, x_4) = x_1x_3(x_2\bar{x}_4 + \bar{x}_2x_4) + \bar{x}_1\bar{x}_3(\bar{x}_2\bar{x}_4 + x_2x_4) \quad (17.21)$$

$$= x_1x_3\varphi(x_2, x_4) + \bar{x}_1\bar{x}_3\overline{\varphi(x_2, x_4)} \quad (17.22)$$

$$= F[\varphi(x_2, x_4), x_1, x_3] \quad (17.23)$$

where

$$\varphi = x_2\bar{x}_4 + \bar{x}_2x_4 \quad (17.24)$$

This corresponds to the partitioning $x_1, x_3 \mid x_2, x_4$. ∎

Nothing was said in the previous example as to how it was determined that a simple disjoint partition of the form $x_1, x_3 \mid x_2, x_4$ could actually be obtained.

Let us explore the possibility of a general method of determining whether or not a particular decomposition exists. Consider, for example, the function

$$f(x_1, x_2, x_3, x_4) = \sum m(4, 5, 6, 7, 8, 13, 14, 15) \quad (17.25)$$

which may be decomposable into the form $F[\varphi(x_1, x_4), x_2, x_3]$. As a first

x_1x_4 x_2x_3	00	01	11	10
00	$0^{[0]}$	$0^{[1]}$	$0^{[9]}$	$1^{[8]}$
01	$0^{[2]}$	$0^{[3]}$	$0^{[11]}$	$0^{[10]}$
11	$1^{[6]}$	$1^{[7]}$	$1^{[15]}$	$1^{[14]}$
10	$1^{[4]}$	$1^{[5]}$	$1^{[13]}$	$0^{[12]}$

FIGURE 17.7. *Partition map.*

step, we can collect the minterms containing each of the four combinations of x_2x_3 and then factor these terms out to give the form

$$f(x_1, x_2, x_3, x_4) = \bar{x}_2\bar{x}_3\alpha(x_1, x_4) + \bar{x}_2x_3\beta(x_1, x_4)$$
$$+ \, x_2x_3\lambda(x_1, x_4) + x_2\bar{x}_3\delta(x_1, x_4) \qquad (17.26)$$

This is the step that produced Equation 17.21 above and Equation 17.17 in the parity function example. This factoring can be done for any function whether it is decomposable or not. Now we shall make a special map of the function, called a *partition map* (Fig. 17.7). This is simply a K-map with the free variables running down the side and the bound variables across the top. Note that the minterm numbers are still those based on the original ordering of the variables specified in Equation 17.25.

Comparing the partition map with Equation 17.26, we can see that the first row must correspond to the function $\alpha(x_1, x_4)$, the second row to $\beta(x_1, x_4)$, the third row to $\lambda(x_1, x_4)$, and the fourth row to $\delta(x_1, x_4)$. Again, this is still completely general. Now recall that for a simple disjoint function, F must be a function of x_2, x_3 and a *single* function $\varphi(x_1, x_4)$. In the above, let the first row function α be the function φ. Then, for the condition of a *single* function to be met, the remaining functions must be either α, $\bar{\alpha}$, 0, or 1. This in turn means that the remaining rows must be either identical, complementary, all 0's, or all 1's. In this example, the conditions are met and the function can be read off the partition matrix as

$$f(x_1, x_2, x_3, x_4) = \bar{x}_2\bar{x}_3\varphi(x_1, x_4) + x_2\bar{x}_3\overline{\varphi(x_1, x_4)} + x_2x_3$$
$$= F[\varphi(x_1, x_4), x_2, x_3] \qquad (17.27)$$

where $\varphi(x_1, x_4) = x_1\bar{x}_4$.

We now have a rule for detecting a simple disjoint decomposition on the partition map. However, this rule may be rather difficult to apply in large partition maps. As a further aid, let us define *column multiplicity* as the number of different column patterns of 1's and 0's. For example, in the partition of Fig. 17.7, there are two distinct column patterns,

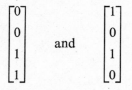

The following theorem offers a simple method by which it can be determined if a partition map corresponds to a simple disjoint decomposition.

THEOREM 17.6.   *A partition map corresponds to a simple disjoint decomposition if and only if the column multiplicity is less than or equal to two.*

Proof: If a function is decomposable into row functions, the functional values of each row of the Karnaugh map must form one of the four functions, 0, 1, φ, or $\bar{\varphi}$. Rows corresponding to the functions 0 and 1 have the same entry in every column and thus have no effect on column multiplicity. If a column has a 1 in some φ-row, it must have a 0 in every $\bar{\varphi}$-row and a 1 in every φ-row. Similarly, a 0 in a φ-row implies a 0 in every φ-row and a 1 in

x_2		0				1		
x_3x_4	00	01	11	10	00	01	11	10
x_1 0	0	1	3	2	4	5	7	6
1	8	9	11	10	12	13	15	14

x_1		0				1		
x_3x_4	00	01	11	10	00	01	11	10
0	0	1	3	2	8	9	11	10
x_2 1	4	5	7	6	12	13	15	14

x_1		0				1		
x_2x_4	00	01	11	10	00	01	11	10
0	0	1	5	4	8	9	13	12
x_3 1	2	3	7	6	10	11	15	14

x_1		0				1		
x_2x_3	00	01	11	10	00	01	11	10
0	0	2	6	4	8	10	14	12
x_4 1	1	3	7	5	9	11	15	13

x_3x_4	00	01	11	10
00	0	1	3	2
01	4	5	7	6
x_1x_2 11	12	13	15	14
10	8	9	11	10

x_2x_4	00	01	11	10
00	0	1	5	4
01	2	3	7	6
x_1x_3 11	10	11	15	14
10	8	9	13	12

x_2x_3	00	01	11	10
00	0	2	6	4
01	1	3	7	5
x_1x_4 11	9	11	15	13
10	8	10	14	12

FIGURE 17.8.   *4-variable decomposition charts.*

every $\bar{\varphi}$-row. Thus, there are only two distinct types of columns possible. If we assume a column multiplicity of two, we may similarly deduce the existence of a simple disjoint decomposition.   Q.E.D.

With the above, we have the basic rule needed to find decompositions. However, the amount of work is still rather extensive, since there are many possible partitions of the variables into bound and free variables. To reduce the amount of labor involved, we introduce *decomposition charts*, which are simply K-maps showing the position of the minterms for all possible combinations of bound and free variables. The decomposition charts for 4-variable functions are shown in Fig. 17.8. To use the charts, we simply circle the minterm numbers of the function on all charts and then look for column multiplicities of 2 or 1. Note that the 4 × 4 charts should be viewed sideways to reverse the order of bound and free variables. Thus, for example, the leftmost 4 × 4 chart will indicate decomposition of the form $F\varphi(x_3, x_4)$, x_1, x_2 when viewed normally, and decompositions of the form $F\varphi(x_1, x_2), x_3, x_4$ when viewed sideways. The 2 × 8 charts should only be viewed normally, since decompositions with only one bound variable are trivial.

Example 17.6

Use decomposition charts to find all simple disjoint decompositions of

$$f(x_1, x_2, x_3, x_4) = \sum m(4, 5, 6, 7, 8, 13, 14, 15) \tag{17.28}$$

Solution: The decomposition charts for this function are shown in Fig. 17.9. The maps of Fig. 17.9b and 17.9e viewed normally indicate the partitionings $x_2 \mid x_1x_3x_4$ and $x_1x_2 \mid x_3x_4$. The maps of Fig. 17.9f and 17.9g viewed sideways indicate the partitionings $x_2x_4 \mid x_1x_3$ and $x_2x_3 \mid x_1x_4$. The resultant decompositions can then be read from the charts as:

$$f(x_2 \mid x_1x_3x_4) = \bar{x}_2\varphi(x_1, x_3, x_4) + x_2\overline{\varphi(x_1, x_3, x_4)}$$
$$= x_2(x_1\bar{x}_3\bar{x}_4) + x_2\overline{(x_1\bar{x}_3\bar{x}_4)} \tag{17.29}$$

where $\varphi(x_1, x_3, x_4) = x_1\bar{x}_3\bar{x}_4$.

$$f(x_1x_2 \mid x_3x_4) = \bar{x}_1x_2 + x_1x_2\varphi(x_3, x_4) + x_1\bar{x}_2\overline{\varphi(x_3, x_4)}$$
$$= \bar{x}_1x_2 + x_1x_2(x_3 + x_4) + x_1\bar{x}_2\overline{(x_3 + x_4)} \tag{17.30}$$

where $\varphi(x_3, x_4) = x_3 + x_4$.

$$f(x_2x_4 \mid x_1x_3) = x_2x_4 + x_2\bar{x}_4\varphi(x_1, x_3) + \bar{x}_2\bar{x}_4\overline{\varphi(x_1, x_3)}$$
$$= x_2x_4 + x_2\bar{x}_4(\bar{x}_1 + x_3) + \bar{x}_2\bar{x}_4\overline{(\bar{x}_1 + x_3)} \tag{17.31}$$

where $\varphi(x_1, x_3) = \bar{x}_1 + x_3$.

$$f(x_2x_3 \mid x_1x_4) = x_2x_3 + x_2\bar{x}_3\varphi(x_1, x_4) + \bar{x}_2\bar{x}_3\overline{\varphi(x_1, x_4)}$$
$$= x_2x_3 + x_2\bar{x}_3(\bar{x}_1 + x_4) + \bar{x}_2\bar{x}_3\overline{(\bar{x}_1 + x_4)} \tag{17.32}$$

where $\varphi(x_1, x_4) = \bar{x}_1 + x_4$. ■

x_2	0				1			
x_3x_4	00	01	11	10	00	01	11	10
x_1 0	0	1	3	2	(4)	(5)	(7)	(6)
x_1 1	(8)	9	11	10	12	(13)	(15)	(14)

(a)

x_1	0				1			
x_3x_4	00	01	11	10	00	01	11	10
0	0	1	3	2	(8)	9	11	10
x_2 1	(4)	(5)	(7)	(6)	12	(13)	(15)	(14)

(b)

x_1	0				1			
x_2x_4	00	01	11	10	00	01	11	10
0	0	1	(5)	(4)	(8)	9	(13)	12
x_3 1	2	3	(7)	(6)	10	11	(15)	(14)

(c)

x_1	0				1			
x_2x_3	00	01	11	10	00	01	11	10
0	0	2	(6)	(4)	(8)	10	(14)	12
x_4 1	1	3	(7)	(5)	9	11	(15)	(13)

(d)

x_3x_4	00	01	11	10
00	0	1	3	2
x_1x_2 01	(4)	(5)	(7)	(6)
11	12	(13)	(15)	(14)
10	(8)	9	11	10

(e)

x_2x_4	00	01	11	10
00	0	1	(5)	(4)
x_1x_3 01	2	3	(7)	(6)
11	10	11	(15)	(14)
10	(8)	9	(13)	12

(f)

x_2x_3	00	01	11	10
00	0	2	(6)	(4)
x_1x_4 01	1	3	(7)	(5)
11	9	11	(15)	(13)
10	(8)	10	(14)	12

(g)

FIGURE 17.9. *Decomposition charts, Example 17.5.*

The same procedure can be applied to functions of more variables. The decomposition charts for 5 variables are shown in Fig. 17.10. The method can, in theory, be extended to any number of variables, but the number and size of the charts rapidly becomes impractical beyond 5 variables. Curtis[5] gives the charts for 6 variables but there are 29* of them, making the practicality of the method for 6 variables rather doubtful. However, the generality of the method provides the basis for computer mechanization of the process for large numbers of variables.

Functions with don't-cares can be handled very simply on the decomposition charts. Circle the 1's of the function in the usual manner and indicate the don't-cares by lines through the corresponding minterms. Then assign the don't-cares to achieve a column multiplicity of two or less.

Example 17.7

Determine a simple disjoint decomposition of

$$f(x_1, x_2, x_3, x_4, x_5) = \sum m(1, 2, 7, 9, 10, 18, 19, 25, 31) + d(0, 15, 20, 23, 26)$$

* Curtis also gives a scheme for reducing the number of charts to 16, but the patterns are rather more difficult to recognize on the condensed charts.

x_2x_3	00				01				11				10			
x_4x_5	00	01	11	10	00	01	11	10	00	01	11	10	00	01	11	10
1	0	1	3	2	4	5	7	6	12	13	15	14	8	9	11	10
x_1 1	16	17	19	18	20	21	23	22	28	29	31	30	24	25	27	26

x_1x_3	00				01				11				10			
x_4x_5	00	01	11	10	00	01	11	10	00	01	11	10	00	01	11	10
0	0	1	3	2	4	5	7	6	20	21	23	22	16	17	19	18
x_2 1	8	9	11	10	12	13	15	14	28	29	31	30	24	25	27	26

x_1x_2	00				01				11				10			
x_4x_3	00	01	11	10	00	01	11	10	00	01	11	10	00	01	11	10
0	0	1	3	2	8	9	11	10	24	25	27	26	16	17	19	18
x_3 1	4	5	7	6	12	13	15	14	28	29	31	30	20	21	23	22

x_1x_2	00				01				11				10			
x_3x_5	00	01	11	10	00	01	11	10	00	01	11	10	00	01	11	10
0	0	1	5	4	8	9	13	12	24	25	29	28	16	17	21	20
x_4 1	2	3	7	6	10	11	15	14	26	27	31	30	18	19	23	22

x_1x_2	00				01				11				10			
x_3x_4	00	01	11	10	00	01	11	10	00	01	11	10	00	01	11	10
0	0	2	6	4	8	10	14	12	24	26	30	28	16	18	22	20
x_5 1	1	3	7	5	9	11	15	13	25	27	31	29	17	19	23	21

x_3	0				1			
x_4x_5	00	01	11	10	00	01	11	10
00	0	1	3	2	4	5	7	6
x_1x_2 01	8	9	11	10	12	13	15	14
11	24	25	27	26	28	29	31	30
10	16	17	19	18	20	21	23	22

x_2	0				1			
x_4x_5	00	01	11	10	00	01	11	10
00	0	1	3	2	8	9	11	10
x_1x_3 01	4	5	7	6	12	13	15	14
11	20	21	23	22	28	29	31	30
10	16	17	19	18	24	25	27	26

FIGURE 17.10. *Decomposition charts for 5-variable functions.*

x_2	0				1			
x_3x_5	00	01	11	10	00	01	11	10
x_1x_4 00	0	1	5	4	8	9	13	12
01	2	3	7	6	10	11	15	14
11	18	19	23	22	26	27	31	30
10	16	17	21	20	24	25	29	28

x_2	0				1			
x_3x_4	00	01	11	10	00	01	11	10
x_1x_5 00	0	2	6	4	8	10	14	12
01	1	3	7	5	9	11	15	13
11	17	19	23	21	25	27	31	29
10	16	18	22	20	24	26	30	28

x_1	0				1			
x_4x_5	00	01	11	10	00	01	11	10
x_2x_3 00	0	1	3	2	16	17	19	18
01	4	5	7	6	20	21	23	22
11	12	13	15	14	28	29	31	30
10	8	9	11	10	24	25	27	26

x_1	0				1			
x_3x_5	00	01	11	10	00	01	11	10
x_2x_4 00	0	1	5	4	16	17	21	20
01	2	3	7	6	18	19	23	22
11	10	11	15	14	26	27	31	30
10	8	9	13	12	24	25	29	28

x_1	0				1			
x_3x_4	00	01	11	10	00	01	11	10
x_2x_5 00	0	2	6	4	16	18	22	20
01	1	3	7	5	17	19	23	21
11	9	11	15	13	25	27	31	29
10	8	10	14	12	24	26	30	28

x_1	0				1			
x_2x_5	00	01	11	10	00	01	11	10
x_3x_4 00	0	1	9	8	16	17	25	24
01	2	3	11	10	18	19	27	26
11	6	7	15	14	22	23	31	30
10	4	5	13	12	20	21	29	28

x_1	0				1			
x_2x_4	00	01	11	10	00	01	11	10
x_3x_5 00	0	2	10	8	16	18	26	24
01	1	3	11	9	17	19	27	25
11	5	7	15	13	21	23	31	29
10	4	6	14	12	20	22	30	28

x_1	0				1			
x_2x_3	00	01	11	10	00	01	11	10
x_4x_5 00	0	4	12	8	16	20	28	24
01	1	5	13	9	17	21	29	25
11	3	7	15	11	19	23	31	27
10	2	6	14	10	18	22	30	26

FIGURE 17.10   (continued)

Solution: The decomposition chart for the partition $x_3x_4 \mid x_1x_2x_5$ is shown in Fig. 17.11a, with the minterms circled and the don't-cares lined out. We see that the chart has a column multiplicity of four in both directions. If the don't-cares at 15 and 26 are assigned as 1's, a column multiplicity of two is achieved with the chart viewed in the normal direction, as shown in Fig. 17.11b. From this chart, interpreted as a K-map, we can read off the realization

$$(x_3x_4 \mid x_1x_2x_5) = (\bar{x}_3\bar{x}_4 + x_3x_4)\varphi(x_1, x_2, x_5) + \bar{x}_3x_4\overline{\varphi(x_1, x_2, x_5)} \quad (17.33)$$

where $\varphi(x_1, x_2, x_5) = x_2x_5 + \bar{x}_1x_5$. ∎

x_1	0				1			
x_2x_5	00	01	11	10	00	01	11	10
x_3x_4 00	0̶	①	⑨	8	16	17	㉕	24
01	②	3	11	⑩	⑱	⑲	27	2̶6̶
11	6	⑦	1̶5̶	14	22	2̶3̶	㉛	30
10	4	5	13	12	2̶0̶	21	29	28

(a)

x_1	0				1			
x_2x_3	00	01	11	10	00	01	11	10
x_3x_4 00	0̶	①	⑨	8	16	17	㉕	24
01	②	3	11	⑩	⑱	⑲	27	㉖
11	6	⑦	⑮	14	22	2̶3̶	㉛	30
10	4	5	13	12	2̶8̶	21	29	28

(b)

FIGURE 17.11.   *Decomposition charts, Example 17.7.*

17.6   Complex Disjoint Decomposition

It is apparent that there must be many forms of decomposition other than simple disjoint. In this section, we consider decompositions resulting from the partitioning of the variables into three or more disjoint subsets.

Definition 17.3.   Let $X = x_1, x_2, \ldots, x_n$ be a set of n switching variables and let $A_1, A_2, \ldots, A_m$ be disjoint subsets of X such that $A_1 \cup A_2 \cup \cdots \cup A_m = X$. Decompositions of the form

$$f(X) = F[\varphi_1(A_1), \varphi_2(A_2), \ldots, \varphi_m(A_m)]$$

or

$$f(X) = F[\varphi_1(A_1), \varphi_2(A_2), \ldots, \varphi_{m-1}(A_{m-1}), A_m] \qquad (17.34)$$

shall be known as multiple disjoint decompositions. Decompositions of the form

$$f(X) = F(\lambda[\varphi(A), B], C) \qquad (17.35)$$

or, in general, of the form

$$f(X) = F[\varphi_{m-1}(\varphi_{m-2}[\ldots (\varphi_1(\varphi_0(A_0), A_1], A_2) \ldots A_{m-2}], A_{m-1}), A_m] \quad (17.36)$$

shall be known as iterative disjoint decompositions. Finally, combinations of these forms, such as,

$$f(X) = F[\gamma[\varphi(A), B], \lambda(C), D] \tag{17.37}$$

shall be known as *complex disjoint* decompositions. In all cases, the decompositions are nontrivial only if all the bound sets contain at least 2 variables, since functions of a single variable are trivial.

Ashenhurst[4] has given a set of theorems relating the above types of decompositions to simple disjoint decompositions. These theorems enable us to extend the use of the decomposition chart to the determination of complex decompositions.

Rather than present the somewhat tedious proofs of these theorems, we shall content ourselves with an example of the application of each theorem.

THEOREM 17.7.   *Let $f(X)$ be a function for which there exist two simple disjoint decompositions:*

$$\begin{aligned} f(X) &= F[\lambda(A, B), C] \\ &= G[\varphi(A), B, C] \end{aligned} \tag{17.38}$$

then there exists an iterative disjoint decomposition

$$f(X) = F[\rho[\varphi(A), B], C] \tag{17.39}$$

where

$$\rho[\varphi(A), B] = \lambda(A, B)$$

As an example of the application of this theorem, consider the function

$$f(x_1, x_2, x_3, x_4, x_5) = \sum m(5, 10, 11, 14, 17, 21, 26, 30)$$

Two decomposition charts of this function are shown in Fig. 17.12. The first indicates the existence of a decomposition

$$f(X) = F[\lambda(x_1, x_3, x_5), x_2, x_4] \tag{17.40}$$

while the second, viewed sideways, shows a decomposition

$$f(X) = G[\varphi(x_1, x_3), x_2, x_4, x_5] \tag{17.41}$$

Thus, the conditions of Theorem 17.7 are met with $A = (x_1 x_3)$, $B = (x_5)$, and $C = (x_2, x_4)$. From Fig. 17.12a, we read off the realization

$$f(x) = \bar{x}_2 \bar{x}_4 \lambda(x_1, x_3, x_5) + x_2 x_4 \overline{\lambda(x_1, x_3, x_5)} \tag{17.42}$$

where $\lambda(x_1, x_3, x_5) = x_3 x_5 + x_1 x_5$.

x_1	0				1			
x_3x_5	00	01	11	10	00	01	11	10
00	0	1	⑤	4	16	⑰	㉑	20
01	2	3	7	6	18	19	23	22
x_2x_4 11	⑩	⑪	15	⑭	㉖	27	31	㉚
10	8	9	13	12	24	25	29	28

(a)

x_2	0				1			
x_4x_5	00	01	11	10	00	01	11	10
00	0	1	3	2	8	9	⑪	⑩
01	4	⑤	7	6	12	13	15	⑭
x_1x_3 11	20	㉑	23	22	28	29	31	㉚
10	16	⑰	19	18	24	25	27	㉖

(b)

FIGURE 17.12. *Decomposition charts indicating iterative decomposition.*

Thus, from Theorem 17.7, $\lambda(x_1, x_3, x_5)$ can be decomposed according to the partition $x_1 x_3 \mid x_5$. By inspection of the equation for λ, we see that the decomposition is given by

$$\rho[\varphi(A), B] = x_5(x_3 + x_1)$$

where $\varphi(A) = x_3 + x_1$.

This example is trivial in the sense that we did not need Theorem 17.7 to tell us that $\lambda(x_1, x_3, x_5)$ could be factored in the above manner. However, the theorem tells us that, no matter how large the number of variables, the charts will enable us to locate decompositions in which the subfunctions can in turn be further decomposed.

THEOREM 17.8. *Let $f(X)$ be a function for which there exist two simple disjoint decompositions*

$$f(X) = F[\lambda(A), B] \tag{17.43}$$

$$= G[\varphi(B), A] \tag{17.44}$$

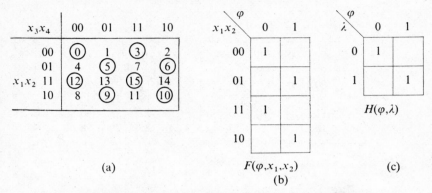

FIGURE 17.13.   *Multiple decompositions of parity function.*

then there exists a multiple disjoint decomposition

$$f(X) = H[\lambda(A),\ \varphi(B)] \qquad (17.45)$$

The conditions of Equations 17.43 and 17.44 will be indicated by a single decomposition chart, which satisfies the column multiplicity requirements when viewed both normally and sideways. An example is the parity function considered earlier. The decomposition chart corresponding to the map of Fig. 17.2 is shown in Fig. 17.13a. Clearly, we have a column multiplicity of two in both directions. Viewed normally, the chart indicates the decomposition

$$f(x_1 x_2 \mid x_3 x_4) = (\bar{x}_1 \bar{x}_2 + x_1 x_2)\overline{\varphi(x_3 x_4)} + (\bar{x}_1 x_2 + x_1 \bar{x}_2)\varphi(x_3, x_4) \qquad (17.46)$$

where $\varphi(x_3 x_4) = x_3 \bar{x}_4 + \bar{x}_3 x_4$.

Viewed sideways, the chart indicates

$$f(x_3 x_4 \mid x_1 x_2) = (\bar{x}_3 \bar{x}_4 + x_3 x_4)\overline{\lambda(x_1, x_2)} + (\bar{x}_3 x_4 + x_3 \bar{x}_4)\lambda(x_1, x_2) \qquad (17.47)$$

where $\lambda(x_1 x_2) = \bar{x}_1 x_2 + x_1 \bar{x}_2$.

The theorem indicates the existence of a multiple decomposition of the form $H[\varphi(x_3, x_4), \lambda(x_1, x_2)]$. To determine H, we make a K-map of $F[\varphi, x_1, x_2]$ as given by Equation 17.46 (Fig. 17.13b). The values of $\lambda(x_1, x_2)$ are indicated on this map, immediately enabling us to construct the map of $H(\varphi, \lambda)$ as shown in Fig. 17.13c. From this map, we read off

$$H(\varphi, \lambda) = \bar{\varphi}\bar{\lambda} + \varphi\lambda \qquad (17.48)$$

where $\varphi = x_3 \oplus x_4$ and $\lambda = x_1 \oplus x_2$.

This will be recognized as the same result as obtained by factoring. Again the example is perhaps trivial, but the method extends to functions of any number of variables.

THEOREM 17.9. *Let $f(X)$ be a function for which there exist two simple disjoint decompositions*

$$f(X) = F[\lambda(A), B, C] \tag{17.49}$$

$$= G[\varphi(B), A, C] \tag{17.50}$$

then there exists a multiple disjoint decomposition

$$f(X) = H[\lambda(A), \varphi(B), C] \tag{17.51}$$

As an example, consider the parity function of five variables $S^5_{1,3,5}(x_1x_2x_3x_4x_5)$, for which two decomposition charts are shown in Fig. 17.14. Chart (a) viewed sideways gives the decomposition

$$f(x_3x_4x_5 \mid x_1x_2) = (\bar{x}_3\bar{x}_4\bar{x}_5 + \bar{x}_3x_4x_5 + x_3\bar{x}_4x_5 + x_3x_4\bar{x}_5)\lambda(x_1, x_2) + (\bar{x}_3\bar{x}_4x_5 + \bar{x}_3x_4\bar{x}_5 + x_3\bar{x}_4\bar{x}_5 + x_3x_4x_5)\overline{\lambda(x_1, x_2)} \tag{17.52}$$

where $\lambda(x_1, x_2) = \bar{x}_1x_2 + x_1\bar{x}_2$.

	x_3	0				1			
	x_4x_5	00	01	11	10	00	01	11	10
x_1x_2	00	0	(1)	3	(2)	(4)	5	(7)	6
	01	(8)	9	(11)	10	12	(13)	15	(14)
	11	24	(25)	27	(26)	(28)	29	(31)	30
	10	(16)	17	(19)	18	20	(21)	23	(22)

(a)

	x_1	0				1			
	x_2x_5	00	01	11	10	00	01	11	10
x_3x_4	00	0	(1)	9	(8)	(16)	17	(25)	24
	01	(2)	3	(11)	10	18	(19)	27	(26)
	11	6	(7)	15	(14)	(22)	23	(31)	30
	10	(4)	5	(13)	12	20	(21)	29	(28)

(b)

FIGURE 17.14. *Decomposition chart for $S^5_{1,3,5}(x_1, x_2, x_3, x_4, x_5)$.*

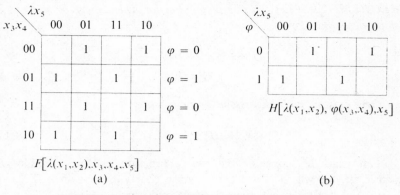

FIGURE 17.15.   *Construction of composite function.*

Chart (b), viewed sideways, gives the decomposition

$$f(x_1 x_2 x_5 \mid x_3 x_4) = (\bar{x}_1 \bar{x}_2 \bar{x}_5 + \bar{x}_1 x_2 x_5 + x_1 \bar{x}_2 x_5 + x_1 x_2 \bar{x}_5)\varphi(x_3, x_4)$$
$$+ (\bar{x}_1 \bar{x}_2 x_5 + \bar{x}_1 x_2 \bar{x}_5 + x_1 \bar{x}_2 \bar{x}_5 + x_1 x_2 x_5)\overline{\varphi(x_3, x_4)}$$

$$(17.53)$$

where $\varphi(x_3, x_4) = \bar{x}_3 x_4 + x_3 \bar{x}_4$.

The conditions of Theorem 17.9 are thus satisfied with $A = \{x_1, x_2\}$, $B = \{x_3, x_4\}$, and $C = \{x_5\}$. To determine the composite function, we construct a K-map of Expression 17.52, as shown in Fig. 17.15a, indicating the corresponding values of $\varphi(x_3, x_4)$. From these, the map of $H[\lambda(x_1, x_2), \varphi(x_3, x_4), x_5]$ may be constructed as shown in Fig. 17.15b. From the map, we read off the function

$$H[\lambda, \varphi, x_5] = (\bar{\lambda}\bar{x}_5 + \lambda x_5)\varphi + (\bar{\lambda}x_5 + \lambda \bar{x}_5)\bar{\varphi} \qquad (17.54)$$

This completes the decomposition corresponding to Theorem 17.9, but we note that the final map of H (Fig. 17.14b) is also decomposable. Thus, we can define

$$\rho(\lambda, x_5) = \bar{\lambda}x_5 + \lambda \bar{x}_5$$
$$= \lambda \oplus x_5$$

Then the function becomes:

$$f(X) = \bar{\rho}\varphi + \rho\bar{\varphi} \qquad (17.55)$$

where $\rho = \lambda \oplus x_5$, $\varphi = x_3 \oplus x_4$, and $\lambda = x_1 \oplus x_2$.

It is apparent that functions might also be decomposed in a nondisjoint fashion, and Curtis[5] has developed the theory quite extensively. However, the methods involved are so complex as to make them of doubtful value.

It is apparent that there is little point trying to decompose a function unless the possibility of significant economy exists. As we indicated earlier, many of the functions whose two level realizations are most expensive are subject to disjoint decomposition.

The reader may have noted in the example illustrating Theorem 17.9 that the two decomposition charts (Fig. 17.14) are identical. Since the value of a symmetric function is dependent only on the *number* of variables taking on the value 1, it is apparent that this condition (identical decomposition charts) will always hold for symmetric functions. Thus, we may conclude that one or more of the theorems given here will be applicable to most (if not all) symmetric functions. This observation, in turn, suggests that the methods of disjoint decomposition will generally be adequate for the simplification of symmetric functions.

It is clear that the value of decomposition will increase as the number of variables increases. However, the design procedures for functions of more than six variables get so complex that it is generally more efficient to break the problem into smaller parts at the *specification* stage.

17.7 Iterative Networks

An iterative network is a highly repetitive form of a combinational logic network. This repetitive structure makes it possible to describe iterative networks utilizing techniques already developed for sequential circuits. We shall limit our discussion to one-dimensional iterative networks represented by the cascade of identical cells given in Fig. 17.16a. A typical cell with appropriate input and output notation is given in Fig. 17.16b. Notice the two distinct types of inputs, primary inputs from the outside world and secondary inputs from the previous cell in the cascade. Similarly, there are two types of outputs, primary, to the outside world, and secondary, to the next cell in the cascade. At the left of the cascade are a set of boundary inputs which we shall denote in the same manner as secondary inputs. In some cases, these inputs will be constant values. A set of boundary outputs emerges from the rightmost cell in the cascade. Although these outputs are to the outside world, they will be labeled in the same manner as secondary outputs. In a few cases, the boundary outputs will be the only outputs of the iterative network.

The next three examples will serve to illustrate the analysis and synthesis of a few practical iterative networks. We shall find that it is possible to use all the techniques associated with the synthesis of clock mode sequential circuits to synthesize iterative networks. As with sequential circuits, we

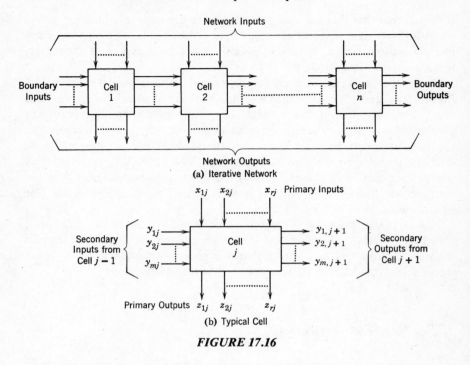

Network Inputs

Boundary Inputs

Cell 1

Cell 2

Cell *n*

Boundary Outputs

Network Outputs
(a) Iterative Network

x_{1j} x_{2j} x_{rj} Primary Inputs

Secondary Inputs from Cell $j-1$

y_{1j}
y_{2j}
y_{mj}

Cell j

$y_{1,j+1}$
$y_{2,j+1}$
$y_{m,j+1}$

Secondary Outputs from Cell $j+1$

Primary Outputs z_{1j} z_{2j} z_{rj}

(b) Typical Cell

FIGURE 17.16

shall first analyze a network already laid out in iterative form. By first translating the network of Example 17.8 to a form very similar to a state table, we hope to make the reverse process more meaningful.

Example 17.8

Although the reader is accustomed to thinking of counters as self-contained sequential circuits, such need not be the case. Very often it is desirable to include a counting capability in a general purpose arithmetic register in computer. Going one step farther to include the possible counting capability in a bank of several arithmetic registers provides a rationale for the network of Fig. 17.17.

The inputs $x_1 \cdots x_n$ in Fig. 17.17 come from the output of a data bus. Depending on control inputs to the bus this information vector might be the contents of any one of several data registers. The outputs of the iterative network $z_1, z_2, \ldots, z_n$ represent the new contents of the same register after incrementing. A control pulse will cause the new information to be triggered back into the same register. The arrangement in Fig. 17.17 makes it possible to share the incrementing logic over the entire bank of registers rather than including it separately in each register. Admittedly, in an MSI environment, the economic advantage would be important only if the number of registers in the bank was quite large. Another criticism of the network is the possible need for the effect of x_1 to propagate through $n-1$ levels of AND gates. A practical realization would no doubt use

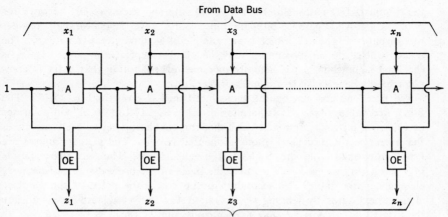

FIGURE 17.17.   Increment network.

(a)

y_j	y_{j+1}		z_j	
	$x_j = 0$	$x_j = 1$	$x_j = 0$	$x_j = 1$
0	0	0	0	1
1	0	1	1	0

x_j	0	1
A	$A, 0$	$A, 1$
B	$A, 1$	$B, 0$

q_{j+1}, z_j

(b)   (c)

FIGURE 17.18.   Development of state table.

multi-input AND gates. Finally, the AND gate and exclusive-OR network for bit 1 and the AND gate for bit n are unnecessary. Nevertheless we shall consider the format of Fig. 17.17, which is an example of a simple iterative network. The reader will easily satisfy himself that the number representing $x_1 \cdots x_n$ (x_1 is the least significant bit) is actually incremented by noting that a bit is complemented whenever all less significant bits are 1.

We base our analysis on the typical cell, cell j, given in Fig. 17.18a. The functional values of y_{j+1} and z_j are tabulated in Fig. 17.18b for all combinations of values of inputs x_j and y_j. Notice that this tabulation very closely resembles the *transition table* of clock mode sequential circuits. The primary inputs are listed horizontally as are sequential circuit inputs. The secondary inputs $y_{1_j} \cdots y_{m_j}$ (in this case only y_j) take the place of the present states of memory elements in the table. The secondary outputs take the place of the memory element next states. By defining $y = 0$ as state A and $y = 1$ as state B, we translate the transistion table to the *state table* form of Fig. 17.18c. ■

As was the case with sequential circuits, the first step in the synthesis of an iterative network will be to reduce an English language description to state table form. From there, all of the techniques of Chapters 10 to 12 may be used to obtain an iterative network from a state table. The following two examples will serve to illustrate the synthesis process.

Example 17.9

A set of bits $x_1 \cdots x_n$ may be considered as received information in a communications system. A particular decoding scheme requires a check for odd parity over the first bit, the first two bits, the first three bits, etc. Design an iterative network which will accomplish this function.

Solution: The reader will notice that the desired outputs are the symmetric functions S_1^1, S_1^2, $S_{1,3}^3$, $S_{1,3}^4$, etc. In general, sets of symmetric functions can quite easily be implemented as iterative networks.

A state table may be constructed directly by letting the *secondary input state* to block j be A if parity over primary inputs $x_1, x_2, \ldots, x_{j-1}$ is odd. If parity over these bits is even, the secondary input state will be B. The output z_j is to be one if there is odd parity over bits $x_1, x_2, \ldots, x_j$. Thus, $z_j = 1$ when $x_j = 0$ and the input state is A and when $x_j = 1$ and the input state is B. The secondary output will be A precisely when $z_j = 1$ and will be B when $z_j = 0$. Thus, we have the state table of Fig. 17.19. ■

x_j	0	1
B	$B, 0$	$A, 1$
A	$A, 1$	$B, 0$

q_{j+1}, z_j **FIGURE 17.19.**   *State table for iterative parity checker.*

The next example will be the first to require more than two secondary input states. We shall continue the design process in this case to develop the network realization of a single cell from the state table.

Example 17.10

Bits $x_1, x_2, \ldots, x_n$ are interrupt signals sent to a computer central processor from n peripheral equipments to request service from the processor. At any given time, the computer can communicate with no more than two of these peripherals. Peripheral 1 (corresponding to interrupt x_1) has the highest priority, with peripheral 2 second-highest, etc. It is, therefore, desired to construct a network with outputs $z_1, z_2, \ldots, z_n$ such that no more than two of the outputs will be 1 at a given time. The 1 outputs will correspond to the two highest priority inputs with active interrupts. (Interrupt i is active if $x_i = 1$.) If only one interrupt is active, only one output line will be 1 and no output will be 1 in the case of no active interrupts.

Solution: We define three secondary input states for cell j. The secondary input state will be A if no interrupts of higher priority than peripheral j are active. The state is B if one higher priority interrupt is active and the state is C if two or more higher priority interrupts are active. Thus, the output $z_j = 1$ only if $x_j = 1$ and the input state is A or B. If $x_j = 0$, the secondary output state is the same as the input state. If $x_j = 1$, an input state B must generate an output state C and an input state A must generate an output state B. This information is tabulated in Fig. 17.20a.

A convenient state assignment defines $y_2 y_1 = 00$ as A, $y_2 y_1 = 01$ as B and $y_2 y_1 = 11$ as C. This leads to the transition table of Fig. 17.20b. The equations for the secondary outputs or next state equations (17.56, 17.57) may be obtained directly by treating the transition table as a Karnaugh map. The

$$y_{2,j+1} = y_{2j} + x_j \cdot y_{1j} \tag{17.56}$$

$$y_{1,j+1} = y_{1j} + x_j \tag{17.57}$$

$$z_j = x_j \cdot \bar{y}_{2j} \tag{17.58}$$

primary output, z_j, is similarly obtained from the transition table. The network for a single cell of the priority network may be obtained by implementing these equations. ∎

	$x_j = 0$	$x_j = 1$		$y_{2j} y_{1j}$	$x_j = 0$	$x_j = 1$	$x_j = 0$	$x_j = 1$
A	$A, 0$	$B, 1$		A	00	00	01	0	1
B	$B, 0$	$C, 1$		B	01	01	11	0	1
C	$C, 0$	$C, 0$		C	11	11	11	0	0
					10				

q_{j+1}, z_j — $y_{2,j+1}, y_{1,j+1}$ — z_j

(a)   (b)

FIGURE 17.20. *Tables for cell in priority network.*

Two additional examples of iterative circuits of considerable interest are the adder and the magnitude comparator. (See Problems 17.11 and 17.12.) Both of these networks have two inputs per cell but are otherwise of the same order of difficulty as the above example. The comparator is a special case in that there are no primary outputs. The only outputs of the overall network are the secondary outputs for the right most cell. As indicated in Fig. 17.16a, these outputs are called boundary outputs.

Although iterative networks were first considered in the days when the primary switching element was the relay, they have taken an increased significance with the advent of LSI. In this context, iterative circuits are something of a research topic. The repetitive structure offers the advantage of easier optimization of chip layout. Variations of one-and-two dimensional iterative networks may provide the language through which complex functions can be systematically organized or single chips. Iterative networks have also been considered as a modeling technique in the generation of test sequences for LSI sequential circuits.[9,10]

17.8 Ordering Relations

Another interesting special class of functions is the class of *unate* Boolean functions. These functions will be particularly important to our discussion of threshold logic in the next chapter. Before we proceed to define *unate* rigorously, consider the following two Boolean functions:

$$f_1 = x_1 x_2 x_3 + x_1 x_2 x_4 + \bar{x}_3 \bar{x}_4 \tag{17.59}$$

$$f_2 = x_1 (x_2 + \bar{x}_3 \bar{x}_4) \tag{17.60}$$

Notice that there is no possible way to express f_1 without, for example, both the literal x_3 and its complement $\bar{x}_3$ appearing in the expression. The function f_2 is expressed in such a way that, if a literal appears in the expression, then its complement does not appear. Functions which can be expressed this way are unate.

In order to formalize the definition of unate, we first consider the concept of an *ordering* relation. In Chapter 10, the concept of a relation was introduced, although only equivalence relations were illustrated. Certain relations which do not satisfy the symmetric law are also of interest.

Definition 17.4.   A relation, $\le$ is a *partial-ordering relation* if and only if it is reflexive and transitive, that is

$$x \le x$$

and

$$x \leq y \quad \text{and} \quad y \leq z \quad \text{imply } x \leq z$$

and it is anti-symmetric, that is,

$$x \leq y \quad \text{and} \quad y \leq x \quad \text{imply } x = y \qquad (17.61)$$

Example 17.11

The set of all real numbers is partially ordered. In addition, either or both of the following must be true where a and b are real numbers.

$$a \leq b \quad \text{or} \quad b \leq a \qquad (17.62)$$

A partially ordered set of objects, any two of which satisfy Equation 17.62, is said to be *totally ordered*. ∎

Example 17.12

An example of a partially ordered set which is not totally ordered may be defined as follows. Let S consist of the four sets A, B, $A \cup B$, and $A \cap B$, where $A \neq B$ and neither A nor B is the null set. If we define the ordering relation to be set inclusion, we see in Fig. 17.21 that the definition of partial ordering is satisfied for any two elements, but that

$$A \nsubseteq B \quad \text{and} \quad B \nsubseteq A \qquad (17.63)$$

Therefore, S is not totally ordered. ∎

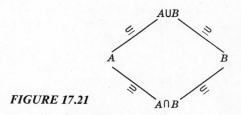

FIGURE 17.21

Definition 17.5. Let S be a set which is partially ordered by $\leq$. An element $u \in S$ is said to be an *upper bound* of the subset, X, $(X \subseteq S)$ if every $x \subseteq X$ satisfies $x \leq u$. The element u is a least *upper bound* of X, if $u \leq v$ where v is any upper bound of X. A *lower bound* and a *greatest lower bound* are similarly defined.

Definition 17.6. A lattice is defined as a set, L, partially ordered by $\leq$, such that every two elements of L have a least upper bound and a greatest lower bound.

Example 17.13

Let S be the set of switching functions of 3 variables x_1, x_2, x_3. Let a relation be defined on S as:

$$x \leq y \quad \text{if} \quad x + y = y$$

Show that S, ordered by $\leq$, forms a lattice.

Solution: Let a, b, and c be any three elements of S. Clearly, $\leq$ is reflexive. If $a + b = b$ and $b + c = c$, then

$$b + c = (a + b) + c = a + (b + c) = a + c \qquad (17.64)$$

Therefore, $\leq$ is transitive. If $a + b = b$ and $b + a = a$, then $a = b$. Thus, $\leq$ is antisymmetric and constitutes a partial ordering of S.

Clearly, $a \leq a + b$ and $b \leq a + b$, so that $a + b$ is an upper bound of $\{a, b\}$. Let u be any upper bound of $\{a, b\}$. Therefore,

$$u = u + a \qquad \text{and} \qquad u = u + b$$

Therefore,

$$u = u + u = (u + a) + (u + b) = u + (a + b)$$

and

$$a + b \leq u \qquad (17.65)$$

Similarly, $ab \leq b$ and $ab \leq a$ so that ab is a lower bound of $\{a, b\}$. Also,

$$a = l + a \qquad \text{and} \qquad b = l + b$$

so that

$$ab = (l + a)(l + b) = l + ab$$

where l is any lower bound of $\{a, b\}$. Thus, $l \leq ab$, ab is the greatest lower bound of $\{a, b\}$, and the set S ordered by $\leq$ is a lattice.   ■

17.9   *Unate Functions*

To permit the definition of a unate function, it is first necessary to define a partial ordering of the vertices of the Boolean hypercube.

Definition 17.7.   A vertex $(x_1, x_2, \ldots, x_n)$ is $\leq$ to a vertex $(y_1, y_2, \ldots, y_n)$ if and only if, for every i, $x_i \leq y_i$. We define $x_i \leq y_i$ if $y_i = 1$ whenever $x_i = 1$.

Example 17.14

Each x_i may take on only the values 0 and 1. Thus, $(x_1, x_2, \ldots, x_n)$ is $\leq$ to $(y_1, y_2, \ldots, y_n)$ if and only if y_i is 1 if x_i is 1. For example,

$$(0, 1, 0) \leq (0, 1, 1)$$

whereas $(0, 1, 0)$ and $(1, 0, 1)$ are not ordered by $\leq$. The complete lattice of all vertices of the three-dimensional Boolean cube is shown in Fig. 17.22. A vertex X is $\leq$ to any vertex Y above it in the diagram, to which it is connected by a line.   ■

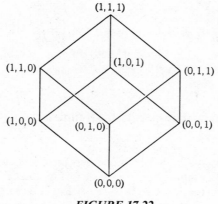

FIGURE 17.22

Definition 17.8.   A switching function of n variables is monotone increasing if and only if $f(\mathbf{X}) \leq f(\mathbf{Y})$ wherever $\mathbf{X} \leq \mathbf{Y}$ where $\mathbf{X}$ represents the vertex $(x_1, x_2, \ldots, x_n)$ and $\mathbf{Y}$ the vertex $(y_1, y_2, \ldots, y_n)$.

The lattice of three-dimensional vertices is repeated in Fig. 17.23a with each vertex, for which the function $x_1 + x_2 x_3 = 1$, indicated by a black circle. The vertices for which $x_1 + \bar{x}_1 \bar{x}_2 \bar{x}_3 = 1$ are similarly indicated in Fig. 17.23b. We see that $x_1 + x_2 x_3$ is monotone increasing. Note however, that $x_1 + \bar{x}_1 \bar{x}_2 \bar{x}_3$ is 1 for $(x_1, x_2, x_3) = (0, 0, 0)$, but is 0 for $(0, 1, 0)$, $(0, 0, 1)$, and $(0, 1, 1)$. As $(0, 0, 0) \leq (0, 1, 0)$, $(0, 0, 1)$, and $(0, 1, 1)$, the conditions of Definition 17.8 are clearly not satisfied; and $x_1 + \bar{x}_1 \bar{x}_2 \bar{x}_3$ is not monotone increasing.

A monotone increasing function is a special case of what we shall now define to be a *unate* function.

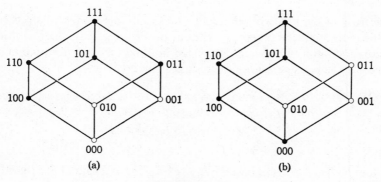

FIGURE 17.23

Definition 17.9.   Let $X^J = (x_1{}^{j_1}, x_2{}^{j_2}, \ldots, x_n{}^{j_n})$ where

$$x_i{}^{j_i} = x_i \quad \text{if} \quad j_i = 0 \quad \text{and} \quad x_i{}^{j_i} = \bar{x}_i \quad \text{if} \quad j_i = 1 \qquad (17.66)$$

A switching function, f, is *unate* if and only if there exists an n-tuple $J = (j_1, j_2, \ldots, j_n)$ such that $f(X) \leq f(Y)$, whenever

$$X^J \leq Y^J \qquad (17.67)$$

We say that f is unate with respect to the n-tuple J.

Example 17.15

The function $f = x_1 + \bar{x}_2 x_3$ is not monotone increasing as indicated by the functional values superimposed on the lattice of Fig. 17.24a. In Fig. 17.24b, the vertices are ordered in the lattice so as to satisfy Equation 17.67 for $J = (0, 1, 0)$. The functional values are the same. In Fig. 17.24b, we see that wherever $X^J \leq Y^J$ that $f(X) \leq f(Y)$. Thus,

$$x_1 + \bar{x}_2 x_3 \qquad (17.68)$$

is unate.   ■

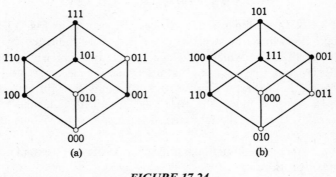

FIGURE 17.24

The truth of the following theorem may seem more evident than its proof. The theorem might be accepted, without proof, as part of the definition of a unate function by the reader interested in obtaining only a minimal background for the consideration of threshold logic.

THEOREM 17.10.   *A function $f(x_1, x_2, \ldots x_n)$ is unate with respect to the n-tuple J if and only if it can be expressed as a sum of products of only the literals $x_1{}^{j_1}, x_2{}^{j_2}, \ldots, x_n{}^{j_n}$ where $J = (j_1, j_2, \ldots, j_n)$.*

Proof:

   A. Let $f(x_1, x_2, \ldots, x_n)$ be a function expressed as a sum of products of the literals $x_1{}^{j_1}, \ldots, x_n{}^{j_n}$. Let X^J and Y^J be vertices such that $X^J \leq Y^J$.

If $f(\mathbf{X}) = 0$, then certainly $f(\mathbf{X}) \leq f(\mathbf{Y})$. Suppose $f(\mathbf{X}) = 1$. Then some $P(\mathbf{X}) = 1$, where P is one of the products of literals making up the expression for f. Since $\mathbf{X^J} \leq \mathbf{Y^J}$, at least as many of the literals $x_1^{j_1}$, $x_2^{j_2}, \ldots, x_n^{j_n}$ must be 1 when evaluated at $\mathbf{Y}$ as when evaluated at $\mathbf{X}$. Therefore, $P(\mathbf{Y}) = 1$ and $f(\mathbf{X}) \leq f(\mathbf{Y})$. Therefore, $f(x_1, x_2, \ldots, x_n)$ is unate.

B. Let $f(x_1, x_2, \ldots, x_n)$ be unate with respect to $\mathbf{J} = (j_1, j_2, \ldots, j_n)$. Let $\mathbf{Y}$ be any vertex for which $f(\mathbf{Y}) = 1$. Let $\mathbf{U} = (u_1, u_2, \ldots, u_n)$ be a vertex such that $f(\mathbf{U^J}) = 1$, $\mathbf{U^J} \leq \mathbf{Y^J}$ and such that there exists no other vertex $\mathbf{V^J}$ where $f(\mathbf{V^J}) = 1$ and $\mathbf{V^J} \leq \mathbf{U^J}$. Let $P(\mathbf{X})$ be the product of all literals $x_i^{j_i}$ for which $u_i^{j_i} = 1$. Therefore, $P(\mathbf{U^J}) = 1$. Since $f(x_1, x_2, \ldots, x_n)$ is unate, $f(\mathbf{Z}) = 1$ for all $\mathbf{Z}$ where $\mathbf{U^J} \leq \mathbf{Z^J}$.

However, if $P(\mathbf{Z}) = 1$, then $\mathbf{U^J} \leq \mathbf{Z^J}$. Therefore, if $P(\mathbf{X}) = 1$, then $f(\mathbf{X}) = 1$; and $P(\mathbf{X})$ may be a term in a sum-of-products expression for f.

A similar $P(\mathbf{X})$ may be found corresponding to any vertex $\mathbf{Y}$, for which $f(\mathbf{Y}) = 1$. Thus, f may be expressed as the sum of all such *distinct* products, $P(\mathbf{X})$.

Example 17.16

Determine a sum-of-products expression for the function, tabulated in Fig. 17.25a, which is unate with respect to $\mathbf{J} = (0, 1, 0, 1)$.

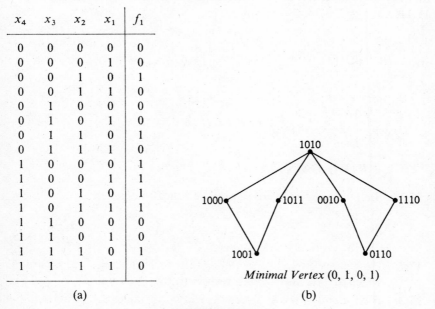

x_4	x_3	x_2	x_1	f_1
0	0	0	0	0
0	0	0	1	0
0	0	1	0	1
0	0	1	1	0
0	1	0	0	0
0	1	0	1	0
0	1	1	0	1
0	1	1	1	0
1	0	0	0	1
1	0	0	1	1
1	0	1	0	1
1	0	1	1	1
1	1	0	0	0
1	1	0	1	0
1	1	1	0	1
1	1	1	1	0

Minimal Vertex (0, 1, 0, 1)

(a)   (b)

FIGURE 17.25

Solution: The minimal element in the lattice is $(0, \bar{0}, 0, \bar{0}) = (0, 1, 0, 1)$. A partial lattice showing only the vertices for which $f_1 = 1$ is shown in Fig. 17.25b. It will be left to the reader to construct the complete lattice and verify that f_1 is unate. According to Theorem 17.10, only the two minimal vertices for which the function is 1 need be considered. As in the proof of the theorem, we form the functional representation from $x_4{}^{j_4}$, $x_3{}^{j_3}$, and $x_2{}^{j_2}$, $x_1{}^{j_1}$, since $(1, 0, 0, 1)^J = (1, \bar{0}, 0, \bar{1}) = (1, 1, 0, 0)$ and $(0, 1, 1, 0)^J = (0, \bar{1}, 1, \bar{0}) = (0, 0, 1, 1)$. Therefore,

$$f_1 = x_4\bar{x}_3 + x_2\bar{x}_1 \tag{17.69}$$

The reader will notice that the sum-of-products representation obtained in the previous example was the unique minimal sum-of-products representation. The following theorem is useful in determining whether a given Boolean function is unate. Its proof will be left to the reader.

THEOREM 17.11. *A switching function, f, is unate if and only if (1) the set of essential prime implicants of f constitutes a unique minimal cover of f and (2) this unique cover of f results in a sum-of-products representation in which only the literals $x_1{}^{j_1}$, $x_2{}^{j_2}$, $\dots$, $x_n{}^{j_n}$ appear for some $\mathbf{J} = (j_1, j_2, \dots, j_n)$.*

It is evident from Fig. 17.26 that the expression

$$x_1 + \bar{x}_2\bar{x}_3 + x_2x_3$$

is a unique minimal cover composed of essential prime subcubes. This unique minimal cover does not satisfy the second condition of Theorem 17.11, so $x_1 + \bar{x}_2\bar{x}_3 + x_2x_3$ is not unate.

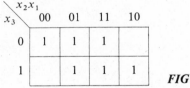

FIGURE 17.26

PROBLEMS

17.1. Simplify the following symmetric function expression.
    (a) $[S_{1,3,5,7}^{7}(x_1, \dots, x_7)] \cdot [S_{4,5,6,7}^{7}(x_1, \dots, x_7)]$
    (b) $S_{2,3,5}^{5}(x_1, \dots, x_5) + S_3{}^{5}(x_1, \dots, x_5)$
    (c) $[S_{1,3,5}^{5}(\bar{x}_1, \bar{x}_2, \dots, \bar{x}_5)] \cdot [S_{2,3,4,5}^{5}(x_1, x_2, \dots, x_5)]$

17.2. How many different functions of the form $f(x_1, x_2, \dots, x_6)$ are there which are symmetric with respect to the six literals $x_1, x_2, \dots, x_6$?

17.3. How many of the 16 functions of two variables are symmetric (with respect to any of the four possible pairs of literals)?

17.4. Consider the ten bits $x_1, x_2, \ldots, x_{10}$ arranged in ascending order. There are eight possible ways of selecting three successive bits from this arrangement. The function which is 1 when a majority of some such set of three bits is 1 is, of course, a symmetric function.

   (a) Write, as a function of symmetric functions, an expression which will be 1 whenever the majorities of bits in an odd number of the sets of three bits are 1.

   (b) Is the resulting expression a symmetric function? Why?

   (c) Repeat part (a), this time considering all $_{10}C_3 = 120$ possible combinations of 3 of the 10 bits. Simplify.

   (d) Is the function in part (c) symmetric? Why?

17.5. Determine a simple disjoint decomposition for each of the following functions for which such a decomposition exists.

   (a) $f(x_1, x_2, x_3, x_4) = \sum m(0, 9, 10, 11, 5, 6, 7, 12)$

   (b) $f(x_1, x_2, x_3, x_4) = \sum m(0, 3, 4, 7, 12, 15)$

   (c) $f(x_1, x_2, x_3, x_4) = \sum m(0, 2, 4, 9, 6, 13)$

   (d) $f(x_1, x_2, x_3, x_4, x_5) = \sum m(3, 10, 14, 17, 18, 22, 23, 24, 27,$
   $$28, 31)$$

17.6. Repeat Problem 17.5 for the following two functions.

   (a) $f(x_1, x_2, x_3, x_4, x_5) = \sum m(1, 3, 5, 7, 9, 10, 13, 14, 21, 23, 26,$
   $$27, 29, 30)$$

   (b) $f(x_1, x_2, x_3, x_4, x_5) = \sum m(1, 3, 9, 10, 17, 19, 24, 27)$

17.7. There exists an iterative disjoint decomposition for the following function of five variables. Determine this decomposition and express it in the form of Equation 17.39.
$$f(x_1, x_2, x_3, x_4, x_5) = \sum m(2, 6, 8, 12, 16, 20, 26, 30)$$

17.8. Determine a multiple disjoint decomposition of the 4-variable function.
$$f(x_1, x_2, x_3, x_4) = \sum m(0, 3, 4, 7, 9, 10, 13, 14)$$

17.9. Develop an original 5-variable example of the multiple disjoint decomposition given by Equation 17.51. Let the variables be partitioned as given by
$$f(x_1, x_2, x_3, x_4, x_5) = H[\lambda(x_1, x_2), \varphi(x_3, x_4), x_5]$$

17.10. Show that a Boolean algebra is a lattice. Which postulates of a Boolean algebra are not satisfied by every lattice?

17.11. Each cell of an iterative network is to have one primary input and one primary output. Let the primary input of cell i be x_i and let the primary output be z_i. Determine the cell table for cell i such that $z_i = 1$ if and only if $x_{i-2} = x_{i-1} = x_i$. Determine a network realization of the typical cell i.

17.12. The *n*-bit binary adder of Fig. 8.7 may be regarded as an iterative network. A single cell of the network is given in a circuit form as Fig. 8.3. Determine the state table of a single cell of the adder.

17.13. A circuit which will compare the magnitudes of two 18-bit numbers is to be formulated as an iterative network. Each cell is to have two secondary outputs and no primary outputs. The input numbers are given as vectors *X1* and *X2* (See Chapter 15). The secondary outputs of cell *j* should indicate whether the binary number represented by $\alpha^j/X1$ is greater than, equal to, or less than the binary number represented by $a^j/X2$. Determine the state table of a single cell and a circuit realization of that cell. Sketch the complete network. What will be the output of the network?

17.14. Which of the following sets are partially ordered? Totally ordered?
(a) Integers.
(b) Real numbers.
(c) Complex numbers (define $\leq$).
(d) A set of all equivalent incompletely specified sequential machines, where a machine S is $\leq$ a machine T if and only if every state in S is compatible with some state in T.
(e) The schools in the Big Ten Conference, where $A \leq B$ if and only if B defeated A in football this fall.

17.15. Which of the sets described in Problem 17.14 form lattices?

17.16. Which of the following switching functions are unate?
(a) $f(x_1, x_2, x_3, x_4) = \sum m(3, 5, 7, 10, 11, 12, 13, 14, 15)$
(b) $f(x_1, x_2, x_3, x_4) = \sum m(4, 8, 10, 12, 13, 14)$
(c) $x_1 \oplus x_2$
(d) $f(x_1, x_2, x_3, x_4) = \sum m(2, 3, 4, 10, 12, 13, 14)$

17.17. Which of the switching functions in Problem 17.5 are unate?

BIBLIOGRAPHY

1. Marcus, M. P. "The Detection and Identification of Symmetric Switching Functions with the use of Tables of Combinations," *IRE Trans. on Electronic Computers*, **EC-5**: 4, 237–239 (Dec. 1956).

2. Marcus, M. P. *Switching Circuits for Engineers*, Prentice-Hall, Englewood Cliffs, N.J., 1962.

3. Maley, G. A. and J. Earle. *The Logic Design of Transistor Digital Computers*. Prentice-Hall, Englewood Cliffs, N.J., 1963.

4. Ashenhurst R. L. "The Decomposition of Switching Functions," *Proceedings of the Symposium on the Theory of Switching*, April 2–5, 1957, Vol. 29 of *Annals of Computers*, Harvard University, 74–116, 1959.

5. Curtis, H. A. *Design of Switching Circuits*, Van Nostrand, Princeton, N.J., 1962.

6. Harrison, Michael. *Introduction to Switching and Automata Theory*, McGraw-Hill, New York, 1965.

7. Birkhoff, G. and S. MacLane. *Survey of Modern Algebra* (3rd ed.), Macmillan, New York, 1965.

8. Shannon, C. E. "A Symbolic Analysis of Relay and Switching Circuits," *Trans. AIEE*, **57**, 713–723 (1938).

9. Hennie, F. C. *Finite State Models for Logical Machines*, Wiley, New York, 1968.

10. Givone, D. D. *Introduction to Switching Circuit Theory*, McGraw-Hill, New York, 1970.

Chapter 18 *Threshold Logic*

18.1 *Generalized Resistor Transistor Logic Circuit*

One early version of a NOR gate was a resistor-transistor device much like Fig. 18.1. The reader may wonder about the performance of such a circuit if the input resistors were not all equal. In Fig. 18.1 we see a generalized version of a resistor transistor logic circuit.

To simplify the analysis of this circuit, we shall neglect the very small base-to-emitter input voltage. In this case, we have

$$Vx_0G_0 + Vx_1G_2 + \cdots + Vx_{n-1}G_{n-1} - I_t = I_{\text{in}} \qquad (18.1)$$

where the product Vx_i indicates that, if the Boolean variable x_i is one, a voltage V is applied across the conductance G_i. This expression can be written

$$\sum_{i=0}^{n-1} G_i x_i - \frac{I_t}{V} = \frac{I_{\text{in}}}{V} \qquad (18.2)$$

If $\sum_{i=0}^{n-1} G_i x_i$ is sufficiently greater than I_t/V, there will be enough current into the base to switch on the transistor and reduce the output voltage to zero. On the other hand, if $\sum_{i=0}^{n-1} G_i x_i < I_t/V$, the transistor output will be V volts. Thus, by adjusting the conductances $\{G_i\}$ and I_t, the output can be made to take on various functional values for the 2^n different combinations of input variables.

545

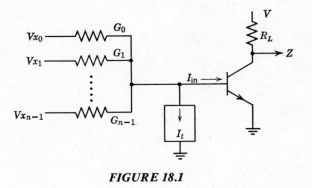

FIGURE 18.1

Example 18.1

Suppose that a circuit of the form in Fig. 18.1 is constructed for four input variables with conductances $G_3 = 0.003$, $G_2 = 0.002$, $G_1 = 0.001$ and $G_0 = 0.001$. Let $V = 10$ volts and $I_t = 0.025$ amp. Tabulate the output voltage for the sixteen possible combinations of input values.

Solution: Let us establish the output for the input values in the order tabulated in Fig. 18.2.

We first calculate $I_t/V = 0.0025$.

Clearly, for the four cases where $x_3 = x_2 = 0$, $G_1x_1 + G_2x_2 + G_3x_3 + G_0x_0 \leq 0.002$. Therefore, the output is 1 for the first four rows of the table.

x_3	x_2	x_1	x_0	f
0	0	0	0	
0	0	0	1	
0	0	1	0	
0	0	1	1	
0	1	0	0	
0	1	0	1	
0	1	1	0	
0	1	1	1	
1	0	0	0	
1	0	0	1	
1	0	1	0	
1	0	1	1	
1	1	0	0	
1	1	0	1	
1	1	1	0	
1	1	1	1	

FIGURE 18.2

x_3	x_2	x_1	x_0	f
0	0	0	0	1
0	0	0	1	1
0	0	1	0	1
0	0	1	1	1
0	1	0	0	1
0	1	0	1	0
0	1	1	0	0
0	1	1	1	0
1	0	0	0	0
1	0	0	1	0
1	0	1	0	0
1	0	1	1	0
1	1	0	0	0
1	1	0	1	0
1	1	1	0	0
1	1	1	1	0

FIGURE 18.3

For (0100), we have $G_0 \cdot 0 + G_1 \cdot 0 + G_2 \cdot 1 + G_3 \cdot 0 = 0.002$, and again the output is 1. For (0101) and (0110), $\sum_{i=0}^{3} G_i x_i = 0.003$, which is greater than I_t/V. In these cases,

$$I_{in} = 10(0.003 - 0.0025) = 5 \text{ ma}$$

A typical transistor will easily be switched on by this amount of current so the output will be 0 for these two cases. For (0111), the input current will be still greater, so the output will again be 0.

For all eight cases in which $x_3 = 1$, we have

$$\sum_{i=0}^{3} G_i x_i \geq 0.003$$

so the output is similarly 0 for all eight cases.

The above results are tabulated in Fig. 18.3. The function in Fig. 18.3 may be expressed as

$$f = \overline{x_3 + x_2(x_1 + x_0)} = \bar{x}_3(\bar{x}_2 + \bar{x}_1\bar{x}_0) \quad \blacksquare \tag{18.3}$$

Because of the inherent inverting property of transistors, a logic circuit of the form discussed above will be called an inverting threshold element. In the next section, we shall turn our attention to identifying the class of functions which may be realized by these devices. It is most convenient, however, to first relate our discussion to elements which do not possess this inverting property.

18.2   Linear Separability*

A Boolean function, $f(x_0, x_1, \ldots, x_{n-1})$, is said to be linearly separable if and only if there exists an n-tuple of ones and zeros $(j_0, j_1, \ldots, j_{n-1})$, a set of weights $N_0, N_1, \ldots, N_{n-1}$, and a threshold value T such that

$$S = \sum_{i=0}^{n-1} N_i x_i^{j_i} \quad \begin{cases} < T \text{ whenever } f = 0 \\ \geq T \text{ whenever } f = 1 \end{cases} \tag{18.4}$$

where

$$x_i^{j_i} = \begin{cases} x_i & \text{if } j_i = 0 \\ \bar{x}_i & \text{if } j_i = 1 \end{cases}$$

As was the case in Section 18.1, $N_i x_i^{j_i}$ is to be thought of as the standard algebraic product of a real number and a Boolean literal.

Example 18.2

It is easily seen that $f = x_3 + x_2(x_1 + x_0)$ given in 18.3 is linearly separable. If weights and threshold value are chosen proportional to the conductance values and I_t/V given in Example 18.1,

$$\sum_{i=0}^{3} N_i x_i \geq T$$

if and only if

$$\sum_{i=0}^{3} G_i x_i \geq \frac{I_t}{V}.$$

The function $\overline{x_3 + x_2(x_1 + x_0)}$, rather than $x_3 + x_2(x_1 + x_0)$, was realized in Example 18.1 because the transistor served to invert the functional values as defined in 18.4. A convenient set of weights is $N_3 = 3$, $N_2 = 2$, and $N_1 = 1$ with threshold $T = 2.5$.

The function $f = \bar{x}_3(\bar{x}_2 + \bar{x}_1 \bar{x}_0)$ is also linearly separable with n-tuple (1111), the same weights as above, and threshold $T = 5$. Verification of this fact will be left to the reader.

THEOREM 18.1.   *If a set of weights can be found satisfying 18.4 for a threshold $T_1 > 0$, then a set can be found satisfying 18.4 for any $T_2 > 0$.*

Proof: Each new weight is obtained as follows

$$N_i' = N_i \frac{T_2}{T_1}$$

$$S' = \sum_{i=0}^{n-1} N_i' x_i^{j_i} \geq T_2$$

---

* Linearly separable functions are commonly called *threshold* functions.

if and only if

$$S' = \sum_{i=0}^{n-1} N_i \left(\frac{T_2}{T_1}\right) x_i^{j_i} = \frac{T_2}{T_1} \sum_{i=0}^{n-1} N_i x_i^{j_i} \geq T_2$$

if and only if

$$S = \sum_{i=0}^{n-1} N_i x_i^{j_i} \geq T_2\left(\frac{T_1}{T_2}\right) = T_1 \tag{18.5}$$

THEOREM 18.2. *If a function f is linearly separable, an n-tuple* $(j_0', j_1', \ldots, j_{n-1}')$ *and a set of weights* $\{N_i'\}$ *can be found so that each* $N_i' \geq 0$.

Proof: Assume that f is separable in terms of $\{N_i\}$, T, and $(j_0 \ldots j_{n-1})$. For every $N_i < 0$, let $j_i' = 1 - j_i$ to form a new n-tuple (that is, if $x_i^{j_i} = x_i$ let $x_i^{j_i'} = \bar{x}_i$ and vice versa). Form a new set of weights given by

$$N_i' = |N_i|$$

Therefore,

$$S' = \sum_{i=0}^{n-1} N_i' x_i^{j_i'}$$

Let I be the set of all i such that $N_i < 0$. Therefore,

$$\sum_{i=0}^{n-1} N_i' x_i^{j_i'} = \sum_{i=0}^{n-1} |N_i| \, x_i^{j_i'} = \sum_{i \in I} |N_i| \, x_i^{1-j_i} + \sum_{i \notin I} N_i x_i^{j_i}$$

$$= \sum_{i \in I} N_i x_i^{j_i} + \sum_{i \in I} |N_i| + \sum_{i \notin I} N_i x_i^{j_i}$$

$$= \sum_{i=0}^{n-1} N_i x_i^{j_i} + \sum_{i \in I} |N_i|$$

Therefore,

$$\sum_{i=0}^{n-1} N_i' x_i^{j_i'} \geq T + \sum_{i \in I} |N_i|$$

if and only if

$$\sum_{i=0}^{n-1} N_i x_i^{j_i} \geq T$$

Therefore, define

$$T' = T + \sum_{i \in I} |N_i| \tag{18.6}$$

Example 18.3

Determine the Boolean function separated by the following summation and threshold, $T = 1$.

$$S = 2x_2 - x_1 + x_0 \tag{18.7}$$

Then find a set of positive weights and a threshold which separates the same function.

x_2	x_1	x_0	S	f (for $T = 1$)
0	0	0	0	0
0	0	1	1	1
0	1	0	-1	0
0	1	1	0	0
1	0	0	2	1
1	0	1	3	1
1	1	0	1	1
1	1	1	2	1

FIGURE 18.4

Solution: Tabulating the summation for all possible combinations of values results in Fig. 18.4. The function in the right-hand column of Fig. 18.4, which can be expressed as

$$f = x_2 + \bar{x}_1 x_0$$

is the only function separated by the given weights and threshold.

Now, applying Theorem 18.2, we have $N_2' = N_2 = 2$, $N_1' = |N_1| = 1$, $N_0' = N_0 = 1$, and

$$T' = T + |N_1| = 1 + 1 = 2 \qquad (18.8)$$

It can be easily verified by the reader that

$$S' = 2x_2 + \bar{x}_1 + x_0 \qquad (18.9)$$

is greater than or equal to 2 for precisely those combinations of values for which the function in Fig. 18.4 is 1. ■

THEOREM 18.3. *A function, $\bar{f}$, is linearly separable if and only if f is linearly separable.*

Proof: Let f be linearly separable. Then there exists a set of weights $\{N_i\}$, a threshold, T, and an n-tuple $(j_0, j_1, \ldots, j_{n-1})$ such that

$$\sum_{i=0}^{n-1} N_i x_i^{j_i} \begin{cases} \geq T & \text{when } f = 1 \\ < T & \text{when } f = 0 \end{cases}$$

Because either $x_i^{j_i}$ or $x_i^{1-j_i}$ must always be 1, we have

$$\sum_{i=0}^{n-1} N_i x_i^{j_i} + \sum_{i=0}^{n-1} N_i x_i^{1-j_i} = \sum_{i=0}^{n-1} N_i \qquad (18.10)$$

Suppose $\bar{f} = 1$. Therefore, $f = 0$ and

$$\sum_{i=0}^{n-1} N_i x_i^{j_i} < T$$

If so, there exists some δ such that

$$\sum_{i=0}^{n-1} N_i x_i^{j_i} < T - \delta$$

Therefore, from 18.10,

$$\sum_{i=0}^{n-1} N_i x_i^{1-j_i} \geq \sum_{i=0}^{n-1} N_i - (T - \delta)$$

We have established a set of weights, $\{N_i' = N_i\}$, a threshold, $\sum_{i=0}^{n-1} N_i - (T - \delta)$, and an n-tuple, $(1 - j_0, 1 - j_1, \ldots, 1 - j_{n-1})$, satisfying 18.4, whenever $\bar{f}$ is 1. If $\bar{f} = 0$, we have $f = 1$ and

$$\sum_{i=0}^{n-1} N_i x_i^{j_i} \geq T > T - \delta$$

Therefore, from 18.10,

$$\sum_{i=0}^{n-1} N_i x_i^{1-j_i} < \sum_{i=0}^{n-1} N_i - (T - \delta)$$

We have shown that, for any Boolean function, f, if f is linearly separable, so is $\bar{f}$. Therefore, if $\bar{f}$ is linearly separable, so is $\bar{\bar{f}} = f$. This completes the proof.

Example 18.4

Determine a set of weights, an n-tuple, and a threshold separating f, where f is tabulated in Fig. 18.4.

Solution: Fortunately, the proof of Theorem 18.3 is constructive, thereby providing a method of calculating the required quantities. From 18.7, we have

$$(j_0, j_1, j_2) = (0, 0, 0)$$

Therefore,

$$(1 - j_0, 1 - j_1, 1 - j_2) = (1, 1, 1)$$

The weights are unchanged, and the threshold T' is given by

$$T' = (2 - 1 + 1) - (1 - 0.5) = 1.5$$

It will be left to the reader to verify that

$$S = 2\bar{x}_2 - \bar{x}_1 + \bar{x}_0 \begin{cases} \geq 1.5 & \text{when } f = 1 \\ < 1.5 & \text{when } f = 0 \end{cases} \quad \blacksquare$$

The importance of Theorems 18.2 and 18.3 lies in their implication that a function can be realized by a resistor-transistor threshold element if and only if it is linearly separable. This is not strictly true, of course, as increased function complexity will eventually result in a current difference between the 0 and 1 cases insufficient to switch the transistor. For simple functions, however, Theorem 18.2 assures us that the natural restriction to positive weights in a practical threshold element will not stand in the way of the realization of a linearly separable function. Similarly, Theorem 18.3 assures us that any linearly separable function can be realized by an inverting element.

THEOREM 18.4. *If a function, f, is linearly separable, there exists a set of weights and a threshold, T, so that the function satisfies Expression 18.4 in terms of the n-tuple* $(0, 0, \ldots, 0)$.

The proof of this theorem is similar to that of Theorem 18.2 and will be left to the reader. Negative weights as well as a negative threshold value may be required.

In terms of the weights promised by Theorem 18.4, the summation in Expression 18.4 reduces to

$$S = \sum_{i=0}^{n-1} N_i x_i \qquad (18.11)$$

Equating the summation to the threshold, T, we obtain the equation of a hyperplane,

$$\sum_{i=0}^{n-1} N_i x_i = T \qquad (18.12)$$

Vertices on the Boolean hypercube, for which the function is 1, will be found on one side of this plane, and vertices for which the function is 0 will be found on the other. It may be necessary to adjust T slightly so that the plane will not pass through any vertices.

Example 18.5

It is easily verified that the plane

$$-2x_2 - x_0 + x_1 = -1.5$$

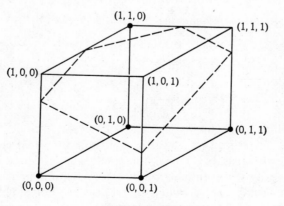

FIGURE 18.5

separates the function

$$f = \bar{x}_2 + x_1 \bar{x}_0$$

In Fig. 18.5, we see the intersection of this plane with the three-dimensional Boolean cube. Notice that the vertices for which $f = 1$ (darkened circles) are on one side of the plane and the vertices for which $f = 0$ are on the other.

Although higher dimensional cases cannot be so easily visualized, no further conceptual problems are involved. ∎

18.3 Conditions for Linear Separability

So far nothing has been said about determining whether a given Boolean function is linearly separable. Presumably, not all of them are separable. In this section, a number of necessary conditions on linearly separable functions will be developed. As we shall see, none of these conditions will be sufficient. In fact no set of necessary and sufficient conditions is known to provide a simple verification of linear separability. To determine the linear separability of a function, we first consider it in terms of the necessary conditions, in order of their complexity. Should the function satisfy all these criteria, we then attempt to determine a set of weights by some trial-and-error algorithm.

THEOREM 18.5. *A linearly separable function is unate.*

Proof: By Theorem 18.2, there exists a positive set of weights which will separate any linearly separable function in terms of some n-tuple $(j_0, j_1, \ldots, j_{n-1})$. Let us consider this n-tuple as the least element in a lattice of the vertices of the Boolean hypercube. Suppose for some vertex $(a_0, a_1, \ldots, a_{n-1})$, $f(a_0, a_1, \ldots, a_{n-1}) = 1$. Therefore,

$$\sum_{i=0}^{n-1} N_i a_i^{j_i} = \sum_{a_i \neq j_i} N_i \geq T$$

Now suppose $(b_0, b_1, \ldots, b_{n-1})$ is some vertex in the lattice such that if

$$a_i^{j_i} = 1 \qquad \text{then} \qquad b_i^{j_i} = 1$$

Therefore,

$$\sum_{i=0}^{n-1} N_i b_i^{j_i} \geq \sum_{i=0}^{n-1} N_i a_i^{j_i} \geq T$$

and

$$\sum_{i=0}^{n-1} N_i b_i \geq T$$

and $f(b_0, b_1, \ldots, b_{n-1}) = 1$. Thus, by Theorem 17.10, the function f must be unate in terms of $(x^{j_0}, x^{j_1}, \ldots, x^{j_{n-1}})$.

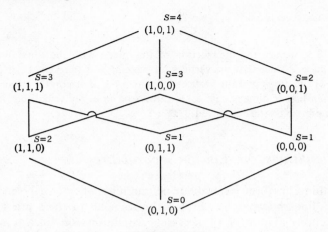

FIGURE 18.6

Example 18.6

Verify that the linearly separable function which is separated by the summation

$$S = 2x_2 + \bar{x}_1 + x_0$$

with threshold, $T = 2$, is unate.

Solution: That the above is true follows from Theorem 18.5. As an illustration of the theorem, let us construct the appropriate lattice for the function. The above summation using positive weights tells us that the least element of the lattice must be $(0, 1, 0)$. The lattice can then be constructed as shown in Fig. 18.6.

The values of S can then be computed for each vertex. These values are also shown in Fig. 18.6. A comparison of the values of S with the threshold, $T = 2$, tells us that only the three vertices for which $S = 1$, or $S = 0$, have functional values 0. As expected, the functional values are nondecreasing along any path from the bottom to the top of the lattice. Definition 17.9 is, therefore, satisfied in this case. Notice that the functional values are those tabulated in Fig. 18.4. The function is, therefore, $f = x_2 + \bar{x}_1 x_0$, which we can easily see is unate. ■

Example 18.7

Verify that the unate function

$$f = x_0 x_1 + x_2 x_3 \tag{18.13}$$

is not linearly separable.

Solution: In order that $f(0, 0, 1, 1) = f(1, 1, 0, 0) = 1$, we must have

$$N_0 + N_1 \geq T \tag{18.14}$$

and

$$N_2 + N_3 \geq T \tag{18.15}$$

On the other hand, $f(0, 1, 0, 1)$ and $f(1, 0, 1, 0)$ must be zero. Therefore,

$$N_0 + N_2 < T \tag{18.16}$$

and

$$N_1 + N_3 < T \tag{18.17}$$

Adding expressions 18.14 and 18.15 yields 18.18, and adding 18.16 and 18.17 yields 18.19.

$$N_0 + N_1 + N_2 + N_3 \geq 2T \tag{18.18}$$

$$N_0 + N_1 + N_2 + N_3 < 2T \tag{18.19}$$

This contradiction tells us that a set of weights and threshold cannot be obtained for this function in terms of n-tuple $(0, 0, \ldots, 0)$. Therefore, by Theorem 18.4, the function is not linearly separable. ∎

To facilitate the proofs of the following theorems, the 2^n vertices of the Boolean hypercube will be designated as vectors $\mathbf{v}_0, \mathbf{v}_1, \ldots, \mathbf{v}_{2n-1}$. These vectors will be subject to the standard laws of vector addition and matrix multiplication.

THEOREM 18.6. *If there exist k vertices $\mathbf{v}_1, \mathbf{v}_2, \ldots, \mathbf{v}_k$ for which $f = 1$ and k vertices $\mathbf{u}_1, \mathbf{u}_2, \ldots, \mathbf{u}_k$ for which $f = 0$, such that the vector sums are equal as given by 18.20, then f is not linearly separable.*

$$\sum_{j=1}^{k} \mathbf{v}_j = \sum_{j=1}^{k} \mathbf{u}_j \tag{18.20}$$

Proof: Let $\mathbf{v}_j$ and $\mathbf{u}_j$ be columns vectors and let the weights be arranged in a row vector $N = (N_0 N_1 \cdots N_{n-1})$. We can now conveniently write, for example, $S(\mathbf{v}_j) = \mathbf{N} \cdot \mathbf{v}_j$, which is a produ t of a $1 \times n$ matrix and an $n \times 1$ matrix. It may be equally well regarded as a vector dot product.

Assume that there exist k vectors $\{\mathbf{v}_j\}$ for which $f = 1$ and k vectors $\{\mathbf{u}_j\}$ for which $f = 0$ such that

$$\sum_{j=1}^{k} \mathbf{v}_j = \sum_{j=1}^{k} \mathbf{u}_j$$

Therefore,

$$\mathbf{N} \cdot \left(\sum_{j=1}^{k} \mathbf{v}_j \right) = \mathbf{N} \cdot \left(\sum_{j=1}^{k} \mathbf{u}_j \right)$$

Or equally well,

$$\sum_{j=1}^{k} \mathbf{N} \cdot \mathbf{v}_j = \sum_{j=1}^{k} \mathbf{N} \cdot \mathbf{u}_j \tag{18.21}$$

However, if f were linearly separable, we would have

$$\mathbf{N} \cdot \mathbf{v}_j \geq T \quad \text{and} \quad \mathbf{N} \cdot \mathbf{u}_j < T$$

and

$$\sum_{j=1}^{k} \mathbf{N}\mathbf{v}_j \geq kT > \sum_{j=1}^{k} \mathbf{N}\mathbf{u}_j \qquad (18.22)$$

Since expressions 18.21 and 18.22 are obviously contradictory, we conclude that f cannot be linearly separable; and the theorem is proved.

Example 18.8

Let us apply Theorem 18.6 to the simple unate function given in Example 18.7, which we have already demonstrated to be not linearly separable.

We note that $f = 1$ for the vertices

$$\begin{bmatrix} x_0 \\ x_1 \\ x_2 \\ x_3 \end{bmatrix} = \begin{bmatrix} 0 \\ 0 \\ 1 \\ 1 \end{bmatrix} \quad \text{and} \quad \begin{bmatrix} 1 \\ 1 \\ 0 \\ 0 \end{bmatrix}$$

Similarly, $f = 0$ for

$$\begin{bmatrix} 0 \\ 1 \\ 0 \\ 1 \end{bmatrix} \quad \text{and} \quad \begin{bmatrix} 1 \\ 0 \\ 1 \\ 0 \end{bmatrix}$$

Therefore, there exist vectors such that

$$\sum_{j=1}^{2} \mathbf{v}_j = \begin{bmatrix} 0 \\ 0 \\ 1 \\ 1 \end{bmatrix} + \begin{bmatrix} 1 \\ 1 \\ 0 \\ 0 \end{bmatrix} = \begin{bmatrix} 1 \\ 1 \\ 1 \\ 1 \end{bmatrix} = \begin{bmatrix} 0 \\ 1 \\ 0 \\ 1 \end{bmatrix} + \begin{bmatrix} 1 \\ 0 \\ 1 \\ 0 \end{bmatrix} = \sum_{j=1}^{2} \mathbf{u}_j$$

Thus, by Theorem 18.6, we again conclude that

$$f = x_0 x_1 + x_2 x_3$$

is not linearly separable. ∎

For a function of about six or fewer variables, the normal form expression may suggest sets of variables which might be fixed to verify that the function is not linearly separable. We would not expect to apply either of these theorems exhaustively, except as a digital computer program. Even if applied exhaustively, these theorems still do not constitute a sufficient condition for linear separability. E. F. Moore has worked out a 12-variable counter example.

The following theorem is a further specialization which is probably most convenient to use when manipulating a function of a small number of variables by hand. It will also prove to be of assistance in determining weights.

THEOREM 18.7. *Suppose that a function, f, is linearly separable in terms of the positive weights $N_0, N_1, \ldots, N_{n-1}$. Let a_i be the number of vertices for which $f = 1$ and $x_i^{j_i} = 1$. Then the following four conditions hold.*

(1) *If $N_i \geq N_k$, then $a_i \geq a_k$.*
(2) *If $N_i = N_k$, then $a_i = a_k$.*
(3) *If $a_i > a_k$, then $N_i > N_k$.*
(4) *If $a_i = a_k$, then the weights may be chosen such that $N_i = N_k$.*

The above theorem may be proved either by using Theorem 18.6 or by referring directly to the definition of linear separability. Both proofs are straightforward, although somewhat awkward, and will be omitted. Instead, let us illustrate the application of the theorem in the following example.

Example 18.9
Determine whether or not the following two functions are linearly separable and, if so, determine an appropriate set of weights and threshold values.

$$f_1 = x_2 x_1 x_0 + x_4 x_3 + x_4 x_2 + x_3 x_2 + x_3 x_0 + x_4 x_1$$
$$f_2 = x_4 x_3 \bar{x}_0 + x_4 \bar{x}_2 + x_3 \bar{x}_2 + x_3 \bar{x}_1 + x_4 x_3 x_0 + x_4 \bar{x}_1 + x_4 x_0 + \bar{x}_2 \bar{x}_1 x_0$$

Solution: Considering f_1 first, we note that this function is unate. Rather than applying Theorem 18.6 and 18.7 exhaustively, let us attempt to determine an order for the variables. Toward this end, we first map the function as illustrated in Fig. 18.7.

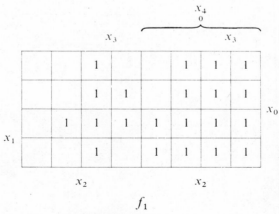

$$f_1$$
FIGURE 18.7. f_1.

Here we see that $a_4 = 14$, $a_3 = 14$, $a_1 = 12$, and $a_0 = 12$. From Theorem 18.7, we conclude that

$$N_4 = N_3 > N_2 > N_1 = N_0 \tag{18.23}$$

Knowledge of the order of the literals suggests the following factorization:

$$x_4(x_3 + x_2 + x_1) + x_3(x_2 + x_0) + x_2 x_1 x_0$$

It is now more apparent, perhaps, that $x_4 x_0$ is not included in the function, but that $x_3 x_0$ is. Therefore, we must have

$$N_4 + N_0 < T \quad \text{and} \quad N_3 + N_0 \geq T$$

Consequently,

$$N_4 < N_3 \tag{18.24}$$

The contradiction between expressions 18.23 and 18.24 tells us that f_1 is not linearly separable.

Notice that f_2 appears as a unate function after making the simplification

$$x_4 x_3 \bar{x}_0 + x_4 x_3 x_0 = x_4 x_3$$

The Karnaugh map of this function appears in Fig. 18.8. A count of the vertices reveals that $a_4 = 15$, $a_3 = 14$, $a_2 = 13$, $a_1 = 13$, and $a_0 = 12$. This suggests the factorization

$$f_2 = x_4(x_3 + \bar{x}_2 + \bar{x}_1 + x_0) + x_3(\bar{x}_2 + \bar{x}_1) + \bar{x}_2 \bar{x}_1 x_0 \tag{18.25}$$

Since there must be four distinct weights, with only $N_2 = N_1$, we try first the

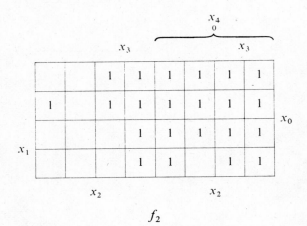

FIGURE 18.8

simplest combination,

$$N_4 = 4, \qquad\qquad N_3 = 3$$
$$N_2 = N_1 = 2, \qquad N_0 = 1$$

Notice that the smallest summation of weights corresponding to a Boolean product in 18.25 is precisely equal to 5. On the other hand, the summation is less than 5 for any product of the appearing literals, which is not included in f_2. Therefore, this simple set of weights is sufficient to separate f_2 if $T = 5$. ∎

In this presentation, we have contented ourselves with finding weights by trial and error. This approach is probably sufficient for functions which might be realized by a practical threshold device. Algorithms for the determination of weights for linearly separable functions of a large number of variables are, however, of considerable theoretical interest.* These techniques usually take the form of computer programs and have in some cases been extended to form the basis of adaptive pattern recognition techniques.

As has been amply demonstrated, not all Boolean functions are linearly separable. After considering a few examples, the reader should be able to convince himself that only a small portion of all switching functions are linearly separable. For example, R. O. Winder[3] has succeeded in verifying the number of seven variable threshold functions to be 8,378,070,864.

This appears to be a large number of functions, but it is only $2.4(10)^{-27}\%$ of the total number of seven variable switching functions. On the other hand, almost 3% of the four variable functions are linearly separable.

Since such a small portion of functions of a large number of variables are linearly separable, we immediately begin to look for ways to realize nonseparable functions with two or more threshold devices. There have been numerous approaches to this problem.[4,5,6] If switching speed is an important consideration, it might be desirable to limit consideration to two-level realizations. Under these circumstances, there is much to be said for learning to recognize threshold functions of a few variables on a Karnaugh map. Alternatively, various of the programmed computational methods have been extended to more than one threshold device. No general method has been found to be completely successful for functions requiring more than two or three threshold devices. Additionally, the complex sets of weights resulting from these multi-threshold techniques make the achievement of a physical realization prohibitive.

A third approach is to restrict consideration to special cases of threshold functions (such as majority or voting functions), which are separable in terms of an elementary set of weights. The remainder of this chapter is mostly devoted to this approach. In particular, the realization of symmetric functions will be considered.

* See, for example, "Synthesis of Linear Input Logic by the Simplex Method," R. C. Minnick, *Denver Research Institute Computer Symposium*, July 30, 1959 (1).

18.4   *Magnetic Threshold Logic Devices*

Before moving on to the consideration of combinations of threshold devices, let us consider another type of realization, in which the principal component is a magnetic core similar to those used in computer memories. The core is made of a ferrite material, which exhibits two distinct states of remanent magnetism and a relatively square hysteresis loop (Fig. 18.9a). The two remanent states of the core are ϕ_1 and ϕ_2. Assume that the core is initially in state ϕ_1. As long as the mmf applied to the core is less than F_s, the core will operate on the lower portion of the curve, with only very small changes in ϕ as the mmf changes. If the mmf exceeds F_s by even a fairly small amount, the cores switches from ϕ_1 to ϕ_2 and stays at ϕ_2 as long as the mmf exceeds $-F_s$. When the mmf becomes less than $-F_s$, the core switches back to ϕ_1. We can see that we have the basis for a threshold device here, since a relatively small change in input can cause large discrete change between two operating conditions.

The construction of a magnetic core threshold device is shown in Fig. 18.9b. For each input x_i there is a separate winding on the core with N_i turns. There will be a current of I amperes in a winding if $x_i = 1$, no current if $x_i = 0$. The N_i correspond to the weights; and the mmf, F_s, corresponds to the threshold, T. Thus, if

$$\sum_{i=0}^{n-1} I N_i x_i < F_s$$

the core will be in state ϕ_1. If this summation is greater than F_s, the core will be in state ϕ_2.

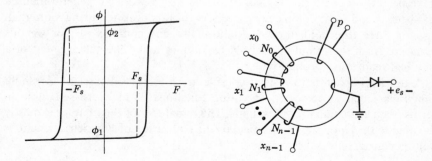

FIGURE 18.9

It is also necessary to provide some means of detecting which state the core is in. For this purpose, we provide a pulse winding and a sense winding. To determine the state of the core, a pulse of current is applied to the p winding. This pulse provides a sufficiently negative mmf to place the core in the ϕ_1 state, regardless of the values of the input variables. If the core is in the ϕ_2 state when the pulse arrives, there will be a change in flux from ϕ_2 to ϕ_1, which will induce a voltage in the sense winding. If the core is in the ϕ_1 state, there will be no change of flux and no induced voltage. The diode in the sense line prevents a pulse when the core switches from ϕ_1 to ϕ_2.

Example 18.10

Design core circuit to implement

$$f(x_1, x_2, x_3) = x_1 + x_2x_3 \tag{18.26}$$

using a core with the hysteresis loop shown in Fig 18.10a.

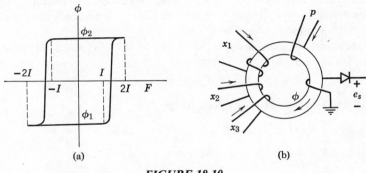

(a)   (b)

FIGURE 18.10

Solution: From the loop, we need an mmf of at least $2I$ to switch the core. If I is the signal current, and $N_1 = 2$ turns and $N_2 = N_3 = 1$ turn, then the mmf will be given by

$$\sum_{i=0}^{n-1} IN_ix_i = (2x_i + x_2 + x_3)I \tag{18.27}$$

which will be at least $2I$ for

$$f = x_1 + x_2x_3 = 1 \tag{18.28}$$

When all three variables equal 1, the mmf will be $4I$, so the current in the pulse winding must be at least $6I$ in the direction shown in Fig. 18.10b to ensure driving the core to ϕ_1. ∎

18.5   The Realization of Symmetric Functions Using More than One Threshold Device

The characteristics of a physical device are never constant. Usually, for example, they vary with temperature. This feature in conjunction with the usual gain limitation places a limit to the complexity of linearly separable functions realizable in terms of a single linear threshold device. This situation is illustrated in Problems 18.8 and 18.10 for the case of the resistor-transistor circuits. For the most part, this has meant that threshold elements have found application only for a few special classes of functions.

One promising application of threshold logic lies in the realization of symmetric functions. As pointed out in Problem 18.9, the tolerance problem is less severe in realizing $_nS_{\geq k}$ than in the realization of any other function requiring at least k of n inputs executed to switch the transistor. We shall see presently that arbitrary symmetric functions are easily expressed in terms of symmetric functions of this form. Therefore, physically realizable configurations involving more than one linear threshold device can be obtained for symmetric functions.

The positive weights, n-tuple, and threshold, in terms of which the symmetric function

$$_nS_{\geq k}(x_0^{j_0}, x_1^{j_1}, \ldots, x_{n-1}^{j_{n-1}}) \tag{18.29}$$

is linearly separable, are easily seen to be

$$N_i = 1 \qquad (j_0, j_1, \ldots, j_{n-1}) \qquad \text{and} \qquad T = k$$

Thus, if p of the n literals are 1, we have

$$\sum_{i=0}^{n-1} N_i x_i^{j_i} = p < T$$

where $p < k$, and

$$\sum_{i=0}^{n-1} N_i x_i^{j_i} = p \geq T$$

where $p \geq k$.

It is in terms of functions of the form of Expression 15.29 that our multi-threshold realizations of arbitrary symmetric functions will be based.

Example 18.11

Devise a resistor-transistor realization of the symmetric function $_5S_{0,1,3,4}$ $(x_0, \bar{x}_1, x_2, x_3, x_4)$.

Solution: First we note that

$$_5S_{0,1,3,4} = \overline{_5S_{2,5}} = \overline{(_5S_{\geq 2})_5S_{\geq 3}} + {_5S_{\geq 5}} \tag{18.30}$$

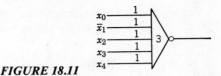

FIGURE 18.11

We have succeeded in expressing $_5S_{2,5}$ in terms of three symmetric functions, all of which are linearly separable. If each of these functions were generated, they in turn could be related by a threshold device leading to a 4-element realization. There is a much better way to proceed, however.

Let us generate first only $_5\overline{S_{\geq 3}}(x_0, \bar{x}_1, x_2, x_3, x_4)$. The resistor-transistor realization of this function is illustrated in Fig. 18.11. The symbol depicted is similar to that of a NAND gate as given in Chapter 2, except that more than one input is present and the weights and threshold value are given. Only one more threshold element is required. This element can best be thought of as an inverting realization of $_5S_{\geq 5}$ with an input of weight 3 from the element in Fig. 18.11. This is illustrated in Fig. 18.12. When the output of the first level element is 1, it can be visualized as reducing the threshold of the second level element from 5 to 2. In that case, only two of the literal inputs to the second level element need be 1 to turn on the transistor. The reader should satisfy himself completely that the output z conforms to Equation 18.30 by following the signal for each of 0, 1, 2, 3, 4, and 5 one inputs. In Problem 18.12, the reader will find that many more conventional logic elements would be required for realization of this function. ∎

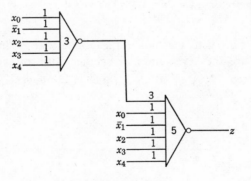

FIGURE 18.12

Perhaps the most important symmetric function is the parity checking function. In the realization of this function, the notion of using the output of threshold elements to control the thresholds of higher level elements, as illustrated in Example 18.11, is carried to its logical conclusion.

Let us restrict ourselves to the odd and even parity check functions of $2^k - 1$ literals where k is a positive integer

$$_{2^k-1}S_{0,2,4,\ldots,2^k-2}$$

and

$$_{2^k-1}S_{1,3,5,\ldots,2^k-1} \tag{18.31}$$

Let us illustrate the procedure by developing a realization of the even parity check function for $k = 3$. The extension of this procedure to a higher number of variables will be apparent. Again, we must provide for the final inversion.

$$_{7}S_{0,2,4,6} = \overline{_{7}S_{1,3,5,7}} \tag{18.32}$$

From the right side of Equation 18.32, we see that the final element must be controlled so as to have apparent thresholds at 1, 3, 5, and 7. The threshold must be one when the number of excited inputs is 0 or 1; it must be three whenever the number of excited inputs is 2 or 3, etc.

That this can be achieved will be demonstrated by forming the following expressions which are equivalent to 18.32:

$$\overline{_{7}S_{1,3,5,7}} = \overline{_{7}S_1 + {_{7}S_3} + {_{7}S_5} + {_{7}S_7}} \tag{18.33}$$

$$= \overline{_{7}S_{\geq 1}(_{7}S_{0,1,4,5})_{7}\overline{S}_{\geq 4} + {_{7}S_3}\ _{7}\overline{S}_{\geq 4} + {_{7}S_{\geq 5}}(_{7}S_{0,1,4,5}) + {_{7}S_7}} \tag{18.34}$$

$$= \overline{_{7}S_{\geq 1}(\overline{_{7}S_{2,3,6,7}})_{7}\overline{S}_{\geq 4} + {_{7}S_{\geq 3}}\ _{7}\overline{S}_{\geq 4} + {_{7}S_{\geq 5}}\ _{7}\overline{S_{2,3,6,7}} + {_{7}S_7}} \tag{18.35}$$

Suppose now we try to mechanize the above with a final level transistor whose inputs are the seven literals together with an input from a representation of each $\overline{_{7}S_{2,3,6,7}}$ and $_{7}S_{\geq 4}$. Such a configuration is illustrated in Fig. 18.13.

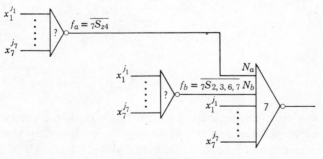

FIGURE 18.13

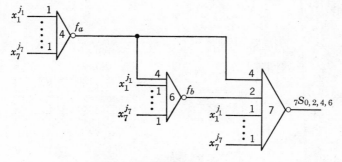

FIGURE 18.14

Clearly, the threshold must be 7 when $f_a = f_b = 0$. From Equation 18.35, however, we see that the effective threshold must be reduced to 5 when $f_b = 1$ and $f_a = 0$. Therefore, from

$$T' = 5 = 7 - N_b \tag{18.36}$$

we see that $N_b = 2$. Similarly, when $f_b = 0$ and $f_a = 1$, we have

$$T'' = 3 = 7 - N_a \tag{18.37}$$

Therefore, $N_a = 4$. For $f_a = f_b = 1$,

$$T''' = 1 = 7 - N_a - N_b \tag{18.38}$$

must be satisfied. Fortunately, the values of $N_a = 4$ and $N_b = 2$, already determined, satisfy Equation 18.38.

We notice immediately that $f_a = {}_7S_{\geq 4}$ is realizable by a single linear threshold device. This is not the case, however, with f_b:

$$f_b = \overline{{}_7S_{2,3} + {}_7S_{6,7}} = \overline{{}_7S_{\geq 2}\,\overline{{}_7S_{\geq 4}}} + {}_7S_{6,7} \tag{18.39}$$

So we see that f_b is realizable with only the literals and f_a, which must already be generated, as inputs. The complete circuit is illustrated in Fig. 18.14. The reader should verify the output for each possible number of excited inputs. In Fig. 18.15 the apparent thresholds and the outputs of each element are tabulated for every input sum.

Notice that, in order to achieve the seven transitions in output, it was necessary for the three transistors to assume all eight possible combinations of output. In general, $N + 1$ distinct states will be required where N is the number of output transitions. As we shall soon see, it will not always be

$\sum_{i=1}^{7} x_i^{j_i}$	T_a'	f_a	T_b'	f_b	T_c'	f_c
0	4	1	2	1	1	1
1	4	1	2	1	1	0
2	4	1	2	0	3	1
3	4	1	2	0	3	0
4	4	0	6	1	5	1
5	4	0	6	1	5	0
6	4	0	6	0	7	1
7	4	0	6	0	7	0

FIGURE 18.15

possible to take advantage of every state of the threshold elements required in the realization of a symmetric function. That this is possible in the case of the parity check function results from the opportunity to binary code the bias weights.

As was mentioned earlier, an optimal parity checker can be worked out for any $2^k - 1$ inputs. The 15-variable device is illustrated in Fig. 18.16. The further extension of the configuration should be evident.

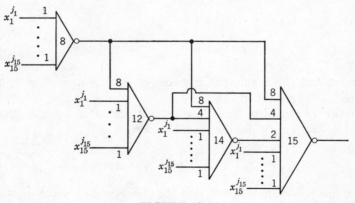

FIGURE 18.16

Example 18.12

Devise a realization of $_9S_{0,3,5,7}$.

Solution: Attempting to proceed in the same manner as with the parity check function, we obtain

$$_9S_{0,3,5,7} = {}_9S_{1,2} + {}_9S_4 + {}_9S_6 + {}_9S_{8,9}$$

$$= {}_9S_{\geq 1} f_a f_b + {}_9S_{\geq 4} f_a + {}_9S_{\geq 6} f_b + {}_9S_{\geq 8}$$

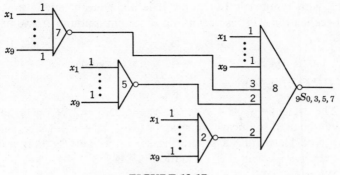

FIGURE 18.17

We note that the actual threshold of the final level device must be 8. In order that the threshold be properly adjusted for the three cases where fewer than 8 inputs must turn on the transistor, the following equations must be satisfied:

$$6 = 8 - N_b$$
$$4 = 8 - N_a$$
$$1 = 8 - N_a - N_b$$

This is clearly impossible. It is, therefore, necessary to utilize 3 functional inputs to find the final threshold element such that

$$_9S_{0,3,5,7} = \overline{_9S_{\geq 1}f_c + _9S_{\geq 4}f_a + _9S_{\geq 6}f_b + _9S_{\geq 8}}$$

In this case, we may choose functions which are linearly separable in terms of the input literals as follows:

$$f_c = \overline{_9S_{\geq 3}} \qquad f_a = \overline{_9S_{\geq 5}} \qquad f_6 = \overline{_9S_{\geq 7}}$$

In this case, then, nothing is gained by utilizing any of the three functions in the generation of the others. This will not always be so. The complete realization is illustrated in Fig. 18.17. ∎

Although there is considerable interest in threshold logic, this interest has has been primarily theoretical. None of the threshold devices so far devised have been sufficiently fast to achieve widespread use. Interest in the already cited computer algorithms[4,b] for realizing very complex functions stems from the broad possibilities inherent in the simulation of threshold elements as adaptive pattern recognition devices. This extension of linear input logic into the field of artificial intelligence is an extremely interesting subject, but one which we do not have space to consider here.[7,8,9]

For arbitrary functions of up to about six variables, it is possible to obtain economical realizations by following a procedure similar to that employed with symmetric functions. It is not guaranteed, however, that a minimal realization will result without a certain amount of enumeration of possibilities.

PROBLEMS

18.1. The weights $N_1 = 4, N_2 = 3, N_3 = 3, N_4 = 2, N_5 = 1$, and $T = 6$ can be used to linearly separate $x_1(\bar{x}_2 + x_3 + \bar{x}_4) + \bar{x}_2 x_3 + \bar{x}_2 \bar{x}_4 x_5 + x_3 \bar{x}_4 x_5$. Determine a separating set of weights so that no inverted literals appear in the summation.

18.2. What function is linearly separated by the weights $N_1 = 3, N_2 = 1$, $N_3 = -1, \ N_4 = -2, \ T = 1$? Determine a set of nonnegative weights which will separate this function.

18.3. Prove Theorem 18.4.

18.4. Determine whether or not each of the following functions is linearly separable.

$$f_1 = x_1 x_2 x_4 + x_2 x_3 x_4 + x_1 x_2 x_3 x_5 + x_1 x_4 x_5$$

$$f_2 = \bar{x}_3 x_4 + x_1 \bar{x}_3 \bar{x}_4 + x_1 x_3 x_4 + x_3 x_4 x_5 + x_1 \bar{x}_2 x_3 x_5 + \bar{x}_3 x_5 + \bar{x}_1 \bar{x}_2 \bar{x}_3 x_4$$

18.5. Determine weights and threshold values for the functions in Problem 18.4 that are linearly separable.

18.6. List all linearly separable functions of three variables.

18.7. Show that the following function satisfies the orderability criteria in Theorem 18.7, but is not linearly separable.

$$f = x_8[x_7 + x_6 + x_5(x_4 + x_3 + x_2 + x_1) + x_4 x_3 x_2 x_1]$$

$$+ x_7 x_6[x_5(x_4 x_3 x_2 + x_4 x_3 x_1 + x_4 x_2 x_1 + x_3 x_2 x_1) + x_4 x_3 x_2 x_1]$$

18.8. In the circuit in Fig. 18.1, suppose the minimum sum of conductances corresponding to excited inputs is G_{0t} when the circuit output is to be 0. Also, let $G_{\min}$ be the smallest conductance. Suppose that the voltage V and the threshold current I_t can be assumed to be precisely constant. Suppose, however, that temperature fluctuations and manufacturing tolerances caused the value of a conductance, G_i, to take on values in the interval

$$G_i(1 - \delta) \le G_i \le G_i(1 + \delta)$$

(a) If a current of I_0 amperes is required to reliably turn on the transistor, show that a limit on function complexity can be expressed by

$$2G_{0t} \leq \frac{G_{\min}(1 + \delta) - I_0/V}{\delta}$$

(b) Why is this a limit on function complexity?

18.9. Suppose that K is the minimum number of inputs which must be excited to turn on the transistor in Fig. 18.1

(a) Using the results of Problem 18.8, show that

$$K \leq \frac{G_{\min}(1 + \delta) - I_0/V}{2G_{\min}\delta}$$

(b) Show that only in the case where $f = {}_nS_{\geq k}$ does the above bound become an equality.

18.10. With δ defined as in Problem 18.8, determine the maximum tolerable δ in terms of I_0/V for the realization of the function, f_2, given in Example 18.8.

18.11. How many functions of five variables are linearly separable?

18.12. Devise a minimal multilevel realization of ${}_5S_{0,1,3,4}$ using NAND gates.

18.13. Devise and verify a 3-transistor realization of ${}_7S_{1,3,5,7}(x_1, x_2, x_3, x_4, x_5, x_6, x_7)$. *Hint:* Use as input literals $\bar{x}_1, \bar{x}_2, \ldots, \bar{x}_7$.

18.14. It is desired that an analog-digital converter be designed, whose input could vary between 0 and 16 volts at a low frequency. It is desired that 4-binary-bit accuracy be available at the output. Design a circuit similar to Fig. 18.15 which can perform this function.

18.15. Suppose a function $f(x_0, x_1, \ldots, x_{n-1})$ is linearly separable. Show that:

(a) $f(x_0, x_1, \ldots, x_{n-1}) + x_n$ is linearly separable.

(b) $[f(x_0, x_1, \ldots, x_{n-1})]x_n$ is linearly separable.

18.16. Determine a minimal multi-threshold realization of

$${}_{13}S_{0,1,3,5,6,8,10,11}(x_1, x_2, \ldots, x_{13})$$

18.17. Let $f(x_0, x_1, \ldots, x_{n-1})$ be any Boolean function. Determine a set of weights $\{N_i\}$ such that in no case will

$$\sum_{i=0}^{n-1} N_i a_i = \sum_{i=0}^{n-1} N_i b_i$$

where $f(a_0 \cdots a_{n-1}) = 1$ and $f(b_0 \cdots b_{n-1}) = 0$.

18.18. Determine an economical multi-threshold realization of

$$F = \sum m(0, 1, 4, 7, 9, 10, 11, 16, 17, 23, 24, 25, 27, 30, 31)$$

BIBLIOGRAPHY

1. Minnick, R. C. "Synthesis of Linear Input Logic by the Simplex Method," *Denver Research Institute Computer Symposium*, July 30, 1959.

2. Chao, S. C. "A Generalized Resistor Transistor Logic Circuit and Some Applications," *IRE Trans.*, **PGEC-8:** 1, 8–12 (March 1959).

3. Winder, R. O. "Enumeration of Seven-Argument Threshold Functions," *IEEE Trans.*, **EC-14:** 3, 315 (June 1965).

4. Hopcroft, J. E. and R. L. Mattson. *IEEE Trans. on Electronic Computers*, **EC-14:** 4, 552 (Aug. 1965).

5. Ho, Y. C. and R. L. Kashyap. "An Algorithm for Linear Inequalities and Its Applications," *IEEE Trans. on Electronic Computers*, **EC-14:** 5, 683 (Oct. 1965).

6. Stram, O. B. "Arbitrary Boolean Functions on N Variables Realizable in Terms of Threshold Devices," *Proceedings IRE*, **49:** 1, 210–220 (Jan. 1961).

7. Nilson, Nils. *Learning Machines*, McGraw-Hill, New York, 1965.

8. Rosenblatt, Frank. *Principles of Neurodynamics: Perceptrons and the Theory of Brain Mechanisms*, Spartan Books, Washington, D.C., 1962.

9. Widrow, Bernard. "Generalization and Information Storage in Networks of Adaline Neurons," *Self Organizing Systems*, Spartan Books, Washington, D.C., 1962: 435.

10. Kautz, W. H. "The Realization of Symmetric Switching Functions with Linear Input Logic Elements," *IRE Trans.*, **PGEC-10:** 3 (Sept. 1961).

11. Chow, P. E. "Boolean Functions Realizable with Single Threshold Devices," *Proceedings of the IRE*, **49**, 1, 370–371 (Jan. 1961).

12. Sheng, C. L. and H. R. Hwa. "Testing and Realization of Threshold Functions by Successive Higher Ordering of Incremental Weights," *IEEE Trans. on Electronic Computers*, **EC-15:** 1, 212–219 (April 1966).

13. Haring, D. R. "Multi-Threshold Threshold Elements," *IEEE Trans. on Electronic Computers*, **EC-15:** 1, 45–65 (Feb. 1966).

A Selection of Minimal Closed Covers

In Chapter 12, after finding the set of all maximal compatibles, the selection of a minimal cover was carried out by an essentially trial-and-error process. This was practical since the number of maximal compatibles was reasonable and there were few implications. In very complex cases, with many compatibles and interacting implications, a more systematic procedure would be highly desirable. Unfortunately, no procedure exists, other than complete enumeration, which guarantees finding the minimal equivalent of a state table in every case. However, a paper published by Grasselli and Luccio[1] provides the basis for a straightforward procedure which will reduce the number of alternatives considerably in most problems.

As a starting point, it is necessary to formalize the procedure of combining sets of compatibility pairs into larger compatibility classes. We have already seen that this can be done, but there is some question as to the effect of this procedure on closure. Theorem A.1 provides the answer to this question.

THEOREM A.1. *Let* $\{q_1, q_2, \ldots, q_n\} = \{q_i\}$ *be a set of states which are pairwise compatible in a single closed collection. Then* $\{q_i\}$ *is a compatibility class within some closed collection.*

Proof: Since the initial collection is closed, the next states $\delta(q_1, X_j)$ and $\delta(q_2, X_j)$ for any input X_j are a compatible pair in the same collection. The same holds true for the next states for any other pair in $\{q_i\}$. Thus, all the next states of $\{q_i\}$ for a given input X_j are pairwise compatible and can form a single compatibility class $\{\delta[q_i, X_j]\}$. Therefore, the requirements of Definition 12.7 are satisfied with respect to $\{q_i\}$. Similarly, all the next states of $\{\delta[q_1, X_j]\}$ are pairwise compatible in the original closed collection and can form a single compatibility class, satisfying Definition 12.7 for $\{\delta[q_i, X_j]\}$. This process continues until a complete closed collection is formed. Since only a finite number of distinct classes can be formed from a finite set of states, the process must terminate. Q.E.D.

THEOREM A.2. *The union of all distinct maximal compatibles formed from the states of a circuit S is a closed collection of compatibility classes.*

Proof: Starting with any maximal compatible, generate a closed collection by the process described in the proof of Theorem A.1. Every class in this collection must either be a maximal compatible or else be included in a maximal compatible. Thus, Definition 12.7 is satisfied. Q.E.D.

The reader can easily verify that the collections of maximal compatibles found in the examples of Chapter 12 are indeed closed collections. These collections of maximal compatibles may be used as a starting point in the process of finding a minimal cover. In these examples, we were able to find minimal covers without great difficulty, simply by inspection. In general, this may not be possible. First, in a problem of any complexity, there will be a large number of maximal compatibles. Second, all possible subsets of the maximal compatibles are also candidates for inclusion in a minimal cover. The first step, then is to reduce the number of candidates.

If C_α and C_β are compatibility classes and P is a collection of classes, we mean by $C_\alpha \subseteq C_\beta$ that all the states in C_α are included in C_β, and by $C_\beta \in P$, that C_β is a member of P.

Definition A.1. The *class set* P_α implied by a compatibility class C_α is the set $\{D_{\alpha_1}, D_{\alpha_2}, \ldots, D_{\alpha_n}\} = \{D_{\alpha_i}\}$, of all classes implied by C_α, such that
1. D_{α_i} has more than one element.
2. $D_{\alpha_i} \not\subseteq C_\alpha$.
3. $D_{\alpha_i} \not\subseteq D_{\alpha_j}$ if $D_{\alpha_j} \in P_\alpha$.

Less formally, the class set is made up of the compatibility classes, other than singletons, implied by a given set and not included in the given set. Figure A.1 shows a small implication table, the determination of the maximal compatibles, and the class sets implied by the maximal compatibles. Note that the class (*abde*) does not imply a class set since all the pairs implied by this class are subsets of the class. Class (*bcd*), however, implies three pairs which are not included in the class.

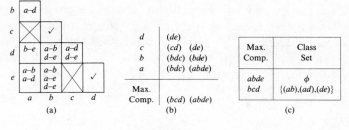

(a) (b) (c)

FIGURE A.1

Condition 3 of Definition A.1 may be applicable when some class of three or more states implies all possible pairs of states from another set of three or more states. Consider the two partial state tables shown in Fig. A.2, and assume that in both cases (abc) is a compatibility class. In both cases, (abc) implies the classes (de), (df), and (ef). For the table of Fig. A.2, the three classes are implied for different inputs; hence they do not combine into a single implied class. For the table of Fig. A.2b, however, all three classes are implied by a single input, so they combine into a single implied class (def). By Condition 3, only this larger set need be included in the class set. Note that this situation cannot be found with the implication table alone, but requires reference to the state table. In the case of Fig. A.1, it was not necessary to refer to the state table, since the pairs appearing did not make up all possible combinations of some larger set of states.

Definition A.2. A compatibility class C_α *excludes* a compatibility class C_β if and only if
1. $C_\beta \subseteq C_\alpha$ and
2. $P_\alpha \subseteq P_\beta$

where P_α and P_β are the class sets implied by C_α and C_β, respectively.

A class which is not excluded will be called a *prime* compatibility class.

	$q^{\nu+1}$		
q^ν	$X = 0$	$X = 1$	$X = 2$
a	d	f	$-$
b	e	$-$	e
c	$-$	d	f

(a)

	$q^{\nu+1}$	
q^ν	$X = 0$	$X = 1$
a	d	f
b	e	$-$
c	f	d

(b)

FIGURE A.2

Suppose a minimal set of compatibility classes covering a given state table has been found, including some class C_β which is not prime. Let C_α be a class which excludes C_β. Since C_α includes all the states covered by C_β and does not imply any classes not implied by C_β, both cover and closure will be maintained if C_α replaces C_β. We have thus proved the following theorem.

THEOREM A.3.   *At least one minimal covering will consist solely of prime compatibility classes.*

The concept of "exclusion" is used to eliminate classes from consideration in much the same manner that dominance is used to eliminate prime implicants. An excluded class covers only states that are included in some larger class, as specified by condition 1 of Definition A.2. Condition 2 ensures that closure is not violated by the substitution of the larger class for the smaller one. It is important to note that not all minimal covers consist of prime classes and that not all covers consisting only of prime classes are minimal. Theorem A.3 simply allows us to restrict our attention to prime classes with the assurance that a minimal cover can still be found.

Example A.1
Determine the prime compatibility classes for the circuit partially described by the implication table of Fig. A.3.

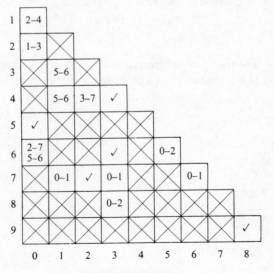

FIGURE A.3

8	(89)								
7	(89)								
6	(89)	(67)							
5	(89)	(67)	(56)						
4	(89)	(67)	(56)						
3	(89)	(367)	(56)	(38)	(34)				
2	(89)	(367)	(56)	(38)	(34)	(27)	(24)		
1	(89)	(367)	(56)	(38)	(134)	(27)	(24)	(137)	
0	(89)	(367)	(056)	(38)	(134)	(27)	(24)	(137)	(02)	(01)
M.C.	(056)	(134)	(137)	(367)	(01)	(02)	(24)	(27)	(38)	(89)

FIGURE A.4.   *Determination of maximal compatibles, Example A.1.*

Solution: The first step is the determination of the maximal compatibles, as shown in Fig. A.4.

The process of determining the prime compatibles starts with the maximal compatibles, which are themselves prime. These are listed in Fig. A.5a, together with the implied class sets.

The remaining prime classes are found by deleting one state at a time from each of the maximal compatibles. If the implied class set is smaller than that of the maximal compatible or any other prime class containing the subset, then it is prime. After this has been done for all of the maximal compatibles, the procedure is repeated for the resultant prime classes, until no further decomposition takes place.

For example, the maximal compatible (134) calls for a check of (13), (14), and (34). We see that (13) and (14) imply (56), just as does (134), so they are

Max Comp.	056	134	137	367	01	02	24	27	38	89
Class Sets	27 02	56	56 01	01	24	13	37	ϕ	02	ϕ

(a)

Other Prime Classes	05	06	34	17	36	1	56
Class Sets	ϕ	27 56	ϕ	01	ϕ	ϕ	02

(b)

FIGURE A.5.   *Determination of prime classes, Example A.1.*

not prime. Class (34), however, implies the empty set and must be included in the list of Fig. 11.15b. Note that a subset must be checked against *all* prime classes which contain it. For example, subset (37), obtained from maximal compatible (137), is not excluded by (137) but is excluded by (367).

Also note that a prime class which implies the empty set excludes all its subsets and thus need not be checked for further prime classes. In this case, only (06), (17) and (56) need be checked further. Class (17) calls for a check of (1) and (7). We see that (7) is excluded by (27), but (1) is included in no prime class implying the empty set, so (1) must be added to the list of prime classes. ■

We are now at much the same position as in the Quine-McCluskey method when the determination of the prime implicants has been completed. Just as the prime implicants are the only products that need be considered in realizing a combinational function, so only the prime classes need be considered in finding a cover for a state table. The first step in selecting a cover from the prime classes is to determine any essential prime classes, that is, prime classes which are the only ones including one or more states. Again, the parallel to the Quine-McCluskey method is noted.

Next, we delete the covered states from the unselected classes. If the resultant reduced classes are then excluded, they may be removed from further consideration. This step is roughly comparable to reduction of the prime implicant table. We then check again for classes made essential by this second step. This process may be continued until no further reduction is possible. We shall see in the completion of this example how the choice of prime compatibles can often be further narrowed through intuitive treatment of a pictorial model.

We shall see in the next few examples how the choice of prime compatibles can often be further narrowed through intuitive treatment of a pictorial model.

Example A.1 (continued)

Find a minimal cover from the set of prime compatibility classes.

We first note that (89) is essential to cover (9). We then delete (8) from (38) leaving (3), which is excluded by (34). So (38) is discarded. This does not make any other sets essential, so we set up a pictorial representation of the implications of the remaining classes (Fig. A.6), which we shall refer to as the *implication graph*. Each class which has been selected or is still under consideration (not excluded) is represented by a circle. Selected classes are indicated by an asterisk. Every class implied by a class C_α is indicated by an arrow from C_α to that class. If only one class can satisfy an implication, we indicate a *strict implication* by a solid arrow. If an implication can be satisfied by any of two or more classes, we indicate an *optional implication* by dashed arrows.

A good place to start the reduction process is with classes which imply other classes but satisfy no implications themselves—for example, class (17). Note

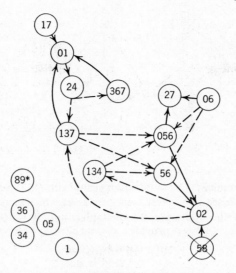

FIGURE A.6. *Implication graph for Example A.1.*

that the chain of implications set up by the selection of (17),

$$(137)$$

$$(17) \rightarrow (01) \rightarrow (24)$$

$$(367)$$

includes classes which cover both (1) and (7). Therefore, (17) may be discarded. By the same argument, (06) may be discarded. Each time a class is discarded, we should check to see if any of the remaining classes have been made essential as a result. This has not yet occurred, so we continue.

Next, let us compare classes which overlap, such as (056) and (56). Both satisfy the same implications, and both imply (02), thus providing a cover for (0). Thus, (056) can accomplish nothing that (56) cannot, and it implies more classes. Therefore, (056) may be discarded. Even after discarding (056), no classes have become essential. To facilitate the discussion, we redraw the graph, incorporating the above simplifications as shown in Fig. A.7.

Let us make a tentative selection of (137), to see what sort of cover results. The strict implication of (137) requires that (01), (24), (56), and (02) also be selected. With the addition of the essential class (89), we have a 6-state closed cover,

$$(137)(01)(02)(24)(56)(89)$$

Because of the strict implication, it is evident that this 6-state cover is the smallest that can be found by using (137). Next, let us eliminate (137) and see

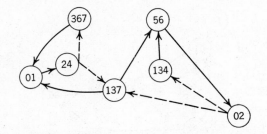

FIGURE A.7.   *Simplified implication graph for Example A.1.*

what possibilities remain. The implication graph with (137) removed is shown in Fig. A.8. Here we see two *cycles*, sets of classes which must be selected as a group because of the cyclic pattern of implication. If we select cycle A, the addition of (05) and (89) will provide a 5-state closed cover,

$$(01)(24)(367)(05)(89) \tag{A.1}$$

If we select cycle B, we also get a five-state closed cover,

$$(02)(134)(56)(27)(89) \tag{A.2}$$

The reader can quickly verify that any cover not including one of the cycles must have at least six states. Thus, the covers of (A.1) and (A.2) are minimal. ■

It should be emphasized that the above procedure is not a complete or exact algorithm. Rather, it consists of the application of a few "common sense" rules in order to reduce the number of candidates for inclusion in a cover to a point where a trial-and-error check of all possibilities is practical. Also, the implication graphs are not an essential feature of the method. The same information is contained in the list of prime classes and class sets. However, the graphs, like Karnaugh maps, make it easier to recognize significant patterns and relationships. The reader who desires practice with this method should rework some of the more difficult exercises of Chapter 12.

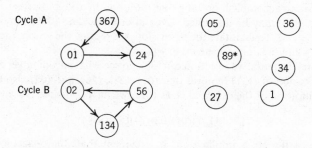

FIGURE A.8.   *Final implication graph for Example A.1.*

BIBLIOGRAPHY

1. Grasselli, A. and F. Lucio. "A Method for Minimizing the Number of Internal States in Incompletely Specified Sequential Networks," *IEEE Transactions on Electronic Computers*, EC-14: 3, 330–359 (June 1965).

Appendix B Relay Circuits

B.1 Basic Characteristics of Relay Circuits

As was mentioned in Chapter 1, switching theory was originally developed to deal with the problems of designing switch and relay circuits for dial telephone systems. The earliest digital computers were also constructed wholly or in part of relays. Today, the relay has been largely replaced by electronic circuits in the majority of digital systems, so that the emphasis of this book has been on electronic realizations. However, there are still situations where relays are the most practical means of realizing logic circuits. This is particularly true in fields such as industrial control where logic circuits are used to control the application of large amounts of power. For this reason, we present here a brief outline of the means by which the theories and techniques developed in this book can be applied to relay circuit design. The reader interested in more detail is referred to any of several excellent books.[1,3,4]

A relay consists of an electromagnet and a set of switch contacts. When the relay coil is energized, the switch operates. The three basic types of contacts are shown in Fig. B.1. In each case, the switch contacts, which are generally referred to as *springs*, are shown in the position they take when the relay coil is not energized; this position is known as the *normal* position. Thus, the first relay is normally open (NO), and the contacts close when the coil is energized. The second relay is normally closed (NC),

580

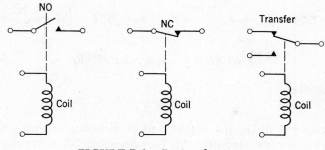

FIGURE B.1.   *Basic relay types.*

and the contacts open when the relay is energized. The third relay has a
transfer contact, a combination of NO and NC contacts sharing a common
movable spring.

Although we have shown only one set of contacts on each relay in Fig. B.1,
practical relays will generally have several sets of contacts, often including
all three types.

The standard form of a relay circuit is shown in Fig. B.2. The inputs are
signals applied to the coils, which take on the value 1 if the relay is energized,
the value 0 if it is not. In this particular form, a coil is energized by connecting
the corresponding input terminal to ground, which will probably be
accomplished by another switch or relay circuit. But we are not concerned
about this any more than we are concerned with the origin of the inputs to an
electronic gate network.

The contact network is made up of the contacts of the relays interconnected in some fashion. The outputs are known as the *transmissions*, T_i, and
take on the value 1 if the output is connected to ground through the network,
the value 0 if it is not.

Consider the trivial relay circuits of Fig. B.3, where the contacts are
operated by the coil shown. In Fig. B.3a, the output will be connected to
ground $(T = 1)$ only if the relay is operated $(X = 1)$. Therefore, this circuit

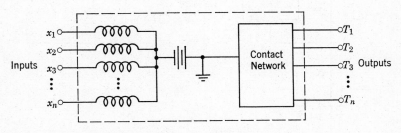

FIGURE B.2.   *Standard form of relay circuit.*

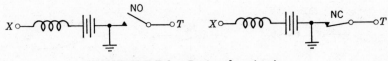

FIGURE B.3.   *Basic relay circuits.*

can be described by

$$T = X$$

Similarly, in Fig. B.3b, the output will be connected to ground $(T = 1)$ if the relay is not operated $(X = 0)$. This circuit can thus be described by

$$T = \overline{X}$$

We thus see that, in writing the equation for transmission, a normally open contact may be represented by the corresponding input variable, a normally closed contact by its complement. Once this is understood, there is nothing to be gained by drawing out all the detail of Fig. B.3. Instead, we draw the circuit as shown in Fig. B.4, where we replace the NO contact

o——X——o   o——$\overline{X}$——o   **FIGURE B.4.**   *Schematic representation of*
$T = X$ $T = \overline{X}$ *circuits of Fig. B.3.*

with the variable X and the NC contact with $\overline{X}$. It is understood that the transmission is 1 when the terminals are connected by the relay contacts. To clarify this, we show the complete representations for AND and OR circuits, along with the corresponding schematic notation, in Fig. B.5.

We have now seen that AND, OR, and NEGATION can all be realized with relays, which tells us that any switching function can be realized with

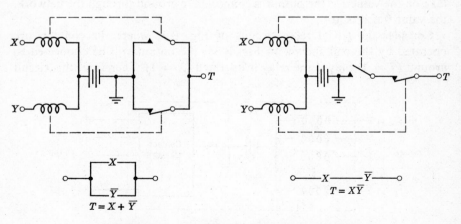

FIGURE B.5.   *Relay realizations of OR and AND.*

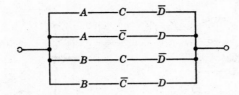

FIGURE B.6.   *Relay circuit for Equation B.1.*

relays. For example, consider the function of Equation 6.6, for which the minimal second-order form is

$$f(A, B, C, D) = AC\bar{D} + A\bar{C}D + BC\bar{D} + B\bar{C}D \qquad (B.1)$$

The relay form of this equation can be written directly, as shown in Fig. B.6. The simple and direct translation from the sum-of-products form to a series-parallel relay circuit is most striking. Although the algebraic form is minimal second-order, it is apparent that the relay circuit is more complicated than it need be. First, we see that the two A contacts and the two B contacts can be combined, as shown in Fig. B.7a. This simplification corresponds to the factoring

$$A\bar{C}D + AC\bar{D} + B\bar{C}D + BC\bar{D} = A(\bar{C}D + C\bar{D}) + B(\bar{C}D + C\bar{D}) \quad (B.2)$$

Noting that the terms in parentheses are identical, we can immediately make the further simplification shown in Fig. B.7b. This corresponds to the form

$$f(A, B, C, D) = (A + B)(C\bar{D} + \bar{C}D) \qquad (B.3)$$

the 3-level gate realization which is shown in Fig. 6.21.

The above example points out that, in relay circuits, unlike electronic gate circuits, there is nothing particularly desirable about the second-order form. The circuit of Fig. B.7b is certainly not slower than that of Fig. B.6. In a contact network, the only delay is in the propagation of a signal down a wire, a delay that is completely negligible compared to the operating speed

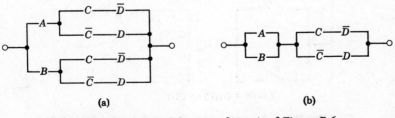

(a)   (b)

FIGURE B.7.   *Simplification of circuit of Figure B.6.*

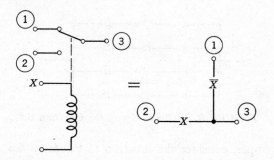

FIGURE B.8.   *Schematic representation of transfer contacts.*

of the relays. The speed of relay circuits is thus completely independent of the form of the contact networks.

In simplifying relay circuits, the basic objective is to reduce the number of springs. We also would like to eliminate relays if possible, but the number of relays is determined by the number of variables required, a factor not always under our control. One way to reduce the number of springs is to combine NO and NC contacts on the same relay into transfer contacts. The relation between a transfer contact and the schematic representation is shown in Fig. B.8. In the circuit of Fig. B.7b, we see immediately that transfer contacts can be used for the C and D variables. Thus, in terms of actual contact configuration, the circuit of Fig. B.7b will appear as shown in Fig. B.9.

Another method of simplification peculiar to relay circuits makes use of the bilateral characters of relay contacts. Contacts conduct current in both directions, which electronic gates do not, and sometimes this fact can be used to advantage. Assume that we wish to realize the function

$$T(A, B, C, D, E) = A\bar{B} + \bar{A}C + \bar{A}\bar{B}E + ACE \tag{B.4}$$

The direct realization of this equation is shown in Fig. B.10a. By inspection, we can simplify to the form of Fig. B.10b. We will leave it to the reader to

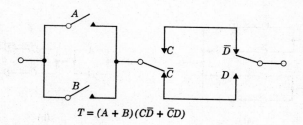

$$T = (A + B)(C\bar{D} + \bar{C}D)$$

FIGURE B.9.   *Contact configuration of circuit of Fig. B.7b.*

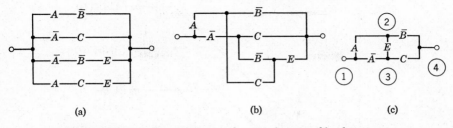

| (a) | (b) | (c) |

FIGURE B.10.   *Simplification of circuit by use of bridge circuit.*

satisfy himself that the two $\bar{B}$ contacts and the two C contacts cannot be combined into single contacts without introducing undesired transmissions. If E could conduct current in only one direction, nothing further could be done to simplify this circuit. But the bilateral character of the contact makes it possible to simplify to the *bridge* circuit of Fig. B.10c. Path ① → ③ → ④ provides the term $\bar{A}C$, path ① → ② → ④ the term $A\bar{B}$, path ① → ② → ③ → ④ the term ACE, and path ① → ③ → ② → ④ the term $\bar{A}\bar{B}E$.

With all these refinements, how do we go about designing relay circuits? First, the techniques presented in this book should be used to find the simplest second-order forms, and the direct realization of these forms should be drawn. Then, by an essentially intuitive combination of visual inspection and algebraic manipulation, contacts are eliminated or combined into transfer contacts wherever possible. Just because there is no single best form, such as second-order, there is no single procedure for finding a minimal circuit or even for determining if a given circuit is minimal. Good relay design is primarily the product of experience.

B.2   Relay Realizations of Symmetric Functions

Symmetric functions are economically realized by using contact networks. Notice that the five outputs of the networks in Fig. B.11 are the five elementary symmetric functions of four variables. Thus, any symmetric function may be realized by connecting together some combination of outputs. If outputs are connected together, some of the contacts become redundant and may be eliminated. For discussion of the procedure involved, the reader is referred to Marcus.[1] Clearly, this realization may be extended to symmetric functions of any number of literals.

Another special network which is conveniently realized by using relays is the standard tree network shown in Fig. B.12. Notice that each possible 3-variable minterm is realized at one of the terminals of the tree. Again the extension to n variables is obvious. Networks of this form are useful as selection networks.

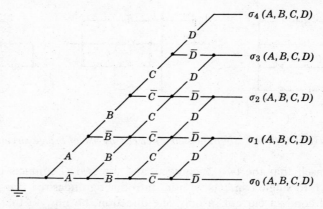

FIGURE B.11. *Relay realization of symmetric functions.*

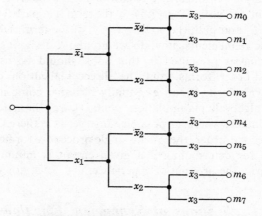

FIGURE B.12. *Relay tree circuit.*

B.3 Relays in Sequential Circuits

Relays called latching relays are available, which may be held by a spring in either of two positions when no input is applied. These devices behave in much the same fashion as flip-flops, since when a latching relay is set to one position by the energizing of one of two coils, it remains there until cleared by an energizing of the other coil. Similarly, two ordinary relays can be cross-coupled to form a memory element, as can two NOR or NAND gates.

Systems featuring latching relays may be considered as synchronous or pulse-mode logic circuits, although this point of view is not particularly common. More often, feedback is added to combinational networks of

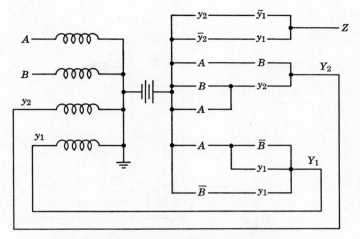

FIGURE B.13.   *Relay fundamental mode circuit.*

relays to form fundamental mode circuits. In Fig. B.13, we see a fundamental mode relay circuit which will have a 1 output if the most recent input transition has taken place on line A, and a 0 output if the most recent transition has been on line B.

Note that the delay involved in switching a relay is of several orders of magnitude longer than the delays in typical electronic switches. The delay in the coils will affect excitation uniformly. Therefore, the switching of contacts due to input transitions will always be complete before further transitions are caused by changes in the states of the secondaries. Because of this delay in the feedback loop, essential hazards are not a problem in relay circuits.

The primitive flow table corresponding to the circuit in Fig. B.13 may be found in Fig. B.14a. The implication table, the maximal compatibles, and the minimal state table are found in Figs. B.14b, c, and d, respectively. The excitation map is shown in Fig. B.14e, and the excitation and output expressions are given in Fig. B.14f. As the outputs are dependent only on the history of input transitions, the minimal state table remains a Moore table, and a hazard-free circuit results without requiring any special precautions.

As a final example, let us consider the following design problem.

Example B.1

An elevator is to operate between two floors, A and B, where B is the top floor. A controller is to be designed for the elevator. The controller is to have two outputs. One of these outputs, Z_2, it to be 1 whenever the drive motor is to be in operation. The other output, Z_1, is to indicate whether the elevator should be pulled up or let down. If the elevator is stopped, Z_1 will indicate whether or not

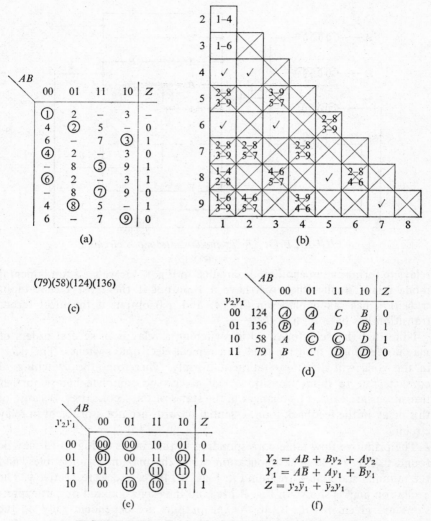

FIGURE B.14. *Tables describing circuit of Fig. B.13.*

the door is to be open. The required outputs are tabulated in Fig. B.15. The door-open cycle is separately timed, and it must terminate when a door-close signal, C, is received from the timer. C will go to 0 after the door has closed. The door-drive motor will cause the door to close and remain closed when $Z_2Z_1 \neq 01$ and will cause it to open and remain open as long as $Z_2Z_1 = 01$. The remaining inputs include the two operator inputs A and B and a signal I, which is 1 when the elevator is within a certain interval between floors. The input A will be 1 whenever the call button is pushed at the bottom floor or whenever the

	Z_2	Z_1
Stopped, Door Closed	0	0
Stopped, Door Open	0	1
Go, Up	1	1
Go, Down	1	0

FIGURE B.15

floor A button is pushed in the elevator. The input B is similarly 1 when it is desired that the elevator proceed to the second floor.

Solution: The general operation of the circuit is indicated in Fig. B.16. Notice that, for the most part, the elevator merely proceeds clockwise through the series of eight conditions when so instructed by the inputs. The only alternative actions are that the door must open when the elevator is called at the floor at which it is already located. No decelerating state is indicated for the end of a transition between floors. It is assumed, for example, that satisfactory operation will take place if the output goes to $Z_2 Z_1 = 01$ when the state of the controller goes from U to BO. That is, the drive will stop and the door-open cycle will begin when the elevator leaves the interval between floors. It is assumed that inertia will carry the elevator into place and delay in the door-open mechanism will prevent the door opening until the elevator has stopped.

Because of the complexity of the problem, a primitive flow table would have an impractical number of rows. Instead, we work directly for a final flow table as shown in Fig. B.17. We assume a state for each one of the eight distinct operating conditions indicated in Fig. B.16.

To fill in the table, we start at some arbitrary condition and fill in the various sequences. For example, let us start with the elevator stopped at the A floor with the door closed, no signals present, that is, stable AC at $ABCI = 0000$. Now, if

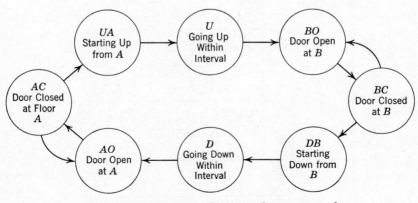

FIGURE B.16.   *State diagram, elevator control.*

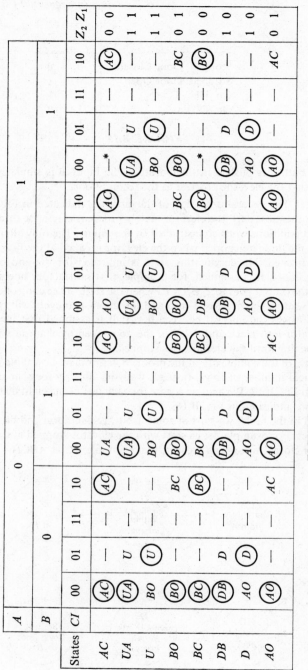

FIGURE B.17. State table, elevator control.

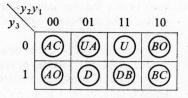

FIGURE B.18.   *State assignment, elevator control.*

the *B* signal arrives, the circuit goes through unstable *UA* to stable *UA* at *ABCI* = 0100. If the *B* signal goes off before the *I* signal comes on, the circuit moves to stable *UA* at *ABCI* = 0000. Then, when the *I* signal comes on, the circuit goes through unstable *U* to stable *U* at *ABCI* = 0001. When the *I* signal goes off, the circuit moves to *BO* at *ABCI* = 0000, and so forth.

Most of the don't-care conditions arise from the fact that a door-close signal will not occur when the elevator is between floors, and *I* will be 0 when the elevator is at rest. The only exceptions are the two don't-cares indicated by asterisks. In these cases, the idle elevator is summoned simultaneously at both floors. It makes no difference whether the elevator is preempted by one of the floors or whether it remains at rest until someone releases a button. Consider the five input conditions for which the controller remains at *AC* in the first column. Clearly,

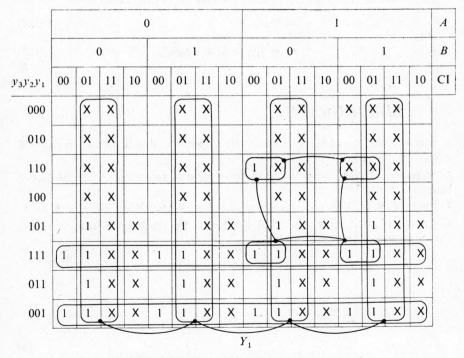

FIGURE B.19.   *Excitation map, elevator control.*

the elevator must remain at rest if there are no inputs. Similarly, it must remain at rest if the door-close signal, C, is 1. The elevator must not move with the door open.

The reader will be left to deduce the remaining entries in the table.

Due to the cyclic nature of Fig. B.16, there is no difficulty in assigning the eight stable states to three secondaries without critical races. The assignment chosen is listed in Fig. B.18. Using this assignment, three separate excitation maps were easily compiled for Y_3, Y_2, and Y_1. Although the maps involve seven variables, it was not difficult to obtain minimal hazard-free expressions. The simplest of the three excitation maps is shown in Fig. B.19. From this, the excitation expression is easily seen to be

$$Y_1 = I + \bar{y}_3\bar{y}_2y_1 + y_3y_2y_1 + y_3y_2A\bar{C} \tag{B.5}$$

The remaining excitation expressions are found to be

$$Y_2 = y_2\bar{C}I + \bar{y}_3I + y_3y_2I + \bar{A}\bar{B}CIy_3 + \bar{y}_3y_2y_1 + ACy_2 + BIy_2 \tag{B.6}$$

$$Y_3 = y_3y_1 + y_3\bar{y}_2\bar{C} + \bar{y}_2\bar{y}_1A\bar{C} + y_2\bar{B}C\bar{I} + y_2ABC \\ + y_3\bar{B}\bar{C} + y_3A\bar{B}C\bar{I} + y_3y_2C \tag{B.7}$$

The only product included solely to avoid a hazard is $\bar{y}_3y_2y_1$ in Y_2. The minimal hazard-free expressions for Z_2 and Z_1 are very easily determined to be

$$Z_2 = y_1 \tag{B.8}$$

and

$$Z_1 = \bar{y}_3y_2 + \bar{y}_3y_1 + y_3\bar{y}_2\bar{y}_1 \tag{B.9}$$

■

BIBLIOGRAPHY

1. Marcus, M. P. *Switching Circuits for Engineers*, Prentice-Hall, Englewood Cliffs, N.J., 1962.

2. Harrison, Michael. *Introduction to Switching and Automata Theory*, McGraw-Hill, New York, 1965.

3. Caldwell, S. H. *Switching Circuits and Logical Design*, Wiley, New York, 1958.

4. Keister, W., A. Z. Ritchie, and S. H. Washburn. *The Design of Switching Circuits*, Van Nostrand, New York, 1951.

Index

Adder, 6, 175
  full, 177
  BCD, 133
Addition, 14
Address, 6
Algebra, Boolean, 45
Applicable, sequence, 333
AND, 24
Arithmetic, 6, 14
Asynchronous, 237

Base, 9
Binary, 1, 10
  adder, 180
  connectives, 26
Binary-coded decimal (BCD), 18, 183
Bistable, 213
Boole, George, 67
Boolean
  algebra, 45
  hypercube, 140
Branch, 442
  network, 450, 469
Bundling, 515

Catenation, 456
Class set, 572
Clock, 216, 220, 227, 233
  controlled, 439
  direct, 438
  skew, 420
Clock-mode, 227, 437, 495
Closed collection, 339, 572
Code, 180
  conversion, 189, 193
  error-correcting, 202
  error-detecting, 202
  excess-3, 183
  Gray, 183, 184
  Hamming, 204

Code (*continued*)
  reflected, 183
  seven-segment, 185, 190
  2-out-of-5, 183
Combinational logic, 7
  subroutine, 458
Comparator, 476
Compatible, 333, 571
  maximal, 337, 572
  pair, 337
Compatibility, 332
  class, 337, 339, 571
  prime, 573
Completely specified, 258
Compression, 456
Computer, 5
Connection,
  matrix, 497
  statement, 452, 458
Connectives, 24
  sufficient sets, 35
Contact,
  bounce, 414
  normally closed, 70, 582
  normally open, 70, 582
  transfer, 584
Control delay, 374
Control model, 438
Control sequence, 443
Control sequencer, 450, 480
Controlled clocking, 439
Core, magnetic, 560
Cost, 156, 223
  design, 485
  incremental, 488
  wiring, 485
Counter, 221, 323
  up-down, 254, 290
Cover, 339, 572
  collection, 340

Cover (*continued*)
  minimal, 574
Critical race, 380
Cubical representation, 139
Custom, 485
Cycle, 379, 578

Data transfer, 441, 452
Decimal, 13
Decoder, 186, 465
  tree, 190
Decomposition, 515–529
  charts, 519, 521
  complex disjoint, 523
  simple disjoint, 515
Delay, 77
DeMorgan's Law, 36, 53
Design, 244
Destination set, 387
Digital, 1
  digital system, 7
Diode, 71
  diode-transistor logic, 75
Direct clocking, 438
Dominance, 155
Don't-Cares, 129, 329
  incidental, 329
Duality, 48

Economics, 484
Equivalence
  circuit, 264
  class, 257, 272
  relation, 257
Error
  correcting, 202
  detecting, 202
Essential prime implicant, 149
Excitation-variables, 226, 235, 365
Excitation maps, 280, 376
Exclusive-or, 25
Extended state table, 440

Factoring, 513
Fault detection, 501
Finite-memory, 249
Flip-flop, 213
  D, 217, 432
  JK, 216, 242
  RS, 214
  SC, 214

Flip-flop (*continued*)
  T, 217
  master-slave, 230, 421
Flow table, 367
Flux, magnetic, 561
Frequency-division, 232
Full adder, 177
Fundamental-mode, 235, 365

Gate, 30
Gray code, 183, 184

Hamming code, 204
Hazards, 394
  dynamic, 398
  essential, 400
  static, 396
Hexadecimal, 13
Huffman-Mealy method, 262
Hybrid MSI, 485
Hypercube, 139, 552, 554
Hysteresis loop, 560
Huntington, E. V., 45, 67

If-then, 27
Implication
  graph, 577
  strict, 576
  table, 267
Implied class, 340
  set, 268, 572
Inclusive-or, 25
Incompletely specified
  functions, 129
  sequential circuits, 312
Increment, 460
Indistinguishable, 259
Integrated circuit logic, 79, 475
  CMOS, 85
  DTL, 80
  ECL, 83
  MOS, 489
  RTL, 81
  TTL, 81
  comparison of types, 86
  large scale, 473
  medium scale, 473
Interface, 481
Inverter, 30, 32, 33, 74
Iterative circuit, 199, 529–534

JK flip-flop, 216

Karnaugh map, 103, 162
K-cubes, 140

Latch, 238
LED, 185
Level-mode, 234, 363–417
Linearly-separable, 548
Logic, 24
  magnetic, 560
  symbols, 32
  vector, 456
LSI, 473

MOS, 489
Magnetic-logic, 560
Master-slave flip-flop, 230, 421
  clocked, 233
  edge-sensing, 423
Matrix
  connection, 497
Maximal compatible, 337, 575
Maxterm, 101
McCluskey, E. J., 143
Mealy circuits, 309
Mealy-Moore translation, 317
Memory, 5, 212
  finite, 250
Memory element, 213
  input equations, 275
Microprogram, 195
Minimum distance, 202
Minterm, 101, 144
Moore circuit, 309
Moore, E. F., 556
MOS, 485
  shift register, 491, 493
  static, 494
MOSFET, 491
MSI, 473
  hybrid, 485
Multiple-output, 160
Multiplication, 14
Multiplicity, column, 517

NAND LOGIC, 35, 88, 194
Negative logic, 72
Negation, 33
NOR, 28, 76, 90
NOT, 24

N-tuple, 551

Octal, 13
OR, 25
Ordering, 534
Overflow, 18, 38

Parallel, 3, 225
Parity, 199, 219, 224, 292, 513
Partition, 257, 261, 279, 438, 451
Petrick's method, 155
Positive logic, 71
Postulates, Huntington's, 45
Prime compatability class, 573
Prime implicant, 143, 147
  essential, 150
Product-of-sums, 100
Program, 442
Pulse-mode, 227, 307–328

Quine-McCluskey minimization, 143, 576

Race, 380
Radix, 9
Read-only-memory (ROM), 190, 486
Realization, 30
Reduction, 457
Reflexive, 257
Register, 456, 481
Regular expression, 245
Relation, 256
  equivalence, 257
  ordering, 534
Relay, 70
  circuits, 580–592
Reset state, 246, 255
Resistor-transistor logic, 546, 562
Rotate, 457
RS flip-flop, 214

SCEPTRE, 496
SC flip-flop, 214
Secondary variable, 365
Sequential circuit, 7, 211
  clock-mode, 227, 241, 437
  completely-specified, 258
  finite-memory, 250
  general-model, 226
  incompletely specified, 329
  level-mode, 235
  pulse-mode, 227, 307

Serial, 3, 225
  parity checker, 220
Seven-segment-code, 185
Sheffer, H. M., 28, 67
Shift register, 221, 229, 457
  2-phase, 491
  dynamic, 491
  4-phase, 493
  MOS, 491
Simulation, 495
  level mode, 499
Standard forms, 98
State
  assignment, 245, 275, 285, 384
  diagram, 244, 245
  minimization, 258
  table, 244, 265, 273, 440
Statement variable, 24
Strongly connected, 356
Strict implication, 576
Subroutine
  combinational logic, 458
Sum of products, 98, 149
Switch logic, 69
Symbols
  logic, 32
Symmetric, 257
  function, 508, 562, 585
Synchronizing sequence, 255
Synchronous, 228
Synthesis, 443
Speed, 223

Test generation, 500
Threshold logic, 545
  magnetic, 560

Threshold logic (continued)
  resistor-transistor, 546
Tic-tac-toe, 219
Timing, 230
Transfer, 441
Transistor, 73
Transistor inverter, 74
Transition
  diagram, 382
  list, 277
  table, 242, 279, 293, 352, 376
Transitive, 257, 338
Transmissions, relay, 581
Tree-decoder, 19
Truth function, 23, 28
Truth-functional compound, 24, 29
Truth table, 25
Truth value, 24

Unate, 534, 553, 558
  functions, 536–540

Variable
  bound, 516
  excitation, 226
  free, 516
  statement, 24
Vector, 454, 456
Venn diagram, 54
Vertex, 139, 552, 554
  counting, 557

Weights, 557
Wired-OR, 91
  design on K-map, 128
Wiring cost, 485